Operator Theory
Advances and Applications
Vol. 106

Contributions to Operator Theory in Spaces with an Indefinite Metric

The Heinz Langer Anniversary Volume

A. Dijksma
I. Gohberg
M.A. Kaashoek
R. Mennicken
Editors

Springer Basel AG

Editors:
Aalt Dijksma, Department of Mathematics, University of Groningen,
P.O. Box 800, 9700 AV Groningen, The Netherlands
email: a.dijksma@math.rug.nl

Israel Gohberg, School of Mathematical Sciences, Tel Aviv University,
Ramat Aviv 69978, Israel
email: gohberg@math.tau.ac.il

Marinus A. Kaashoek, Department of Mathematics, Free University,
De Boelelaan 1081a,1081 HV Amsterdam, The Netherlands
email: kaash@cs.vu.nl

Reinhard Mennicken, NWF 1, Mathematics, University of Regensburg,
Universitaetsstr. 31, D-93053 Regensburg, Germany
email: reinhard.mennicken@mathematik.uni-regensburg.de

1991 Mathematics Subject Classification 47-xx,

Picture of Heinz Langer
courtesy Paul A. Fuhrmann

A CIP catalogue record for this book is available from the
Library of Congress, Washington D.C., USA

Deutsche Bibliothek Cataloging-in-Publication Data
Dijksma, Aalt:
Contributions to operator theory in spaces with an indefinite metric :
the Heinz Langer anniversary volume / A. Dijksma ... – Basel ;
Boston ; Berlin : Birkhäuser, 1998
(Operator theory ; Vol. 106)
ISBN 978-3-0348-9782-2 ISBN 978-3-0348-8812-7 (eBook)
DOI 10.1007/978-3-0348-8812-7

Originally published by Birkhäuser Verlag, Basel in 1998
Softcover reprint of the hardcover 1st edition 1998

Printed on acid-free paper produced from chlorine-free pulp. TCF ∞
Cover design: Heinz Hiltbrunner, Basel

ISBN 978-3-0348-9782-2

9 8 7 6 5 4 3 2 1

Contents

Heinz Langer

Operator Theory:
Advances and Applications, Vol. 106

Heinz Langer and his work

AAD DIJKSMA AND ISRAEL GOHBERG

1. Introduction

This volume is dedicated to Professor Heinz Langer to honor him for his outstanding contributions to mathematics. His results in spectral analysis and its applications, in particular in spaces with an indefinite metric, are fundamental. Five main themes emerge in Heinz Langer's work, some of them are closely connected or have much in common:

(1) Spectral theory of operators in spaces with indefinite inner product.
(2) Pencils of linear operators (nonlinear eigenvalue problems).
(3) Extension theory of operators in spaces with indefinite inner product.
(4) Block operator matrices.
(5) One-dimensional Markov processes.

Heinz has written more than 130 research papers with 45 coauthors from 11 countries. He advised about 25 Ph.D. students and always enjoyed cooperation with colleagues, students and friends. As a teacher, Heinz has the ability to clarify connections and to point out the important. His work has numerous followers and great influence in the world centers of operator theory.

The occasion marking the origin of this book is Heinz Langer's sixtieth birthday. Two of his collaborators Martin Blümlinger and Fritz Vogl, with the help of Gabi Schuster, organized a two day Colloquium on Thursday and Friday, 12 and 13 October 1995, at the Technical University of Vienna. Friends and colleagues from all over the world attended. At the end of the conference it was decided to prepare this anniversary volume.

2. Biography of Heinz Langer

Heinz Langer was born on August 8, 1935 in Dresden. He went to school and Gymnasium there and attended the Technical University of Dresden from 1953 to 1958. Originally he wanted to study physics, but the selection principles in the then eastern part of divided Germany were against him. Thanks to some personal connections of a friend to a professor in mathematics, he was enrolled in mathematics. In 1958, after the diploma (with a thesis on perturbation theory of linear operators), staying at the TU Dresden as an assistant, he began studying linear operators in indefinite inner product spaces.

It was his teacher, Professor P. H. Müller, who recommended this topic: At a conference in Hungary he had listened to a lecture of János Bognár, and had encountered such questions in his investigations of operator polynomials. Heinz studied the fundamental papers by I. S. Iohvidov and M. G. Kreĭn, and, in the fall of 1959, he showed his results to Professor Szőkefalvi-Nagy from Szeged, who visited Dresden on his first trip abroad after the 1956 revolution in Hungary. Sz.-Nagy reacted positively but, since he did not consider himself a specialist in this field, recommended to send it to Professor M. G. Kreĭn. Heinz sent a handwritten (Kreĭn mentioned this later not only once!) manuscript to him. Its main result was a generalization of the Theorem of L. S. Pontrjagin about the existence of a maximal nonnegative invariant subspace of a self-adjoint operator in a Pontrjagin space to a Kreĭn space. This result attracted Kreĭn's attention, and so he invited Heinz to stay for one year in Odessa.

There existed exchange programs for graduate students between the German Democratic Republic (GDR) and the Soviet Union (mainly used by students from the GDR), and Heinz applied. Before being admitted one had to undergo a preparatory course at a special faculty of the University of Halle, which lasted one month in 1960. Heinz attended this course in the summer of 1960, but after two weeks he was sent home: He was found ideologically unsuitable for a longer stay in the Soviet Union.

In the fall of 1960 Heinz received his Ph.D. at the TU Dresden. In September 1961, after someone from the ministry had encouraged him to apply again for a stay in Odessa and the preparatory course had been shortened from one month to two weeks, students and graduate students separated, he was finally admitted to go to the University of Odessa on a post doc fellowship for one year. However, until he arrived in Odessa, Heinz did not know that M. G. Kreĭn did not work at the University of Odessa, but held the chair of Theoretical Mechanics at the Odessa Civil Engineering Institute.

At that time in Odessa each week there was a special lecture of M. G. Kreĭn and about three seminars at different institutions (regularly at the Civil Engineering Institute, the Pedagogical Institute and monthly at the House of Scientists). In the second part of his Ph.D.Thesis, Heinz had applied the Livšic-Brodskiĭ model for operators with finite-dimensional imaginary part in order to obtain a model for self-adjoint operators inPontryagin spaces. So he was quite well acquainted not only with indefinite inner product spaces, but also with some of the other main topics of interest of the Odessa school of functional analysis when he arrived, and he could actively take part in it.

The intense mathematical life in the circle around M.G. Kreĭn to which his former students belonged, among them M. S. Brodskiĭ, M. L. Brodskiĭ, I. S. Iohvidov, I. S. Kac, Ju. P. Ginzburg, Ju. L. Smul'jan, but also V. P. Potapov and

L. A. Sakhnovic, deeply impressed Heinz and had a great influence on his entire carrier and interests. In the Introduction to his Habilitationsschrift Heinz describes M. G. Kreĭn's influence as follows: 'Wer das Glück hatte, eine längere Zeit in der Umgebung von Professor M. G. Krein in Odessa arbeiten zu können, weiß, welche Fülle von Gedanken und Anregungen er ständig ausstrahlt. Diese habe ich in wesentlich größerem Maße ausgenutzt, als in der Einleitung zum Ausdruck gebracht werden konnte.' Heinz's high regard for M. G. Kreĭn was reciprocated: M. G. Kreĭn considered Heinz one of his most brilliant students and collaborators.

This fruitful collaboration lasted for almost twenty five years and ended with the death of M. G. Kreĭn. M. G. Kreĭn has worked with many mathematicians. Of his joint publications, most are written with I. Gohberg, next in number come those with Heinz. In Odessa Heinz also met I. Gohberg for the first time. The friendship with him and with the doctoral students of M. G. Kreĭn of that time (V. M. Adamjan, D. S. Arov, V. A. Javrjan and S. N. Saakjan) lasts until today!

During this year in Odessa Heinz completed the main result of his thesis by adding a statement about the location of the spectrum of the restriction of the selfadjoint operator in this invariant subspace which was now the full generalization of Pontryagin's theorem. Jointly with M. G. Kreĭn he proved the existence of a spectral function (with certain critical points) for a selfadjoint operator in a Pontryagin space. In the following years (until 1965) Heinz showed the existence of a spectral function for the more general situation of a definitizable operator in a Kreĭn space. Thus, besides the existence of a maximal nonnegative invariant subspace, a second cornerstone of the spectral theory of selfadjoint and other classes of operators in Kreĭn spaces was laid.

In 1955 R. J. Duffin proved a result in connection with network theory about strongly damped selfadjoint second order matrix pencils $\lambda^2 I+\lambda B+C$, which M. G. Kreĭn ingeneously interpreted as the existence of a solution Z of the quadratic matrix equation $Z^2 + BZ + C = 0$. Heinz realized that the main result of his Ph.D. thesis, which was proved just as an abstract generalization without any application in mind, could easily be applied in order to get a corresponding result for the infinite dimensional case. This was worked out by M. G. Kreĭn and Heinz in the summer of 1962, and thus finally the spectral theory of selfadjoint operators in Kreĭn spaces found an important application. It should be added that at about the same time M. G. Kreĭn was working with I. Gohberg on the two books on nonselfadjoint operators and quite a few results of this theory also turned out to be useful for second order pencils.

The main results of the years 1961–1965 were summed up in Heinz's Habilitationsschrift. This Habilitationsschrift became well-known among people interested in spaces with an indefinite inner product and had a big impact on the development of the spectral theory of operators in such spaces. The results about the spectral function for definitizable operators were published without proof only in 1971 in [22], and the original proofs were published only seventeen years later in the Lecture Notes [63].

At this time, Heinz received an offer to work at the Mathematical Institute of the Academy of Sciences of the GDR, but he preferred to remain at the Technical University of Dresden as this also involved teaching and working with Ph.D. and post doctoral students, which he always liked and still likes to do.

The next important period in Heinz's development was his one year stay in Canada during the academic year 1966/67. After having met him at an operator theory conference in Balatonföldvar in Hungary in 1964, Professor I. Halperin invited Heinz to spend the academic year 1965/66 with a fellowship of the National Research Council of Canada at the University of Toronto. As was to be expected in these years of the Berlin wall, the authorities of the GDR did not allow Heinz to accept this invitation. However, Halperin insistently renewed it for the following year, and so Heinz could spend the academic year 1966/67 at the University of Toronto. Shortly before, Heinz had married, and in May 1967 his only daughter Henriette was born. In Canada he also met Peter Lancaster, with whom he shared interests not only in operator pencils, but also personal ones like skiing and hiking in the mountains.

After returning in 1967, Heinz was appointed Professor at the Technical University of Dresden. Shortly afterwards in 1968, with the '3rd Hochschulreform' in the GDR, research at the universities was reorganized. It turned out that officially there should be no research group in analysis at the Technical University of Dresden, but only groups in 'Numerik', 'Mathematische Kybernetik und Rechentechnik' and 'Stochastik'. Heinz joined the last group and became interested in semigroup theory and Markov processes, in particular one-dimensional Markov processes. A fairly wide class of such processes, which contains diffusion processes and birth and death processes, can be described by a second order generalized or Kreĭn-Feller derivative, which Heinz had come across in Odessa. Together with his students he considered in particular processes with nonlocal boundary conditions and the time reversal of such processes. Nevertheless, he continued to be interested in operators in indefinite inner product spaces. The disadvantage of the situation was that he could not lecture about the topics he liked best. Instead, he lectured on 'Semigroups', 'Spectral theory of Kreĭn-Feller differential operators', 'Markov processes' etc and the topics for graduate students also had to have a probabilistic touch. At this point, it turned out to be useful that Heinz was known abroad: From 1970 on some mathematicians from other countries came to Dresden to do their Ph.D. work with him (Pekka Sorjonen, Björn Textorius, Karim Daho, and later Branko Ćurgus, Muhamed Borogovac), which was, of course, in operator theory, the topic which always was his favourite.

In 1969 Heinz again visited M. G. Kreĭn for one month. By then the famous papers of Adamjan, Arov and Kreĭn were finished. Because of these results, it seemed to be necessary and promising to generalize the extension theory for symmetric operators in Hilbert space to Pontryagin spaces. In fact, it was clear that this would give an operator theoretic approach to the Adamjan, Arov, Kreĭn results.

This turned out to be an interesting and fruitful program: Already the generalization of the classical von Neumann-Kreĭn-Naimark extension theory showed new and interesting features in the indefinite case. In the following years also M. G. Kreĭn's theory of generalized resolvents, resolvent matrices and entire operators was extended to the indefinite situation.

In the seventies and early eighties Heinz visited Odessa quite regularly, sometimes officially, sometimes not quite officially, and then it was difficult to get the permission for the stay in Odessa. Sometimes the Rector of the Odessa Civil Engineering Institute could help. Another time, Heinz, without permission, just stayed at the apartment of Ju. L. Smul'jan, which was of course completely illegal and certainly a risk for the Smul'jan family.

During these years, besides the abstract lines of extension theory of symmetric operators, applications to indefinite moment problems, interpolation problems, to the continuation problem for hermitian functions with a finite number of negative squares, and to boundary eigenvalue problems were also studied. In the abstract results as well as in the applications the classical Hilbert space results were sometimes also completed, that is, the role of the Q–function in extension theory was worked out. These were the main topics of Heinz's work in these years, often done jointly with students and friends, mostly from outside Dresden or even outside the GDR. In addition, he also returned to work on operator pencils in the seventies. One of the main results he proved was the equivalence of a factorization of an n-th order pencil with the existence of a properly supported invariant subspace of the companion operator. At that time operator pencils were also studied by A. S. Markus and V. M. Matsaev. While Heinz used results from operator theory in indefinite inner product spaces as a tool, Markus and Matsaev applied results from factorization theory of analytic operator functions by I. Gohberg and J. Leiterer. It happened that on the same day in Kreĭn's seminar both methods were presented in a kind of friendly competition. However, only after Markus and Matsaev had moved to Israel and Heinz to Vienna, did the three of them start working together.

In the seventies Heinz could travel to the West almost every year, keeping the number of trips in balance with trips to the East. He lectured at the Universities of Jyväskylä, Uppsala, Linköping, Antwerp, and the KTH Stockholm during stays of a few weeks. Nevertheless, his applications for journeys to the West were not always successful, and it was impossible until the middle of the eighties to go to conferences there. In one case the permission to attend was granted only after the conference was over. While abroad he usually contacted Israel Gohberg and other friends (which was forbidden by the GDR authorities). Once Israel sent a letter for Heinz to Sweden, which arrived only after he had left. It was forwarded by the secretary to Dresden. Luckily someone gave him the open letter before it was read by some department officials (which was the rule for mail from abroad), and so it did not cause Heinz any problems.

In the eighties an intense collaboration developed with Aad Dijksma and Henk de Snoo from Groningen. They studied thoroughly the classes of analytic functions which arose in the extension theory of symmetric operators in spaces with indefinite inner product and applied this theory in order to get a unified treatment for selfadjoint boundary eigenvalue problems of ordinary differential operators containing the eigenvalue parameter in the boundary condition. On the invitation of Rien Kaashoek, he also regularly visited the Vrije Universiteit Amsterdam. All these contacts with colleagues from operator theory, which usually grew into friendship, were very important for Heinz's work. Also at that time, a result of R. Beals appeared about the half range completeness of Sturm-Liouville operators with an indefinite weight. Under the influence of Åke Pleijel, Heinz had considered such problems with Karim Daho already in the seventies in the context of the Kreĭn space generated by the weight function. So he understood that Beals' result could be interpreted as the regularity of the critical point infinity of the spectral function of the selfadjoint operator which can be associated with the problem in this Kreĭn space. This was further elaborated in cooperation with Branko Ćurgus.

In January 1988 Heinz was allowed for the first time to accept an invitation to West Germany: Reinhard Mennicken had invited him to spend one month at the University of Regensburg. They started joint work on the connections of operator pencils and special functions. Following an idea of F. W. Schäfke, certain systems of special functions were interpreted as eigenfunction systems of pencils of differential operators. They also began studying block operator matrices, which has been another main topic of Heinz's interests since then. The problem is to express the spectral properties of an operator, acting in the product of two spaces and given as a block operator matrix, by the properties of the entries of this matrix.

At the beginning of October 1989, shortly before the fall of the Berlin wall, Heinz fled from the GDR and went to West Germany. His first contact point was Regensburg. The decision to leave had ripened for a longer time and was certainly hard: a secure position at the university, pupils, friends and a part of his life had to be left behind. However, the pressure was stronger. Thanks to the assistance and intercession of Albert Schneider, Heinz obtained first a one year position as a professor at the University of Dortmund, and then, with the support of Reinhard Mennicken, a professorship at the University of Regensburg. Since August 1991 Heinz has held a chair in 'Anwendungsorientierte Analysis' at the Technical University of Vienna. Released from the psychological tension of life in the GDR, Heinz's life and work has come to a new blossoming. Within the last seven years he has organized three workshops on operator theory and its applications in Vienna, one of them in cooperation with the Schrödinger Institute, and he enjoys attending conferences all over the world. He has created a center of active research in operator theory in Vienna, attracting visitors from many countries, and still keeping a nice balance between those coming from the West and those from the East.

3. Some main results

In this section we explain some of Heinz's main results in detail. We focus on some theorems from the first three themes mentioned in the Introduction and relate them to the work of others, but first we recall some definitions.

An inner product space $(\mathcal{K}, [\cdot,\cdot])$ is called a Kreĭn space if $\mathcal{K}$ is a complex linear space which has a fundamental decomposition with respect to the inner product $[\cdot,\cdot]$, that is, a decomposition of the form

$$\mathcal{K} = \mathcal{K}_+[\dot{+}]\mathcal{K}_-,$$

where $[\dot{+}]$ denotes the direct $[\cdot,\cdot]$-orthogonal sum and $(\mathcal{K}_\pm, \pm[\cdot,\cdot])$ are Hilbert spaces. The fundamental decomposition induces a Hilbert space inner product on $\mathcal{K}$, given by

$$(x,y) = [x_+, y_+] - [x_-, y_-], \quad x = x_+ + x_-, \ y = y_+ + y_-, \quad x_\pm, y_\pm \in \mathcal{K}_\pm.$$

The operator $J = P_+ - P_-$, where $P_\pm$ is the $(\cdot,\cdot)$-orthogonal projection onto $\mathcal{K}_\pm$, is called the fundamental symmetry corresponding to the fundamental decomposition. Note $(x,y) = [Jx,y], x,y \in \mathcal{K}$. Although the fundamental decomposition is not unique, different ones generate equivalent Hilbert space norms. Topological notions refer to this Hilbert space topology. For example, a subspace of $\mathcal{K}$ is a linear manifold in $\mathcal{K}$ which is closed and continuity of an operator means continuity with respect to this norm topology, etc. We denote by $\mathbf{L}(\mathcal{K})$ the set of bounded linear operators on $\mathcal{K}$.

The numbers $\dim \mathcal{K}_\pm$, each either a nonnegative integer or infinity, do not depend on the fundamental decomposition $\mathcal{K} = \mathcal{K}_+[\dot{+}]\mathcal{K}_-$ of $\mathcal{K}$. If $\dim \mathcal{K}_+ = 0$ $\mathcal{K}$ is sometimes called an anti-Hilbert space. The Kreĭn space $(\mathcal{K}, [\cdot,\cdot])$ is called a π_κ-space or a Pontryagin space of index κ if $\kappa := \min(\dim \mathcal{K}_+, \dim \mathcal{K}_-) < \infty$. In the sequel we consider only Pontryagin spaces for which $\kappa = \dim \mathcal{K}_-$.

A linear subset is called nonnegative if its elements x have a nonnegative self inner product: $[x,x] \geq 0$; a nonpositive subset is defined in a similar way.

The linear operators, which we consider in the Kreĭn space $\mathcal{K}$, will in general be densely defined and closed or closable. If A is a densely defined operator then its adjoint A^+ is defined as follows: $\operatorname{dom} A^+$ is the set of all $u \in \mathcal{K}$ for which there exists a $v \in \mathcal{K}$ with

$$[Ax, u] = [x, v] \text{ for all } x \in \operatorname{dom} A,$$

and in this case $A^+u = v$. We have $A^+ = JA^*J$, where A^* is the adjoint of A with respect to the Hilbert space inner product $[J\cdot,\cdot]$. The operator A in the Kreĭn space $\mathcal{K}$ is called selfadjoint if $A = A^+$, symmetric if $A \subseteq A^+$, unitary if $A^+A = AA^+ = I$, isometric if $A^+A = I$, contractive if $[Ax, Ax] \leq [x,x]$ and expansive if $[Ax, Ax] \geq [x,x]$, for all $x \in \mathcal{K}$.

As mentioned before Heinz started his work in indefinite inner product spaces by studying the two articles by I. S. Iokhvidov and M. G. Kreĭn that had appeared in 1956 and 1959. Later these papers were incorporated in the joint work [57] with Heinz. It is a clear and comprehensive introduction to spectral theory in Pontryagin spaces and contains the basic results, not only on hermitian and isometric operators, but also on contractive and expansive operators in Pontryagin spaces. The material about the latter classes of operators has recently been reconsidered and completed in the joint publication [105] with Tomas Azizov.

From this early period also dates the Hilbert space result called "Langer's Lemma" (terminology from N.K. Nikol'skiĭ, Treatise on the shift operator, Grundlehren der mathematischen Wissenschaften 273, Springer-Verlag, Berlin, 1985) about the orthogonal decomposition of a Hilbert space contraction into a unitary part and a completely nonunitary part; see [2]. The same result was obtained independently by B. Sz.-Nagy and C. Foiaş.

(1) The maximal nonnegative invariant subspace theorem in Kreĭn spaces proved by Heinz reads as follows: If A is a selfadjoint operator in the Kreĭn space $\mathcal{K}$ and for some fundamental symmetry $J = P_+ - P_-$, $\operatorname{ran} P_- \subset \operatorname{dom} A$ and P_+AP_- is compact, then A has a maximal nonnegative invariant subspace. In particular, a (bounded or unbounded) selfadjoint operator A in a Pontryagin space has a maximal nonnegative and a maximal nonpositive invariant subspace. Since the latter is finite dimensional, A has eigenvalues with corresponding nonpositive eigenvectors. This statement can be made more precise by considering multiplicities and the location of the eigenvalues with respect to the real axis, see [57].

The spectral theory of definitizable operators in Kreĭn spaces developed in the Habilitationsschrift "Spektraltheorie linearer Operatoren in J-Räumen und einige Anwendungen auf die Schar $L(\lambda) = \lambda^2 + \lambda B + C$", TU-Dresden, 1965, is a cornerstone in the operator theory in spaces with an indefinite metric. The spectral function of a definitizable operator and the description of its behavior in the critical points are powerful tools in the abstract theory as well as for the investigation of operators in function spaces, such as differential operators. We review the main definitions and results:

The selfadjoint operator A in the Kreĭn space $\mathcal{K}$ is called definitizable if the resolvent set $\rho(A)$ of A is nonempty and there exists a polynomial p such that

$$[p(A)x, x] \geq 0, \quad x \in \operatorname{dom} A^k,$$

where k is the degree of p; p is called a definitizing polynomial of A. The set $c(A)$ of critical points of A is the set of all $\lambda \in \mathbf{R}$ such that $p(\lambda) = 0$ for each definitizing polynomial p of A, and $\infty \in c(A)$ if one (and hence each) definitizing polynomial is of odd degree and $\sigma(A)$ contains arbitrarily large positive and negative numbers. It can be shown that $c(A) \subseteq \sigma(A)$, the spectrum of A. We denote by $\mathcal{R}_A$ the Boolean algebra generated by all intervals of $\mathbf{R} \cup \{\infty\}$ whose endpoints are not in $c(A)$. Heinz proved that there exists a unique mapping $E : \mathcal{R}_A \to \mathbf{L}(\mathcal{K})$ with the following properties (Δ and Δ' are arbitrary elements of $\mathcal{R}_A$):

(a) $E(\Delta) = E(\Delta)^+$, $E(\emptyset) = 0$.
(b) $E(\Delta \cap \Delta') = E(\Delta)E(\Delta')$ (in particular, $E(\Delta)$ is a projection).
(c) $E(\Delta \cup \Delta') = E(\Delta) + E(\Delta') - E(\Delta)E(\Delta')$.
(d) $E(\mathbf{R}) = 1_{\mathcal{K}} - E_0$, where E_0 is the Riesz-Dunford projection associated with the nonreal spectrum $\sigma(A) \setminus \mathbf{R}$ of A.
(e) If $p|_\Delta > 0$ (or $p|_\Delta < 0$) for some definitizing polynomial p of A then $E(\Delta)\mathcal{K}$ is a positive (negative, respectively) subspace.
(f) $E(\Delta) \in \{(A - z)^{-1}\}''$, the double commutant of the resolvent $(A - z)^{-1}$ of A, $z \in \rho(A)$.
(g) If Δ is bounded then $E(\Delta)\mathcal{K} \subseteq \operatorname{dom} A$; if Δ is unbounded then $E(\Delta)\mathcal{K} \cap \operatorname{dom} A$ is dense in $E(\Delta)\mathcal{K}$. In both cases $\sigma(A|_{E(\Delta)\cap\mathcal{K}}) \subset \Delta^c$.
(h) If A is bounded then

$$p(A) = \int_{\mathbf{R}} p(\lambda)E(d\lambda) + N,$$

where N is a bounded nonnegative operator in $\mathcal{K}$ with $N^2 = 0$. (The integral here is improper with respect to the points $c(A)$ as at these points E is not defined.)

The mapping E is called the spectral function (with critical points) of the definitizable operator A. The statement (e) implies that $\mathcal{K}_\Delta = E(\Delta)\mathcal{K}$ is a Hilbert space or anti–Hilbert space if p is positive or negative on $\Delta \cap \sigma(A)$ for some definitizing polynomial p. This space $\mathcal{K}_\Delta$ reduces A and $A|_{\mathcal{K}_\Delta}$ is a bounded or unbounded and densely defined Hilbert space selfadjoint operator in $\mathcal{K}_\Delta$. Thus, with the exception of the (finitely many) points in $c(A)$, the definitizable operator A has locally the same spectral properties as a selfadjoint operator in a Hilbert space. The critical points $\lambda \in c(A)$ can be characterized as follows:

$$\Delta \in \mathcal{R}_A,\ \lambda \in \Delta \Rightarrow \text{ the inner product } [\cdot,\cdot] \text{ is indefinite on } \mathcal{K}_\Delta.$$

In [22] and [12] the results from the Habilitationsschrift in a completed form were published. It was proved in [22] among other more general results that a bounded definitizable selfadjoint operator in a Kreĭn space has a maximal dual pair of invariant subspaces. The question of uniqueness of such pairs was considered in the note [18].

The paper [12] concerns selfadjoint operators in Kreĭn spaces that arise from a fundamentally reducible selfadjoint operator by perturbations of Matsaev class. Under some additional assumptions it was proved that these operators possess a local spectral function. This paper is the starting point for the study of locally definitizable operators in papers by Peter Jonas and in the joint work [125] with Alexander Markus and Vladimir Matsaev. In the latter paper the sign classification of the spectrum of a selfadjoint operator in a Kreĭn space is obtained with the help of approximating eigensequences. This new approach is applied to the study of bounded and compact perturbations of selfadjoint operators in Kreĭn spaces. The main result in [125] on compact perturbations contains the result from [12] and has applications to the block operator matrices.

The papers [49] with Peter Jonas and [65] with Branko Najman deal with the perturbation theory of definitizable operators. In the first paper, for example, it is shown that within the class of selfadjoint operators, finite-rank perturbations of the resolvent preserve definitizability and that the emerging new critical points are of a special type. In the second paper, stability properties of the spectral function and its critical points are studied. In the paper [103] with Peter Jonas and Björn Textorius a model for an arbitrary selfadjoint operator in a Pontrjagin space is established. The model is closely related to selfadjoint differential operators with inner singularities arising in mathematical physics, presently under investigation with Aad Dijksma and Yuri Shondin.

(2) The spectral theory of selfadjoint operator pencils, of which Heinz is co-founder, is closely connected with indefinite metrics. The maximal nonnegative invariant subspace therorem has a large number of applications, for example, to the existence of an operator root and hence to the factorization of operator pencils. The joint papers [8] and [7] contain new ideas and methods which determined the development of this area for decades, and lead to new publications in spectral theory and in applications to mechanics and physics. To be more specific, with the pencil $L(\lambda) = \lambda^2 + \lambda B + C$ there is associated the quadratic operator equation

$$Z^2 + BZ + C = 0$$

and M.G. Kreĭn and Heinz looked for a root Z whose spectrum coincides with a specified part of the spectrum of the pencil. This problem is closely connected with the problem of factorizing the pencil, that is, the problem of representing it in the form

$$L(\lambda) = (\lambda I - Y)(\lambda I - Z).$$

This approach can be used even when the pencil is not selfadjoint. But in the selfadjoint case $B = B^*$, $C = C^*$, they proved that the quadratic equation has a root with the help of the invariant subspace theorem mentioned above applied to a certain companion matrix for the pencil. In the seventies Heinz proved general yet strong results on the factorization of operator polynomials of arbitrary degree; see [27], [30], [33], and [38]. For example, the operator polynomial

$$L(\lambda) = \lambda^n I + \lambda^{n-1} A_{n-1} + \cdots + \lambda A_1 + A_0$$

with operators A_j in a Hilbert space $(\mathcal{H}, (\cdot,\cdot))$ admits a factorization

$$L(\lambda) = N(\lambda)M(\lambda), \quad M(\lambda) = \lambda^k I + \sum_{j=0}^{k-1} \lambda^j B_j, \quad N(\lambda) = \lambda^{n-k} I + \sum_{j=0}^{n-k-1} \lambda^j C_j,$$

if and only if the companion operator

$$\tilde{A} = \begin{pmatrix} -A_{n-1} & \dots & -A_1 & -A_0 \\ 1 & & 0 & 0 \\ \vdots & & \vdots & \vdots \\ 0 & & 1 & 0 \end{pmatrix}$$

acting in $\tilde{\mathcal{K}} = \mathcal{H}^n$ has a specific invariant subspace. This result has been applied by many authors both for the selfadjoint and for the nonselfadjoint case. When the operators A_j are selfadjoint, the companion operator is selfadjoint with respect to the $\tilde{G}$–inner product $(\tilde{G}\cdot, \cdot)$ on $\tilde{\mathcal{K}}$, where

$$\tilde{G} = \begin{pmatrix} 0 & 0 & \dots & 0 & 1 \\ 0 & 0 & & 1 & A_{n-1} \\ \vdots & \vdots & & & \vdots \\ 0 & 1 & & & A_2 \\ 1 & A_{n-1} & \dots & A_2 & A_1 \end{pmatrix}.$$

In examples the operators $A_0, \dots, A_{n-1}$ are often unbounded. Sometimes $\tilde{A}$ can also be considered in this situation, sometimes by a simple transformation the given pencil can be transformed into one with bounded operators.

Because of the formula

$$L(\lambda)^{-1} = Q(\tilde{A} - \lambda)^{-1}P, \quad P = \begin{pmatrix} I \\ 0 \\ \vdots \\ 0 \end{pmatrix}, \; Q = \begin{pmatrix} 0 & \dots & 0 & I \end{pmatrix},$$

where P is a mapping from $\mathcal{H}$ into $\tilde{\mathcal{K}}$ and Q maps $\tilde{\mathcal{K}}$ into $\mathcal{H}$, the companion matrix $\tilde{A}$ is sometimes called the linearization of the pencil $L(\lambda)$. In the lecture series [120] other eigenvalue problems whose linearization lead to selfadjoint operators in Kreĭn spaces are discussed.

We also mention the following natural and beautiful result, which has a simple formulation but a complicated proof. We use the same notation as above. If $L(\lambda)$ is a selfadjoint polynomial and for some segment $[a, b]$ on the real axis,

$$L(a) \ll 0, \quad L(b) \gg 0, \quad L'(\lambda) \gg 0 \; (a < \lambda < b),$$

then $L(\lambda)$ admits the above factorization with $k = 1$, $M(\lambda) = \lambda I - Z$ and the spectrum of the operator Z lies in (a, b). The operator Z not only has a real spectrum, but it is also similar to a selfadjoint operator.

Finally, in 1971–1973 Heinz studied the important class of weakly hyperbolic selfadjoint operator polynomials of arbitrary degree (or polynomials with real zeros) and proved theorems about their so-called spectral zones and factorizations.

For the quadratic case this class is the class of "strongly damped pencils" and was considered earlier jointly with M.G. Kreĭn in [7] and [8].

(3) In the four "Fortsetzungsprobleme" papers [35], [40], [54], and [75] M.G. Kreĭn and Heinz formulate and study indefinite analogues of interpolation, moment and continuation problems. These papers contain a wealth of interesting results, which have subsequently been generalized by many authors. The indefiniteness comes in by requiring that certain kernels have κ negative squares, $\kappa \in \{0, 1, \ldots\}$. A kernel K on a nonempty set Ω is a function $K : \Omega \times \Omega \to \mathbf{C}$ which is hermitian: $K(z, w) = \overline{K(w, z)}$. It has κ negative squares on Ω if for every natural number n and arbitrary points $z_1, z_2, \ldots, z_n \in \Omega$, the hermitian matrix $(K(z_i, z_j))_{i,j=1}^n$ has at most and for at least one choice of $n, z_1, \ldots, z_n$ exactly κ negative eigenvalues counting multiplicities. Special kernels yield special classes of functions; we mention two examples from [35] and [75]:

(a) A function Q belongs to the class N_κ of generalized Nevanlinna functions if it is meromorphic on $\mathbf{C}^+$ and the kernel

$$N_Q(z, w) = \frac{Q(z) - \overline{Q(w)}}{z - \bar{w}}$$

has κ negative squares. For $\kappa = 0$, the class N_0 coincides with the class of Nevanlinna functions; by definition these functions are holomorphic on $\mathbf{C}^+$ and have a nonnegative imaginary part there. By N_κ^+ we denote the set of $Q \in N_\kappa$ for which $zQ(z) \in N_0$.

Like Nevanlinna functions, the functions in class N_κ have an operator and an integral representation, they are given in [35]. The latter is rather complicated because N_κ-functions have singularities which account for the negative squares; those at a nonreal point are just poles, but the ones on the real axis may be embedded.

(b) A function f belongs to the class $\mathcal{P}_\kappa$ if it is defined and continuous on $\mathbf{R}$, $f(t) = \overline{f(-t)}$, and the kernel $H_f(s, t) = f(s - t)$ has κ negative squares.

From the many interpolation, moment and continuation problems studied by M.G. Kreĭn and Heinz, we single out the following two. The Stieltjes moment problem:

Given a sequence $(s_j)_{j=0}^\infty$ of complex numbers such that of the Hankel forms

$$\sum_{j,k} s_{j+k} x_j \bar{x}_k, \quad \sum_{j,k} s_{j+k+1} x_j \bar{x}_k$$

the first has κ negative squares and the second is nonnegative, find all $Q \in N_\kappa^+$ such that

$$Q(z) \sim -\frac{s_0}{z} - \frac{s_1}{z^2} - \cdots, \quad z = iy, \ y \to \infty,$$

and the continuation problem:

Given the continuous function $f : [-2a, 2a] \to \mathbf{C}$ such that $f(t) = \overline{f(-t)}, t \in [-2a, 2a]$ and $H_f(s,t)$ has κ negative squares on $[-a, a]$, find all $\tilde{f} \in \mathcal{P}_\kappa$ such that

$$\tilde{f}(t) = f(t), \quad t \in [-2a, 2a].$$

For $\kappa = 0$, these problems where studied before by A.I. Akhiezer and M.G. Kreĭn, but even when restricted to this case some of the results in the Fortsetzungsprobleme were new.

The conditions on the data are necessary and sufficient for the existence of a solution, and there is either one solution or there are infinitely many solutions.

If the moment problem has infinitely many solutions, a 2×2 matrix function $W(z) = (w_{ij}(z))_{i,j=1}^2$ exists such that the formula

$$Q(z) = \frac{w_{11}(z)N(z) + w_{12}(z)}{w_{21}(z)N(z) + w_{22}(z)}$$

gives a one-to-one correspondence between all solutions $Q(z)$ and all functions $N(z) \in N_0^+ \cup \{\infty\}$.

A similar result holds for the continuation problem, but extra assumptions on the function f are needed: Assume that (i) f has an accelerant, that is, there is a hermitian function $H \in L^2(-2a, 2a)$ such that

$$f(t) = f(0) - \frac{1}{2}|t| - \int_0^t (t-s)H(s)ds, \quad t \in [-2a, 2a],$$

and that (ii) -1 does not belong to the spectrum of the integral operator $\mathbf{H}$ on $L^2(0, 2a)$ defined by

$$\mathbf{H}\varphi(t) = \int_0^{2a} H(t-s)\varphi(s)ds, \quad t \in [0, 2a].$$

Then if the continuation problem has infinitely many solutions, a 2×2 matrix function $\tilde{W}(z) = (\tilde{w}_{ij}(z))_{i,j=1}^2$ exists such that the formula

$$-i\int_0^\infty e^{-izt}\tilde{f}(t)dt = \frac{w_{11}(z)N(z) + w_{12}(z)}{w_{21}(z)N(z) + w_{22}(z)}, \quad \operatorname{Im} z \leq -\gamma,$$

for some $\gamma \geq 0$ gives a one-to-one correspondence between all solutions $\tilde{f}$ and all functions $N(z) \in N_0 \cup \{\infty\}$.

Suitably normalized, the resolvent matrices $W(z)$ and $\tilde{W}(z)$ are unique; their entries are entire and have finite order. The matrix $W(z)$ coincides essentially with the transmission matrix of a string that can be associated with the moment problem. The string has a special structure: besides positive masses, it also has a finite number of negative masses and certain new elements called dipoles; see [54], part II. Under certain conditions on the accelerant, the matrix $\tilde{W}(z)$ is a solution of a Hamiltonian system of differential equations; see [75]. These results are closely related to a theorem of Louis de Branges in his theory of Hilbert spaces of entire functions.

The method M.G. Kreĭn and Heinz used to obtain the above fractional linear transformation representation of the solutions is based on the extension theory of a symmetric operator or isometric operator in a Pontryagin space, developed in for example [20],[21], [28], and [40]: The data of the problem at hand give rise to a symmetric operator S in a Pontrjagin space $\mathcal{P}$ of index κ and an element u from $\mathcal{P}$. The solutions correspond 1–1 to the u-resolvents $[(\tilde{A}-z)^{-1}u,u]$ of S, where $\tilde{A}$ runs through the class of selfadjoint extensions of S with nonempty resolvent set acting in spaces of the form $\tilde{\mathcal{P}}=\mathcal{P}\oplus\mathcal{H}$, $\mathcal{H}$ a Hilbert space. These u-resolvents can be written as a fractional linear transformation over the functions from the class $N_0^+\cup\{\infty\}$ or $N_0\cup\{\infty\}$, depending on the problem.

Extension theory also entails the study of Straus extensions of a symmetric operator and the description of these involves unitary colligations and characteristic functions. For the indefinite case this has been worked out in, for example, [79], [80] and [82] with Aad Dijksma and Henk de Snoo and [91] also with Branko Ćurgus, and applied to the study of nonstandard boundary eigenvalue problems associated with Sturm-Liouville and Hamiltonian systems of differential operators; see [68], [85], [87], and [104]. Here nonstandard means that the boundary conditions contain the eigenvalue parameter. Earlier on eigenfunction expansions were obtained for the Hilbert space case in [62], [69], [74] with Björn Textorius using Kreĭn's method of directing functionals. Basis properties of the eigenfunctions for certain classes of boundary eigenvalue problems have been obtained recently with Reinhard Mennicken and Christiane Tretter in [121], [128] and [129].

We thank Paul A. Fuhrman for the picture of Heinz, Peter Jonas, Alex Markus and Wilfried Schenk for their valuable contributions to this section, and Christa Binder who in part helped us with the bibliography below.

List of publications of Heinz Langer

[1] On J–Hermitian operators, Doklady Akad. Nauk SSSR 134, 2 (1960), 263–266 (Russian); English transl.: Soviet Math. Dokl. 1 (1960), 1052–1055.

[2] Ein Zerspaltungssatz für Operatoren im Hilbertraum, Acta Math. Acad. Sci. Hung. XII, 3/4 (1961), 441–445.

[3] Zur Spektraltheorie J–selbstadjungierter Operatoren, Math. Ann. 146 (1962), 60–85.

[4] Über die Wurzeln eines maximalen dissipativen Operators, Acta Math. Acad. Sci. Hung. XIII, 3/4 (1962), 415–424.

[5] Eine Verallgemeinerung eines Satzes von L.S. Pontrjagin, Math. Ann. 152 (1963), 434–436.

[6] The spectral function of a selfadjoint operator in a space with indefinite metric, Doklady Akad. Nauk SSSR 152, 1 (1963), 39–42 (Russian); English transl.: Soviet Math. Dokl. 4 (1963), 1236–1239 (with M.G. Kreĭn).

[7] A contribution to the theory of quadratic pencils of selfadjoint operators, Doklady Akad. Nauk SSSR 154, 6 (1964), 1258–1261 (Russian); English transl.: Soviet Math. Dokl. 5 (1964), 266–269 (with M.G. Kreĭn).

[8] On some mathematical principles in the linear theory of damped oscillations of continua, Proc. Int. Sympos. on Applications of the Theory of Functions in Continuum Mechanics, Tbilissi, 1963, Vol. II: Fluid and Gas Mechanics, Math. Methods, Moscow, 1965, 283–322 (Russian); English transl.: Integral Equations Operator Theory 1 (1978), 364–399 and 539–566 (with M.G. Kreĭn).

[9] Eine Erweiterung der Spurformel der Störungstheorie, Math. Nachr. 30, 1/2 (1965), 123–135.

[10] Invariant subspaces of linear operators on a space with indefinite metric, Doklady Akad. Nauk SSSR 169, 1 (1966), 12–15 (Russian); English transl.: Soviet Math. Dokl. 7 (1966), 849-852.

[11] Einige Bemerkungen über dissipative Operatoren im Hilbertraum, Wiss. Zeitschrift der Techn. Universität Dresden 15, 4 (1966), 669–673 (with V. Nollau).

[12] Spektralfunktionen einer Klasse J–selbstadjungierter Operatoren, Math. Nachr. 33, 1/2 (1967), 107–120.

[13] Über stark gedämpfte Scharen im Hilbertraum, J. Math. Mech. 17, 7 (1968), 685–706.

[14] Über Lancaster's Zerlegung von Matrizenscharen, Arch. Rat. Mech. Anal. 29, 1 (1968), 75–80.

[15] Über einen Satz von M.A. Neumark, Math. Ann. 175 (1968), 303–314.

[16] Über die schwache Stabilität linearer Differentialgleichungen mit periodischen Koeffizienten, Math. Scand. 22 (1968), 203–208.

[17] A remark on invariant subspaces of linear operators in Banach spaces with an indefinite metric, Matem. Issledovanija Kišinev 4, 1 (1969), 27–34 (Russian).

[18] On maximal dual pairs of invariant subspaces of J-selfadjoint operators, Matem. Zametki 7 (1970), 443–447 (Russian).

[19] Über die Methode der richtenden Funktionale von M.G. Kreĭn, Acta. Sci. Math. Hung. 21, 1/2 (1970), 207–224.

[20] Über die verallgemeinerten Resolventen und die charakteristische Funktion eines isometrischen Operators in Raume Π_κ, Colloquia Math. Soc. Janos Bolyai, Tihany (Hungary), 5. Hilbert Space Operators, 1970, 353–399 (with M.G. Kreĭn).

[21] Defect subspaces and generalized resolvents of an hermitian operator in the space Π_κ, Funkcional. Anal. i Priložen. 5,2 (1971), 59–71; 5,3 (1971), 54–69 (Russian); English transl.: Functional Analysis Appl. (1971), 136–146; (1972) 217–228 (with M.G. Kreĭn).

[22] Invariante Teilräume definisierbarer J-selbstadjungierter Operatoren, Ann. Acad. Sci. Fenn. A I, 475 (1971), 1–23.

[23] Generalized coresolvents of a π-isometric operator with unequal defect numbers, Funkcional. Anal. i Priložen. 5,4 (1971), 73–75 (Russian); English transl.: Functional Analysis Appl., 5 (1971), 329–331.

[24] Verallgemeinerte Resolventen eines J-nichtnegativen Operators mit endlichem Defekt, J. Functional Analysis 8,2 (1971), 287–320.

[25] Zur Spektraltheorie verallgemeinerter gewöhnlicher Differentialoperatoren zweiter Ordnung mit einer nichtmonotonen Gewichtsfunktion, Universität Jyväskylä (Finland), Mathematisches Institut, Bericht 14 (1972), 1–58.

[26] Über verallgemeinerte gewöhnliche Differentialoperatoren mit nichtlokalen Randbedingungen und die von ihnen erzeugten Markov–Prozesse, Publ. Res. Inst. Math. Sci. (Kyoto) 7,3 (1972), 655–702 (with L. Partzsch and D. Schütze).

[27] Über eine Klasse polynomialer Scharen selbstadjungierter Operatoren im Hilbertraum, J. Functional Analysis 12,1 (1973), 13–29.

[28] Über die Q-Funktion eines π-hermitschen Operators im Raume Π_κ, Acta Sci. Math. Szeged 34 (1973), 191–230 (with M.G. Kreĭn).

[29] Über eine Klasse nichtlinearer Eigenwertprobleme, Acta Sci. Math. Szeged 35 (1973), 73–86.

[30] Über eine Klasse polynomialer Scharen selbstadjungierter Operatoren im Hilbertraum, II, J. Functional Analysis 16,2 (1974), 221–234.

[31] Verallgemeinerte Resolventen hermitescher und isometrischer Operatoren im Pontrjaginraum, Ann. Acad. Sci. Fenn. A I, 561 (1974), 1–45 (with P. Sorjonen).

[32] Über indexerhaltende Erweiterungen eines hermiteschen Operators im Pontrjaginraum, Math. Nachr. 64 (1974), 289–317 (with M. Grossman).

[33] Zur Spektraltheorie polynomialer Scharen selbstadjungierter Operatoren im Hilbertraum, Math. Nachr. 65 (1975), 301–319.

[34] Invariant subspaces for a class of operators in spaces with indefinite metric, J. Functional Analysis 19, 2 (1975), 232–241.

[35] Über einige Fortsetzungsprobleme, die eng mit der Theorie hermitescher Operatoren im Raume Π_κ zusammenhängen, Teil I: Einige Funktionenklassen und ihre Darstellungen, Math. Nachr. 77 (1977), 187–236 (with M.G. Kreĭn).

[36] A class of infinitesimal generators of one–dimensional Markov processes, J. Math. Soc. Japan 28, 2 (1976), 242–249.

[37] Zu einem Satz über Verteilungen quadratischer Formen in Hilberträumen, Math. Nachr. 61 (1974), 175–179 (with G. Maibaum and P.H. Müller).

[38] Factorization of operator pencils, Acta Sci. Math. Szeged 38, 1/2 (1976), 83–96.

[39] On the indefinite power moment problem, Doklady Akad. Nauk SSSR 226, 2 (1976), 261–264; English transl.: Soviet Math. Doklady 17 (1976), 90–93 (with M.G. Kreĭn).

[40] Über einige Fortsetzungsprobleme, die eng mit der Theorie hermitescher Operatoren im Raume Π_κ zusammenhängen, Teil II: Verallgemeinerte Resolventen, u-Resolventen und ganze Operatoren, J. Functional Analysis 30, 3 (1978), 390–447 (with M.G. Kreĭn).

[41] Spektralfunktionen einer Klasse von Differentialoperatoren zweiter Ordnung mit nichtlinearem Eigenwertparameter, Ann. Acad. Sci. Fenn. A I, Vol. 2 (1976), 269–301.

[42] Absolutstetigkeit der Übergangsfunktion einer Klasse eindimensionaler Fellerprozesse, Math. Nachr. 75 (1976), 101–112.

[43] Generalized resolvents and Q-functions of closed linear relations (subspaces) in Hilbert space, Pacific J. Math. 72, 1 (1977), 135–165 (with B. Textorius).

[44] Sturm–Liouville operators with an indefinite weight function, Proc. Royal Soc. Edinburgh, A 78 (1977), 161–191 (with K. Daho).

[45] Some remarks on a paper by W.N. Everitt, Proc. Royal Soc. Edinburgh, A 78 (1977), 71–79 (with K. Daho).

[46] Sturm–Liouville problems with indefinite weight function and operators in spaces with indefinite metric, Differential Equations Proc. Uppsala 1977, Int. Conference, Symp. Univ. Uppsala 7, 1977, 114–124.

[47] A generalization of M.G. Kreĭn's method of directing functionals to linear relations, Proc. Royal Soc. Edinburgh, A 81 (1978), 237–246 (with B. Textorius).

[48] Singular generalized second order differential operators with accessible or entrance boundaries, Preprint TU Dresden 07–10–78.

[49] Compact perturbations of definitizable operators, J. Operator Theory 2 (1979), 63–77 (with P. Jonas).

[50] Random spectral functions of a random string, Preprint TU Dresden 07–18–79.

[51] A factorization theorem for operator pencils, Integral Equations Operator Theory 2 (1979), 344-363 (with K. Harbarth).

[52] Szökefalvi–Nagy, Bela 65 éves, Matematika Lapok 27, 1–2 (1976–79), 7–24.

[53] A class of infinitesimal generators of one–dimensional Markov processes II. Invariant measures, J. Math. Soc. Japan 31, 1 (1980), 1–18 (with W. Schenk).

[54] On some extension problems which are closely connected with the theory of hermitian operators in a space Π_κ, III. Indefinite analogues of the Hamburger and Stieltjes moment problems, Beiträge zur Analysis 14 (1979), 25–40; 15 (1980), 27–45 (with M.G. Kreĭn).

[55] A class of infinitesimal generators of one–dimensional Markov processes III, Math. Nachr. 102 (1981), 25–44 (with W. Schenk).

[56] Generalized resolvents of contractions, Acta Sci. Math. Szeged 44 (1982), 125–131 (with B. Textorius).

[57] Introduction to the Spectral Theory of Operators in Spaces with an Indefinite Metric, Mathematical Research, Vol. 9, Akademie Verlag, Berlin, 1982 (with I.S. Iohvidov and M.G. Kreĭn).

[58] Some propositions on analytic matrix functions related to the theory of operators in the space Π_κ, Acta Sci. Math. Szeged 43 (1981), 181–205 (with M.G. Kreĭn).

[59] Continuous analogues of orthogonal polynomials with respect to an indefinite weight on the unit circle, and extension problems associated with them, Doklady Akad. Nauk SSSR 258, 3 (1981), 537–540 (Russian); English transl.: Soviet Math. Dokl. 23 (1981), 553–557 (with M.G. Kreĭn).

[60] Generalized resolvents and spectral functions of a matrix generalization of the Kreĭn–Feller second order derivative, Math. Nachr. 100 (1981), 163–186 (with L.P. Klotz).

[61] Generalized resolvents of dual pairs of contractions, Operator Theory: Adv. Appl., Vol. 6, Birkhäuser Verlag, Basel, 1982, 103–118 (with B. Textorius).

[62] L–Resolvent matrices of symmetric linear relations with equal defect numbers; applications to canonical differential relations, Integral Equations Operator Theory 5 (1982), 208–243 (with B. Textorius).

[63] Spectral functions of definitizable operators in Kreĭn spaces, Proc. Graduate School "Functional Analysis", Dubrovnik 1981. Lecture Notes in Math. 948, Springer Verlag, Berlin, 1982, 1–46.

[64] Some questions in the perturbation theory of J–nonnegative operators in Kreĭn spaces, Math. Nachr. 114 (1983), 205–226 (with P. Jonas).

[65] Perturbation theory for definitizable operators in Kreĭn spaces, J. Operator Theory 9 (1983), 297–317 (with B. Najman).

[66] On measurable hermitian–indefinite functions with a finite number of negative squares, Acta Sci. Math. Szeged 45 (1983), 281–292.

[67] Knotting of one–dimensional Feller processes, Math. Nachr. 113 (1983), 151–161 (with W. Schenk).

[68] Selfadjoint π_κ–extensions of symmetric subspaces: An abstract approach to boundary problems with spectral parameters in the boundary conditions, Integral Equations Operator Theory 7 (1984), 459–515 (with A. Dijksma and H.S.V. de Snoo).

[69] Spectral functions of a symmetric linear relation with a directing mapping. I, Proc. Royal Soc. Edinburgh, 97 A (1984), 165–176 (with B. Textorius).

[70] Spectral properties of selfadjoint differential operators with an indefinite weight function, Proc. 1984 Workshop "Spectral Theory of Sturm–Liouville differential operators", ANL–84.47, Argonne National Laboratory, Argonne, Illinois (1984), 73–80 (with B. Ćurgus).

[71] Matrix functions of the class $N_\kappa^{n\times n}$, Math. Nachr. 120 (1985), 275–294 (with K. Daho).

[72] A characterization of generalized zeros of negative type of functions of the class N_κ, Operator Theory: Adv. Appl., Vol. 17, Birkhäuser Verlag, Basel, 1985, 201–212.

[73] Some interlacing results for hermitian indefinite matrices, Linear Algebra Appl. 69 (1985), 131–154 (with B. Najman).

[74] Spectral functions of a symmetric linear relation with a directing mapping. II, Proc. Royal Soc. Edinburgh, 101 A (1985), 111–124 (with B. Textorius).

[75] On some continuation problems which are closely related to the theory of Hermitian operators in spaces Π_κ. IV: Continuous analogues of orthogonal polynomials on the unit circle with respect to an indefinite weight and related continuation problems for some classes of functions, J. Operator Theory 13 (1985), 299–417 (with M.G. Kreĭn).

[76] A model for π–selfadjoint operators in π_1–spaces and a special linear pencil, Integral Equations Operator Theory 8 (1985), 13–35 (with P. Jonas).

[77] Duality of a class of one–dimensional Markov processes, Math. Nachr. 125 (1986), 69–81 (with W. Schenk).

[78] Some spectral properties of operators which are related to one–dimensional Markov processes, Math. Nachr. 127 (1986), 51–63 (with B. Zagany).

[79] Unitary colligations in π_κ–spaces, characteristic functions and Štraus extensions, Pacific J. Math. 125, 2 (1986), 347–362 (with A. Dijksma and H.S.V. de Snoo).

[80] Representations of holomorphic functions by means of resolvents of unitary or selfadjoint operators on Kreĭn spaces, Operator Theory: Adv. Appl., Vol. 24, Birkhäuser Verlag, Basel, 1987, 123–143 (with A. Dijksma and H.S.V. de Snoo).

[81] Characteristic functions of unitary operator colligations in π_κ–spaces, Operator Theory: Adv. Appl., Vol. 19, Birkhäuser Verlag, Basel, 1986, 125–194 (with A. Dijksma and H.S.V. de Snoo).

[82] Kreĭn space extensions of an isometric operator in Hilbert space, characteristic functions and related questions, Proc. Graduate School "Functional analysis II", Dubrovnik 1985. Lecture Notes in Math. 1242, Springer Verlag, Berlin, 1987, 1–42 (with A. Dijksma and H.S.V. de Snoo).

[83] Sturm–Liouville operators with an indefinite weight function: The periodic case, Radovi Matematicki 2 (1986), 165–188 (with K. Daho).

[84] Time reversal of quasidiffusions, Lecture Notes in Control and Information Sciences, Vol. 96, Springer Verlag, Berlin, 1987, 156–163 (with W. Schenk).

[85] Symmetric Sturm–Liouville operators with eigenvalue depending boundary conditions, in: Oscillation, Bifurcation and Chaos, Canad. Math. Soc.–Amer. Math. Soc. Conference Proc., Vol. 8 (1987), 87–116 (with A. Dijksma and H.S.V. de Snoo).

[86] Generalized zeros of negative type of matrix functions of the class N_κ, Operator Theory: Adv. Appl., Vol. 28, Birkhäuser Verlag, Basel, 1988, 17–26 (with M. Borogovac).

[87] Hamiltonian systems with eigenvalue depending boundary conditions, Operator Theory: Adv. Appl., Vol. 35, Birkhäuser Verlag, Basel, 1988, 37–84 (with A. Dijksma and H.S.V. de Snoo).

[88] Remarks on the perturbation of analytic matrix functions. II, Integral Equations Operator Theory 12 (1989), 392–407 (with B. Najman).

[89] A Kreĭn space approach to symmetric ordinary differential operators with an indefinite weight function, J. Differential Equations 79, 1 (1989), 31–61 (with B. Ćurgus).

[90] Time reversal of transient gap diffusions, Mathematical Research, Vol. 54, Akademie-Verlag, Berlin, 1989, 104–114 (with W. Schenk).

[91] Characteristic functions of unitary colligations and of bounded operators in Kreĭn spaces, Operator Theory: Adv. Appl., Vol. 41, Birkhäuser Verlag, Basel, 1989, 125–152 (with B. Ćurgus, A. Dijksma, and H.V.S. de Snoo).

[92] Generalized second–order differential operators, corresponding gap diffusions and superharmonic transformations, Math. Nachr. 148 (1990), 7–45 (with W. Schenk).

[93] Definitizing polynomials of unitary and hermitian operators in Pontrjagin spaces, Math. Annalen 288 (1990), 231–243 (with Z. Sasvari).

[94] A transformation of right-definite S-hermitian systems to canonical systems, Differential and Integral Equations 3,5 (1990), 901–908 (with R. Mennicken).

[95] Generalized coresolvents of standard isometric operators and generalized resolvents of standard symmetric relations in Kreĭn spaces, Operator Theory: Adv. Appl., Vol. 48, Birkhäuser Verlag, Basel, 1990, 261–274 (with A. Dijksma and H.S.V. de Snoo).

[96] A second order differential operator depending nonlinearly on the eigenvalue parameter, Operator Theory: Adv. Appl., Vol. 48, Birkhäuser Verlag, Basel, 1990, 319–332 (with R. Mennicken and M. Möller).

[97] Linearization of boundary eigenvalue problems, Integral Equations Operator Theory 14 (1991), 105–119 (with M. Möller).

[98] On spectral properties of regular quasidefinite pencils $F - \lambda G$, Results in Mathematics 19 (1991), 89–109 (with A. Schneider).

[99] Remarks on the perturbation of analytic matrix functions.III, Integral Equations Operator Theory 15 (1992), 796–806 (with B. Najman).

[100] Perturbation of the eigenvalues of quadratic matrix polynomials, SIAM J. Matrix Anal. Appl. 13, 2 (1992), 474-489 (with B. Najman and K. Veselic).

[101] Some remarks about polynomials which are orthogonal with respect to an indefinite weight, Results in Mathematics 21 (1992), 152–164 (with A. Schneider).

[102] On Floquet eigenvalue problems for first order differential systems in the complex domain, J. Reine Angew. Math. 425 (1992), 87–121 (with R. Mennicken and M. Möller).

[103] Models and unitary equivalence of cyclic selfadjoint operators in Pontrjagin spaces, Operator Theory: Adv. Appl., Vol. 59, Birkhäuser Verlag, Basel, 1992, 252–284 (with P. Jonas and B. Textorius).

[104] Eigenvalues and pole functions of Hamiltonian systems with eigenvalue depending boundary conditions, Math. Nachr. 161 (1993), 107–154 (with A. Dijksma and H.V.S. de Snoo).

[105] Some spectral properties of contractive and expansive operators in indefinite inner product spaces, Math. Nachr. 162 (1993), 247–259 (with T.Ja. Azizov).

[106] Leading coefficients of the eigenvalues of perturbed analytic matrix functions, Integral Equations Operator Theory 16 (1993), 600–604 (with B. Najman).

[107] Expansions of analytic functions in series of Floquet solutions of first-order differential systems, Math. Nachr. 162 (1993), 279–314 (with R. Mennicken and M. Möller).

[108] Expansions of analytic functions in products of Bessel functions, Results in Math. 24 (1993), 129–146 (with R. Mennicken, M. Möller, and A. Sattler).

[109] Sturm-Liouville problems with coefficients which depend analytically on the eigenvalue parameter, Acta Sci.Math. (Szeged) 57 (1993), 25–44 (with F.V. Atkinson and R. Mennicken).

[110] On an elliptic boundary value problem arising in magnetohydrodynamics, Quaestiones mathematicae 17 (1994), 141–159 (with M. Faierman, R. Mennicken, and M. Möller).

[111] On the papers by M.G. Kreĭn in the theory of spaces with indefinite metric, Ukrainskiĭ Matem. Žurnal 46, 1/2 (1994), 5–17 (Russian, with T.Ya. Azizov and Yu.P. Ginzburg).

[112] Eigenvalues of a Sturm-Liouville problem depending rationally on the eigenvalue parameter, Mathematical Research, Vol. 79: Systems and Networks, Vol. II (1994), 589–594 (with V. Adamyan and R. Mennicken).

[113] The essential spectrum of some matrix operators, Math. Nachr. 167 (1994), 5–20 (with F.V. Atkinson, R. Mennicken, and A. Shkalikov).

[114] Generalized matrix functions of the class $N_\kappa^{m\times m}$, in: Linear and Complex Analysis, Problem Book 3, part I (V.P. Havin and N.K. Nikolski (eds)), Lecture Notes in Mathematics 1573, Springer Verlag, Berlin, 1994, 201–204 (with A. Dijksma and H.S.V. de Snoo).

[115] Spectral properties of a class of rational operator valued functions, J. Operator Theory 33 (1995), 259–277 (with V.M. Adamjan).

[116] Selfadjoint extensions of a closed linear relation of defect one in a Kreĭn space, Operator Theory: Adv. Appl., Vol. 80, Birkhäuser Verlag, Basel, 1995, 176–203 (with P. Jonas).

[117] Selfadjoint extensions for a class of symmetric operators with defect numbers (1,1), Topics in Operator Theory, Operator Algebras and Applications, Vol. 1 (1995), 115–145 (with S. Hassi and H.S.V. de Snoo).

[118] Spectral components of selfadjoint block operator matrices with unbounded entries, Math. Nachr. 178 (1996), 43–80 (with V. Adamjan, R. Mennicken, and J. Saurer).

[119] The essential spectrum of a non-elliptic boundary value problem, Math. Nachr. 178 (1996), 233–248 (with M. Möller).

[120] Operator Theory and Ordinary Differential Operators, in: Lectures on Operator Theory and its Applications, Fields Institute Monographs, Vol. 3, Amer. Math. Soc, Providence, RI, 1996, 75–139 (with A. Dijksma).

[121] A selfadjoint linear pencil $Q - \lambda P$ of ordinary differential operators, Methods of Functional Analysis and Topology 2, 1 (1996), 38–54 (with R. Mennicken and C. Tretter).

[122] Spectral properties of a compactly perturbed span of projections, Integral Equations Operator Theory 26 (1996), 353–366 (with V. Pivovarcik and C. Tretter).

[123] Elliptic problems involving an indefinite weight, Operator Theory: Adv. Appl., Vol. 87, Birkhäuser Verlag, Basel, 1996, 105–124 (with M. Faierman).

[124] Instability of singular critical points of definitizable operators, Integral Equations Operator Theory 28 (1997), 60–71 (with B. Najman).

[125] Locally definite operators in indefinite inner product spaces, Math. Annalen 308 (1997), 405–424 (with A. Markus and V. Matsaev).

[126] Notes on a Nevanlinna-Pick interpolation problem for generalized Nevanlinna functions, Operator Theory: Adv. Appl., Vol. 95, Birkhäuser Verlag, Basel, 1997, 69–91 (with A. Dijksma).

[127] Nonnegative solutions of algebraic Riccati equations, Linear Alg. Appl. 261 (1997), 317–352 (with A.C.M. Ran and D. Temme).

[128] Spectral properties of the Orr-Sommerfeld problem, Proc. Royal Soc. Edinburgh 127A (1997), 1245–1261 (with C. Tretter).

[129] Spectral decomposition of some nonselfadjoint block operator matrices, J. Operator Theory 39, 2 (1998), 339–359 (with C. Tretter).

[130] Resolvents of symmetric operators and the degenerated Nevanlinna-Pick problem, Operator Theory: Adv. Appl., Vol. 103, Birkhäuser Verlag, Basel, 1998, 233–261 (with H. Woracek).

[131] Classical Nevanlinna-Pick interpolation with real interpolation points, Operator Theory: Adv. Appl. (with D. Alpay and A. Dijksma).

[132] Variational principles for real eigenvalues of selfadjoint operator pencils, Integral Equations Operator Theory (with D. Eschwé and P. Binding)

[133] The spectral shift function for certain block operator matrices, Math. Nachr. (with V. Adamjan)

[134] Direct and inverse spectral problems for generalized strings, Integral Equations Operator Theory (with H. Winkler)

Operator Theory:
Advances and Applications, Vol. 106

On the spectra of some classes of quadratic operator pencils

V. Adamyan and V. Pivovarchik

Dedicated to Heinz Langer on the occasion of his 60th birthday

This paper is devoted to the description of the spectra of some unbounded quadratic operator pencils in Hilbert spaces arising in the theory of damped oscillations of continua.

1. Introduction

The renowned work of Mark Krein and Heinz Langer on strongly damped quadratic operator pencils [KL] was a fundamental contribution to the study of spectral problems which are nonlinear in the spectral parameter. The influence of [KL] may be traced in later developments of abstract spectral theory of operator pencils and its applications to mechanics of continua. In this paper we go again through a very simple aspect of the theory worked out in [KL], which is related to the description of the spectra of quadratic operator pencils. However we do this under some weaker demands on the operator coefficients than in [KL].

Like in other investigations on this topic let us start with the following provocative example. Consider the equation

$$\beta\frac{\partial^4 u}{\partial x^4}+\frac{\partial^3}{\partial t\partial x^2}\left(\alpha(x)\frac{\partial^2 u}{\partial x^2}\right)+\frac{\partial}{\partial x}\left(g(x)\frac{\partial u}{\partial x}\right) + k(x)\frac{\partial u}{\partial t} + \tag{1.1}$$
$$- \rho(x)\frac{\partial^2 u}{\partial t^2}=0,\quad x\in[0,1],$$

which describes small transverse vibrations of a viscoelastic thin beam of unit length with external and internal damping (the so-called Kelvin-Voigt material [Pal]). Here $x\in[0,1]$ is the coordinate along the beam, $t\geq 0$ is the time, $u(x,t)$ is the transverse displacement at the position x and time t. The function $\alpha(x)\geq 0$ describes the internal damping, $k(x)\geq 0$ is the coefficient of the external damping (viscous friction), and $g(x)$ is responsible for the forces of contraction or tension. The function $\rho\geq 0$ is the density. Let us assume that ρ is equal to zero only on a set of measure zero. For the sake of simplicity we assume also that $k(x)$ is continuous, $g(x)$ is continuously differentiable, and $\alpha(x)$ is twice continuously differentiable.

As to ρ, we only assume that it belongs to $\mathbf{L}^2(0,1)$. Let us choose the boundary conditions corresponding to clamped ends of the beam:

$$u(0,t) = \frac{\partial u(x,t)}{\partial x}\,|_{x=0} = u(1,t) = \frac{\partial u(x,t)}{\partial x}\,|_{x=1} = 0. \tag{1.2}$$

Substituting $u(x,t) = y(\lambda,x)e^{\lambda t}$ into (1.1) and (1.2) we obtain the following boundary problem

$$\begin{aligned} &y^{iv} + (g(x)y')' + \lambda\left((\alpha(x)y'')'' + k(x)y\right) - \lambda^2\rho(x)y = 0,\\ &y(\lambda,0) = y'(\lambda,0) = y(\lambda,1) = y'(\lambda,1) = 0. \end{aligned} \tag{1.3}$$

This problem with $\alpha = constant$ was considered in [ZKM], [Pi1], [Pi2], [LS] (and in [Pi3], [GS] on a half-axis). Introduce the following operators acting in $\mathbf{L}^2(0,1)$:

$$\begin{aligned}
&(1.4)\quad (Ay)(x) &=&\ y^{iv}(x),\\
&\qquad\quad \mathcal{D}(A) &=&\ \{y : y \in \mathbf{W}_4^2(0,1),\ y(0) = y'(0) = y(1) = y'(1) = 0\},\\
&(1.5)\quad (By)(x) &=&\ (g(x)y'(x))',\\
&\qquad\quad \mathcal{D}(B) &=&\ \{y : y \in \mathbf{W}_2^2(0,1),\ y(0) = y(1) = 0\},\\
&(1.6)\quad (Cy)(x) &=&\ (\alpha(x)y''(x))'' + k(x)y, \quad \mathcal{D}(C) = \mathcal{D}(A),\\
&(1.7)\quad (Dy)(x) &=&\ \rho(x)y(x), \quad \mathcal{D}(D) = \{y : \rho y \in \mathbf{L}^2(0,1)\}.
\end{aligned}$$

Then the spectral problem (1.1), (1.2) is reduced to that for the quadratic operator pencil

$$L_r(\lambda) = \lambda^2 D + \lambda C + A + B. \tag{1.8}$$

The equation analogous to (1.1) for small damped vibrations of a thin pipeline conveying the stationary flow of an incompressible fluid contains also an addendum

$$\gamma\frac{\partial^2 u}{\partial t\partial x}, \quad \gamma = constant,$$

to account for gyroscopic forces. In this case we face spectral problems for the pencil

$$L(\lambda) = L_r(\lambda) + i\lambda G,$$

where G is the selfadjoint operator

$$(Gy)(x) = -i\gamma y'(x), \quad \mathcal{D}(G) = \{y : y \in \mathbf{W}_1^2(0,1),\ y(0) = y(1)\}. \tag{1.9}$$

These problems give good grounds for studying the abstract spectral problems for non-selfadjoint quadratic pencils of the form

$$L(\lambda) = \lambda^2 D + \lambda(C + iG) + A + B, \tag{1.10}$$

where all coefficients are in general unbounded operators acting in some Hilbert space, A is positive and invertible, B, C, D, and G are selfadjoint operators with domains containing the domain of A, so that by definition $\mathcal{D}(L(\lambda)) = \mathcal{D}(A)$ for all $\lambda \in \mathbf{C}$. Observe that even the definitions of the resolvent set and the spectra of such pencils demand more care than in the case of operator pencils with bounded coefficients. For example (see [Pi4]), for the operator pencil

$$M(\lambda) = \lambda^2 I + \lambda A + A,$$

where I is the identity operator and A is an unbounded positive invertible operator with compact inverse, we see that $M(-1) = I \mid_{\mathcal{D}(A)}$ is not closed. Consequently, the number -1 should not be considered as a regular point of $M(\lambda)$. In fact, -1 is an accumulation point of the eigenvalues of $M(\lambda)$.

From now on we assume that $C \geq 0$ and $D > 0$. For the special case, where (i) the negative spectrum of $A + B$ consists of normal eigenvalues, (ii) for some $\alpha > 0$

$$C = \alpha A + C_1 \tag{1.11}$$

and (iii) C_1 is subordinate to A in the sense that there exist constants $l > 1$, $a > 0$, and $b > 0$ such that for $y \in \mathcal{D}(A)$,

$$\max\left\{\|By\|^l,\ \|C_1 y\|^l\right\} \leq a\|y\|^l + b\|Ay\|\|y\|^{l-1},$$

it was proved in particular in [Pi1], [Pi2] that the part of the spectrum of $L_r(\lambda)$ in the open right half-plane is real and consists of semisimple normal eigenvalues and that the total multiplicity of the spectrum of $L_r(\lambda)$ in the right half-plane (possibly infinite) is equal to the total multiplicity of the negative spectrum of $A + B$.

An interesting attempt to describe the spectral structure of $L(\lambda)$ in the general case was undertaken in [S] on the basis of a generalized notion of the spectrum of $L(\lambda)$, which is understood loosely speaking as the spectrum of the properly defined pencil $\widehat{L}(\lambda) = A^{-\frac{1}{2}} L(\lambda) A^{-\frac{1}{2}}$. In this way the beautiful index formula and important inequalities concerning the set of normal eigenvalues of $L(\lambda)$ were established. However, from the results presented in [S] it is hard to see directly what is the essential spectrum of the given $L(\lambda)$.

In the present paper under some assumptions we try to describe the spectrum of $L(\lambda)$ remaining in the framework of standard definitions. In Section 2 the essential and discrete spectrum of an abstract quadratic pencil with operator coefficients acting in a Hilbert space are singled out. Here the assumptions on coefficients are taken such that they include the above examples. In Section 3 for the case of pencils with selfadjoint coefficients we prove under the same assumptions that the number of all eigenvalues of the pencil in the right half-plane counted with respect to their multiplicities coincides with that for the negative eigenvalues of $A + B$. In Section 4 we show that the results obtained for abstract operator pencils are indeed applicable to the above examples. It is worth mentioning that in this paper we do not use any linearization in the spectral parameter.

2. The essential and discrete spectrum of $L(\lambda)$

Let A and D be selfajoint operators and let B, C, and G be densely defined symmetric operators in the Hilbert space $\mathcal{H}$. Throughout the paper we assume that the following conditions are satisfied:

(a) $A \gg 0$ and A^{-1} is compact,

(b) $\mathcal{D}(A) \subset \mathcal{D}(C)$ and there exist a bounded operator $K \geq 0$ and a compact operator R such that
$$C = (K + R)A,$$

(c) $\mathcal{D}(A) \subset \mathcal{D}(B)$, $\mathcal{D}(A) \subset \mathcal{D}(D)$, $\mathcal{D}(A) \subset \mathcal{D}(G)$, and there exist compact operators X, Y, and Z such that
$$B = XA, \quad D = YA, \quad G = ZA,$$

(d) $D > 0$ and $C \geq 0$.

This work is devoted to study of spectra in the right half-plane of quadratic pencils
$$L(\lambda) = \lambda^2 D + \lambda(C + iG) + A + B$$
on the λ-independent linear manifold $\mathcal{D}(L) = \mathcal{D}(A)$.

First we introduce some notations. Let $T(\lambda)$ be a holomorphic operator function on some open connected set $\mathbf{\Omega} \subset \mathbf{C}$, the values of which are closed operators acting in $\mathcal{H}$ with a λ-independent dense domain $\mathcal{D}(T)$. Suppose that $T(\lambda)x$ for each $x \in \mathcal{D}(T)$ is a holomorphic vector function on $\mathbf{\Omega}$. If $T(\lambda)$ is bijective and $T(\lambda)^{-1}$ is bounded, we say as usual that $T(\lambda)$ has a bounded inverse. We call
$$\rho(T) := \{\lambda \in \mathbf{\Omega} : T(\lambda) \text{ has a bounded inverse }\}$$
the *resolvent set* of T and
$$\sigma(T) := \mathbf{\Omega} \setminus \rho(T)$$
the *spectrum* of T. The point spectrum of T is defined by
$$\sigma_p(T) := \{\lambda \in \mathbf{\Omega} : T(\lambda) \text{ is not injective }\}.$$
A point $\lambda_0 \in \sigma_p(T)$ is called an *eigenvalue* of T, and a vector $x_0 \neq 0$ such that
$$T(\lambda_0)x_0 = 0$$
is called an *eigenvector* of T with respect to the eigenvalue λ_0. The dimension of the kernel $\mathcal{N}(\lambda_0)$ of operator $T(\lambda_0)$ is called the *geometric multiplicity* of the

eigenvalue λ_0. The vectors x_0, x_1, ..., x_{m-1} form a *chain of associated vectors* to the eigenvector $x_0 \neq 0$ if

$$\sum_{s=0}^{k} \frac{1}{s!} \left(\frac{\partial^s}{\partial z^s} T \right) (\lambda_0) x_{k-s} = 0, \ k = 1, ..., m-1.$$

The number m is called the *length* of this chain. Denote by $\mathcal{L}(\lambda_0)$ the minimal subspace containing the vectors of all chains of associated vectors to all eigenvectors with respect to the eigenvalue λ_0. The dimension of $\mathcal{L}(\lambda_0)$ is called the *algebraic multiplicity* of the eigenvalue λ_0. An *isolated* eigenvalue λ_0 of T is called *normal* if λ_0 has finite algebraic multiplicity and the range of the operator $T(\lambda_0)$ is closed and has at most finite co-dimension i.e. if $T(\lambda_0)$ is fredholmian. We will denote by $\sigma_0(T)$ the set of all normal eigenvalues of T. This set forms the *discrete spectrum* of T. The set

$$\sigma_{ess}(T) := \{\lambda \in \sigma(T) : T(\lambda) \text{ is not fredholmian}\}$$

is called the *essential spectrum* of T. By our assumptions (a)–(d),

$$BA^{-1} = X, \quad CA^{-1} = K + R, \quad DA^{-1} = Y, \quad GA^{-1} = Z$$

are bounded operators and BA^{-1}, DA^{-1}, and GA^{-1} are compact. Let

$$\widetilde{L}_{\pm}(\lambda) := \lambda^2 Y + \lambda(K + R \pm iZ) + I + X. \tag{2.1}$$

For each $\lambda \in \mathbf{C}$, $\widetilde{L}_{\pm}(\lambda)$ a bounded operator. Since A has a bounded inverse, the equality

$$L(\lambda) = \widetilde{L}_{+}(\lambda)A \tag{2.2}$$

shows that $\sigma(L) = \sigma(\widetilde{L}_{+})$. Moreover, $\lambda_0 \in \mathbf{C}$ is a normal eigenvalue of $L(\lambda_0)$ or $L(\overline{\lambda}_0)^*$ respectively if and only if λ_0 is a normal eigenvalue of $\widetilde{L}_{+}(\lambda_0)$ or $\widetilde{L}_{-}(\lambda_0)$ respectively and $L(\lambda)$ is not fredholmian if and only if $\widetilde{L}_{+}(\lambda)$ is not fredholmian. It follows that

$$\sigma_0(L) = \sigma_0(\widetilde{L}_{+}), \quad \sigma_0(L^*) = \sigma_0(\widetilde{L}_{-}), \quad \sigma_{ess}(L) = \sigma_{ess}(\widetilde{L}_{+}). \tag{2.3}$$

For each $x \in \mathcal{D}(A)$ we have

$$L(\lambda)^* x = (\overline{\lambda}^2 D + \overline{\lambda}(C - iG) + A + B)x = \widetilde{L}_{-}(\overline{\lambda})Ax.$$

Hence

$$\overline{\widetilde{L}_{-}(\overline{\lambda})A} \subseteq L(\lambda)^*, \quad \lambda \in \mathbf{C},$$

and moreover, since A is an invertible operator, we have for $\overline{\lambda} \in \rho(\widetilde{L}_{-})$,

$$L(\lambda)^* = \widetilde{L}_{-}(\overline{\lambda})A.$$

Therefore

$$\{\lambda : \overline{\lambda} \in \rho(\widetilde{L}_-)\} \subseteq \rho(L^*), \quad \sigma_{ess}(L^*) \subset \{\lambda : \overline{\lambda} \in \sigma_{ess}(\widetilde{L}_-)\},$$
$$\{\lambda : \overline{\lambda} \in \sigma_p(\widetilde{L}_-)\} \subseteq \sigma_p(L^*) \tag{2.4}$$

Let us denote by $\widetilde{L}_0(\lambda)$ the linear bounded pencil $I + \lambda K$. Evidently,

$$\sigma(\widetilde{L}_0) \subset (-\infty, -\|K\|^{-1}].$$

Theorem 2.1. *Let A, B, C, D, and G satisfy the assumptions* (a)–(d) *above. Then for the quadratic pencil $L(\lambda)$,*

$$\sigma_{ess}(L) = \sigma_{ess}(\widetilde{L}_0) \subset (-\infty, -\|K\|^{-1}] \quad \text{and} \quad \sigma(L) \setminus \sigma_{ess}(\widetilde{L}_0) = \sigma_0(L).$$

Proof. According to (2.1) and the assumptions (a)–(c),

$$\widetilde{L}_+(\lambda) = I + \lambda K + Q(\lambda) = \widetilde{L}_0(\lambda) + Q(\lambda), \tag{2.5}$$

where

$$Q(\lambda) := \lambda^2 Y + \lambda(R + iZ) + X$$

is a quadratic pencil with compact coefficients. It follows from (2.5) that $\widetilde{L}(\lambda)$ is not fredholmian if and only if $I + \lambda K$ is not fredholmian. Hence

$$\sigma_{ess}(L) = \sigma_{ess}(\widetilde{L}) = \sigma_{ess}(\widetilde{L}_0).$$

Let

$$F(\lambda) := (I + \lambda K)^{-1} Q(\lambda), \quad \lambda \notin \sigma(\widetilde{L}_0).$$

It follows from (2.2) and the representation

$$\widetilde{L}_+(\lambda) = (I + \lambda K)(I + F(\lambda)) = \widetilde{L}_0(\lambda)(I + F(\lambda)), \quad \lambda \notin \sigma(\widetilde{L}_0), \tag{2.6}$$

that $\lambda_0 \in \mathbf{C} \setminus \sigma(\widetilde{L}_0)$ is an eigenvalue of $\widetilde{L}_+(\lambda)$ (and hence of $L(\lambda)$) of geometric multiplicity n if and only if the equation

$$x + F(\lambda_0)x = 0, \quad x \in \mathcal{H}, \tag{2.7}$$

has n linearly independent solutions. Note that $F(\lambda)$ is a holomorphic operator function in $\mathbf{C} \setminus \sigma(\widetilde{L}_0)$ with compact values and that $\mathbf{C} \setminus \sigma(\widetilde{L}_0)$ is a connected set. Therefore by a theorem of I. Gohberg, for all $\lambda \in \mathbf{C} \setminus \sigma(\widetilde{L}_0)$ except for some isolated points the number $n(\lambda)$ of linearly independent solutions of (2.7) is a constant n_0 and at the mentioned isolated points: $n_0 < n(\lambda) < \infty$ (see [GK], [GGK], Ch. XI, Corollary 8.4). Besides the algebraic multiplicity of each eigenvalue $\lambda \in \mathbf{C} \setminus \sigma(\widetilde{L}_0)$ of $I + F$ is finite (see [M]). In particular, if $n(\lambda_0) = 0$ for at least one $\lambda_0 \in \mathbf{C} \setminus \sigma(\widetilde{L}_0)$, then $n_0 = 0$ and the operator $I + F(\lambda)$ is invertible for almost all $\lambda \in \mathbf{C} \setminus \sigma(\widetilde{L}_0)$ except for a set of isolated points, which consists of all normal eigenvalues of $I + F(\lambda)$. Let us prove that for given $F(\lambda)$ we indeed have $n_0 = 0$. To this end it suffices to show that there exist points $\lambda \in \mathbf{C} \setminus \sigma(\widetilde{L}_0)$, which are not eigenvalues of L. The last claim is evident for the auxiliary symmetric pencil

$$L_r(\lambda) := \lambda^2 D + \lambda C + A + B, \tag{2.8}$$

which is obtained from $L(\lambda)$ by taking $G = 0$. Indeed, by the assumptions (a)–(d), we have for any $\lambda = \mu + i\tau$, $\mu > 0$, $\tau \neq 0$ and all $x \in \mathcal{D}(A)$ with $x \neq 0$,

$$\|L_r(\lambda)x\| \geq |\operatorname{Im}(L_r(\lambda)x, x)| \geq |\tau| \, ((2\mu D + C)x, x) > 0.$$

Hence *the non-real points of the open right half-plane are not eigenvalues of* $L_r(\lambda)$ *i.e. the eigenvalues of* $L_r(\lambda)$ *in the open right half-plane only belong to the positive half-axis.* Applying the above reasoning for the special case $G = 0$ we see that all such eigenvalues are normal and that they do not accumulate anywhere in $[0, +\infty)$. Since for $\lambda > 0$ we have

$$\operatorname{Re}(L(\lambda)x, x) = (L_r(\lambda)x, x), \ x \in \mathcal{D}(A),$$

it remains to prove that $L_r(\lambda) > 0$ for some $\lambda > 0$. This property follows from the next lemma.

Lemma 2.2. *Let A, B, C, and D satisfy the assumptions* (a)–(d) *above. Then for* $\lambda \geq 0$,
a) *values of L_r are self-adjoint operators,*
b) *there exists an $\alpha > 0$ such that $L_r(\lambda) \gg 0$ for $\lambda \geq \alpha$, and*
c) *$L_r(\lambda)^{-1}$ converges strongly to zero as $\lambda \to +\infty$.*

Proof. By assumption, B is an A-compact symmetric operator. Therefore *$A + B$ is a self-adjoint operator* [K]. Since A is a self-adjoint operator and $A \gg 0$ we conclude that the A-bound of B is infinitesimal and as follows that *$A + B$ is bounded below* [K]. The A-compactness of B entails also

$$\sigma_{ess}(A + B) = \sigma_{ess}(A). \tag{2.9}$$

Set

$$\beta := \inf \sigma_{ess}(A + B) = \inf \sigma_{ess}(A).$$

By the assumptions $\beta > 0$. We see that *the number of all eigenvalues of $A + B$ less then any $\beta' < \beta$ counted with respect to their multiplicities is finite.*

To prove that $L_r(\lambda)$ on $\mathcal{D}(A)$ is a self-adjoint operator for each $\lambda > 0$ let us assume first that $A + B \geq 0$. Then $L_r(\lambda) > 0$ for $\lambda > 0$ and it follows that the bounded pencil

$$\widetilde{L}_r(\lambda) := \widetilde{L}_{r,+}(\lambda) = (I + (\lambda^2 Z + \lambda R + X)(I + \lambda K)^{-1})(I + \lambda K)$$

is invertible. Hence for each $\lambda > 0$, the range of the symmetric operator $L_r(\lambda)$ is all of $\mathcal{H}$. Thus, by a theorem of von Neumann, this operator is self-adjoint. Let N_- be the number of negative eigenvalues of $A+B$ counted with respect to their multiplicity. To deal with the case $N_- > 0$, we note that $N_- < \infty$ and therefore there exists a non-negative perturbation B_1 of finite rank such that $A+B+B_1 \geq 0$. Evidently, B_1 is a bounded operator. By the above arguments for the case $N_0 = 0$, the values of the pencil

$$L_{r,1}(\lambda) := L_r(\lambda) + B_1$$

are self-adjoint operators for $\lambda > 0$. Since $L_r(\lambda)$ differs from $L_{r,1}(\lambda)$ by a λ-independent bounded self-adjoint operator we conclude that $L_r(\lambda)$ is self-adjoint for each $\lambda > 0$.

Let $e_1, ..., e_s, ..., e_\ell$ denote some orthonormal basis of the invariant subspace $\mathcal{H}_-(\frac{\beta'}{2})$ of the self-adjoint operator $A+B$ generated by its spectrum in $(-\infty, \frac{\beta'}{2})$ and let $P_- := \sum_s(\cdot, e_s)e_s$ be the orthogonal projector on $\mathcal{H}_-(\frac{\beta'}{2})$. Since $(-\infty, \frac{\beta'}{2}] \cap \sigma_{ess}(A+B) = \emptyset$, we have $\ell = \operatorname{rank} P_- < \infty$. Set

$$\widehat{L}_r(\lambda) := L_r(\lambda) + (|\nu_0| + \frac{\beta'}{2})P_-,$$

where ν_0 is the smallest eigenvalue of $A+B$. By construction,

$$A + \widehat{B} := A + B + (|\nu_0| + \frac{\beta'}{2})P_- \geq \frac{\beta'}{2} > 0.$$

Whence $\widehat{L}_r(\lambda)$ is a self-adjoint operator for $\lambda > 0$ and we have

$$\widehat{L}_r(\lambda) \geq \lambda^2 D + \frac{\beta'}{2} \gg 0, \quad \lambda > 0. \tag{2.10}$$

Let $\mathbf{E}(t)$ be the spectral function of D. As $D > 0$ and taking into account (2.10) we obtain for arbitrary $x \in \mathcal{H}$,

$$(\widehat{L}_r(\lambda)^{-1}x, x) \leq ((\lambda^2 D + \frac{\beta'}{2})^{-1}x, x) = \int_0^\infty \frac{2}{2\lambda^2 t + \beta'} d(\mathbf{E}(t)x, x). \tag{2.11}$$

Since by our assumption $\mathbf{E}(+0) = \mathbf{E}(0)$, (2.11) clearly forces

$$\lim_{\lambda > 0, \lambda \to +\infty} (\widehat{L}_r(\lambda)^{-1}x, x) = 0, \quad x \in \mathcal{H}. \tag{2.12}$$

Combining the Cauchy inequality with (2.12), we can assert that $\widehat{L}_r(\lambda)^{-1}$ tends weakly to zero as $\lambda \to +\infty$. But $\widehat{L}_r(\lambda)^{-1}$ is a non-negative and non-increasing operator function on $[0, \infty)$. Hence $\widehat{L}_r(\lambda)^{-1}$ tends strongly to zero as $\lambda \to +\infty$.

For $\lambda \in \rho(\widehat{L}_r) \cap \rho(L_r)$,

$$L_r(\lambda)^{-1} = \widehat{L}_r(\lambda)^{-1} + \sum_{s,l=1}^{N_-} (\Gamma(\lambda)^{-1})_{s,l} \, (\cdot, \widehat{L}_r(\lambda)^{-1}e_l) \, \widehat{L}_r(\lambda)^{-1}e_s, \tag{2.13}$$

where $\Gamma(\lambda)$ is a holomorphic matrix function in the open right half-plane with elements

(2.14) $$\Gamma(\lambda)_{s,l} := \delta_{s,l} + (\widehat{L}_r(\lambda)^{-1}e_s, e_l),$$

$\delta_{s,l}$ being the Kronecker symbol. Since $\widehat{L}_r(\lambda)^{-1} \geq 0$, $\lambda > 0$, it follows from (2.13) that $L_r(\lambda)^{-1} \geq 0$ for those $\lambda > 0$ for which the second term on the right hand side of (2.13) is a non-negative operator i.e. for which $\Gamma(\lambda)^{-1} \gg 0$ and hence $\Gamma(\lambda) \gg 0$. But by the above

$$\lim_{\lambda\to+\infty} (\widehat{L}_r(\lambda)^{-1}e_s, e_l) = 0.$$

Therefore there exists an $\alpha > 0$ such that for $\lambda \geq \alpha$ we have

(2.15) $$\Gamma(\lambda) > \frac{1}{2}I \gg 0.$$

Whence $L_r(\lambda)^{-1}$ and consequently $L_r(\lambda)$ is a non-negative operator for $\lambda \geq \alpha$. By (2.13) and (2.15), we have for $\lambda \geq \alpha$,

$$L_r(\lambda)^{-1} \leq \widehat{L}_r(\lambda)^{-1} + 2\widehat{L}_r(\lambda)^{-1}P_-\widehat{L}_r(\lambda)^{-1} \leq (\frac{2}{\beta'} + \frac{4}{\beta'^2})\, I.$$

Hence

$$s - \lim_{\lambda\to+\infty} L_r(\lambda)^{-1} = 0$$

and

$$L_r(\lambda) \geq \frac{\beta'^2}{2(2+\beta')} \gg 0.$$

□

Returning to the proof of Theorem 2.1 let us suppose now that $\lambda_0 \notin \sigma_{ess}(\widetilde{L}_0)$ is a normal eigenvalue of $\widetilde{L}_0(\lambda)$. Evidently, $\mathrm{Im}\lambda_0 = 0$. We write P_0 for the orthogonal projector onto the null-space of $\widetilde{L}_0(\lambda_0)$. By our assumptions, P_0 has finite rank. Set

$$\widetilde{\widetilde{L}}_0(\lambda) = I + \lambda K + P_0.$$

The operator function $\widetilde{\widetilde{L}}_0(\lambda)$ is invertible at λ_0 and at each point of some neighborhood of λ_0. According to (2.1) and the assumptions (a)–(d),

(2.16) $$\widetilde{L}_+(\lambda) = \widetilde{\widetilde{L}}_0(\lambda) + Q(\lambda) - P_0 = \widetilde{\widetilde{L}}_0(\lambda)(I + \widetilde{G}(\lambda)),$$

where $\widetilde{F}(\lambda) = \widetilde{\widetilde{L}}_0(\lambda)^{-1}(Q(\lambda) - P_0)$ is an operator function with compact values, which is holomorphic in some neighborhood of λ_0. Using the representation (2.16) we can now proceed by analogy with the above to show that λ_0 is not an accumulation point for eigenvalues of $L(\lambda)$. □

3. The spectrum of $L_r(\lambda)$ in the right half-plane

In the sequel we will denote by $\{\mu_j\}$ some enumeration of the distinct eigenvalues of $L(\lambda)$ in the open right half-plane and n_j will stand for the algebraic multiplicity of μ_j.

Remark 3.1. *Geometric multiplicities of positive eigenvalues of $L_r(\lambda)$ coincide with their algebraic multiplicities.*

Proof. Indeed, let $\mu_j > 0$ be an eigenvalue of $L_r(\lambda)$ and let $x_0, x_1 \in \mathcal{D}(A)$ satisfy the system of equations

$$L_r(\mu_j)x_0 = 0,\ L_r(\mu_j)x_1 + L_r'(\mu_j)x_0 = 0. \tag{3.1}$$

Since $L_r(\mu_j)$ is a self-adjoint operator, (3.1) yields

$$(L_r'(\mu_j)x_0, x_0) = 2\mu_j(Dx_0, x_0) + (Cx_0, x_0) = 0.$$

By assumption (d) above and since $\mu_j > 0$, we have $x_0 = 0$. □

Theorem 3.2. *Let A, B, C and D satisfy the assumptions* (a)–(d) *above. Then all eigenvalues $\{\mu_j\}$ of $L_r(\lambda)$ in the open right half-plane are real, the algebraic multiplicity n_j of each μ_j coincides with its geometric multiplicity, the number $\sum_j n_j$ is finite and*

$$\sum_j n_j = N_-, \tag{3.2}$$

where N_- is the number of all negative eigenvalues of self-adjoint operator $A+B$ counted with respect to their multiplicity.

Proof. For $\lambda > 0$, let $n_-(\lambda)$ denote the sum of the multiciplicities of all negative eigenvalues of the self-adjoint operator $L_r(\lambda)$. As the operator function $L_r(\lambda)$ is non-decreasing in $[0,\infty)$ and taking into account the mini-max properties of semi-bounded operators, we see that the function $n_-(\lambda)$ is non-increasing in $[0,\infty)$. Hence $n_-(\lambda) \le N_-$.

Note that $n_-(\lambda)$ is equal to the dimension of each maximal negative subspace with respect to the quadratic form $(L_r(\lambda)x, x)$, $x \in \mathcal{D}(A)$. Since $L_r(\lambda_1) \le L_r(\lambda_2)$ for $0 < \lambda_1 < \lambda_2$, a maximal negative subspace for $L_r(\lambda_2)$ is equal to or can be enlarged to a maximal negative subspaces for $L_r(\lambda_1)$. Let $\mu_j > 0$ be an eigenvalue of $L_r(\lambda)$ and let $\mathcal{M}_j \subset \mathcal{D}(A)$ be a maximal non-positive subspace in the quadratic form $(L_r(\mu_j)x, x)$, $x \in \mathcal{D}(A)$. Evidently,

$$\dim \mathcal{M}_j = n_-(\mu_j) + n_j,$$

where n_j is the multiplicity of μ_j. For any $y \in \mathcal{M}_j$ and $\lambda < \mu_j$, $\lambda > 0$ we have

$$(L_r(\lambda)y, y) < (L_r(\mu_j)y, y) \le 0.$$

Hence for $0 < \lambda < \mu_j < \lambda'$,

$$n_-(\lambda') < n_-(\mu_j) < n_-(\mu_j) + n_j \leq n_-(\lambda) \leq N_-, \tag{3.3}$$

and it follows that

$$\sum_{\mu_j > \lambda > 0} n_j \leq N_-. \tag{3.4}$$

On the other hand, let $\lambda_0 > 0$ be a regular point of $L_r(\lambda)$. Then λ_0 is a regular point of $\widetilde{L}_r(\lambda) := \lambda^2 Y + \lambda(K+R) + X + I$ and

$$L_r(\lambda_0)^{-1} = A^{-1}\widetilde{L}_r(\lambda_0)^{-1}.$$

Since X, Y, $K+R$, and $\widetilde{L}_r(\lambda_0)^{-1}$ are bounded operators and

$$L_r(\lambda) = (I + ((\lambda^2 - \lambda_0^2)Y + (\lambda - \lambda_0)(K+R))\widetilde{L}_r(\lambda_0)^{-1})\widetilde{L}_r(\lambda_0)A,$$

there exists some $\delta > 0$ such that the disk $\mathbf{D}_\delta = \{\lambda : |\lambda - \lambda_0| < \delta\}$ belongs to $\rho(L_r)$. Moreover, the operator functions $L_r(\lambda)^{-1}$ and $\widetilde{L}_r(\lambda)^{-1}$ are bounded and holomorphic in $\mathbf{D}_\delta$. From now we assume that $N_- > 0$ and denote by $-\nu_0$, $\nu_0 > 0$, the smallest eigenvalue of $A+B$. Since $L_r(\lambda) > A+B$ for $\lambda > 0$, $-\nu_0$ does not exceed of the smallest eigenvalue of $L_r(\lambda)$.

Choosing a number $\tau > 2\nu_0$ we define in $(\lambda_0 - \delta, \lambda_0 + \delta)$ the self-adjoint contractive and invertible operator function

$$Q(\lambda) := L_r(\lambda)(\tau + L_r(\lambda))^{-1} = (I + \tau L_r(\lambda)^{-1})^{-1}.$$

$Q(\lambda)$ is differentiable in $(\lambda_0 - \delta, \lambda_0 + \delta)$ with respect to the operator norm and

$$Q'(\lambda) = \tau Q(\lambda)A^{-1}\widetilde{L}_r(\lambda)^{-1}(2\lambda Y + K + R)\widetilde{L}_r(\lambda)^{-1}Q(\lambda).$$

Therefore $Q(\lambda)$ is continuous in $(\lambda_0 - \delta, \lambda_0 + \delta)$ with respect to the operator norm.

Let us denote by $\mathcal{H}_-(\lambda)$ the invariant subspace of $Q(\lambda)$ generated by its negative spectrum. Observe that $\mathcal{H}_-(\lambda)$ coincides with such a subspace for $L_r(\lambda)$ and $\dim \mathcal{H}_-(\lambda) = n_-(\lambda)$. Since $Q(\lambda_0)$ has a bounded inverse, there exists a positive number $\eta > 0$ such that $(-\eta, \eta) \subset \rho(Q(\lambda_0))$. Obviously,

$$\begin{aligned} (Q(\lambda_0)x, x) &\leq -\eta \|x\|^2, \quad x \in \mathcal{H}_-(\lambda_0), \\ (Q(\lambda_0)x, x) &\geq \eta \|x\|^2, \quad x \in \mathcal{H}_-(\lambda_0)^\perp. \end{aligned}$$

The continuity of $Q(\lambda)$ with respect to the operator norm implies that there exists a δ', $0 < \delta' < \delta$, such that for $\lambda \in [\lambda_0 - \delta', \lambda_0 + \delta']$

$$\begin{aligned} (Q(\lambda)x, x) &\leq -\frac{\eta}{2} \|x\|^2, \quad x \in \mathcal{H}_-(\lambda_0), \\ (Q(\lambda)x, x) &\geq \frac{\eta}{2} \|x\|^2, \quad x \in \mathcal{H}_-(\lambda_0)^\perp. \end{aligned} \tag{3.5}$$

Observe that $n_-(\lambda)$ coincides with the dimension of the maximal negative subspace in the quadratic form $(Q(\lambda)x, x)$, $x \in \mathcal{H}_-$. It follows from (3.5) that $n_-(\lambda) = n_-(\lambda_0)$ for $\lambda \in [\lambda_0 - \delta', \lambda_0 + \delta']$.

We proved that the jumps of the non-increasing function $n_-(\lambda)$ occur exactly in the eigenvalues of $L_r(\lambda)$ in $(0, +\infty)$ and that the size of the jump of $n_-(\lambda)$ at $\lambda = \mu_j$ is equal to n_j. Hence the equality (3.2) holds if (and only if) there exists a finite number $\alpha > 0$ such that $n_-(\lambda) = 0$ for $\lambda \geq \alpha$. This is equivalent to the inequality $L_r(\lambda) > 0$ for $\lambda \geq \alpha > 0$. By Lemma 2.2, such an $\alpha > 0$ indeed exists.

□

4. Example

Let us return now to the problem (1.1), (1.2). The inverse of the operator A given by (1.4) is the nuclear integral operator

$$(A^{-1}y)(x) = \int_0^1 W(x,\xi)y(\xi)d\xi,$$

with continuous kernel $W(x, \xi)$. Note that for fixed ξ, $W(x, \xi)$ is twice continuously differentiable in x and

$$\frac{\partial^3}{\partial x^3}W(x,\xi) = \begin{cases} -1 + 3\xi^2 - 2\xi^3, & 0 \leq x < \xi \leq 1, \\ 3\xi^2 - 2\xi^3, & 0 \leq \xi < x \leq 1. \end{cases}$$

In the spectral analysis above the operator

$$\widetilde{L}_+(\lambda) = L(\lambda)A^{-1} = \lambda^2 Y + \lambda(K + R + iZ) + I + X$$

occurred; see (2.1). For the problem (1.1), (1.2) the operator coefficients take the following form: K is the multiplication operator in $\mathbf{L}^2(0,1)$ by the continuous function $\alpha(x)$ and R, Z, Y, and X are integral operators with kernels

$$2\alpha'(x)\frac{\partial^3}{\partial x^3}W(x,\xi) + \alpha(x)\frac{\partial^2}{\partial x^2}W(x,\xi) + k(x)W(x,\xi), \quad \rho(x)W(x,\xi),$$
$$-\gamma\frac{\partial}{\partial x}W(x,\xi), \quad \text{and} \quad \frac{\partial}{\partial x}\left(g(x)\frac{\partial}{\partial x}W(x,\xi)\right),$$

respectively. By the assumptions, all kernels in here are evidently square integrable on $(0,1) \times (0,1)$. Whence R, Z, Y, and X are compact operators and of Hilbert-Schmidt class. We see that for the pencil generated by the boundary problem (1.1), (1.2) assumptions (a)–(d) of Section 2 hold.

Let us denote

$$a = \min_{x\in[0,1]}\alpha(x), \quad b = \max_{x\in[0,1]}\alpha(x).$$

By our assumption $\alpha(x) \geq 0$ and therfore $a \geq 0$. Applying Theorem 2.1 we obtain

Theorem 4.1. *The essential spectrum of the quadratic pencil* $L_r(\lambda) + i\lambda G$, *where* $L_r(\lambda)$ *is the pencil generated by the boundary problem* (1.1), (1.2) *and* G *is defined by* (1.9), *coincides with the interval* $[-\frac{1}{a}, -\frac{1}{b}]$. *The spectrum of* $L_r(\lambda)+i\lambda G$ *outside this interval consists of normal eigenvalues.*

Theorem 4.2. *The spectrum of the quadratic pencil* $L_r(\lambda)$ *generated by the boundary problem* (1.1), (1.2) *in the open right half-plane consists of a finite number of positive normal eigenvalues, which do not possess associated vectors. The number of these eigenvalues counted with respect to their multiplicities coincides with the number of negative eigenvalues of the selfadjoint operator* $A + B$ *counted with respect to their multiplicities, where* A *and* B *are defined by* (1.4) *and* (1.5).

Acknowledgements

The authors wish to express their thanks to Professor Aad Dijksma for his attention to this work and important remarks.
The research described in this publication was made possible in part by Award #UM 1-298 of the Government of Ukraine and US Civilian Research and Development Foundation for the former Soviet Union and in part by support of International Soros Science and Educational Program through grants SPU 001 and #APV 071080.

References

[GGK] GOHBERG I.C., GOLDBERG S., KAASHOEK M.A.: Classes of Linear Operators, Volume I, Operator Theory: Adv. and Appl., vol. **49**, Birkhäuser-Verlag Basel-Boston-New York,1996

[GK] GOHBERG I.C., KREIN M.G.: Introduction to the Theory of Linear Nonselfadjoint Operators, Amer. Math. Soc., Providence, 1988

[GS] GRINIV R.O., SHKALIKOV A.A.: On Operator Pencils Arising in the Problem of Beam Oscillations with Internal Damping (in Russian), Matem. Zametki, vol. **56** , #2, (1994), 114–131; English Transl. in Math. Notes, vol. **56**(1994)

[K] KATO T.: Perturbation Theory for Linear Operators, Springer-Verlag Berlin-Heidelberg-New York, 1966

[KL] KREIN M.G., LANGER H.: On Some Mathematical Principles in the Linear Theory of Damped Oscillations of Continua I, II, Integral Eq. and Operator Theory, Vol.1 (1978), 364–399, 539–566

[LS] LANCASTER P., SHKALIKOV A.A.: Damped Vibrations of Beams and Related Spectral Problems, Canadian Appl.Math. Quart., vol. **2**, #4(1994), 45–90

[M] MARKUS A.: On Holomorphic Operator Functions, Doklady Akad. Nauk SSSR, vol. **119** , #6(1958), 1099–1102

[PaI] PAÏDUSSIS M.P., ISSIDN.T.: Dynamic Stability of Pipes Conveying Fluid, J. Sound and Vibrations, vol. **33**, #3 (1974), 267–294

[Pi1] PIVOVARCHIK V.N.: Problem Connected with Oscillations of Elastic Beams with Internal and Viscous Damping (in Russian), Moscow Univ. Bulletin, vol. **42**, (1987), 68–71

[Pi2] PIVOVARCHIK V.N.: On the Spectrum of Certain Quadratic Pencils of Unbounded Operators(in Russian), Function. Anal. i ego Prilozhen., vol. **23**, #1(1989), 80–81

[Pi3] PIVOVARCHIK V.N.: On Oscillatiions of a Semiinfinite Beam with Internal and External Damping(in Russian), Prikladnaya Math. and Mech., vol. **52**, #5 (1988), 829–836; English Transl. in J. Appl. Math. and Mech. (1989)

[Pi4] PIVOVARCHIK V.N.: On Closednessof the Approximative Spectrum of a Polynomial Operator Pencil(in Russian), Mathem. Zametki, vol. **47**, #6 (1990), 147–148

[S] SHKALIKOV A.A.: Operator Pencils Arising in Elasticity and Hydrodynamics: the Instability Index Formula, Operator Theory: Adv. and Appl.: vol. **87** , (1996), 358–385

[ZKM] ZEFIROV V.N., KOLESOV V.V., MILOSLAVSKIIA.I.: Investigation of Characteristic Freaquences ofinear Pipe (in Russian), Izv. Akad. Nauk SSSR, Multifrequency Tone Telegraphy, #1, (1985), 179–188

Department of Theoretical Physics
Odessa State University
ul. Dvorjanskaja 2
270026 Odessa, Ukraine

Department of Higher Mathematics
Odessa State Academy of Civil Engineering and Architecture
ul. Didrikhsona 4
270028 Odessa, Ukraine

1991 Mathematics Subject Classification. Primary 47A56, 47A10

Operator Theory:
Advances and Applications, Vol. 106

Special realizations for Schur upper triangular operators

D. Alpay and Y. Peretz

Dedicated to Heinz Langer on the occasion of his 60th birthday

Spaces introduced by L. de Branges and J. Rovnyak provide isometric, coisometric and unitary realizations of Schur functions. In this paper we show that similar realizations exist in the "nonstationary setting", i.e. when one considers upper triangular contractions (which appear in time–variant system theory as "transfer functions" of dissipative systems). Two approaches are used: complementation theory and an approach based on the theory of relations in Hilbert spaces.

1. Introduction

In this paper we consider realization problems for the nonstationary analogue of the Schur class of analytic functions. Let us recall that the classical Schur class consists of all functions S analytic inside the open unit disk $\mathbb{D}$ and bounded by 1 in modulus. It occurs prominently in interpolation theory, invariant subspace theory and in the theory of linear dissipative systems (see [34], [42], [38], [2]). Let $\mathcal{H}$ and $\mathcal{G}$ be two Hilbert spaces. We denote by $\mathcal{L}(\mathcal{H},\mathcal{G})$ and $\mathcal{L}(\mathcal{H})$ the sets of all bounded linear operators from $\mathcal{H}$ into $\mathcal{G}$ and from $\mathcal{H}$ into itself respectively, and recall that a $\mathcal{L}(\mathcal{H})$–valued function $K(z,\omega)$ analytic in z and $\bar{\omega}$ in some region Ω is called nonnegative if it is Hermitian, i.e. $K(z,\omega) = K(\omega,z)^*$ and if, for every choice of positive integer n, $z_1, z_2, \ldots, z_n \in \Omega$ and vectors $\xi_1, \xi_2, \ldots, \xi_n \in \mathcal{H}$ the matrix $(\langle K(z_j,z_i)\xi_i, \xi_j\rangle_{\mathcal{H}})_{i,j=1}^n$ is nonnegative. To every nonnegative function K is associated a uniquely defined Hilbert space $\mathcal{H}(K)$ of $\mathcal{H}$–valued analytic functions with the following two properties:

1. $K(\cdot,\omega)\xi \in \mathcal{H}(K)$ for any $\xi \in \mathcal{H}$ and any $\omega \in \Omega$.
2. $\langle f, K(\cdot,\omega)\xi\rangle_{\mathcal{H}(K)} = \langle f(\omega), \xi\rangle_{\mathcal{H}}$ for any $\xi \in \mathcal{H}$, any $\omega \in \Omega$ and any $f \in \mathcal{H}(K)$.

The space $\mathcal{H}(K)$ is called the reproducing kernel Hilbert space with reproducing kernel K. See [14], [47], [50]. The operator–valued Schur class $\mathcal{S}(\mathcal{H},\mathcal{G})$ consists of all functions S analytic in $\mathbb{D}$ whose values are contractions from $\mathcal{H}$ into $\mathcal{G}$. For every Schur function $S \in \mathcal{S}(\mathcal{H},\mathcal{G})$ the associated conjugate function

$$\tilde{S}(z) = S(\bar{z})^* \tag{1.1}$$

belongs to the class $\mathcal{S}(\mathcal{G},\mathcal{H})$ and the kernels

$$K_S(z,\omega)=\frac{I_{\mathcal{G}}-S(z)S(\omega)^*}{1-z\bar{\omega}},\quad K_{\tilde{S}}(z,\omega)=\frac{I_{\mathcal{H}}-\tilde{S}(z)\tilde{S}(\omega)^*}{1-z\bar{\omega}} \tag{1.2}$$

and

$$D_S(z,\omega)=\begin{pmatrix} K_S(z,\omega) & \frac{S(z)-S(\bar{\omega})}{z-\bar{\omega}} \\ \frac{\tilde{S}(z)-\tilde{S}(\bar{\omega})}{z-\bar{\omega}} & K_{\tilde{S}}(z,\omega)\end{pmatrix} \tag{1.3}$$

are nonnegative in $\mathbb{D}$. (Here $I_{\mathcal{G}}$ denotes the identity operator of $\mathcal{G}$). This result originates with the work of L. de Branges and J. Rovnyak (see [25], [13], [11]) who introduced the reproducing kernel Hilbert spaces $\mathcal{H}(S)$, $\mathcal{H}(\tilde{S})$ and $\mathcal{D}(S)$ associated to the kernels (1.2) and (1.3) respectively. These spaces appear in the theory of canonical models for contraction operators and provide coisometric, isometric and unitary realizations of S. For instance, one has

$$S(z)=D+zC(I_{\mathcal{H}(S)}-zA)^{-1}B, \tag{1.4}$$

where

$$\begin{pmatrix} A & B \\ C & D\end{pmatrix}:\begin{pmatrix}\mathcal{H}(S)\\ \mathcal{F}\end{pmatrix}\longrightarrow\begin{pmatrix}\mathcal{H}(S)\\ \mathcal{G}\end{pmatrix} \tag{1.5}$$

is the backwards–shift realization defined by

$$(Au)(z) = \frac{u(z)-u(0)}{z} \tag{1.6}$$

$$Bf(z) = \frac{S(z)-S(0)}{z}f \tag{1.7}$$

$$Cu = u(0) \tag{1.8}$$

$$Df = S(0)f. \tag{1.9}$$

Furthermore, the matrix (1.5) is coisometric, and the realization (1.4) is closely outerconnected in the sense that $\cap_{n=0}^{\infty}\mathrm{Ker}\, CA^n=\{0\}$. In general, it is not minimal but the closely outerconnectedness property insures uniqueness up to a similarity operator which moreover is unitary; see [32], [33], [28], [11].

Besides the works of de Branges and Rovnyak mentioned above, we refer to [17], [44], [48], [49], [11] for more on the relationships between these spaces and operator models.

Schur functions are transfer functions of time–invariant dissipative linear systems (see [37, p. 185]). The need to consider time–variant systems motivated another generalization of the Schur class, namely the class $\mathcal{S}$ of all upper triangular contraction operators S; see [29], [31], [8], [39]; the precise definition is given in the sequel. In [7] and [8] was defined a point evaluation for upper triangular operators (called there the "W–transform") which allowed to define and investigate analogues of classical interpolation problems (see [39]). The nonstationary analogue of the reproducing kernel Hilbert space with reproducing kernel K_S appears first in [8, Section 7].

In the present paper we use this space and the analogue of the space with reproducing kernel D_S to give coisometric, isometric and unitary realizations for Schur class upper triangular operators, using two approaches: the classical approach using the notion of complementation due to L. de Branges [25] (see also T. Ando [13]) and a more recent approach using the notion of linear relations and developed in [11] (for the case where the kernels (1.2) and (1.3) have a finite number of negative squares and where $\mathcal{F}$ and $\mathcal{G}$ are Pontryagin spaces of same index; see [10] for a summary of the method and of the results). One of the difficulties in obtaining the realizations is that there is no natural analogue of the conjugate function (1.1) for upper triangular operators and we have to consider also lower triangular operators.

We mention that nonstationary analogues of (1.4) appear in the litterature; see for instance [18, (5.7) p. 44] and [30, (5.17) p. 200]. There, the complex variable z is replaced by a bilateral shift in an appropriate space. Unitary systems associated to an upper triangular contraction were already considered in the 1988 paper [16] (see also [27, Chapter 2]) and are also considered in the preprint [15]. Formulas that we obtain here are of a different kind. We consider S as a function defined on the diagonal operators (i.e. the W–transform of S) and replace the complex numbers by diagonal operators. The various links between these different works remain to be done. Some of these relationships are explored in the preprint [4].

The paper consists of six sections besides the introduction. In Section 2 we review the nonstationary setting to be used throughout the paper. In the third section we define the state spaces which will provide coisometric realizations for the operator S. The results of this are basically taken from [8]. The coisometric and isometric realizations are studied in Section 4 using complementation theory and in Section 5 using the theory of relations. The main results are Theorem 4.1 (the backwards shift colligation) and Theorem 4.5. In the sixth section we introduce the analogue of the two by two kernel (1.3). The reproducing kernel property is exposed in Lemma 6.3, which is the main result of the section. The corresponding reproducing kernel space is the state space of unitary realization of S, which is computed in the last section; see Theorems 7.1 and 7.2.

In the case of the Schur functions one can also consider coisometric, isometric and unitary realizations centered at an arbitrary point $\alpha \in \mathbb{D}$; see [5], [6]. The same problem holds here; one can define coisometric, isometric and unitary realizations for an upper triangular contraction centered at an arbirary diagonal operator W of norm strictly less than 1. This is done in the preprint [12]. The case of upper triangular operators with positive real part is considered in the preprint [9].

Quite a number of problems in the stationary framework (such as operator models, $\mathbf{H}_2$–interpolation, the Bergman and Dirichlet kernels) still make sense in the setting of upper triangular operators. Using the results of the present paper, we will discuss these aspects in future publications.

2. Preliminaries

In this section we briefly review the nonstationary framework in which we will work. We closely follow the analysis and notations of [8] and [30], and the knowledgeable reader can proceed directly to the third section. A summary in form of a table is given at the end of the next section for the convenience of the reader.

Let $\mathcal{M}$ and $\mathcal{N}$ be two separable Hilbert spaces, "the coefficient spaces". As in [30, Section 1], the set of bounded linear operators from the space $\ell^2_{\mathcal{N}}$ of square summable sequences with components in $\mathcal{N}$ into the space $\ell^2_{\mathcal{M}}$ of square summable sequences with components in $\mathcal{M}$ is denoted by $\mathcal{X}\left(\ell^2_{\mathcal{N}};\ell^2_{\mathcal{M}}\right)$ (or $\mathcal{X}\left(\ell^2_{\mathcal{N}}\right)$, when $\mathcal{M}=\mathcal{N}$, or $\mathcal{X}$ when the coefficient spaces are understood from the context). The spaces $\ell^2_{\mathcal{N}}$, $\ell^2_{\mathcal{M}}$ are taken with the standard inner products. Let $Z_{\mathcal{N}}$ be the bilateral backward shift operator

$$(Z_{\mathcal{N}}f)_i = f_{i+1}, \qquad i=\ldots,-1,0,1,\ldots$$

where $f=(\ldots,f_{-1},f_0,f_1,\ldots)\in\ell^2_{\mathcal{N}}$. The operator $Z_{\mathcal{N}}$ is unitary on $\ell^2_{\mathcal{N}}$ i.e. $Z_{\mathcal{N}}Z^*_{\mathcal{N}}=Z^*_{\mathcal{N}}Z_{\mathcal{N}}=I_{\mathcal{N}}$, and

$$\pi^*_{\mathcal{N}}Z^j_{\mathcal{N}}\pi_{\mathcal{N}}=\begin{cases} I_{\mathcal{N}} & \text{if} \quad j=0\\ 0_{\mathcal{N}} & \text{if} \quad j\neq 0.\end{cases}$$

where π denote the injection map

$$\pi_{\mathcal{N}}: u\in\mathcal{N}\rightarrow f\in\ell^2_{\mathcal{N}} \quad \text{where} \quad \begin{cases} f_0=u\\ f_i=0, \quad i\neq 0\end{cases}.$$

We define the space of upper triangular operators by

$$\mathcal{U}\left(\ell^2_{\mathcal{N}};\ell^2_{\mathcal{M}}\right)=\left\{A\in\mathcal{X}\left(\ell^2_{\mathcal{N}};\ell^2_{\mathcal{M}}\right) \mid \pi^*_{\mathcal{M}}Z^i_{\mathcal{M}}AZ^{*j}_{\mathcal{N}}\pi_{\mathcal{N}}=0 \quad \text{for} \quad i>j\ \right\},$$

and the space of lower triangular operators by

$$\mathcal{L}\left(\ell^2_{\mathcal{N}};\ell^2_{\mathcal{M}}\right)=\left\{A\in\mathcal{X}\left(\ell^2_{\mathcal{N}};\ell^2_{\mathcal{M}}\right) \mid \pi^*_{\mathcal{M}}Z^i_{\mathcal{M}}AZ^{*j}_{\mathcal{N}}\pi_{\mathcal{N}}=0 \quad \text{for} \quad i<j\ \right\}.$$

The space of diagonal operators $\mathcal{D}\left(\ell^2_{\mathcal{N}};\ell^2_{\mathcal{M}}\right)$ consists of the operators which are both upper and lower triangulars. As for the space $\mathcal{X}$, we usually denote these spaces by $\mathcal{U}$, $\mathcal{L}$ and $\mathcal{D}$ when the sequences spaces are understood from the context. Similarly, we write Z instead of $Z_{\mathcal{M}}$ and I instead of $I_{\mathcal{M}}$.

Let $A^{(j)}=Z^{*j}_{\mathcal{M}}AZ^j_{\mathcal{N}}$ for $A\in\mathcal{X}$ and $j=\ldots,-1,0,1,\ldots$; note that $\left(A^{(j)}\right)_{st}=A_{s-j,t-j}$ and that the maps $A\mapsto A^{(j)}$ take the spaces $\mathcal{L}$, $\mathcal{D}$, $\mathcal{U}$ into themselves. Clearly, for $A\in\mathcal{X}(\ell^2_{\mathcal{B}};\ell^2_{\mathcal{M}})$ and $B\in\mathcal{X}(\ell^2_{\mathcal{N}};\ell^2_{\mathcal{B}})$ ($\mathcal{B}$ being some Hilbert space) we have that $(AB)^{(j)}=A^{(j)}B^{(j)}$ and $A^{(j+k)}=\left(A^{(j)}\right)^{(k)}$.

In [8] (for the case $\mathcal{M} = \mathcal{N}$) it is shown that for every $F \in \mathcal{U}$, there exists a unique sequence of operators $F_{[j]} \in \mathcal{D}$, $j = 0, 1, \ldots$ such that

$$F - \sum_{j=0}^{n-1} Z^j_{\mathcal{M}} F_{[j]} \in Z^n_{\mathcal{M}} \mathcal{U}.$$

In fact, $\left(F_{[j]}\right)_{ii} = F_{i-j,i}$ and we can formally represent $F \in \mathcal{U}$ as the sum of its diagonals

$$F = \sum_{n=0}^{\infty} Z^n_{\mathcal{M}} F_{[n]}.$$

This expression is just formal, and in general there is no notion of convergence.

We now introduce the operator left transform

$$F^{\wedge}(W) = \sum_{n=0}^{\infty} W^{[n]} F_{[n]} = \sum_{n=0}^{\infty} \left(W Z^*_{\mathcal{M}}\right)^n Z^n_{\mathcal{M}} F_{[n]}$$

with

$$W^{[0]} = I_{\mathcal{M}} \quad \text{and} \quad W^{[j+1]} = W\left(W^{[j]}\right)^{(1)} \quad \text{for} \quad j = 0, 1, \ldots$$

for any $W \in \mathcal{X}\left(\ell^2_{\mathcal{M}}\right)$ for which $\ell_{W,\mathcal{M}} = \lim_{n\uparrow\infty} \|W^{[n]}\|^{1/n} < 1$ (see [8]), where the last limit is the spectral radius $r_{sp}\left(W Z^*_{\mathcal{M}}\right)$ of $W Z^*_{\mathcal{M}}$. Similarly we have a right transform defined by

$$F^{\triangle}(W) = \sum_{n=0}^{\infty} \left\{F_{[n]}\right\}^{(-n)} \left\{W^{[n]}\right\}^{(1-n)} = \sum_{n=0}^{\infty} Z^n_{\mathcal{M}} F_{[n]} \left(Z^*_{\mathcal{N}} W\right)^n$$

which exists for any $W \in \mathcal{X}\left(\ell^2_{\mathcal{N}}\right)$ for which $\ell_{W,\mathcal{N}} = r_{sp}\left(Z^*_{\mathcal{N}} W\right)$.

When $\mathcal{M} = \mathcal{N}$, it is easily checked and useful to note for future use that $r_{sp}\left(Z^*_{\mathcal{M}} W\right) = r_{sp}\left(W Z^*_{\mathcal{N}}\right)$.

The maps $W \mapsto F^{\wedge}(W)$ and $W \mapsto F^{\triangle}(W)$ will be considered for $W \in \mathcal{D}\left(\ell^2_{\mathcal{M}}\right)$, $W \in \mathcal{D}\left(\ell^2_{\mathcal{N}}\right)$ respectively and we define

$$\Omega_{\mathcal{M}} = \left\{W \in \mathcal{D}\left(\ell^2_{\mathcal{M}}\right) \mid \ell_{W,\mathcal{M}} < 1\right\}, \qquad \Omega_{\mathcal{N}} = \left\{W \in \mathcal{D}\left(\ell^2_{\mathcal{N}}\right) \mid \ell_{W,\mathcal{N}} < 1\right\}$$

the sets of all diagonal operators W for which the related transforms exist.

When $\mathcal{M} = \mathcal{N}$, we denote by Ω

$$\Omega = \Omega_{\mathcal{M}} = \Omega_{\mathcal{N}}. \tag{2.1}$$

The set Ω contains in particular elements of $\mathcal{D}$ of norm strictly less than 1.

The first transform was introduced, as already mentioned, in [7]. The second was considered in [21]. Continuous analogues are introduced in [19] and [20]. In [8] it is shown that for $F \in \mathcal{U}$, $D \in \mathcal{D}$ the operator $\left(Z_{\mathcal{M}} - W\right)^{-1}(F - D)$ belongs to $\mathcal{U}$

for $W \in \Omega_{\mathcal{M}}$ if and only if $D = F^{\wedge}(W)$. Similarly, $(F - D)(Z_{\mathcal{N}} - W)^{-1}$ belongs to $\mathcal{U}$ for $W \in \Omega_{\mathcal{N}}$ if and only if $D = F^{\triangle}(W)$. Therefore $F^{\wedge}(W)$ and $F^{\triangle}(W)$ are the analogues of the point evaluation of an analytic function in the open unit disk. The sets $\Omega_{\mathcal{M}}$ and $\Omega_{\mathcal{N}}$ are analogues of the open unit disk and $\mathcal{D}(\ell^2_{\mathcal{M}})$ and $\mathcal{D}(\ell^2_{\mathcal{N}})$ are analogues of $\mathbb{C}$.

Let $F \in \mathcal{U}$; we have

$$F^{\wedge}(W^*)^* = \sum_{n=0}^{\infty} F^*_{[n]} Z^{*n} (ZW)^n \quad \text{and} \quad F^{\triangle}(W^*)^* = \sum_{n=0}^{\infty} (WZ)^n F^*_{[n]} Z^{*n}$$

where $W^* \in \Omega_{\mathcal{M}}$ in the first case and $W^* \in \Omega_{\mathcal{N}}$ in the second case.

This gives rise to the analogues of the left and right transforms for lower triangular operators. Let $G \in \mathcal{L}$. Formally $G = \sum_{n=0}^{\infty} G_{[n]} Z^{*n}$. We define

$$G^{\vee}(W) = \sum_{n=0}^{\infty} G_{[n]} Z^{*n} (ZW)^n$$

to be the right transform for $\mathcal{L}$ operators and by

$$G^{\nabla}(W) = \sum_{n=0}^{\infty} (WZ)^n G_{[n]} Z^{*n}$$

the left transform for $\mathcal{L}$ operators. The four transforms are related by conjugation in the sense that for any $F \in \mathcal{U}$, the operator F^* belongs to $\mathcal{L}$ and thus, for $W^* \in \Omega_{\mathcal{M}}$ and $\Omega_{\mathcal{N}}$ respectively

$$F^{\wedge}(W^*)^* = F^{*\vee}(W) \tag{2.2}$$

and

$$F^{\triangle}(W^*)^* = F^{*\nabla}(W). \tag{2.3}$$

The next topic in this preliminary section is the notion of reproducing kernel Hilbert space module, and of two particular instances of such an object, namely the spaces of upper and lower triangular Hilbert–Schmidt operators. We refer to [40] (or, if the reader follows the present work with the paper [8] at hand, to [8, Section 7.1]) for the definition and main properties of Hilbert–Schmidt operators. We will just recall that these operators form a Hilbert space (denoted here by the symbol $\mathcal{X}_2$) when endowed with the inner product

$$\langle A, B\rangle_{HS} = \mathrm{Tr} B^* A$$

(where Tr denotes the operator trace).

We set

$$\mathcal{U}_2 = \mathcal{U} \cap \mathcal{X}_2, \quad \mathcal{L}_2 = \mathcal{L} \cap \mathcal{X}_2, \quad \text{and} \quad \mathcal{D}_2 = \mathcal{U}_2 \cap \mathcal{L}_2 .$$

These spaces are reproducing kernel spaces in a sense to be explained in the sequel. We first need some more definitions. Let K_W be an operator–valued function from Ω to $\mathcal{U}$. Then the operator–valued function

$$K^L(V,W) : \Omega \times \Omega \longrightarrow \mathcal{U} \quad \text{defined by} \quad K^L(V,W) = K_W^{\wedge}(V)$$

is said to be nonnegative if for any $V_i \in \Omega, D_i \in \mathcal{D}_2, i = 1, \ldots, n$, the trace of the (trace class) operator

$$\sum_{i,j=1}^{n} D_j^* K^L(V_j, V_i) D_i$$

is nonnegative. For instance, for the operator–valued function

$$K_W = (I - ZW^*)^{-1}$$

$K^L(V,W)$ is given by

$$K^L(V,W) = \sum_{n=0}^{\infty} V^{[n]} W^{[n]*}$$

and is nonnegative.

Let $\mathcal{H}$ be a Hilbert space of operators in $\mathcal{U}_2$ with an inner product $\langle , \rangle_{\mathcal{H}}^L$ and Hermitian form $[,]_{\mathcal{H}}^L$, which is continuous with respect to $\langle , \rangle_{\mathcal{H}}^L$. Assume that FA belongs to $\mathcal{H}$, for any $F \in \mathcal{H}$ and any $A \in \mathcal{D}$. Then $\mathcal{H}$ will be called a Hilbert space right module. The operator–valued function K_W will be called a left reproducing kernel for $\mathcal{H}$ if

1. $K_W G \in \mathcal{H}$ for any $G \in \mathcal{D}_2$.
2. $[F, K_W G]_{\mathcal{H}}^L = G^* F^{\wedge}(W)$.

In this case $\mathcal{H}$ will be called a left reproducing kernel Hilbert space right module. We send the reader to [8, Sections 7 and 8] for related discussion of reproducing kernel Hilbert spaces.

There is a similar notion of right reproducing kernel Hilbert space left module, defined with respect to the right transform. Let $\mathcal{H}$ be a Hilbert space of operators in $\mathcal{U}_2$ with an inner product $\langle , \rangle_{\mathcal{H}}^R$ and Hermitian form $[,]_{\mathcal{H}}^R$, which is continuous with respect to $\langle , \rangle_{\mathcal{H}}^R$. Assume now that AF belongs to $\mathcal{H}$, for any $F \in \mathcal{H}$ and any $A \in \mathcal{D}$. Then $\mathcal{H}$ will be called a Hilbert space left module. Let K_W be an operator–valued function from Ω to $\mathcal{U}$. It is called a right reproducing kernel for $\mathcal{H}$ if

1. $G K_W \in \mathcal{H}$ for any $G \in \mathcal{D}_2$.
2. $[F, G K_W]_{\mathcal{H}}^R = F^{\triangle}(W) G^*$.

In this case $\mathcal{H}$ will be called a right reproducing kernel Hilbert space left module.

We review here a basic result in this framework.

Theorem 2.1. *Let $K_W : \Omega \longrightarrow \mathcal{U}$ be an operator–valued function such that*

$$K^L(V,W) : \Omega \times \Omega \longrightarrow \mathcal{U}$$

defined by $K^L(V,W) = K_W^{\wedge}(V)$ is nonnegative. Then there exists a unique left reproducing kernel Hilbert space right module $\mathcal{H}$, with $K^L(V,W)$ as a reproducing kernel. The space $\mathcal{H}$ is the closure of the set of all operators F of the form

$$F = \textstyle\sum_{i=1}^n K_{W_i} A_i, \quad W_i \in \Omega, \quad A_i \in \mathcal{D}_2, \quad i, \ldots, n$$

in the inner product given by

$$\langle F, G \rangle_{\mathcal{H}}^L = \mathrm{Tr} \sum_{i=1}^{n} \sum_{j=1}^{m} B_j^* K^L(V_j, W_i) A_i$$

for $F = \sum_{i=1}^n K_{W_i} A_i$ *and* $G = \sum_{j=1}^m K_{V_j} B_j$. *Moreover, the form*

$$[F, G]_{\mathcal{H}}^L = \sum_{i=1}^{n} \sum_{j=1}^{m} B_j^* K^L(V_j, W_i) A_i$$

is Hermitian and is continuous with respect to $\langle , \rangle_{\mathcal{H}}^L$.

The proof of this theorem goes in much the same way as in the stationary case: one can show that the set of all operators of the form $F = \sum_{i=1}^n K_{W_i} A_i$ is a pre–Hilbert space with respect to the inner product defined above. Identifying all the operators that belong to the same equivalence class of Cauchy sequences makes it a Hilbert space. The rest is plain.

The spaces $\mathcal{U}_2$ and $\mathcal{L}_2$ are reproducing kernel Hilbert modules. Let

$$\rho_W^{\wedge -1} = \sum_{n=0}^{\infty} (ZW^*)^n \tag{2.4}$$

and

$$\rho_W^{\triangle -1} = \sum_{n=0}^{\infty} (W^* Z)^n \tag{2.5}$$

where $\rho_W^{\wedge} = I - ZW^*$, and $\rho_W^{\triangle} = I - W^* Z$ respectively. For any $E \in \mathcal{D}_2$ the operators $\rho_W^{\wedge -1} E$,and $E \rho_W^{\triangle -1}$ both belong to $\mathcal{U}_2$ and

$$\langle F, \rho_W^{\wedge -1} E \rangle_{HS} = \mathrm{Tr} E^* \rho_W^{\wedge -*} F = \mathrm{Tr} E^* F^{\wedge}(W) \tag{2.6}$$

and

$$\langle F, E \rho_W^{\triangle -1} \rangle_{HS} = \mathrm{Tr} F \rho_W^{\triangle -*} E^* = \mathrm{Tr} F^{\triangle}(W) E^* \tag{2.7}$$

for any $F \in \mathcal{U}_2$, $E \in \mathcal{D}_2$ and $W \in \Omega_{\mathcal{M}}$ and $\Omega_{\mathcal{N}}$ respectively. The computations are as in [8] and omitted. The functions $\rho_W^{\wedge -1}$ and $\rho_W^{\triangle -1}$ will be called the left and right Cauchy kernels. Formulas (2.6) and (2.7) express the following: $\mathcal{U}_2$ is a left reproducing kernel Hilbert space right module with the kernel (2.4) and with respect to the inner product (2.6). We write $\langle , \rangle^L_{\mathcal{U}_2}$ for its inner product. In the same manner $\mathcal{U}_2$ is a right reproducing kernel Hilbert space left module with the kernel (2.5) and with respect to the inner product (2.7). We denote by $\langle , \rangle^R_{\mathcal{U}_2}$ its inner product. In [8] it is shown that the set of all the operators of the form $\rho_W^{\wedge -1}E$, is dense in $(\mathcal{U}_2, \| \cdot \|^L_{\mathcal{U}_2})$. One can show in the same way that the set of all operators of the form $E\rho_W^{\triangle -1}$ is also dense in $(\mathcal{U}_2, \| \cdot \|^R_{\mathcal{U}_2})$.

Similarly we define two kernels for $\mathcal{L}_2$, namely

$$\rho_W^{\vee -1} = \sum_{n=0}^{\infty} (W^*Z^*)^n, \qquad \rho_W^{\nabla -1} = \sum_{n=0}^{\infty} (Z^*W^*)^n,$$

where $\rho_W^{\vee} = I - W^*Z^*$, and $\rho_W^{\nabla} = I - Z^*W^*$.

Straightforward computations show that

$$\langle G, E\rho_W^{\vee -1}\rangle_{HS} = \mathrm{Tr} G\rho_W^{\vee -*}E^* = \mathrm{Tr} G^{\vee}(W)E^* \tag{2.8}$$

and

$$\langle G, \rho_W^{\nabla -1}E\rangle_{HS} = \mathrm{Tr} E^*\rho_W^{\nabla -*}G = \mathrm{Tr} E^*G^{\nabla}(W) \tag{2.9}$$

for any $G \in \mathcal{L}_2$, $E \in \mathcal{D}_2$ and $W^* \in \Omega_{\mathcal{M}}$ and $\Omega_{\mathcal{N}}$ respectively. The space $\mathcal{L}_2$ with the inner product (2.8), which we denote by $\langle , \rangle^R_{\mathcal{L}_2}$ is a right reproducing kernel Hilbert space left module, and the same space with the inner product (2.9), which we denote by $\langle , \rangle^L_{\mathcal{L}_2}$ is a left reproducing kernel Hilbert space right module. One can show that the sets of all operators of the form $E\rho_W^{\vee -1}$, $\rho_W^{\nabla -1}E$, for $E \in \mathcal{D}_2$ and $W \in \Omega$ are dense in $(\mathcal{L}_2, \| \cdot \|^R_{\mathcal{L}_2})$ and $(\mathcal{L}_2, \| \cdot \|^L_{\mathcal{L}_2})$ respectively.

Note that these kernels are related by conjugation in the sense that

$$\rho_W^{\wedge -*} = \rho_{W^*}^{\vee -1}, \qquad \text{and} \qquad \rho_W^{\triangle -*} = \rho_{W^*}^{\nabla -1}.$$

We refer to [41] and [45] for more information on reproducing kernel Hilbert modules.

As already mentionned, a table summarizing the main feature of the nonstationary case versus the stationary case is given at the end of the next section.

3. The state space $\mathcal{H}_L(S)$

In this section we introduce the spaces which will be shown to be state spaces of a coisometric and closely outer connected realization for a Schur class operator S. We first recall the notion of complementation (due to de Branges; we refer to [24] for a recent account). We use the analysis of Ando [13]. Let T be a continuous linear operator between Hilbert spaces $\mathcal{H}$ and $\mathcal{G}$; its range Ran T endowed with the inner product

$$\langle Tf, Tg\rangle_{\text{Ran }T} = \langle Pf, Pg\rangle_{\mathcal{H}}$$

for any $f, g \in \mathcal{H}$, where P is the orthogonal projection on $(\text{Ker } T)^{\perp}$, defines a unique Hilbert space $\mathcal{M}(T) = (\text{Ran}T, \|\cdot\|_{\text{Ran}T})$, and T is a coisometry from $\mathcal{H}$ to Ran T. A Hilbert space $\mathcal{M}$ included in a Hilbert space $\mathcal{H}$ is said to be contractively included if the inclusion map is a contraction. It is shown in [13] that this is the case if and only if $\mathcal{M} = \mathcal{M}(T)$ for some contraction T, and TT^* is then uniquely determined from $\mathcal{M}$. When $\mathcal{M}$ is contractively included in $\mathcal{H}$, its de Branges complement $\mathcal{M}'$ is defined by

$$\mathcal{M}' = \mathcal{M}\left((I - TT^*)^{\frac{1}{2}}\right) \tag{3.1}$$

where T is any contraction such that $\mathcal{M} = \mathcal{M}(T)$. Remark that since $I - TT^*$ is nonnegative, the square root in (3.1) is a well defined contraction. Thus the de Branges complement $\mathcal{M}'$ is also contractively included in $\mathcal{H}$, and $(\mathcal{M}')' = \mathcal{M}$. When $\mathcal{M}$ is isometrically included in $\mathcal{H}$, its de Branges complement coincides with the orthogonal complement, because $\mathcal{M} = \mathcal{M}(P_{\mathcal{M}})$, where $P_{\mathcal{M}}$ is the orthogonal projection on $\mathcal{M}$. To pursue the analysis we quote the following result ([13, Theorem 3.9]):

Theorem 3.1. *Let $\mathcal{M} = \mathcal{M}(T)$ with $\|T\| \leq 1$. Then*

$$\|f\|^2_{\mathcal{M}'} = \sup_{g\in\mathcal{H}} \left\{\|f + Tg\|^2_{\mathcal{H}} - \|g\|^2_{\mathcal{H}}\right\}, \tag{3.2}$$

in the sense that f belongs to $\mathcal{M}'$ if and only if the right hand side of (3.2) *is finite. When this condition is in force,* (3.2) *defines the norm of f in $\mathcal{M}'$.*

In the classical scalar stationary case, one considers the multiplication operator $M_S : \mathbf{H}_2 \to \mathbf{H}_2$, defined by $(M_S f)(z) = S(z)f(z)$, where $\mathbf{H}_2$ denotes the Hardy space of functions analytic in the open unit disk and where $S(z)$ is an $\mathbf{H}_\infty$ function with $\|S\|_\infty \leq 1$, i.e. a scalar Schur function. The operator M_S is a contraction operator and therefore we can consider now $\mathcal{M}, \mathcal{M}'$ as subspaces of $\mathbf{H}_2$. The space $\mathcal{M}'$ is Ran $\left((I - M_S M_S^*)^{\frac{1}{2}}\right)$, which is often written as $\mathcal{H}(S)$ and was studied by de Branges and Rovnyak [26]. From (3.2) we have the characterization

$$\mathcal{H}(S) = \{f \in \mathbf{H}_2 |\ \kappa(f) < \infty\},$$

where

$$\kappa(f) = \|f\|^2_{\mathcal{H}(S)} = \sup_{g\in\mathbf{H}_2} \left\{\|f + Sg\|^2_{\mathbf{H}_2} - \|g\|^2_{\mathbf{H}_2}\right\}.$$

We refer to [35, Theorem 4.1 p. 275] for more connections between operator ranges and the de Branges–Rovnyak spaces. When S is inner, the contraction operator is in fact an isometry, and $\mathcal{H}(S) = \mathbf{H}_2 \ominus S\mathbf{H}_2$, the orthogonal complement of the

Beurling Lax space (see [43], [46] for the Beurling Lax theorem). Note that if $S(z)$ is a Schur function the conjugate function $\tilde{S}(z) = S(\bar{z})^*$ is also Schur function; this latter gives rise to another de Branges space, namely $\mathcal{H}(\tilde{S})$. In the nonstationary case we do not have a natural analogue of $\tilde{S}$. Indeed, one could think of defining $\tilde{S}$ via

$$\tilde{S} = \sum_{n=0}^{\infty} S_{[n]}^* Z^n, \quad \text{where} \quad S = \sum_{n=0}^{\infty} Z^n S_{[n]},$$

in the sense explained in section 2. The following example shows that $\tilde{S}$ need not be a contraction. Let $\mathcal{M} = \mathbb{C}$ and let $S = S_{[0]} + ZS_{[1]}$ from $\ell^2_{\mathbb{C}}$ into itself be defined by

$$S_{[0]} = \begin{pmatrix} \ddots & & & & \\ & 0 & & & \\ & & 0 & & \\ & & & \boxed{1} & \\ & & & & 0 \\ & & & & & \ddots \end{pmatrix}, \quad \text{and} \quad S_{[1]} = \begin{pmatrix} \ddots & & & & \\ & 0 & & & \\ & & 1 & & \\ & & & \boxed{0} & \\ & & & & 0 \\ & & & & & \ddots \end{pmatrix}.$$

We have

$$S^*S = \begin{pmatrix} \ddots & & & & \\ & 0 & & & \\ & & 1 & & \\ & & & \boxed{1} & \\ & & & & 0 \\ & & & & & \ddots \end{pmatrix},$$

while $\tilde{S} = S_{[0]}^* + S_{[1]}^* Z$, and therefore

$$\tilde{S}^*\tilde{S} = \begin{pmatrix} \ddots & & & & \\ & 0 & & & \\ & & 0 & & \\ & & & \boxed{2} & \\ & & & & 0 \\ & & & & & \ddots \end{pmatrix},$$

which means that $\|S\| \leq 1$, while $\|\tilde{S}\| \not\leq 1$. Thus, in opposition to the classical case we also consider S^* (i.e. "antianalytic", or "anticausal" operator).

In the present context $\mathbf{H}_2$ is replaced by its nonstationary analogue, the space $\mathcal{U}_2$ of upper triangular Hilbert–Schmidt operators, and the space $\mathbf{H}_\infty$ is replaced by its nonstationary analogue, the space $\mathcal{U}$ of upper triangular bounded operators. Remark that $\mathcal{U}_2 \subset \mathcal{U}$ while in the stationary case $\mathbf{H}_\infty \subset \mathbf{H}_2$ (for the unit disk).

We define four de Branges spaces; the first two are connected with S

$$\mathcal{H}_L(S) = \{F \in \mathcal{U}_2 \,|\, \kappa_L(F) < \infty\},$$

where

$$\kappa_L(F) = \sup_{G \in \mathcal{U}_2} \left\{ \|F + SG\|^2_{\mathcal{U}_2} - \|G\|^2_{\mathcal{U}_2} \right\},$$

and

$$\mathcal{H}_R(S) = \{ F \in \mathcal{U}_2 \,|\, \kappa_R(F) < \infty \},$$

where

$$\kappa_R(F) = \sup_{G \in \mathcal{U}_2} \left\{ \|F + GS\|^2_{\mathcal{U}_2} - \|G\|^2_{\mathcal{U}_2} \right\}.$$

The last two are connected with S^*

$$\mathcal{H}_L(S^*) = \{ F \in \mathcal{L}_2 \,|\, \kappa^*_L(F) < \infty \},$$

where

$$\kappa^*_L(F) = \sup_{G \in \mathcal{L}_2} \left\{ \|F + S^*G\|^2_{\mathcal{L}_2} - \|G\|^2_{\mathcal{L}_2} \right\}$$

and

$$\mathcal{H}_R(S^*) = \{ F \in \mathcal{L}_2 \,|\, \kappa^*_R(F) < \infty \},$$

where

$$\kappa^*_R(F) = \sup_{G \in \mathcal{L}_2} \left\{ \|F + GS^*\|^2_{\mathcal{L}_2} - \|G\|^2_{\mathcal{L}_2} \right\}.$$

Let $M^L_S : \mathcal{U}_2 \longrightarrow \mathcal{U}_2$ be defined by

$$M^L_S(F) = SF.$$

Obviously M^L_S is a contraction if and only if S is a Schur upper triangular operator.

Theorem 3.2. *The following are equivalent:*

1. *The operator M^L_S is a contraction from $\mathcal{U}_2$ to $\mathcal{U}_2$.*
2. *The kernel $K^L_S(\cdot, W) = (I - SS^\wedge(W)^*)\, \rho_W^{\wedge -1}$ is nonnegative in $\Omega_{\mathcal{M}}$.*

Proof. First we compute M^{L*}_S on the left kernel of $\mathcal{U}_2$. It is readily seen that

$$\left\langle M^{L*}_S \left(\rho_V^{\wedge -1} E\right), \rho_W^{\wedge -1} G \right\rangle_{\mathcal{U}_2} = \langle S^\wedge(V)^* \rho_V^{\wedge -1} E, \rho_W^{\wedge -1} G \rangle_{\mathcal{U}_2},$$

and therefore

$$M^{L*}_S \left(\rho_V^{\wedge -1} E\right) = S^\wedge(V)^* \rho_V^{\wedge -1} E,$$

so, we have that

$$\left\langle \left(I - M^L_S M^{L*}_S\right) \left(\rho_V^{\wedge -1} E\right), \rho_W^{\wedge -1} G \right\rangle_{\mathcal{U}_2} = \mathrm{Tr} G^* \left((I - SS^\wedge(V)^*)\, \rho_V^{\wedge -1} \right)^\wedge (W) E.$$

and so the claims are equivalent. □

The space $\mathcal{H}_L(S)$ can be characterized as an operator range. We first recall the following result (see [13, Theorem 3.9 p. 27], [3, Section 3]).

Theorem 3.3. *Let $\mathcal{H}$ be a Hilbert space endowed with the inner product $\langle , \rangle_{\mathcal{H}}$, let Γ be a bounded linear nonnegative operator acting on $\mathcal{H}$ and let P be the orthogonal projection onto* Ker Γ. *The space* $\mathrm{Ran}\Gamma^{\frac{1}{2}}$ *equiped with the norm given by*

$$\|\Gamma^{\frac{1}{2}}u\|_{\mathrm{Ran}\Gamma^{\frac{1}{2}}} = \|(I-P)\,u\|_{\mathcal{H}}$$

is a Hilbert space. Moreover, $\mathrm{Ran}\Gamma$ *is dense in* $\mathrm{Ran}\Gamma^{\frac{1}{2}}$ *and* $\langle \Gamma u, \Gamma v\rangle_{\mathrm{Ran}\Gamma^{\frac{1}{2}}} = \langle \Gamma u, v\rangle_{\mathcal{H}}$, *for any $u, v \in \mathcal{H}$.*

As an immediate consequence of this result we have:

Corollary 3.4. *The space $\mathcal{H}_L(S)$ is equal to the operator range*

$$\mathrm{Ran}\left(I - M_S^L M_S^{L*}\right)^{\frac{1}{2}}$$

with the norm $\|\cdot\|_{\mathrm{Ran}\left(I-M_S^L M_S^{L*}\right)^{\frac{1}{2}}}$ and is a left reproducing kernel Hilbert space right module with reproducing kernel $K_S^L(\cdot, W)$. This space is the de Branges complement of the space $S\mathcal{U}_2$, the range of M_S^L.

We associated to S a multiplication operator on the left. Correspondingly we have the operator $M_S^R : \mathcal{U}_2 \longrightarrow \mathcal{U}_2$ defined by $M_S^R(F) = FS$, with associated kernel

$$K_S^R(\cdot, W) = \rho_W^{\triangle -1}\left(I - S^{\triangle}(W)^* S\right).$$

We also have the corresponding operators and kernels for S^* given by $M_{S^*}^L : \mathcal{L}_2 \longrightarrow \mathcal{L}_2$ where $M_{S^*}^L(F) = S^*F$, with associated kernel

$$K_{S^*}^L(\cdot, W) = \left(I - S^* S^{*\nabla}(W)^*\right)\rho_W^{\nabla -1},$$

and $M_{S^*}^R : \mathcal{L}_2 \longrightarrow \mathcal{L}_2$ where $M_{S^*}^R(F) = FS^*$ with associated kernel

$$K_{S^*}^R(\cdot, W) = \rho_W^{\vee -1}\left(I - S^{*\vee}(W)^* S^*\right).$$

The four kernels are related by conjugation in the sense that

$$\left(K_{S^*}^R(\cdot, W^*)\right)^* = K_S^L(\cdot, W) \quad \text{and} \quad \left(K_{S^*}^L(\cdot, W^*)\right)^* = K_S^R(\cdot, W).$$

Obviously we have similar theories for all four kernels, but in the sequel we shall relate especially to the left kernels associated to S.

In the classical case, it is well known that $z \longrightarrow \frac{S(z)-S(\omega)}{z-\omega}\xi$ belongs to $\mathcal{H}(S)$ if S is a $\mathbb{C}^{n\times n}$–valued Schur function and $\omega \in \mathbb{D}$, $\xi \in \mathbb{C}^n$ (see [34, Th 2.3 p. 27]). We conclude this section with a proof of the corresponding fact in the present setting.

The result will be used in the last two sections to define a unitary realization and its state space.

Lemma 3.5. *Let $E \in \mathcal{D}_2$ and let S be any Schur upper triangular operator. Then*
1. *Let $W \in \Omega_{\mathcal{N}}$. The operator*

$$(S - S^{\triangle}(W))\,(Z - W)^{-1}\,E \tag{3.3}$$

belongs to $\mathcal{H}_L(S)$.
2. *Let $W^* \in \Omega_{\mathcal{M}}$. The operator*

$$(S^* - S^{*\vee}(W^*))\,(Z^* - W^*)^{-1}\,E \tag{3.4}$$

belongs to $\mathcal{H}_L(S^*)$.

Proof. It is sufficient to show that

$$\kappa_L\left((S - S^{\triangle}(W))\,(Z - W)^{-1}\,E\right) < \infty$$

and

$$\kappa_L^*\left((S^* - S^{*\vee}(W^*))\,(Z^* - W^*)^{-1}\,E\right) < \infty.$$

For $G \in \mathcal{U}_2$ we have

$$\begin{aligned}
&\|\,(S - S^{\triangle}(W))\,(Z - W)^{-1}\,E + SG\|_{\mathcal{U}_2}^2 \\
&\quad= \|S^{\triangle}(W)\,(I - Z^*W)^{-1}\,E^{(1)}Z^*\|_{\mathcal{X}_2}^2 \\
&\quad\quad -2\mathrm{Re}\left\langle S^{\triangle}(W)\,(I - Z^*W)^{-1}\,E^{(1)}Z^*, S\left((I - Z^*W)^{-1}\,E^{(1)}Z^* + G\right)\right\rangle_{\mathcal{X}_2} \\
&\quad\quad +\|S\left((I - Z^*W)^{-1}\,E^{(1)}Z^* + G\right)\|_{\mathcal{X}_2}^2 \\
&\quad\le \|S^{\triangle}(W)\,(I - Z^*W)^{-1}\,E^{(1)}Z^*\|_{\mathcal{X}_2}^2 \\
&\quad\quad -2\mathrm{Re}\left\langle S^{\triangle}(W)\,(I - Z^*W)^{-1}\,E^{(1)}Z^*, S\left((I - Z^*W)^{-1}\,E^{(1)}Z^*\right)\right\rangle_{\mathcal{X}_2} \\
&\quad\quad +\|\,(I - Z^*W)^{-1}\,E^{(1)}Z^* + G\|_{\mathcal{X}_2}^2 \\
&\quad= \|S^{\triangle}(W)\,(I - Z^*W)^{-1}\,E^{(1)}Z^*\|_{\mathcal{X}_2}^2 \\
&\quad\quad -2\mathrm{Re}\left\langle S^{\triangle}(W)\,(I - Z^*W)^{-1}\,E^{(1)}Z^*, S\left((I - Z^*W)^{-1}\,E^{(1)}Z^*\right)\right\rangle_{\mathcal{X}_2} \\
&\quad\quad +\|\,(I - Z^*W)^{-1}\,E^{(1)}Z^*\|_{\mathcal{L}_2}^2 + \|G\|_{\mathcal{U}_2}^2 \\
&\quad= \mathrm{Tr}ZE^{*(1)}\,(I - W^*Z)^{-1}\,S^{\triangle}(W)^*S^{\triangle}(W)\,(I - Z^*W)^{-1}\,E^{(1)}Z^* \\
&\quad\quad -2\mathrm{Re}\mathrm{Tr}ZE^{*(1)}\,(I - W^*Z)^{-1}\,S^*S^{\triangle}(W)\,(I - Z^*W)^{-1}\,E^{(1)}Z^* \\
&\quad\quad +\mathrm{Tr}ZE^{*(1)}\,(I - W^*Z)^{-1}\,(I - Z^*W)^{-1}\,E^{(1)}Z^* + \|G\|_{\mathcal{U}_2}^2
\end{aligned}$$

$$\begin{aligned}
&= \mathrm{Tr}E^{*(1)}\left(I-W^*Z\right)^{-1}S^{\triangle}(W)^*S^{\triangle}(W)\left(I-Z^*W\right)^{-1}E^{(1)} \\
&\quad -2\mathrm{Re}\mathrm{Tr}\sum_{k,l,m=0}^{\infty}E^{*(1)}\left(W^*Z\right)^k S^*_{[l]}Z^{*l}S^{\triangle}(W)\left(Z^*W\right)^m E^{(1)} \\
&\quad +\mathrm{Tr}ZE^{*(1)}\left(I-W^*Z\right)^{-1}\left(I-Z^*W\right)^{-1}E^{(1)}Z^*+\|G\|^2_{\mathcal{U}_2} \\
&= \mathrm{Tr}E^{*(1)}\left(I-W^*Z\right)^{-1}S^{\triangle}(W)^*S^{\triangle}(W)\left(I-Z^*W\right)^{-1}E^{(1)} \\
&\quad -2\mathrm{Re}\ \mathrm{Tr}\sum_{m=0}^{\infty}E^{*(1)}\left(W^*Z\right)^m\left(\sum_{\ell=0}^{\infty}\left(W^*Z\right)^{\ell}S^*_{[\ell]}Z^{*\ell}S^{\triangle}(W)\right)\left(Z^*W\right)^m E^{(1)} \\
&\quad +\mathrm{Tr}E^{*(1)}\left(I-W^*Z\right)^{-1}\left(I-Z^*W\right)^{-1}E^{(1)}+\|G\|^2_{\mathcal{U}_2} \\
&= \mathrm{Tr}E^{*(1)}\left(I-W^*Z\right)^{-1}\left(I-S^{\triangle}(W)^*S^{\triangle}(W)\right)\left(I-Z^*W\right)^{-1}E^{(1)}+\|G\|^2_{\mathcal{U}_2} \\
&= \mathrm{Tr}E^{*(1)}\rho_W^{\triangle-1}\left(I-S^{\triangle}(W)^*S^{\triangle}(W)\right)\rho_W^{\triangle-*}E^{(1)}+\|G\|^2_{\mathcal{U}_2}.
\end{aligned}$$

Therefore,

$$\begin{aligned}
&\sup_{G\in\mathcal{U}_2}\left\{\|\left(S-S^{\triangle}(W)\right)\left(Z-W\right)^{-1}E+SG\|^2_{\mathcal{U}_2}-\|G\|^2_{\mathcal{U}_2}\right\} \\
&\quad\leq \mathrm{Tr}E^{*(1)}\rho_W^{\triangle-1}\left(I-S^{\triangle}(W)^*S^{\triangle}(W)\right)\rho_W^{\triangle-*}E^{(1)}.
\end{aligned}$$

We conclude that the supremum is finite and therefore the operator

$$\left(S-S^{\triangle}(W)\right)\left(Z-W\right)^{-1}E$$

belongs to $\mathcal{H}_L(S)$. The proof of the second part of the lemma goes in much the same way and will be omitted. □

The next lemma shows the connection between the left and right evaluations.

Lemma 3.6. *Let F, G be any operators in $\mathcal{U}$ and $\mathcal{L}$ respectively, and let $V \in \Omega_{\mathcal{N}}$ and $W \in \Omega_{\mathcal{M}}$. Then,*

$$\left(\left(F-F^{\triangle}(V)\right)\left(Z-V\right)^{-1}\right)^{\wedge}(W) = \left(\left(Z-W\right)^{-1}\left(F-F^{\wedge}(W)\right)\right)^{\triangle}(V)^{(-1)} \tag{3.5}$$

$$\left(\left(Z^*-V^*\right)^{-1}\left(G-G^{\nabla}(V^*)\right)\right)^{\vee}(W^*) = \left(\left(G-G^{\vee}(W^*)\right)\left(Z^*-W^*\right)^{-1}\right)^{\nabla}(V^*)^{(-1)}. \tag{3.6}$$

Proof. The proof is by direct computations.

$$\begin{aligned}
&\left(\left(F-F^{\triangle}(V)\right)\left(Z-V\right)^{-1}\right)^{\wedge}(W)^{(1)} \\
&\quad= Z^*\left(\left(F-F^{\triangle}(V)\right)\left(I-Z^*V\right)^{-1}Z^*\right)^{\wedge}(W)Z
\end{aligned}$$

THE STATIONARY SETTING	THE NONSTATIONARY SETTING
Schur functions	Uppertriangular contractions
The Hardy space $\mathbf{H}_2$	Uppertriangular Hilbert Schmidt operators $\mathcal{U}_2$
The Cauchy kernel $\frac{1}{1-z\omega^*}$	The left and right Cauchy kernels $\rho_W^{\wedge -1} = (I - ZW^*)^{-1}$ and $\rho_W^{\triangle -1} = (I - W^*Z)^{-1}$
Multiplication by z	Multiplication by the bilateral shift Z
The field of complex numbers	The algebra of the diagonal operators
The Taylor series of a function at the origin	The (formal) series $F = \sum_{n=0}^{\infty} Z^n F_{[n]}$
The value of an analytic function $f(\omega)$ at a point ω of the unit disk such that the function $\frac{f(z)-f(\omega)}{z-\omega}$ is analytic	The unique diagonal operators $F^{\wedge}(W)$ and $F^{\triangle}(W)$ such that $(Z-W)^{-1}(F - F^{\wedge}(W))$ and $(F - F^{\triangle}(W))(Z-W)^{-1}$ are uppertriangular operators
Multiplication M_s by s in $\mathbf{H}_2$	Multiplication M_S^L by an uppertriangular contraction S from the left in $\mathcal{U}_2$ Multiplication M_S^R by an upper triangular contraction S from the right in $\mathcal{U}_2$
The de Branges–Rovnyak space Ran $\sqrt{I - M_s M_s^*}$	The operator ranges (de Branges–Rovnyak spaces) Ran $\sqrt{I - M_S^L (M_S^L)^*}$ and Ran $\sqrt{I - M_S^R (M_S^R)^*}$
The reproducing kernel $\frac{1-s(z)s(\omega)^*}{1-z\omega^*}$	The reproducing kernels $(I - SS^{\wedge}(W)^*)\,\rho_W^{\wedge -1}$ and ${\rho_W^{\triangle}}^{-1}(I - S^{\triangle}(W)^* S)$ of the left and right de Branges–Rovnyak spaces
For a Schur function s, the Schur function $\tilde{s}(z) = s(z^*)^*$	The lower triangular operator S^*
The reproducing kernel $\frac{1-\tilde{s}(z)\tilde{s}(\omega)^*}{1-z\omega^*}$	The reproducing kernels $(I - S^* S^{*\nabla}(W)^*)\,\rho_W^{\nabla -1}$ and $\rho_W^{\vee -1}(I - S^{*\vee}(W)^* S^*)$ of the spaces associated to S^*
The "two–by–two" kernel $\begin{pmatrix} K_S(z,\omega) & \frac{S(z)-S(\bar{\omega})}{z-\bar{\omega}} \\ \frac{\tilde{S}(z)-\tilde{S}(\bar{\omega})}{z-\bar{\omega}} & K_{\tilde{S}}(z,\omega) \end{pmatrix}$	The "two–by–two" kernels $\begin{pmatrix} K_S^L(\cdot,W)\,E + (S - S^{\triangle}(W))\,(Z-W)^{-1}\,G^{(-1)} \\ (S^* - S^{*\vee}(W^*))\,(Z^*-W^*)^{-1}\,E^{(1)} + K_{S^*}^L(\cdot,W^*)\,G \end{pmatrix}$ and $\begin{pmatrix} EK_S^R(\cdot,W) + G^{(1)}\,(Z-W)^{-1}\,(S - S^{\wedge}(W)) \\ E^{(-1)}\,(Z^*-W^*)^{-1}\,(S^* - S^{*\nabla}(W^*)) + GK_{S^*}^R(\cdot,W^*) \end{pmatrix}$

Table: analogies between the stationary setting and the nonstationary setting

$$
\begin{aligned}
&= Z^* \left(\sum_{n=1}^{\infty} Z^n F_{[n]} \left(I - (Z^*V)^n \right) \left(I - Z^*V \right)^{-1} Z^* \right)^{\wedge} (W) Z \\
&= Z^* \left(\sum_{n=1}^{\infty} \sum_{k=0}^{n-1} Z^n F_{[n]} \left(Z^*V \right)^k Z^* \right)^{\wedge} (W) Z \\
&= \sum_{n=1}^{\infty} \sum_{k=0}^{n-1} Z^* \left(WZ^* \right)^{n-1-k} Z^n F_{[n]} \left(Z^*V \right)^k \\
&= \left(Z^* \left(I - WZ^* \right)^{-1} \left(F - F^{\wedge}(W) \right) \right)^{\triangle} (V) \\
&= \left((Z - W)^{-1} \left(F - F^{\wedge}(W) \right) \right)^{\triangle} (V),
\end{aligned}
$$

which ends the proof. □

A summary in which the stationary setting is compared with the nonstationary setting is given in the table on the previous page.

4. Realization for Schur operators via complementation

In this section we give two coisometric and two isometric realizations for Schur operators, using complementation theory. The next theorem will lead us a to coisometric realization for S.

Theorem 4.1. *The colligation*

$$
\mathcal{V}_L = \begin{pmatrix} \mathbf{A}_L & \mathbf{B}_L \\ \mathbf{C}_L & \mathbf{D}_L \end{pmatrix} : \begin{pmatrix} \mathcal{H}_L(S) \\ \mathcal{D}_2 \end{pmatrix} \longrightarrow \begin{pmatrix} \mathcal{H}_L(S) \\ \mathcal{D}_2 \end{pmatrix}
$$

defined by

$$
\begin{aligned}
&(4.1) \qquad & \mathbf{A}_L(F) &= \left(F - F_{[0]} \right) Z^{-1}, \\
&(4.2) \qquad & \mathbf{B}_L(E) &= \left(S - S_{[0]} \right) E Z^{-1}, \\
&(4.3) \qquad & \mathbf{C}_L(F) &= F_{[0]}, \\
&(4.4) \qquad & \mathbf{D}_L(E) &= S_{[0]} E
\end{aligned}
$$

is coisometric.

Note that (4.1)–(4.4) define the (left) analogue of the backwards shift realization (1.6)–(1.9). Before turning to the proof of the theorem, we show in the next lemma that the colligation $\mathcal{V}_L$ is well defined.

Lemma 4.2. *Let $F \in \mathcal{H}_L(S)$ and $E \in \mathcal{D}_2$. The operators $\left(F - F_{[0]} \right) Z^{-1}$ and $\left(S - S_{[0]} \right) E Z^{-1}$ both belong to $\mathcal{H}_L(S)$.*

Proof. The proof is adapted from the analysis of [34, Proof of theorem 2.3 p. 27] for the stationary case. It is sufficient to show that

$$\kappa_L\left(\left(F - F_{[0]}\right) Z^{-1}\right) < \infty, \quad \text{and} \quad \kappa_L\left(\left(S - S_{[0]}\right) EZ^{-1}\right) < \infty.$$

Now

$$\begin{aligned}\left\|\left(F - F_{[0]}\right) Z^{-1} + SG\right\|_{\mathcal{U}_2}^2 &= \left\|F - F_{[0]} + SGZ\right\|_{\mathcal{U}_2}^2 \\ &= \|F + SGZ\|_{\mathcal{U}_2}^2 + \left\|F_{[0]}\right\|_{\mathcal{D}_2}^2 - 2\mathrm{Re}\left\langle F_{[0]}, F + SGZ\right\rangle_{\mathcal{U}_2} \\ &= \|F + SGZ\|_{\mathcal{U}_2}^2 - \left\|F_{[0]}\right\|_{\mathcal{D}_2}^2.\end{aligned}$$

Therefore we have

$$\begin{aligned}\left\|\left(F - F_{[0]}\right) Z^{-1} + SG\right\|_{\mathcal{U}_2}^2 - \|G\|_{\mathcal{U}_2}^2 &= \|F + SGZ\|_{\mathcal{U}_2}^2 - \|GZ\|_{\mathcal{U}_2}^2 - \left\|F_{[0]}\right\|_{\mathcal{D}_2}^2 \\ &\leq \|F\|_{\mathcal{H}_L(S)}^2 - \left\|F_{[0]}\right\|_{\mathcal{D}_2}^2.\end{aligned}$$

Thus $\kappa_L\left(\left(F - F_{[0]}\right) Z^{-1}\right) \leq \|F\|_{\mathcal{H}_L(S)}^2 - \left\|F_{[0]}\right\|_{\mathcal{D}_2}^2 < \infty$.

Similarly

$$\begin{aligned}\left\|\left(S - S_{[0]}\right) EZ^{-1} + SG\right\|_{\mathcal{U}_2}^2 &= \left\|-S_{[0]}EZ^{-1} + S\left(G + EZ^{-1}\right)\right\|_{\mathcal{X}_2}^2 \\ &= \left\|-S_{[0]}EZ^{-1}\right\|_{\mathcal{X}_2}^2 + \left\|S\left(G + EZ^{-1}\right)\right\|_{\mathcal{X}_2}^2 \\ &\quad -2\mathrm{Re}\left\langle S_{[0]}EZ^{-1}, S\left(G + EZ^{-1}\right)\right\rangle_{\mathcal{X}_2} \\ &= \left\|S\left(G + EZ^{-1}\right)\right\|_{\mathcal{X}_2}^2 - \left\|S_{[0]}EZ^{-1}\right\|_{\mathcal{X}_2}^2 \\ &\leq \left\|G + EZ^{-1}\right\|_{\mathcal{X}_2}^2 - \left\|S_{[0]}EZ^{-1}\right\|_{\mathcal{X}_2}^2 \\ &= \|G\|_{\mathcal{X}_2}^2 + \left\|EZ^{-1}\right\|_{\mathcal{X}_2}^2 - \left\|S_{[0]}EZ^{-1}\right\|_{\mathcal{X}_2}^2 \\ &= \|G\|_{\mathcal{U}_2}^2 + \|E\|_{\mathcal{D}_2}^2 - \left\|S_{[0]}E\right\|_{\mathcal{D}_2}^2.\end{aligned}$$

Therefore

$$\left\|\left(S - S_{[0]}\right) EZ^{-1} + SG\right\|_{\mathcal{U}_2}^2 - \|G\|_{\mathcal{U}_2}^2 \leq \|E\|_{\mathcal{D}_2}^2 - \left\|S_{[0]}E\right\|_{\mathcal{D}_2}^2.$$

Thus

$$\kappa_L\left(\left(S - S_{[0]}\right) EZ^{-1}\right) \leq \|E\|_{\mathcal{D}_2}^2 - \left\|S_{[0]}E\right\|_{\mathcal{D}_2}^2 < \infty,$$

which ends the proof of the lemma. □

Corollary 4.3. *The operators* $\mathbf{A}_L : \mathcal{H}_L(S) \longrightarrow \mathcal{H}_L(S)$, $\mathbf{B}_L : \mathcal{D}_2 \longrightarrow \mathcal{H}_L(S)$, $\mathbf{C}_L : \mathcal{H}_L(S) \longrightarrow \mathcal{D}_2$, $\mathbf{D}_L : \mathcal{D}_2 \longrightarrow \mathcal{D}_2$ *are contractions.*

We now turn to the proof of Theorem 4.1: We first compute the adjoint of the colligation. Let $E_1, E_2 \in \mathcal{D}_2$. We have

$$\begin{aligned}
\langle \mathbf{D}_L^*(E_1), E_2\rangle_{\mathcal{D}_2} &= \langle E_1, \mathbf{D}_L(E_2)\rangle_{\mathcal{D}_2} \\
&= \left\langle E_1, S_{[0]}E_2\right\rangle_{\mathcal{D}_2} \\
&= \mathrm{Tr} E_2^* S_{[0]}^* E_1 \\
&= \left\langle S_{[0]}^* E_1, E_2\right\rangle_{\mathcal{D}_2},
\end{aligned}$$

and thus $\mathbf{D}_L^*(E) = S_{[0]}^* E$.

Let $E \in \mathcal{D}_2$ and $F \in \mathcal{H}_L(S)$; then,

$$\langle \mathbf{C}_L^*(E), F\rangle_{\mathcal{H}_L(S)} = \langle E, \mathbf{C}_L(F)\rangle_{\mathcal{D}_2} = \left\langle E, F_{[0]}\right\rangle_{\mathcal{D}_2} = \mathrm{Tr} F_{[0]}^* E.$$

Thus for $F = K_S^L(\cdot, W)G$, where $W \in \Omega_{\mathcal{M}}$ and $G \in \mathcal{D}_2$ we get that

$$\begin{aligned}
\mathrm{Tr} G^* \left(\mathbf{C}_L^*(E)\right)^\wedge (W) &= \left\langle \mathbf{C}_L^*(E), K_S^L(\cdot, W)G\right\rangle_{\mathcal{H}_L(S)} \\
&= \mathrm{Tr} G^* \left(I - S^\wedge(W)S_{[0]}^*\right) E \\
&= \mathrm{Tr} G^* \left(\left(I - SS_{[0]}^*\right) E\right)^\wedge (W).
\end{aligned}$$

Therefore

$$\left(\mathbf{C}_L^*(E)\right)^\wedge (W) = \left(\left(I - SS_{[0]}^*\right) E\right)^\wedge (W)$$

for any $W \in \Omega_{\mathcal{M}}$, and we conclude that

$$\mathbf{C}_L^*(E) = \left(I - SS_{[0]}^*\right) E = K_S^L(\cdot, 0)E.$$

Next, let $E \in \mathcal{D}_2$ and $F \in \mathcal{H}_L(S)$. Then,

$$\begin{aligned}
\langle \mathbf{B}_L^*(F), E\rangle_{\mathcal{D}_2} &= \mathrm{Tr} E^* \mathbf{B}_L^*(F) \\
&= \mathrm{Tr} Z E^* Z^* Z \mathbf{B}_L^*(F) Z^*.
\end{aligned}$$

Thus for $F = K_S^L(\cdot, W)G$ where $W \in \Omega_{\mathcal{M}}$ and $G \in \mathcal{D}_2$ we obtain

$$\begin{aligned}
\left\langle \mathbf{B}_L^* \left(K_S^L(\cdot, W)G\right), E\right\rangle_{\mathcal{D}_2} &= \left\langle K_S^L(\cdot, W)G, \mathbf{B}_L(E)\right\rangle_{\mathcal{H}_L(S)} \\
&= \left\langle K_S^L(\cdot, W)G, \left(S - S_{[0]}\right) E Z^{-1}\right\rangle_{\mathcal{H}_L(S)} \\
&= \left\{\mathrm{Tr} G^* \left(\left(S - S_{[0]}\right) E Z^{-1}\right)^\wedge (W)\right\}^* \\
&= \mathrm{Tr} \left(\left(S - S_{[0]}\right) E Z^{-1}\right)^\wedge (W)^* G \\
&= \mathrm{Tr} Z E^* Z^* \left(\left(S - S_{[0]}\right) Z^{-1}\right)^\wedge (W)^* G,
\end{aligned}$$

Therefore

$$Z\mathbf{B}_L^*\left(K_S^L(\cdot,W)G\right)Z^* = \left(\left(S - S_{[0]}\right)Z^{-1}\right)^\wedge (W)^*G,$$

and thus

$$\mathbf{B}_L^*\left(K_S^L(\cdot,W)G\right) = Z^*\left(\left(S - S_{[0]}\right)Z^{-1}\right)^\wedge (W)^*GZ.$$

For the special case of $F = K_S^L(\cdot,W)ZW^*GZ^*$ we get

$$\begin{aligned}\mathbf{B}_L^*\left(K_S^L(\cdot,W)ZW^*GZ^*\right) &= Z^*\left(\left(S - S_{[0]}\right)Z^{-1}\right)^\wedge (W)^*ZW^*G \\ &= \left(S - S_{[0]}\right)^\wedge (W)^*G \\ &= \left(S^\wedge(W)^* - S_{[0]}^*\right)G,\end{aligned}$$

so that

$$\mathbf{B}_L^*\left(K_S^L(\cdot,W)ZW^*GZ^*\right) = \left(S^\wedge(W)^* - S_{[0]}^*\right)G.$$

We now compute $\mathbf{A}_L^*$. We first note the following:

$$\begin{aligned}&\left\langle K_S^L(\cdot,W)ZW^*GZ^*, \left(K_S^L(\cdot,V) - \left(K_S^L(\cdot,V)\right)_{[0]}\right)EZ^{-1}\right\rangle_{\mathcal{H}_L(S)} \\ &= \left\{\mathrm{Tr}ZG^*WZ^*\left(\left(K_S^L(\cdot,V) - \left(K_S^L(\cdot,V)\right)_{[0]}\right)EZ^{-1}\right)^\wedge (W)\right\}^* \\ &= \left\{\mathrm{Tr}ZG^*WZ^*\left(\left(K_S^L(\cdot,V) - \left(K_S^L(\cdot,V)\right)_{[0]}\right)Z^{-1}\right)^\wedge (W)ZEZ^*\right\}^* \\ &= \left\{\mathrm{Tr}ZG^*\left(K_S^L(\cdot,V) - \left(K_S^L(\cdot,V)\right)_{[0]}\right)^\wedge (W)EZ^*\right\}^* \\ &\qquad = \mathrm{Tr}E^*\left(K_S^L(W,V)^* - \left(K_S^L(\cdot,V)\right)_{[0]}^*\right)G \\ &\qquad = \mathrm{Tr}E^*\left(K_S^L(V,W) - K_S^L(V,0)\right)G \\ &\qquad = \left\langle\left(K_S^L(\cdot,W) - K_S^L(\cdot,0)\right)G, K_S^L(\cdot,V)E\right\rangle_{\mathcal{H}_L(S)}\end{aligned}$$

where $E, G \in \mathcal{D}_2$, and $V, W \in \Omega_{\mathcal{M}}$. Therefore

$$\begin{aligned}&\left\langle K_S^L(\cdot,W)ZW^*GZ^*, \left(K_S^L(\cdot,V) - \left(K_S^L(\cdot,V)\right)_{[0]}\right)EZ^{-1}\right\rangle_{\mathcal{H}_L(S)} \\ &\qquad = \left\langle\left(K_S^L(\cdot,W) - K_S^L(\cdot,0)\right)G, K_S^L(\cdot,V)E\right\rangle_{\mathcal{H}_L(S)}.\end{aligned}$$

Thus,

$$\begin{aligned}&\left\langle\mathbf{A}_L^*\left(K_S^L(\cdot,W)ZW^*GZ^*\right), K_S^L(\cdot,V)E\right\rangle_{\mathcal{H}_L(S)} \\ &= \left\langle K_S^L(\cdot,W)ZW^*GZ^*, \mathbf{A}_L\left(K_S^L(\cdot,V)E\right)\right\rangle_{\mathcal{H}_L(S)} \\ &= \left\langle K_S^L(\cdot,W)ZW^*GZ^*, \left(K_S^L(\cdot,V) - \left(K_S^L(\cdot,V)\right)_{[0]}\right)EZ^{-1}\right\rangle_{\mathcal{H}_L(S)} \\ &= \left\langle\left(K_S^L(\cdot,W) - K_S^L(\cdot,0)\right)G, K_S^L(\cdot,V)E\right\rangle_{\mathcal{H}_L(S)}.\end{aligned}$$

And therefore we have got that

$$\mathbf{A}_L^* \left(K_S^L(\cdot, W)ZW^*GZ^*\right) = \left(K_S^L(\cdot, W) - K_S^L(\cdot, 0)\right) G.$$

Now we show that the set of all operators of the form

$$K_S^L(\cdot, W)ZW^*GZ^*$$

is dense in $\mathcal{H}_L(S)$. Let F be any operator in $\mathcal{H}_L(S)$, such that

$$\left\langle F, K_S^L(\cdot, W)ZW^*GZ^*\right\rangle_{\mathcal{H}_L(S)} = 0.$$

Since

$$\begin{aligned}\left\langle F, K_S^L(\cdot, W)ZW^*GZ^*\right\rangle_{\mathcal{H}_L(S)} &= \mathrm{Tr} ZG^*WZ^*F^\wedge(W) \\ &= \mathrm{Tr} Z^* \left\{ZG^*WZ^*F^\wedge(W)\right\} Z \\ &= \mathrm{Tr} G^*(FZ)^\wedge(W),\end{aligned}$$

we get that $(FZ)^\wedge(W) = 0$ for any $W \in \Omega_{\mathcal{M}}$ and so $FZ = 0$ and $F = 0$.

Now we are prepared to show that the colligation $\mathcal{V}_L$ is coisometric. Let $E \in \mathcal{D}_2$:

$$\begin{aligned}\left(\mathbf{C}_L\mathbf{C}_L^* + \mathbf{D}_L\mathbf{D}_L^*\right)(E) &= \mathbf{C}_L\left(\left(I - SS_{[0]}^*\right) E\right) + \mathbf{D}_L\left(S_{[0]}^* E\right) \\ &= \left(I - S_{[0]}S_{[0]}^*\right) E + S_{[0]}S_{[0]}^* E \\ &= E.\end{aligned}$$

Therefore

$$\mathbf{C}_L\mathbf{C}_L^* + \mathbf{D}_L\mathbf{D}_L^* = \mathbf{I}_{\mathcal{D}_2}.$$

Let $E \in \mathcal{D}_2$. Then

$$\begin{aligned}\left(\mathbf{A}_L\mathbf{C}_L^* + \mathbf{B}_L\mathbf{D}_L^*\right)(E) &= \mathbf{A}_L\left(\left(I - SS_{[0]}^*\right) E\right) + \mathbf{B}_L\left(S_{[0]}^* E\right) \\ &= \left(\left(I - SS_{[0]}^*\right) E - \left(I - S_{[0]}S_{[0]}^*\right) E\right) Z^{-1} + \\ &\quad + \left(S - S_{[0]}\right) S_{[0]}^* E Z^{-1} \\ &= \mathbf{0}.\end{aligned}$$

Therefore

$$\mathbf{A}_L\mathbf{C}_L^* + \mathbf{B}_L\mathbf{D}_L^* = \mathbf{0}.$$

Let $G \in \mathcal{D}_2$ and $W \in \Omega_{\mathcal{M}}$. Then,

$$\begin{aligned}&\left(\mathbf{C}_L\mathbf{A}_L^* + \mathbf{D}_L\mathbf{B}_L^*\right)\left(K_S^L(\cdot, W)ZW^*GZ^*\right) \\ &= \mathbf{C}_L\left(\left(K_S^L(\cdot, W) - K_S^L(\cdot, W)\right) G\right) + \mathbf{D}_L\left(\left(S^\wedge(W)^* - S_{[0]^*}\right) G\right)\end{aligned}$$

$$\begin{aligned}
&= \left(\left(K_S^L(\cdot,W)\right)_{[0]} - \left(K_S^L(\cdot,0)\right)_{[0]} \right) G + S_{[0]} \left(S^\wedge(W)^* - S_{[0]}^* \right) G \\
&= \left(\left(I - S_{[0]} S^\wedge(W)^*\right) - \left(I - S_{[0]} S_{[0]}^*\right) \right) G \\
&\quad + S_{[0]} \left(S^\wedge(W)^* - S_{[0]}^* \right) G \\
&= \mathbf{0}.
\end{aligned}$$

The operator $\mathbf{C}_L \mathbf{A}_L^* + \mathbf{D}_L \mathbf{B}_L^*$ is equal to zero on a dense subset of $\mathcal{H}_L(S)$ and so on $\mathcal{H}_L(S)$ itself, i.e.

$$\mathbf{C}_L \mathbf{A}_L^* + \mathbf{D}_L \mathbf{B}_L^* = \mathbf{0}.$$

Let $G \in \mathcal{D}_2$ and $W \in \Omega_{\mathcal{M}}$. Then

$$\begin{aligned}
&(\mathbf{A}_L \mathbf{A}_L^* + \mathbf{B}_L \mathbf{B}_L^*) \left(K_S^L(\cdot,W) Z W^* G Z^*\right) \\
&\quad = \mathbf{A}_L \left(\left(K_S^L(\cdot,W) - K_S^L(\cdot,0)\right) G \right) + \mathbf{B}_L \left(\left(S^\wedge(W)^* - S_{[0]}^* \right) G \right) \\
&\quad = \left(K_S^L(\cdot,W) - \left(K_S^L(\cdot,W)\right)_{[0]} \right) G Z^{-1} - \left(K_S^L(\cdot,0) - \left(K_S^L(\cdot,0)\right)_{[0]} \right) G Z^{-1} \\
&\qquad + \left(S - S_{[0]}\right) \left(S^\wedge(W)^* - S_{[0]}^* \right) G Z^{-1} \\
&\quad = \left(\left(I - S S^\wedge(W)^*\right) \rho_W^{\wedge -1} - \left(I - S_{[0]} S^\wedge(W)^*\right) \right) G Z^{-1} \\
&\qquad - \left(\left(I - S S_{[0]}^*\right) \rho_W^{\wedge -1} - \left(I - S_{[0]} S_{[0]}^*\right) \right) G Z^{-1} \\
&\qquad + \left(S - S_{[0]}\right) \left(S^\wedge(W)^* - S_{[0]}^* \right) G Z^{-1} \\
&\quad = \left(\left(I - S S^\wedge(W)^*\right) \rho_W^{\wedge -1} - \left(I - S S^\wedge(W)^*\right) \right) G Z^{-1} \\
&\quad = \left(I - S S^\wedge(W)^*\right) \rho_W^{\wedge -1} Z W^* G Z^* \\
&\quad = K_S^L(\cdot,W) Z W^* G Z^*.
\end{aligned}$$

The operator $\mathbf{A}_L \mathbf{A}_L^* + \mathbf{B}_L \mathbf{B}_L^*$ coincides with the identity of $\mathcal{H}_L(S)$ on a dense subset and thus on $\mathcal{H}_L(S)$, i.e.

$$\mathbf{A}_L \mathbf{A}_L^* + \mathbf{B}_L \mathbf{B}_L^* = \mathbf{I}_{\mathcal{H}_L(S)}.$$

which ends the proof that $\mathcal{V}_L$ is coisometric.

To state the next result (about the coisometric realization of S) we need a preliminary lemma:

Lemma 4.4. *Let $W \in \Omega_{\mathcal{N}}$ with $\|W\| < 1$. The multiplication operator $\mathbf{M}_W^R : \mathcal{H}_L(S) \longrightarrow \mathcal{H}_L(S)$ defined by*

$$\mathbf{M}_W^R(F) = FW$$

is a strict contraction.

Proof. Let $F = \sum_{i=1}^{n} K_S^L(\cdot, V_i)G_i$. Then, $\mathbf{M}_W^R(F) = \sum_{i=1}^{n} K_S^L(\cdot, V_i)G_iW$, where now $G_iW \in \mathcal{D}_2$ for $i = 1, \ldots, n$ and therefore $\mathbf{M}_W^R(F)$ belongs to $\mathcal{H}_L(S)$.

Let

$$[F, F]_{\mathcal{H}_L(S)} = \sum_{i,j=1}^{n} G_j^* K_S^L(V_j, V_i)G_i.$$

Then, $\|F\|^2_{\mathcal{H}_L(S)} = \mathrm{Tr}[F, F]_{\mathcal{H}_L(S)}$ and

$$\begin{aligned} \|\mathbf{M}_W^R(F)\|^2_{\mathcal{H}_L(S)} &= \mathrm{Tr}W^* \sum_{i,j=1}^{n} G_j^* K_S^L(V_j, V_i)G_iW \\ &= \mathrm{Tr}W^*[F, F]_{\mathcal{H}_L(S)}W. \end{aligned}$$

Note that $[F, F]_{\mathcal{H}_L(S)}$ is a diagonal trace class operator. From the nonnegativity of the kernel $K_S^L(\cdot, V)$ we get

$$\mathrm{Tr}E^*[F, F]_{\mathcal{H}_L(S)}E \geq 0,$$

for every $E \in \mathcal{D}_2$ and therefore $[F, F]_{\mathcal{H}_L(S)} \geq 0$. Let $X = [F, F]_{\mathcal{H}_L(S)}$. The operator $X^{\frac{1}{2}}$ is a Hilbert–Schmidt operator; indeed let $\phi_1, \phi_2, \ldots$ be an orthonormal basis for $\ell^2_{\mathcal{N}}$. Then,

$$\begin{aligned} \sum_{i=1}^{\infty} \|X^{\frac{1}{2}}\phi_i\|^2_{\ell^2_{\mathcal{N}}} &= \sum_{i=1}^{\infty} \langle X^{\frac{1}{2}}\phi_i, X^{\frac{1}{2}}\phi_i\rangle_{\ell^2_{\mathcal{N}}} \\ &= \sum_{i=1}^{\infty} \langle X\phi_i, \phi_i\rangle_{\ell^2_{\mathcal{N}}} \\ &= \mathrm{Tr}X = \mathrm{Tr}[F, F]_{\mathcal{H}_L(S)} < \infty. \end{aligned}$$

Thus $X^{\frac{1}{2}} \in \mathcal{D}_2$. Now

$$\begin{aligned} \|\mathbf{M}_W^R(F)\|^2_{\mathcal{H}_L(S)} &= \mathrm{Tr}W^*[F, F]_{\mathcal{H}_L(S)}W \\ &= \mathrm{Tr}W^* X^{\frac{1}{2}} X^{\frac{1}{2}} W \\ &= \|X^{\frac{1}{2}}W\|^2_{\mathcal{D}_2} \\ &\leq \|W\|^2 \|X^{\frac{1}{2}}\|^2_{\mathcal{D}_2} \\ &= \|W\|^2 \mathrm{Tr}X \\ &= \|W\|^2 \mathrm{Tr}[F, F]_{\mathcal{H}_L(S)} \\ &= \|W\|^2 \|F\|^2_{\mathcal{H}_L(S)}, \end{aligned}$$

and therefore the operator $\mathbf{M}_W^R$ is a strict contraction. □

We now turn to the main theorem of this section.

Theorem 4.5. *Let $S \in \mathcal{S}(\ell^2_{\mathcal{N}}, \ell^2_{\mathcal{M}})$. For any $W \in \Omega_{\mathcal{N}}$ such that $\|W\| < 1$ and any $E \in \mathcal{D}_2(\ell^2_{\mathcal{N}})$ we have that*

$$(SE)^{\triangle}(W) = \left(\mathbf{D}_L + \mathbf{C}_L \mathbf{M}^R_W \left(\mathbf{I}_{\mathcal{H}_L(S)} - \mathbf{A}_L \mathbf{M}^R_W\right)^{-1} \mathbf{B}_L\right)(E).$$

Proof. Let $F \in \mathcal{H}_L(S)$. Then,

$$\mathbf{M}^R_W \mathbf{A}_L(F) = \left(F - F_{[0]}\right) Z^*W = \sum_{n=1}^{\infty} Z^n F_{[n]} Z^* W,$$

and so

$$\mathbf{C}_L \mathbf{M}^R_W \mathbf{A}_L(F) = Z F_{[1]} Z^* W,$$

and in the same manner we get that

$$\mathbf{C}_L \left(\mathbf{M}^R_W \mathbf{A}_L\right)^n (F) = Z^n F_{[n]} \left(Z^* W\right)^n,$$

for $n \geq 1$.

Therefore

$$F^{\triangle}(W) = \mathbf{C}_L \left(\mathbf{I}_{\mathcal{H}_L(S)} - \mathbf{M}^R_W \mathbf{A}_L\right)^{-1}(F),$$

where the indicated inverse exists because $\mathbf{A}_L$ is a contraction and $\mathbf{M}^R_W$ is a strict contraction.

We conclude that for any $E \in \mathcal{D}_2$ and any $W \in \Omega_{\mathcal{N}}$ such that $\|W\| < 1$,

$$\left(\left(S - S_{[0]}\right) E Z^{-1}\right)^{\triangle}(W) = \mathbf{C}_L \left(\mathbf{I}_{\mathcal{H}_L(S)} - \mathbf{M}^R_W \mathbf{A}_L\right)^{-1} \mathbf{B}_L(E).$$

Using the fact that for any $F \in \mathcal{U}$

$$(FZ)^{\triangle}(W) = (FW)^{\triangle}(W) = (F)^{\triangle}(W^{(-1)})W, \tag{4.5}$$

we get that

$$\left(\left(S - S_{[0]}\right) E\right)^{\triangle}(W) = \mathbf{M}^R_W \mathbf{C}_L \left(\mathbf{I}_{\mathcal{H}_L(S)} - \mathbf{M}^R_{W^{(-1)}} \mathbf{A}_L\right)^{-1} \mathbf{B}_L(E).$$

Since $\mathbf{D}_L(E) = S_{[0]}E$, $\mathbf{M}^R_{W^{(-1)}} \mathbf{A}_L = \mathbf{A}_L \mathbf{M}^R_W$ and $\mathbf{M}^R_W \mathbf{C}_L L = \mathbf{C}_L \mathbf{M}^R_W$ we get the right coisometric realization, namely

$$(SE)^{\triangle}(W) = \left(\mathbf{D}_L + \mathbf{C}_L \mathbf{M}^R_W \left(\mathbf{I}_{\mathcal{H}_L(S)} - \mathbf{A}_L \mathbf{M}^R_W\right)^{-1} \mathbf{B}_L\right)(E).$$

□

Similar arguments lead to a coisometric realization in the space $\mathcal{H}_R(S)$. Here we define the colligation $\mathcal{V}_R$

$$\mathcal{V}_R = \begin{pmatrix} \mathbf{A}_R & \mathbf{B}_R \\ \mathbf{C}_R & \mathbf{D}_R \end{pmatrix} : \begin{pmatrix} \mathcal{H}_R(S) \\ \mathcal{D}_2 \end{pmatrix} \longrightarrow \begin{pmatrix} \mathcal{H}_R(S) \\ \mathcal{D}_2 \end{pmatrix}$$

by

$$\begin{aligned} \mathbf{A}_R(F) &= Z^{-1}\left(F - F_{[0]}\right), \\ \mathbf{B}_R(E) &= Z^{-1}E\left(S - S_{[0]}\right), \\ \mathbf{C}_R(F) &= F_{[0]}, \\ \mathbf{D}_R(E) &= ES_{[0]}. \end{aligned}$$

These equations define the (right) analogue of the backwards shift realization (1.6)–(1.9). We skip the proof that $\mathcal{V}_R$ is well defined and coisometric, and proceed directly to show that it provides a realization for S. Let $W \in \Omega_{\mathcal{M}}$. The multiplication operator $\mathbf{M}_W^L : \mathcal{H}_R(S) \longrightarrow \mathcal{H}_R(S)$ defined by

$$\mathbf{M}_W^L(F) = WF,$$

is well defined since $WGK_S^R(\cdot, V)$ belongs to $\mathcal{H}_R(S)$ for any $V \in \Omega_{\mathcal{M}}$, $G \in \mathcal{D}_2$ and since the set of all operators of the form $GK_S^R(\cdot, V)$ is dense in $\mathcal{H}_R(S)$. It is readily seen that $\mathbf{M}_W^L$ is a strict contraction if $\|W\| < 1$.

Let $F \in \mathcal{H}_R(S)$ then

$$\begin{aligned} \mathbf{M}_W^L \mathbf{A}_R(F) &= WZ^*\left(F - F_{[0]}\right) \\ &= \sum_{n=0}^{\infty} WZ^*Z^n F_{[n]}. \end{aligned}$$

Therefore we have that

$$\mathbf{C}_R \mathbf{M}_W^L \mathbf{A}_R(F) = WZ^*ZF_{[1]},$$

and in the same manner we get that

$$\mathbf{C}_R\left(\mathbf{M}_W^L \mathbf{A}_R\right)^n (F) = \left(WZ^*\right)^n Z^n F_{[n]},$$

for $n \geq 1$. So,

$$F^\wedge(W) = \mathbf{C}_R\left(\mathbf{I}_{\mathcal{H}_R(S)} - \mathbf{M}_W^L \mathbf{A}_R\right)^{-1}(F),$$

when the indicated inverse exists since $\mathbf{A}_R$ is a contraction and $\mathbf{M}_W^L$ is a strict contraction. Now, for any $E \in \mathcal{D}_2$ and for any $W \in \Omega_{\mathcal{M}}$ such that $\|W\| < 1$ we have

$$\left(Z^{-1}E\left(S - S_{[0]}\right)\right)^\wedge(W) = \mathbf{C}_R\left(\mathbf{I}_{\mathcal{H}_R(S)} - \mathbf{M}_W^L \mathbf{A}_R\right)^{-1} \mathbf{B}_R(E).$$

Using the fact that for any $F \in \mathcal{U}$

$$(ZF)^\wedge (W) = (WF)^\wedge (W) = W\,(F)^\wedge (W^{(1)}), \tag{4.6}$$

we get

$$(E\,(S - S_{[0]}))^\wedge (W) = \mathbf{M}^L_W \mathbf{C}_R \left(\mathbf{I}_{\mathcal{H}_R(S)} - \mathbf{M}^L_{W^{(1)}} \mathbf{A}_R\right)^{-1} \mathbf{B}_R(E).$$

Since $\mathbf{D}_R(E) = ES_{[0]}$ and $\mathbf{M}^L_{W^{(1)}}\mathbf{A}_R = \mathbf{A}_R\mathbf{M}^L_W$, we get the left coisometric realization, namely

$$(ES)^\wedge (W) = \left(\mathbf{D}_R + \mathbf{M}^L_W \mathbf{C}_R \left(\mathbf{I}_{\mathcal{H}_R(S)} - \mathbf{A}_R \mathbf{M}^L_W\right)^{-1} \mathbf{B}_R\right)(E).$$

To study the question of uniqueness in these representations, we first need to review some definitions. A colligation $\mathcal{V}$

$$\begin{pmatrix} A & B \\ C & D \end{pmatrix} : \quad \begin{pmatrix} \mathcal{H} \\ \mathcal{F} \end{pmatrix} \longrightarrow \begin{pmatrix} \mathcal{H} \\ \mathcal{G} \end{pmatrix} \tag{4.7}$$

is said to be closely inner connected, closely outer connected or closely connected according as

$$\mathcal{H} = \overline{\text{Span}} \left\{ \text{Ran}\,(I_{\mathcal{H}} - \lambda A)^{-1} B \mid \lambda \in \Delta \right\},$$

$$\mathcal{H} = \overline{\text{Span}} \left\{ \text{Ran}\,(I_{\mathcal{H}} - \bar{\mu} A^*)^{-1} C^* \mid \mu \in \Delta \right\},$$

or

$$\mathcal{H} = \overline{\text{Span}} \left\{ \text{Ran}\,(I_{\mathcal{H}} - \lambda A)^{-1} B,\ \text{Ran}\,(I_{\mathcal{H}} - \bar{\mu} A^*)^{-1} C^* \mid \lambda, \bar{\mu} \in \Delta \right\},$$

where Δ is some neighborhood of $\mathbb{D}$ which contains the origin.

These conditions are equivalent to the following:

$$\mathcal{H} = \overline{\text{Span}} \left\{ \text{Ran}\,(A)^m B \mid m \geq 0 \right\},$$

$$\mathcal{H} = \overline{\text{Span}} \left\{ \text{Ran}\,(A^*)^n C^* \mid n \geq 0 \right\},$$

$$\mathcal{H} = \overline{\text{Span}} \left\{ \text{Ran}\,(A)^m B,\ \text{Ran}\,(A^*)^n C^* \mid m, n \geq 0 \right\}.$$

The terms controllable and observable are used in system theory instead of closely inner and closely outer connected; see [23]. In system theory, the main point is on minimality: a finite dimensional colligation which is both controllable and observable (i.e. minimal) is uniquely determined, up to a similarity operator, by its characteristic operator function

$$\theta(\lambda) = D + \lambda C (I_{\mathcal{H}} - \lambda A)^{-1} B.$$

The uniqueness arises here from the metric properties of the colligation. A closely outer connected coisometric colligation is defined uniquely, up to a similarity operator (which is moreover unitary), by its characteristic operator function.

(See Theorem 4.7 for the definition of similarity). Similar statements hold for closely inner connected isometric colligations and closely connected unitary colligations (see [32], [33], [28], [11]).

In the next theorem we show that the colligation $\mathcal{V}_L$ is closely outer connected.

Theorem 4.6. *Let $\mathcal{V}_L$ be as in (4.1)–(4.4). Then*

$$\mathcal{H}_L(S) = \overline{\text{Span}}\left\{\text{Ran}\,(\mathbf{A}_L^*)^n\,\mathbf{C}_L^* \mid n \geq 0\right\},$$

i.e. the colligation $\mathcal{V}_L$ is closely outer connected.

Proof. Let $F \in \mathcal{H}_L(S)$. For $n \geq 0$,

$$\mathbf{C}_L\,(\mathbf{A}_L)^n\,(F) = Z^n F_{[n]} Z^{*n}.$$

Thus, if F belongs to

$$\cap_{n=0}^{\infty}\text{Ker}\;\mathbf{C}_L\,(\mathbf{A}_L)^n\,,$$

we have $Z^n F_{[n]} Z^{*n} = 0$ for any $n \geq 0$. Hence $F_{[n]} = 0$ for any $n \geq 0$, and we conclude that $F = 0$. □

The uniqueness of $\mathcal{V}_L$ is studied in the next result.

Theorem 4.7. *Let $\mathcal{H}$ be a Hilbert space right $\mathcal{D}$–module, and let $\mathcal{V}$ be any left colligation*

$$\mathcal{V} = \begin{pmatrix} \mathbf{A}'_L & \mathbf{B}'_L \\ \mathbf{C}'_L & \mathbf{D}'_L \end{pmatrix} : \begin{pmatrix} \mathcal{H} \\ \mathcal{D}_2 \end{pmatrix} \longrightarrow \begin{pmatrix} \mathcal{H} \\ \mathcal{D}_2 \end{pmatrix},$$

which is coisometric and closely outer connected, such that

$$(SE)^{\triangle}(W) = \left(\mathbf{D}'_L + \mathbf{C}'_L\mathbf{M}^R_W\left(\mathbf{I}_{\mathcal{H}} - \mathbf{A}'_L\mathbf{M}^R_W\right)^{-1}\mathbf{B}'_L\right)(E) \tag{4.8}$$

for any $E \in \mathcal{D}_2$ and any $W \in \Omega_{\mathcal{N}}$ with $\|W\| < \delta$. The colligations $\mathcal{V}_L$ and $\mathcal{V}$ are then unitarily equivalent, i.e. there is a unitary mapping

$$\sigma : \mathcal{H} \to \mathcal{H}_L(S)$$

such that

$$\begin{pmatrix} \sigma^{-1} & 0 \\ 0 & I \end{pmatrix}\begin{pmatrix} \mathbf{A}_L & \mathbf{B}_L \\ \mathbf{C}_L & \mathbf{D}_L \end{pmatrix}\begin{pmatrix} \sigma & 0 \\ 0 & I \end{pmatrix} = \begin{pmatrix} \mathbf{A}'_L & \mathbf{B}'_L \\ \mathbf{C}'_L & \mathbf{D}'_L \end{pmatrix}. \tag{4.9}$$

Proof. By assumption we get that

$$\left(\mathbf{D}'_L + \mathbf{C}'_L\mathbf{M}^R_W\left(\mathbf{I}_{\mathcal{H}} - \mathbf{A}'_L\mathbf{M}^R_W\right)^{-1}\mathbf{B}'_L\right)(E) =$$
$$\left(\mathbf{D}_L + \mathbf{C}_L\mathbf{M}^R_W\left(\mathbf{I}_{\mathcal{H}_L(S)} - \mathbf{A}_L\mathbf{M}^R_W\right)^{-1}\mathbf{B}_L\right)(E),$$

for any $W \in \Omega_{\mathcal{N}}$, $\|W\| < \delta$ and $E \in \mathcal{D}_2$. Let $W = \lambda I$, $|\lambda| < \delta$. Since, as is readily verified,

$$\theta_L(\lambda)(E) = (SE)^{\triangle}(\lambda I)$$

(where θ_L is the characteristic function of the colligation $\mathcal{V}_L$) the related characteristic functions

$$\theta_L(\lambda) = \mathbf{D}_L + \lambda \mathbf{C}_L \left(\mathbf{I}_{\mathcal{H}_L(S)} - \lambda \mathbf{A}_L\right)^{-1} \mathbf{B}_L,$$

$$\theta_L'(\lambda) = \mathbf{D}_L' + \lambda \mathbf{C}_L' \left(\mathbf{I}_{\mathcal{H}} - \lambda \mathbf{A}_L'\right)^{-1} \mathbf{B}_L'$$

coincide in some region of $\mathbb{D}$ containing the origin and are closely outer connected. By the above mentioned result on closely outer connected coisometric colligation, the two colligations differ by a unitary equivalence (see [11]). This ends the proof. □

One should note the following (and the same remark holds for the other uniqueness results in the paper). If we start from a unitary map σ which satisfies (4.9), it need not be true that (4.8) is in force. It will hold when the operators M_W^R commute with σ, i.e. when σ preserves the module structure. This problem is a consequence of the non commutativity. Similar problems were encountered by Ball and Trent [22] in a study of realizations in several variables. There, some additional moment type conditions are added to tackle this question.

The function θ_L is analytic and operator–valued, and uniquely determines S. We note that in [36] the authors also associate to a Schur operator an operator–valued analytic function (different from θ_L), to reduce the study of nonstationary interpolation problems to the case of classical interpolation problem for operator–valued functions.

We now turn to the isometric realizations. Let $\tilde{\mathcal{V}}_L$ be the left colligation associated to $\mathcal{H}_L(S^*)$

$$\tilde{\mathcal{V}}_L = \begin{pmatrix} \tilde{\mathbf{A}}_L & \tilde{\mathbf{B}}_L \\ \tilde{\mathbf{C}}_L & \tilde{\mathbf{D}}_L \end{pmatrix} : \begin{pmatrix} \mathcal{H}_L(S^*) \\ \mathcal{D}_2 \end{pmatrix} \longrightarrow \begin{pmatrix} \mathcal{H}_L(S^*) \\ \mathcal{D}_2 \end{pmatrix},$$

defined by

$$\begin{aligned} \tilde{\mathbf{A}}_L(F) &= \left(F - F_{[0]}\right) Z, \\ \tilde{\mathbf{B}}_L(E) &= \left(S^* - S^*_{[0]}\right) EZ, \\ \tilde{\mathbf{C}}_L(F) &= F_{[0]}, \\ \tilde{\mathbf{D}}_L(E) &= S^*_{[0]} E. \end{aligned}$$

Let $\tilde{\mathcal{V}}_R$ be the right colligation related to $\mathcal{H}_R(S^*)$

$$\tilde{\mathcal{V}}_R = \begin{pmatrix} \tilde{\mathbf{A}}_R & \tilde{\mathbf{B}}_R \\ \tilde{\mathbf{C}}_R & \tilde{\mathbf{D}}_R \end{pmatrix} : \begin{pmatrix} \mathcal{H}_R(S^*) \\ \mathcal{D}_2 \end{pmatrix} \longrightarrow \begin{pmatrix} \mathcal{H}_R(S^*) \\ \mathcal{D}_2 \end{pmatrix},$$

defined by

$$\begin{aligned}
\tilde{\mathbf{A}}_R(F) &= Z\left(F - F_{[0]}\right), \\
\tilde{\mathbf{B}}_R(E) &= ZE\left(S^* - S^*_{[0]}\right), \\
\tilde{\mathbf{C}}_R(F) &= F_{[0]}, \\
\tilde{\mathbf{D}}_R(E) &= ES^*_{[0]}.
\end{aligned}$$

It is readily seen that $\tilde{\mathcal{V}}_L$ and $\tilde{\mathcal{V}}_R$ are coisometric and closely outer connected colligations which supply left and right realizations for S^*. The left realization is given by

$$(4.10)\quad \left(S^*E\right)^\vee(W) = \left(\tilde{\mathbf{D}}_L + \tilde{\mathbf{C}}_L\mathbf{M}^R_W\left(\mathbf{I}_{\mathcal{H}_L(S^*)} - \tilde{\mathbf{A}}_L\mathbf{M}^R_W\right)^{-1}\tilde{\mathbf{B}}_L\right)(E),$$

and the right realization is given by

$$(4.11)\quad \left(ES^*\right)^\nabla(W) = \left(\tilde{\mathbf{D}}_R + \tilde{\mathbf{C}}_R\mathbf{M}^L_W\left(\mathbf{I}_{\mathcal{H}_R(S^*)} - \tilde{\mathbf{A}}_R\mathbf{M}^L_W\right)^{-1}\tilde{\mathbf{B}}_R\right)(E).$$

Theorem 4.8. *Let* $\tilde{\mathcal{V}}_L$ *defined as in* (4.10)*. Then*

$$(SE)^\triangle(W) = \left(\tilde{\mathbf{D}}^*_L + \tilde{\mathbf{B}}^*_L\left(\mathbf{I}_{\mathcal{H}_L(S^*)} - \mathbf{M}^R_{W^{(1)}}\tilde{\mathbf{A}}^*_L\right)^{-1}\mathbf{M}^R_{W^{(1)}}\tilde{\mathbf{C}}^*_L\right)(E),$$

for any $E \in \mathcal{D}_2$ *and any* $W^* \in \Omega_{\mathcal{N}}$ *of norm strictly less than* 1*, and is isometric and closely inner connected left realization for* S*.*

Proof. Let $\mathcal{F}_W : \mathcal{D}_2 \longrightarrow \mathcal{D}_2$ be the operator defined by (4.10), i.e. $G \longrightarrow (S^*G)^\vee(W)$. We have

$$\begin{aligned}
\langle G, \mathcal{F}^*_W(E)\rangle_{\mathcal{D}_2} &= \langle \mathcal{F}_W(G), E\rangle_{\mathcal{D}_2} \\
&= \langle (S^*G)^\vee(W), E\rangle_{\mathcal{D}_2} \\
&= \mathrm{Tr}E^*\left(S^*G\right)^\vee(W) \\
&= \left\{\mathrm{Tr}\left(S^*G\right)^\vee(W)^*E\right\}^* \\
&= \left\{\mathrm{Tr}\left(G^*S\right)^\wedge(W^*)E\right\}^* \\
&= \left\{\mathrm{Tr}\sum_{n=0}^{\infty}\left(W^*Z^*\right)^n G^*Z^nS_{[n]}E\right\}^* \\
&= \left\{\mathrm{Tr}\sum_{n=0}^{\infty} G^*Z^nS_{[n]}E\left(W^*Z^*\right)^n\right\}^* \\
&= \left\{\mathrm{Tr}G^*\left(SE\right)^\triangle\left(W^{*(-1)}\right)\right\}^*.
\end{aligned}$$

Therefore

$$\mathcal{F}_W^*(E) = (SE)^{\triangle}\left(W^{*(-1)}\right),$$

and we conclude that

$$\mathcal{F}_{W^{*(1)}}^*(E) = (SE)^{\triangle}(W).$$

Using the fact that $\left(\mathbf{M}_W^R\right)^* = \mathbf{M}_{W^*}^R$, we get the right realization for S via $\tilde{\mathcal{V}}_L$

$$(SE)^{\triangle}(W) = \left(\tilde{\mathbf{D}}_L^* + \tilde{\mathbf{B}}_L^*\left(\mathbf{I}_{\mathcal{H}_L(S^*)} - \mathbf{M}_{W^{(1)}}^R\tilde{\mathbf{A}}_L^*\right)^{-1}\mathbf{M}_{W^{(1)}}^R\tilde{\mathbf{C}}_L^*\right)(E).$$

The fact that this realization is isometric and closely inner connected is equivalent to the fact that $\tilde{\mathcal{V}}_L$ is coisometric and closely outer connected. □

We remark that similar computations yield the left isometric and closely inner connected realization for S, namely

$$(ES)^{\wedge}(W) = \left(\tilde{\mathbf{D}}_R^* + \tilde{\mathbf{B}}_R^*\left(\mathbf{I}_{\mathcal{H}_R(S^*)} - \mathbf{M}_{W^{(-1)}}^L\tilde{\mathbf{A}}_R^*\right)^{-1}\mathbf{M}_{W^{(-1)}}^L\tilde{\mathbf{C}}_R^*\right)(E).$$

To prove the uniqueness property, one uses now the relationship

$$\theta_R(\lambda)(E) = (ES)^{\wedge}(\lambda).$$

5. Realization for Schur operators via relations

Let $\mathcal{H}_1$, $\mathcal{H}_2$ be two Hilbert spaces. A linear relation in $\mathcal{H}_1\times\mathcal{H}_2$ is a linear subspace $\mathbf{R}$ of $\mathcal{H}_1\times\mathcal{H}_2$. The domain of a linear relation $\mathbf{R}$ is

$$\text{dom }\mathbf{R} = \{f\,|(f,g)\in\mathbf{R}\ \text{ for some } g\in\mathcal{H}_2\},$$

and the range of $\mathbf{R}$ is

$$\text{ran }\mathbf{R} = \{g\,|(f,g)\in\mathbf{R}\ \text{ for some } f\in\mathcal{H}_1\}.$$

We say that $\mathbf{R}$ is contractive if $\langle g,g\rangle_{\mathcal{H}_2}\leq\langle f,f\rangle_{\mathcal{H}_1}$ whenever $(f,g)\in\mathbf{R}$, and isometric if $\langle g,g\rangle_{\mathcal{H}_2}=\langle f,f\rangle_{\mathcal{H}_1}$ whenever $(f,g)\in\mathbf{R}$. A linear relation $\mathbf{R}$ is the graph of a linear operator T if and only if the subspace

$$\{g\in\mathcal{H}_2\,|(0,g)\in\mathbf{R}\},$$

is the zero subspace of $\mathcal{H}_2$. In this case, the domain and range of T coincide with the domain and range of $\mathbf{R}$. The following result allows us to use the notion of relations here. The result is easily shown in the present setting of Hilbert spaces. In the setting of Pontryagin spaces, it is based on a theorem of Y. Shmul'yan [51] and is given in [11].

Theorem 5.1. *Let $\mathbf{R}$ be a linear relation in $\mathcal{H}_1\times\mathcal{H}_2$, where $\mathcal{H}_1$, $\mathcal{H}_2$ are Hilbert spaces, then:*

1. *If* $\mathbf{R}$ *is a densely defined contraction, its closure is the graph of a contraction* $T \in \mathcal{L}(\mathcal{H}_1, \mathcal{H}_2)$.
2. *If* $\mathbf{R}$ *is a densely defined isometry, its closure is the graph of an isometry* $T \in \mathcal{L}(\mathcal{H}_1, \mathcal{H}_2)$.

We now prove Theorem 4.1 using the theory of relations.

Proof of Theorem 4.1: The proof is based on the analysis of [11]. Let $\mathbf{R}$ be the linear relation in $(\mathcal{H}_L(S) \oplus \mathcal{D}_2) \times (\mathcal{H}_L(S) \oplus \mathcal{D}_2)$ spanned by the pairs

$$\left(\begin{pmatrix} K_S^L(\cdot, W) Z W^* G_1 Z^* \\ E_1 \end{pmatrix}, \begin{pmatrix} \left(K_S^L(\cdot, W) - K_S^L(\cdot, 0)\right) G_1 + K_S^L(\cdot, 0) E_1 \\ \left(S^\wedge(W)^* - S_{[0]}^*\right) G_1 + S_{[0]}^* E_1 \end{pmatrix} \right),$$

with $W \in \Omega_{\mathcal{M}}, G_1, E_1 \in \mathcal{D}_2$. Let

$$\left(\begin{pmatrix} K_S^L(\cdot, V) Z V^* G_2 Z^* \\ E_2 \end{pmatrix}, \begin{pmatrix} \left(K_S^L(\cdot, V) - K_S^L(\cdot, V)\right) G_2 + K_S^L(\cdot, 0) E_2 \\ \left(S^\wedge(V)^* - S_{[0]}^*\right) G_2 + S_{[0]}^* E_2 \end{pmatrix} \right),$$

with $V \in \Omega_{\mathcal{M}}, G_2, E_2 \in \mathcal{D}_2$ be another such pair. The reproducing kernel property leads to

$$\begin{aligned}
&\left\langle \begin{pmatrix} \left(K_S^L(\cdot, W) - K_S^L(\cdot, 0)\right) G_1 + K_S^L(\cdot, 0) E_1 \\ \left(S^\wedge(W)^* - S_{[0]}^*\right) G_1 + S_{[0]}^* E_1 \end{pmatrix}, \right. \\
&\qquad \left. \left(\left(K_S^L(\cdot, V) - K_S^L(\cdot, 0)\right) G_2 + \left(S^\wedge(V)^* - S_{[0]}^*\right) G_2 + S_{[0]}^* E_2 \right) \right\rangle_{\mathcal{H}_L(S) \oplus \mathcal{D}_2} \\
= \;& \left\langle \left(K_S^L(\cdot, W) - K_S^L(\cdot, 0)\right) G_1 + K_S^L(\cdot, 0) E_1, K_S^L(\cdot, V) - K_S^L(\cdot, 0) \right\rangle_{\mathcal{H}_L(S)} \\
& + \left\langle \left(K_S^L(\cdot, W) - K_S^L(\cdot, 0)\right) G_1 + K_S^L(\cdot, 0) E_1, K_S^L(\cdot, 0) E_2 \right\rangle_{\mathcal{H}_L(S)} \\
& + \left\langle \left(S^\wedge(W)^* - S_{[0]}^*\right) G_1 + S_{[0]}^* E_1, \left(S^\wedge(V)^* - S_{[0]}^*\right) G_2 + S_{[0]}^* E_2 \right\rangle_{\mathcal{D}_2} \\
= \;& \mathrm{Tr} G_2^* \left(\left(K_S^L(\cdot, W) - K_S^L(\cdot, 0)\right)^\wedge (V) - \left(K_S^L(\cdot, W) - K_S^L(\cdot, 0)\right)^\wedge (0) \right) G_1 \\
& + \mathrm{Tr} G_2^* \left(S^\wedge(V) - S_{[0]}\right) \left(S^\wedge(W)^* - S_{[0]}^*\right) G_1 \\
& + + \mathrm{Tr} E_2^* \left(K_S^L(\cdot, 0)\right)^\wedge (0) E_1 + + \mathrm{Tr} E_2^* S_{[0]} S_{[0]}^* E_1 \\
& + \mathrm{Tr} G_2^* \left(\left(K_S^L(\cdot, 0)\right)^\wedge (V) - \left(K_S^L(\cdot, 0)\right)^\wedge (0) \right) E_1 \\
& + \mathrm{Tr} G_2^* \left(S^\wedge(V) - S_{[0]}\right) S_{[0]}^* E_1 \\
& + \mathrm{Tr} E_2^* \left(K_S^L(0, W) - K_S^L(0, 0)\right) G_1 + \mathrm{Tr} E_2^* S_{[0]} \left(S^\wedge(W)^* - S_{[0]}^*\right) G_1 \\
= \;& \mathrm{Tr} G_2^* V Z^* K_S^L(V, W) Z W^* G_1 + \mathrm{Tr} E_2^* E_1 \\
= \;& \mathrm{Tr} Z G_2^* V Z^* K_S^L(V, W) Z W^* G_1 Z^* + \mathrm{Tr} E_2^* E_1 \\
= \;& \left\langle \begin{pmatrix} K_S^L(\cdot, W) Z W^* G_1 Z^* \\ E_1 \end{pmatrix}, \begin{pmatrix} K_S^L(\cdot, V) Z V^* G_2 Z^* \\ E_2 \end{pmatrix} \right\rangle_{\mathcal{H}_L(S) \oplus \mathcal{D}_2}.
\end{aligned}$$

It follows that $\mathbf{R}$ is an isometric linear relation in $(\mathcal{H}_L(S) \oplus \mathcal{D}_2) \times (\mathcal{H}_L(S) \oplus \mathcal{D}_2)$. The domain of $\mathbf{R}$ is easily seen to be dense in $\mathcal{H}_L(S) \oplus \mathcal{D}_2$. Therefore by Theorem 5.1, the closure of $\mathbf{R}$ is the graph of an isometry

$$\mathcal{V}^* = \begin{pmatrix} \alpha^* & \gamma^* \\ \beta^* & \delta^* \end{pmatrix} : \begin{pmatrix} \mathcal{H}_L(S) \\ \mathcal{D}_2 \end{pmatrix} \longrightarrow \begin{pmatrix} \mathcal{H}_L(S) \\ \mathcal{D}_2 \end{pmatrix}.$$

Let

$$\begin{pmatrix} \alpha & \beta \\ \gamma & \delta \end{pmatrix} \begin{pmatrix} F \\ E \end{pmatrix} = \begin{pmatrix} G \\ J \end{pmatrix}.$$

We have

$$\begin{aligned} &\langle WZ^*G^\wedge(W)Z, G_1\rangle_{\mathcal{D}_2} + \langle J, E_1\rangle_{\mathcal{D}_2} \\ &= \left\langle \begin{pmatrix} G \\ J \end{pmatrix}, \begin{pmatrix} K_S^L(\cdot, W)ZW^*G_1Z^* \\ E_1 \end{pmatrix} \right\rangle_{\mathcal{H}_L(S)\oplus\mathcal{D}_2} \\ &= \left\langle \begin{pmatrix} F \\ E \end{pmatrix}, \begin{pmatrix} \alpha^* & \gamma^* \\ \beta^* & \delta^* \end{pmatrix} \begin{pmatrix} K_S^L(\cdot, W)ZW^*G_1Z^* \\ E_1 \end{pmatrix} \right\rangle_{\mathcal{H}_L(S)\oplus\mathcal{D}_2} \\ &= \left\langle \begin{pmatrix} F \\ E \end{pmatrix}, \begin{pmatrix} \left(K_S^L(\cdot, W) - K_S^L(\cdot, 0)\right) G_1 + K_S^L(\cdot, 0)E_1 \\ \left(S^\wedge(W)^* - S_{[0]}^*\right) G_1 + S_{[0]}^* E_1 \end{pmatrix} \right\rangle_{\mathcal{H}_L(S)\oplus\mathcal{D}_2} \\ &= \langle F^\wedge(W) - F_{[0]}, G_1\rangle_{\mathcal{D}_2} + \langle F_{[0]}, E_1\rangle_{\mathcal{D}_2} \\ &\quad + \langle \left(S^\wedge(W) - S_{[0]}\right) E, G_1\rangle_{\mathcal{D}_2} + \langle S_{[0]}E, E_1\rangle_{\mathcal{D}_2}. \end{aligned}$$

The elements $F \in \mathcal{H}_L(S)$, $E, E_1, G_1 \in \mathcal{D}_2$ and $W \in \Omega_{\mathcal{M}}$ are arbitrary. Choosing $E = 0$ we get

$$\begin{aligned} (GZ)^\wedge(W) &= WZ^*G^\wedge(W)Z \\ &= F^\wedge(W) - F_{[0]} \\ &= \left(F - F_{[0]}\right)^\wedge(W), \end{aligned}$$

and thus

$$\alpha(F) = \left(F - F_{[0]}\right) Z^{-1}.$$

The choice $F = 0$ leads to

$$\begin{aligned} (GZ)^\wedge(W) &= WZ^*G^\wedge(W)Z \\ &= \left(S^\wedge(W) - S_{[0]}\right) E \\ &= \left(\left(S - S_{[0]}\right) E\right)^\wedge(W), \end{aligned}$$

and thus

$$\beta(E) = \left(S - S_{[0]}\right) EZ^{-1}.$$

Still for $E = 0$ we have that $\delta(E) = S_{[0]}E$. We conclude that $\mathcal{V}_L = \mathcal{V}$, i.e.

$$\begin{pmatrix} \alpha & \beta \\ \gamma & \delta \end{pmatrix} = \begin{pmatrix} \mathbf{A}_L & \mathbf{B}_L \\ \mathbf{C}_L & \mathbf{D}_L \end{pmatrix},$$

and we get that the colligation $\mathcal{V}_L$ is well defined and coisometric.

6. The state space $\mathcal{D}_L(S)$

In this section we define the space which will be the state space of a unitary realization of the upper triangular Schur operator S. We suppose that $\mathcal{M} = \mathcal{N}$ and recall that the space Ω was defined by (2.1). The first step is to define the analogue of the two by two kernel (1.3). As we saw in (3.4), the operator $(S^* - S^{*\vee}(W^*))\,(Z^* - W^*)^{-1}\,E$ belongs to $\mathcal{H}_L(S^*)$ for any $W \in \Omega$, and any $E \in \mathcal{D}_2$. This fact allows to define a contraction operator (as in the stationary case) from the space $\mathcal{H}_L(S)$ into the space $\mathcal{H}_L(S^*)$. More precisely, we introduce the operator $\mathbf{\Lambda} : \mathcal{H}_L(S) \longrightarrow \mathcal{H}_L(S^*)$ which acts as

$$\mathbf{\Lambda}\left(K_S^L(\cdot, W)\,E\right) = (S^* - S^{*\vee}(W^*))\,(Z^* - W^*)^{-1}\,E^{(1)}.$$

Lemma 6.1. *The operator $\mathbf{\Lambda} : \mathcal{H}_L(S) \longrightarrow \mathcal{H}_L(S^*)$ is a well defined contraction.*

Proof. We first compute

$$\|\sum_{i=1}^n K_S^L(\cdot, W_i)\,E_i\|^2_{\mathcal{H}_L(S)}$$

using the reproducing property:

$$\begin{aligned}
\|\sum_{i=1}^n K_S^L(\cdot, W_i)\,E_i\|^2_{\mathcal{H}_L(S)} &= \left\langle \sum_{i=1}^n K_S^L(\cdot, W_i)\,E_i, \sum_{j=1}^n K_S^L(\cdot, W_j)\,E_j \right\rangle_{\mathcal{H}_L(S)} \\
&= \operatorname{Tr} \sum_{i,j=1}^n E_j^* K_S^L(W_j, W_i)\,E_i \\
&= \operatorname{Tr} \sum_{i,j=1}^n E_j^* \left((I - SS^\wedge(W_i)^*)\,\rho_{W_i}^{\wedge -1}\right)^\wedge (W_j) E_i \\
&= \operatorname{Tr} \sum_{i,j=1}^n E_j^* \rho_{W_j}^{\wedge -*}\,(I - S^\wedge(W_j) S^\wedge(W_i)^*)\,\rho_{W_i}^{\wedge -1} E_i.
\end{aligned}$$

Therefore

$$\|\sum_{i=1}^n K_S^L(\cdot, W_i)\,E_i\|^2_{\mathcal{H}_L(S)} = \operatorname{Tr} \sum_{i,j=1}^n E_j^* \rho_{W_j}^{\wedge -*}\,(I - S^\wedge(W_j) S^\wedge(W_i)^*)\,\rho_{W_i}^{\wedge -1} E_i.$$

Next we estimate

$$\|\sum_{i=1}^n (S^* - S^{*\vee}(W_i^*))\,(Z^* - W_i^*)^{-1}\,E_i^{(1)}\|^2_{\mathcal{H}_L(S^*)},$$

using the characterization of $\mathcal{H}_L(S^*)$ via complementation:

$$\begin{aligned}
&\|\sum_{i=1}^n (S^* - S^{*\vee}(W_i^*))\,(Z^* - W_i^*)^{-1}\, E_i^{(1)} + S^*G\|_{\mathcal{L}_2}^2 \\
&= \|\sum_{i=1}^n S^{*\vee}(W_i^*)\,(I - ZW_i^*)^{-1}\, E_i Z\|_{\mathcal{X}_2}^2 \\
&- 2\mathrm{Re}\left\langle \sum_{i=1}^n S^{*\vee}(W_i^*)\,(I - ZW_i^*)^{-1}\, E_i Z, S^*\left(\sum_{i=1}^n (I - ZW_i^*)^{-1}\, E_i Z + G\right)\right\rangle_{\mathcal{X}_2} \\
&\quad + \|S^*\left(\sum_{i=1}^n (I - ZW_i^*)^{-1}\, E_i Z + G\right)\|_{\mathcal{X}_2}^2 \\
&\le \|\sum_{i=1}^n S^{*\vee}(W_i^*)\,(I - ZW_i^*)^{-1}\, E_i Z\|_{\mathcal{X}_2}^2 \\
&- 2\mathrm{Re}\left\langle \sum_{i=1}^n S^{*\vee}(W_i^*)\,(I - ZW_i^*)^{-1}\, E_i Z, S^*\left(\sum_{i=1}^n (I - ZW_i^*)^{-1}\, E_i Z + G\right)\right\rangle_{\mathcal{X}_2} \\
&\quad + \|\sum_{i=1}^n (I - ZW_i^*)^{-1}\, E_i Z + G\|_{\mathcal{X}_2}^2 \\
&= \|\sum_{i=1}^n S^{*\vee}(W_i^*)\,(I - ZW_i^*)^{-1}\, E_i Z\|_{\mathcal{X}_2}^2 \\
&- 2\mathrm{Re}\left\langle \sum_{i=1}^n S^{*\vee}(W_i^*)\,(I - ZW_i^*)^{-1}\, E_i Z, S^*\left(\sum_{i=1}^n (I - ZW_i^*)^{-1}\, E_i Z\right)\right\rangle_{\mathcal{X}_2} \\
&\quad + \|\sum_{i=1}^n (I - ZW_i^*)^{-1}\, E_i Z\|_{\mathcal{X}_2}^2 + \|G\|_{\mathcal{L}_2}^2 \\
&= \mathrm{Tr}\sum_{i,j=1}^n Z^*E_j^*\,(I - W_jZ^*)^{-1}\, S^\wedge(W_j)S^\wedge(W_i)^*\,(I - ZW_i^*)^{-1}\, E_i Z \\
&\quad -2\mathrm{Re}\mathrm{Tr}\sum_{i,j=1}^n Z^*E_j^*\,(I - W_jZ^*)^{-1}\, SS^\wedge(W_i)^*\,(I - ZW_i^*)^{-1}\, E_i Z \\
&\quad +\mathrm{Tr}\sum_{i,j=1}^n Z^*E_j^*\,(I - W_jZ^*)^{-1}\,(I - ZW_i^*)^{-1}\, E_i Z + \|G\|_{\mathcal{L}_2}^2 \\
&= \mathrm{Tr}\sum_{i,j=1}^n E_j^*\rho_{W_j}^{\wedge -*} S^\wedge(W_j)S^\wedge(W_i)^*\rho_{W_i}^{\wedge -1} E_i \\
&\quad -2\mathrm{Tr}\sum_{i,j=1}^n E_j^*\rho_{W_j}^{\wedge -*} S^\wedge(W_j)S^\wedge(W_i)^*\rho_{W_i}^{\wedge -1} E_i
\end{aligned}$$

$$
\begin{aligned}
&+\mathrm{Tr} \sum_{i,j=1}^{n} E_j^* \rho_{W_j}^{\wedge -*} \rho_{W_i}^{\wedge -1} E_i + \|G\|_{\mathcal{L}_2}^2 \\
= \ & \mathrm{Tr} \sum_{i,j=1}^{n} E_j^* \rho_{W_j}^{\wedge -*} \left(I - S^{\wedge}(W_j) S^{\wedge}(W_i)^*\right) \rho_{W_i}^{\wedge -1} E_i + \|G\|_{\mathcal{L}_2}^2 .
\end{aligned}
$$

Therefore

$$
\begin{aligned}
&\| \sum_{i=1}^{n} (S^* - S^{*\vee}(W_i^*)) (Z^* - W_i^*)^{-1} E_i^{(1)} \|_{\mathcal{H}_L(S^*)}^2 \\
&\leq \ \mathrm{Tr} \sum_{i,j=1}^{n} E_j^* \rho_{W_j}^{\wedge -*} \left(I - S^{\wedge}(W_j) S^{\wedge}(W_i)^*\right) \rho_{W_i}^{\wedge -1} E_i,
\end{aligned}
$$

and we conclude that $\mathbf{\Lambda}$ is a densely defined contraction, and therefore has a unique extension, which is also a contraction. □

The operator $\mathbf{\Lambda}^*$ is a contraction from $\mathcal{H}_L(S)$ to $\mathcal{H}_L(S^*)$, which we compute in the next lemma.

Lemma 6.2. *The operator* $\mathbf{\Lambda}^*$ *is given by*

$$
\mathbf{\Lambda}^* \left(K_{S^*}^L(\cdot, W^*) E\right) = \left(S - S^{\triangle}(W)\right) (Z - W)^{-1} E^{(-1)}.
$$

Proof. To compute $\mathbf{\Lambda}^*$ we make use of (3.5) and we obtain

$$
\begin{aligned}
&\langle K_S^L(\cdot, V) G, \mathbf{\Lambda}^* \left(K_{S^*}^L(\cdot, W^*) E\right) \rangle_{\mathcal{H}_L(S)} \\
&= \ \langle \mathbf{\Lambda} \left(K_S^L(\cdot, V) G\right), K_{S^*}^L(\cdot, W^*) E \rangle_{\mathcal{H}_L(S^*)} \\
&= \ \mathrm{Tr} E^* \left((S^* - S^{*\vee}(V^*)) (Z^* - V^*)^{-1} G^{(1)} \right)^{\nabla} (W^*) \\
&= \ \mathrm{Tr} E^* \left((Z - V)^{-1} (S - S^{\wedge}(V)) \right)^{\triangle} (W)^* G^{(1)} \\
&= \ \left\{ \mathrm{Tr} G^{*(1)} Z^* \left((S - S^{\triangle}(W)) (Z - W)^{-1} \right)^{\wedge} (V) Z E \right\}^* \\
&= \ \left\{ \mathrm{Tr} Z G^{*(1)} Z^* \left((S - S^{\triangle}(W)) (Z - W)^{-1} \right)^{\wedge} (V) Z E Z^* \right\}^* \\
&= \ \left\{ \mathrm{Tr} G^* \left((S - S^{\triangle}(W)) (Z - W)^{-1} E^{(-1)} \right)^{\wedge} (V) \right\}^* .
\end{aligned}
$$

Therefore

$$
\mathbf{\Lambda}^* \left(K_{S^*}^L(\cdot, W^*) E\right) = \left(S - S^{\triangle}(W)\right) (Z - W)^{-1} E^{(-1)}.
$$

□

To define the two by two kernel and the corresponding space, we use Theorem 3.3. It follows from this theorem that if $\mathcal{H}$ is a left reproducing kernel Hilbert space right module with K_W as a kernel, then $\mathrm{Ran}\Gamma^{\frac{1}{2}}$ is also a left reproducing kernel Hilbert space right module, with reproducing kernel L_W defined by $L_W(G) = \Gamma(K_W G)$.

We apply this analysis to the operator

$$\Gamma = \begin{pmatrix} I_{\mathcal{H}_L(S)} & \mathbf{\Lambda}^* \\ \mathbf{\Lambda} & I_{\mathcal{H}_L(S^*)} \end{pmatrix} : \begin{pmatrix} \mathcal{H}_L(S) \\ \mathcal{H}_L(S^*) \end{pmatrix} \longrightarrow \begin{pmatrix} \mathcal{H}_L(S) \\ \mathcal{H}_L(S^*) \end{pmatrix}, \tag{6.1}$$

(which is a nonnegative operator acting on the space $\mathcal{H}_L(S) \oplus \mathcal{H}_L(S^*)$ thanks to Lemmas 6.1 and 6.2). and denote by $\mathcal{D}_L(S)$ the corresponding left reproducing kernel Hilbert space right module. Furthermore,

$$\begin{pmatrix} I_{\mathcal{H}_L(S)} & \mathbf{\Lambda}^* \\ \mathbf{\Lambda} & I_{\mathcal{H}_L(S^*)} \end{pmatrix} \begin{pmatrix} K_S^L(\cdot, W)\, E \\ K_{S^*}^L(\cdot, W^*)\, G \end{pmatrix}$$
$$= \begin{pmatrix} K_S^L(\cdot, W)\, E + (S - S^{\triangle}(W))\,(Z-W)^{-1}\, G^{(-1)} \\ (S^* - S^{*\vee}(W^*))\,(Z^* - W^*)^{-1}\, E^{(1)} + K_{S^*}^L(\cdot, W^*)\, G \end{pmatrix}.$$

Let $D_S^L(\cdot, W) : \mathcal{D}_2 \oplus \mathcal{D}_2 \longrightarrow \mathcal{H}_L(S) \oplus \mathcal{H}_L(S^*)$, be the "two by two" kernel defined by

$$D_S^L(\cdot, W) \begin{pmatrix} E \\ G \end{pmatrix} = \begin{pmatrix} K_S^L(\cdot, W)\, E + (S - S^{\triangle}(W))\,(Z-W)^{-1}\, G^{(-1)} \\ (S^* - S^{*\vee}(W^*))\,(Z^* - W^*)^{-1}\, E^{(1)} + K_{S^*}^L(\cdot, W^*)\, G \end{pmatrix}.$$

The linear span of all the operators of the form

$$D_S^L(\cdot, W) \begin{pmatrix} E \\ G \end{pmatrix}$$

is dense in $\mathcal{D}_L(S)$. We denote this set by $\overset{\circ}{\mathcal{D}}_L(S)$.

The reproducing property of the kernel $D_S^L(\cdot, W)$ is explicited in the next lemma:

Lemma 6.3. *Let* $\begin{pmatrix} F \\ H \end{pmatrix} \in \mathcal{D}_L(S)$ *and* $\begin{pmatrix} E \\ G \end{pmatrix} \in \mathcal{D}_2 \oplus \mathcal{D}_2$. *Then, it holds that*

$$\left\langle \begin{pmatrix} F \\ H \end{pmatrix}, D_S^L(\cdot, W) \begin{pmatrix} E \\ G \end{pmatrix} \right\rangle_{\mathcal{D}_L(S)} = \mathrm{Tr} \begin{pmatrix} E^* & G^* \end{pmatrix} \begin{pmatrix} F^{\wedge}(W) \\ H^{\triangledown}(W^*) \end{pmatrix}. \tag{6.2}$$

Proof. Let $v = \Gamma^{\frac{1}{2}} \begin{pmatrix} K_S^L(\cdot, W)\, E \\ K_{S^*}^L(\cdot, W^*)\, G \end{pmatrix}$, and let $\begin{pmatrix} F \\ H \end{pmatrix} = \Gamma^{\frac{1}{2}} u$ for some

$$u \in \mathcal{H}_L(S) \oplus \mathcal{H}_L(S^*).$$

Then

$$\begin{aligned}
\langle \Gamma^{\frac{1}{2}}u, \Gamma^{\frac{1}{2}}v\rangle_{\mathcal{D}_L(S)} &= \langle (I-P)\,u, v\rangle_{\mathcal{H}_L(S)\oplus\mathcal{H}_L(S^*)} \\
&= \left\langle (I-P)\,u, \Gamma^{\frac{1}{2}} \begin{pmatrix} K_S^L(\cdot, W)\,E \\ K_{S^*}^L(\cdot, W^*)\,G \end{pmatrix} \right\rangle_{\mathcal{H}_L(S)\oplus\mathcal{H}_L(S^*)} \\
&= \left\langle \Gamma^{\frac{1}{2}}u, \begin{pmatrix} K_S^L(\cdot, W)\,E \\ K_{S^*}^L(\cdot, W^*)\,G \end{pmatrix} \right\rangle_{\mathcal{H}_L(S)\oplus\mathcal{H}_L(S^*)} \\
&= \left\langle \begin{pmatrix} F \\ H \end{pmatrix}, \begin{pmatrix} K_S^L(\cdot, W)\,E \\ K_{S^*}^L(\cdot, W^*)\,G \end{pmatrix} \right\rangle_{\mathcal{H}_L(S)\oplus\mathcal{H}_L(S^*)} \\
&= \operatorname{Tr}\begin{pmatrix} E^* & G^* \end{pmatrix} \begin{pmatrix} F^{\wedge}(W) \\ H^{\nabla}(W^*) \end{pmatrix}.
\end{aligned}$$

On the other hand we get that

$$\Gamma^{\frac{1}{2}}v = \Gamma \begin{pmatrix} K_S^L(\cdot, W)\,E \\ K_{S^*}^L(\cdot, W^*)\,G \end{pmatrix} = D_S^L(\cdot, W) \begin{pmatrix} E \\ G \end{pmatrix},$$

and therefore

$$\langle \Gamma^{\frac{1}{2}}u, \Gamma^{\frac{1}{2}}v\rangle_{\mathcal{D}_L(S)} = \left\langle \begin{pmatrix} F \\ H \end{pmatrix}, D_S^L(\cdot, W) \begin{pmatrix} E \\ G \end{pmatrix} \right\rangle_{\mathcal{D}_L(S)}. \tag{6.3}$$

This proves the reproducing property of the kernel. □

Lemma 6.4. *The set of all operators of the form*

$$D_S^L(\cdot, W) \begin{pmatrix} E \\ (WG)^{(1)} \end{pmatrix}$$

is dense in $\mathcal{D}_S^L$. Similarly, the set of all operators of the form

$$D_S^L(\cdot, W) \begin{pmatrix} (W^*E)^{(-1)} \\ G \end{pmatrix}$$

is dense in $\mathcal{D}_S^L$.

Proof. Let $\begin{pmatrix} F \\ H \end{pmatrix} \in \mathcal{D}_S^L$; then,

$$\begin{aligned}
&\left\langle \begin{pmatrix} F \\ H \end{pmatrix}, D_S^L(\cdot, W) \begin{pmatrix} E \\ (WG)^{(1)} \end{pmatrix} \right\rangle_{\mathcal{D}_L(S)} = \\
&= \operatorname{Tr}\begin{pmatrix} E^* & (WG)^{*(1)} \end{pmatrix} \begin{pmatrix} F^{\wedge}(W) \\ H^{\nabla}(W^*) \end{pmatrix} \\
&= \operatorname{Tr}E^*F^{\wedge}(W) + \operatorname{Tr}(WG)^{*(1)}H^{\nabla}(W^*) \\
&= \operatorname{Tr}E^*F^{\wedge}(W) + \operatorname{Tr}G^*(HZ^*)^{\nabla}(W^*).
\end{aligned}$$

Therefore if $\begin{pmatrix} F \\ H \end{pmatrix}$ is orthogonal to the set of all operators of the form

$$D_S^L(\cdot, W)\begin{pmatrix} E \\ (WG)^{(1)} \end{pmatrix},$$

then choosing $G = 0$ leads to $\mathrm{Tr}E^*F^\wedge(W) = 0$, and we get that $F = 0$. The choice $G = 0$ leads to $\mathrm{Tr}G^*(HZ^*)^\nabla(W^*) = 0$ and we get that $HZ^* = 0$, which is equivalent to say that $H = 0$. A similar argument shows the second part of this lemma. □

Let $M^R_{\mathrm{diag}\ (W,W^{(1)})}$ the operator $\mathcal{H}_L(S) \oplus \mathcal{H}_L(S^*) \longrightarrow \mathcal{H}_L(S) \oplus \mathcal{H}_L(S^*)$ defined by

$$M^R_{\mathrm{diag}\ (W,W^{(1)})}\left(\begin{pmatrix} F \\ H \end{pmatrix}\right) = \begin{pmatrix} FW \\ HW^{(1)} \end{pmatrix}, \tag{6.4}$$

where $E, H \in \mathcal{D}_2$ and $W \in \Omega$ with $\|W\| < 1$. We will use the simpler notation M^R_W, although it creates a notational conflict with the previously defined multiplication operator in $\mathcal{H}_L(S)$. The meaning of the symbol M^R_W will always be clear from the context.

Lemma 6.5. *Let $W \in \mathcal{D}$ of norm strictly less than 1. The operator M^R_W (defined by (6.4)) is a strict contraction from $\mathcal{H}_L(S) \oplus \mathcal{H}_L(S^*)$ into itself.*

Proof.

$$\begin{aligned}\left\|M^R_W\left(\begin{pmatrix} F \\ H \end{pmatrix}\right)\right\|^2_{\mathcal{H}_L(S)\oplus\mathcal{H}_L(S^*)} &= \|FW\|^2_{\mathcal{H}_L(S)} + \|HW^{(1)}\|^2_{\mathcal{H}_L(S^*)} \\ &\leq \|W\|^2\left(\|F\|^2_{\mathcal{H}_L(S)} + \|H\|^2_{\mathcal{H}_L(S^*)}\right) \\ &= \|W\|^2\left\|\begin{pmatrix} F \\ H \end{pmatrix}\right\|^2_{\mathcal{H}_L(S)\oplus\mathcal{H}_L(S^*)},\end{aligned}$$

therefore M^R_W is a strict contraction on $\mathcal{H}_L(S) \oplus \mathcal{H}_L(S^*)$. □

Theorem 6.6. *Let $W \in \mathcal{D}$ and let M^R_W be defined by (6.4). The state space $\mathcal{D}_L(S)$ is M^R_W invariant. Moreover, M^R_W considered as an operator from $\mathcal{D}_L(S)$ to itself is a strict contraction commuting with $\Gamma^{\frac{1}{2}}$, i.e.*

$$M^R_W\Gamma^{\frac{1}{2}} = \Gamma^{\frac{1}{2}}M^R_W.$$

Proof. Let

$$\begin{pmatrix} F \\ H \end{pmatrix} = \begin{pmatrix} K^L_S(\cdot, V)E \\ K^L_{S^*}(\cdot, V^*)G \end{pmatrix},$$

then

$$
\begin{aligned}
& M_W^R \Gamma\left(\begin{pmatrix} F \\ H \end{pmatrix}\right) \\
= \; & M_W^R \Gamma \begin{pmatrix} K_S^L(\cdot, V)E \\ K_{S^*}^L(\cdot, V^*)G \end{pmatrix} \\
= \; & M_W^R \left(\begin{pmatrix} K_S^L(\cdot, V)\,E + \left(S - S^{\triangle}(V)\right)(Z - V)^{-1} G^{(-1)} \\ \left(S^* - S^{*\vee}(V^*)\right)(Z^* - V^*)^{-1} E^{(1)} + K_{S^*}^L(\cdot, V^*)\,G \end{pmatrix}\right) \\
= \; & \begin{pmatrix} K_S^L(\cdot, V)\,EW + \left(S - S^{\triangle}(V)\right)(Z - V)^{-1} G^{(-1)}W \\ \left(S^* - S^{*\vee}(V^*)\right)(Z^* - V^*)^{-1} E^{(1)}W^{(1)} + K_{S^*}^L(\cdot, V^*)\,GW^{(1)} \end{pmatrix} \\
= \; & \begin{pmatrix} K_S^L(\cdot, V)\,EW + \left(S - S^{\triangle}(V)\right)(Z - V)^{-1} \left(GW^{(1)}\right)^{(-1)} \\ \left(S^* - S^{*\vee}(V^*)\right)(Z^* - V^*)^{-1} (EW)^{(1)} + K_{S^*}^L(\cdot, V^*)\,GW^{(1)} \end{pmatrix} \\
= \; & \Gamma \begin{pmatrix} K_S^L(\cdot, V)EW \\ K_{S^*}^L(\cdot, V^*)GW^{(1)} \end{pmatrix} \\
= \; & \Gamma M_W^R \left(\begin{pmatrix} K_S^L(\cdot, V)E \\ K_{S^*}^L(\cdot, V^*)G \end{pmatrix}\right) \\
= \; & \Gamma M_W^R \left(\begin{pmatrix} F \\ H \end{pmatrix}\right),
\end{aligned}
$$

therefore $M_W^R \Gamma = \Gamma M_W^R$ on a dense set of $\mathcal{H}_L(S) \oplus \mathcal{H}_L(S^*)$ and hence on all of $\mathcal{H}_L(S) \oplus \mathcal{H}_L(S^*)$ by continuity. Hence, for every $n \geq 0$

$$M_W^R \Gamma^n = \Gamma^n M_W^R.$$

Let now $\Gamma = \int_0^\ell \lambda dE(\lambda)$ be the spectral decomposition of Γ and let $P_n(\lambda)$ be a sequence of polynomials such that

$$\lim_{n \to \infty} \sup_{\lambda \in [0, \ell]} \left| \lambda^{\frac{1}{2}} - P_n(\lambda) \right| = 0.$$

Then

$$\lim_{n \to \infty} P_n(\Gamma) = \Gamma^{\frac{1}{2}},$$

in the operator norm, and so

$$M_W^R \Gamma^{\frac{1}{2}} = \Gamma^{\frac{1}{2}} M_W^R.$$

Recall that $\mathcal{D}_L(S) = \mathrm{Ran}\Gamma^{\frac{1}{2}}$ and therefore $\mathcal{D}_L(S)$ is M_W^R invariant. It remains to show that M_W^R is a strict contraction acting on $\mathcal{D}_L(S)$. We first show that M_W^R is a bounded operator on $\mathcal{D}_L(S)$.

Indeed it is a everywhere defined and closed operator, for if $F_n \in \mathcal{D}_L(S)$ is a sequence which converges to F and $M_W^R F_n$ converges to H, then

$$\begin{aligned}
\left\langle M_W^R(F), D_S^L(\cdot,V)\begin{pmatrix} E \\ G \end{pmatrix}\right\rangle_{\mathcal{D}_L(S)} &= \left\langle M_W^R\left(\begin{pmatrix} F_1 \\ F_2 \end{pmatrix}\right), D_S^L(\cdot,V)\begin{pmatrix} E \\ G \end{pmatrix}\right\rangle_{\mathcal{D}_L(S)} \\
&= \mathrm{Tr}E^*\left(F_1W\right)^\wedge(V) + \mathrm{Tr}G^*\left(F_2W^{(1)}\right)^\nabla(V) \\
&= \mathrm{Tr}E^*\left(F_1\right)^\wedge(V)W + \mathrm{Tr}G^*\left(F_2\right)^\nabla(V)W^{(1)} \\
&= \mathrm{Tr}WE^*\left(F_1\right)^\wedge(V) + \mathrm{Tr}W^{(1)}G^*\left(F_2\right)^\nabla(V) \\
&= \left\langle F, D_S^L(\cdot,V)\begin{pmatrix} EW^* \\ GW^{(1)*} \end{pmatrix}\right\rangle_{\mathcal{D}_L(S)} \\
&= \lim_{n\to\infty}\left\langle F_n, D_S^L(\cdot,V)\begin{pmatrix} EW^* \\ GW^{(1)*} \end{pmatrix}\right\rangle_{\mathcal{D}_L(S)} \\
&= \lim_{n\to\infty}\left\langle M_W^R\left(F_n\right), D_S^L(\cdot,V)\begin{pmatrix} E \\ G \end{pmatrix}\right\rangle_{\mathcal{D}_L(S)} \\
&= \left\langle H, D_S^L(\cdot,V)\begin{pmatrix} E \\ G \end{pmatrix}\right\rangle_{\mathcal{D}_L(S)}.
\end{aligned}$$

Therefore $M_W^R(F) = H$. By the closed graph theorem we conclude that the operator M_W^R is a bounded operator acting on $\mathcal{D}_L(S)$.

To show that M_W^R is a strict contraction, we use the fact that the operators $\Gamma^{\frac{1}{2}}$ and $I - P$ are commuting, see [1, p. 270] for example.

Let $\begin{pmatrix} F \\ H \end{pmatrix}$ belong to $\mathcal{D}_L(S)$. We can write

$$\begin{pmatrix} F \\ H \end{pmatrix} = \Gamma^{\frac{1}{2}}(I-P)\begin{pmatrix} u \\ v \end{pmatrix},$$

for some $\begin{pmatrix} u \\ v \end{pmatrix}$ in $\mathcal{H}_L(S) \oplus \mathcal{H}_L(S^*)$.

Now,

$$\begin{aligned}
\|M_W^R\begin{pmatrix} F \\ H \end{pmatrix}\|^2_{\mathcal{D}_L(S)} &= \|M_W^R\Gamma^{\frac{1}{2}}(I-P)\begin{pmatrix} u \\ v \end{pmatrix}\|^2_{\mathcal{D}_L(S)} \\
&= \|\Gamma^{\frac{1}{2}}M_W^R(I-P)\begin{pmatrix} u \\ v \end{pmatrix}\|^2_{\mathcal{D}_L(S)} \\
&= \|\left(I-P\right)M_W^R(I-P)\begin{pmatrix} u \\ v \end{pmatrix}\|^2_{\mathcal{H}_L(S)\oplus\mathcal{H}_L(S^*)} \\
&\leq \|M_W^R\left(I-P\right)\begin{pmatrix} u \\ v \end{pmatrix}\|^2_{\mathcal{H}_L(S)\oplus\mathcal{H}_L(S^*)}
\end{aligned}$$

$$\begin{aligned} &\leq \|M_W^R\|^2 \| (I-P) \begin{pmatrix} u \\ v \end{pmatrix} \|^2_{\mathcal{H}_L(S)\oplus\mathcal{H}_L(S^*)} \\ &= \|M_W^R\|^2 \|\Gamma^{\frac{1}{2}} \begin{pmatrix} u \\ v \end{pmatrix} \|^2_{\mathcal{D}_L(S)} \\ &= \|M_W^R\|^2 \| \begin{pmatrix} F \\ H \end{pmatrix} \|^2_{\mathcal{D}_L(S)}, \end{aligned}$$

where the operator norm of M_W^R as an operator acting on $\mathcal{H}_L(S)\oplus\mathcal{H}_L(S^*)$ is strictly less than 1 and therefore M_W^R is a strict contraction on $\mathcal{D}_L(S)$. □

We note that the operator Γ defined in (6.1) satisfies

$$\Gamma^2 \leq 2\Gamma.$$

By [24, Theorem 3], the space $\mathcal{D}_L(S)$ can be identified as follows: it is the unique Hilbert space $\mathcal{G}$ contained continuously in $\mathcal{H}_L(S)\oplus\mathcal{H}_L(S^*)$ such that Γ coincides with the adjoint of the inclusion of $\mathcal{G}$ into $\mathcal{H}_L(S)\oplus\mathcal{H}_L(S^*)$.

7. Unitary realizations

In this section we define a closely connected unitary realization for a Schur upper triangular operator S. Let us recall that unitary systems associated to upper triangular operators were considered in [16] and in [15].

Theorem 7.1. *The colligation* $\mathcal{U}_L = \begin{pmatrix} \alpha_L & \beta_L \\ \gamma_L & \delta_L \end{pmatrix} : \mathcal{D}_L(S)\oplus\mathcal{D}_2 \longrightarrow \mathcal{D}_L(S)\oplus\mathcal{D}_2$ *defined by:*

$$\begin{aligned} \alpha_L \begin{pmatrix} F \\ H \end{pmatrix} &= \begin{pmatrix} (F-F_{[0]})\,Z^{-1} \\ HZ^* - S^*F_{[0]} \end{pmatrix}, \\ \beta_L(E) &= \begin{pmatrix} (S-S_{[0]})\,EZ^{-1} \\ (I-S^*S_{[0]})\,E \end{pmatrix}, \\ \gamma_L \begin{pmatrix} F \\ H \end{pmatrix} &= F_{[0]}, \\ \delta_L(E) &= S_{[0]}E. \end{aligned}$$

is a well defined unitary operator.

Proof. We divide the proof into a number of steps.

Step 1. *Let* $\begin{pmatrix} F \\ H \end{pmatrix} = D_S^L(\cdot, W)\begin{pmatrix} E \\ (WG)^{(1)} \end{pmatrix}$. *Then,*

$$\alpha_L \begin{pmatrix} F \\ H \end{pmatrix} = D_S^L(\cdot, W)\begin{pmatrix} (W^*E)^{(-1)} \\ G \end{pmatrix} - D_S^L(\cdot, 0)\begin{pmatrix} 0 \\ G + S^\wedge(W)^*E \end{pmatrix}. \tag{7.1}$$

Proof of Step 1. We have

$$
\begin{aligned}
&\alpha_L\left(D_S^L(\cdot,W)\begin{pmatrix} E\\ 0\end{pmatrix}\right)\\
&= \begin{pmatrix} \left(K_S^L(\cdot,W)E - \left(K_S^L(\cdot,W)E\right)_{[0]}\right)Z^{-1}\\ (S^*-S^{*\vee}(W^*))(Z^*-W^*)^{-1}E^{(1)}Z^* - S^*\left(K_S^L(\cdot,W)E\right)_{[0]}\end{pmatrix}\\
&= \begin{pmatrix} \left((I-SS^\wedge(W)^*)\rho_W^{\wedge-1}E - \left(I-S_{[0]}S^\wedge(W)^*\right)E\right)Z^{-1}\\ (S^*-S^{*\vee}(W^*))(Z^*-W^*)^{-1}Z^*E - S^*\left(I-S_{[0]}S^\wedge(W)^*\right)E\end{pmatrix}\\
&= \begin{pmatrix} (I-SS^\wedge(W)^*)(I-ZW^*)^{-1}ZW^*EZ^{-1}\\ (S^*-S^{*\vee}(W^*))(I-ZW^*)^{-1}ZW^*E\end{pmatrix}\\
&\quad+\begin{pmatrix} (I-SS^\wedge(W)^*)EZ^{-1} - \left(I-S_{[0]}S^\wedge(W)^*\right)EZ^{-1}\\ (S^*-S^{*\vee}(W^*))E - S^*\left(I-S_{[0]}S^\wedge(W)^*\right)E\end{pmatrix}\\
&= \begin{pmatrix} K_S^L(\cdot,W)(W^*E)^{(-1)}\\ (S^*-S^{*\vee}(W^*))(Z^*-W^*)^{-1}(W^*E)\end{pmatrix}\\
&\quad-\begin{pmatrix} \left(S-S_{[0]}\right)Z^{-1}(S^\wedge(W)^*E)^{(-1)}\\ K_{S^*}^L(\cdot,0)(S^\wedge(W)^*E)\end{pmatrix}\\
&= D_S^L(\cdot,W)\begin{pmatrix}(W^*E)^{(-1)}\\ 0\end{pmatrix} - D_S^L(\cdot,0)\begin{pmatrix}0\\ S^\wedge(W)^*E\end{pmatrix}.
\end{aligned}
$$

and

$$
\begin{aligned}
&\alpha_L\left(D_S^L(\cdot,W)\begin{pmatrix}0\\ (WG)^{(1)}\end{pmatrix}\right) =\\
&\begin{pmatrix} \left((S-S^\triangle(W))(Z-W)^{-1}WG - \left((S-S^\triangle(W))(Z-W)^{-1}WG\right)_{[0]}\right)Z^{-1}\\ K_{S^*}^L(\cdot,W^*)(WG)^{(1)}Z^* - S^*\left((S-S^\triangle(W))(Z-W)^{-1}WG\right)_{[0]}\end{pmatrix}\\
&= \begin{pmatrix} (S-S^\triangle(W))(I-Z^*W)^{-1}Z^*WGZ^{-1} - \left(S^\triangle(W)-S_{[0]}\right)GZ^{-1}\\ (I-S^*S^\triangle(W))(I-Z^*W)^{-1}Z^*WG - S^*\left(S^\triangle(W)-S_{[0]}\right)G\end{pmatrix}\\
&= \begin{pmatrix} (S-S^\triangle(W))(I-Z^*W)^{-1}Z^*WGZ^{-1}\\ (I-S^*S^\triangle(W))(I-Z^*W)^{-1}Z^*WG\end{pmatrix}\\
&\quad+\begin{pmatrix} (S-S^\triangle(W))GZ^{-1} - \left(S-S_{[0]}\right)GZ^{-1}\\ (I-S^*S^\triangle(W))G - \left(I-S^*S_{[0]}\right)G\end{pmatrix}\\
&= \begin{pmatrix} (S-S^\triangle(W))(Z-W)^{-1}G^{(-1)}\\ K_{S^*}^L(\cdot,W^*)G\end{pmatrix}\\
&\quad-\begin{pmatrix} \left(S-S_{[0]}\right)Z^{-1}G^{(-1)}\\ K_{S^*}^L(\cdot,0)G\end{pmatrix}\\
&= D_S^L(\cdot,W)\begin{pmatrix}0\\ G\end{pmatrix} - D_S^L(\cdot,0)\begin{pmatrix}0\\ G\end{pmatrix}.
\end{aligned}
$$

We conclude that

$$\alpha_L\begin{pmatrix} F \\ H \end{pmatrix} = D_S^L(\cdot, W)\begin{pmatrix} (W^*E)^{(-1)} \\ G \end{pmatrix} - D_S^L(\cdot,0)\begin{pmatrix} 0 \\ G+S^{\wedge}(W)^*E \end{pmatrix},$$

and therefore α_L is well defined.

Let $E \in \mathcal{D}_2$:

$$\beta_L(E) = \begin{pmatrix} \left(S - S_{[0]}\right)EZ^{-1} \\ \left(I - S^*S_{[0]}\right)E \end{pmatrix} = D_S^L(\cdot,0)\begin{pmatrix} 0 \\ E \end{pmatrix}. \tag{7.2}$$

Therefore β_L is well defined. Obviously γ_L and δ_L are well defined, and therefore the colligation $\mathcal{U}_L$ is well defined.

Step 2. *The adjoint colligation* $\mathcal{U}_L^* = \begin{pmatrix} \alpha_L^* & \gamma_L^* \\ \beta_L^* & \delta_L^* \end{pmatrix}$ *is given by*

$$\alpha_L^*\begin{pmatrix} F \\ H \end{pmatrix} = \begin{pmatrix} FZ - SH_{[0]} \\ \left(H - H_{[0]}\right)Z \end{pmatrix}, \tag{7.3}$$

$$\gamma_L^*(E) = \begin{pmatrix} \left(I - SS_{[0]}^*\right)E \\ \left(S^* - S_{[0]}^*\right)EZ \end{pmatrix}, \tag{7.4}$$

$$\beta_L^*\begin{pmatrix} F \\ H \end{pmatrix} = H_{[0]}, \tag{7.5}$$

$$\delta_L^*(E) = S_{[0]}^*E. \tag{7.6}$$

Proof of Step 2. We first show that (7.3)–(7.6) are well defined operators. We denote by $\tilde{\mathcal{U}} = \begin{pmatrix} \tilde\alpha & \tilde\gamma \\ \tilde\beta & \tilde\delta \end{pmatrix}$ the expressions (7.3)–(7.6) where $\begin{pmatrix} F \\ H \end{pmatrix}$ belongs to the dense set $\mathring{\mathcal{D}}_L(S)$. We can write:

$$\begin{aligned}
&\tilde\alpha\left(D_S^L(\cdot,W)\begin{pmatrix} (W^*E)^{(-1)} \\ 0 \end{pmatrix}\right) = \\
&= \tilde\alpha\begin{pmatrix} K_S^L(\cdot,W)(W^*E)^{(-1)} \\ (S^* - S^{*\vee}(W^*))(Z^* - W^*)^{-1}W^*E \end{pmatrix} \\
&= \begin{pmatrix} K_S^L(\cdot,W)ZW^*E \\ ((S^* - S^{*\vee}(W^*))(Z^* - W^*)^{-1}W^*EZ \end{pmatrix} \\
&\quad - \begin{pmatrix} S\left((S^* - S^{*\vee}(W^*))(Z^* - W^*)^{-1}W^*E\right)_{[0]} \\ \left((S^* - S^{*\vee}(W^*))(Z^* - W^*)^{-1}W^*E\right)_{[0]}Z \end{pmatrix} \\
&= \begin{pmatrix} (I - SS^{\wedge}(W)^*)(I - ZW^*)^{-1}ZW^*E - S\left(S^{\wedge}(W)^* - S_{[0]}^*\right)E \\ (S^* - S^{*\vee}(W^*))(I - ZW^*)^{-1}ZW^*EZ - \left(S^{\wedge}(W)^* - S_{[0]}^*\right)E \end{pmatrix}
\end{aligned}$$

$$
\begin{aligned}
&= \begin{pmatrix} (I - SS^{\wedge}(W)^*)\,(I - ZW^*)^{-1}\,E \\ (S^* - S^{*\vee}(W^*))\,(I - ZW^*)^{-1}\,EZ \end{pmatrix} \\
&\quad - \begin{pmatrix} (I - SS^{\wedge}(W)^*)\,E - S\left(S^{\wedge}(W)^* - S^*_{[0]}\right)E \\ (S^* - S^{*\vee}(W^*))\,EZ - \left(S^{\wedge}(W)^* - S^*_{[0]}\right)E \end{pmatrix} \\
&= \begin{pmatrix} K^L_S\,(\cdot, W)\,E \\ (S^* - S^{*\vee}(W^*))\,(Z^* - W^*)^{-1}\,E^{(1)} \end{pmatrix} - \begin{pmatrix} K^L_S\,(\cdot, 0)\,E \\ \left(S^* - S^*_{[0]}\right) ZE^{(1)} \end{pmatrix} \\
&= D^L_S(\cdot, W) \begin{pmatrix} E \\ 0 \end{pmatrix} - D^L_S(\cdot, 0) \begin{pmatrix} E \\ 0 \end{pmatrix}
\end{aligned}
$$

and

$$
\begin{aligned}
&\tilde{\alpha}\left(D^L_S(\cdot, W)\begin{pmatrix} 0 \\ G \end{pmatrix}\right) = \\
&= \tilde{\alpha}\begin{pmatrix} (S - S^{\triangle}(W))\,(Z - W)^{-1}\,G^{(-1)} \\ K^L_{S^*}\,(\cdot, W^*)\,G \end{pmatrix} \\
&= \begin{pmatrix} (S - S^{\triangle}(W))\,(Z - W)^{-1} G^{(-1)} Z - S\left(K^L_{S^*}\,(\cdot, W^*)\,G\right)_{[0]} \\ \left(K^L_{S^*}\,(\cdot, W^*)\,G - \left(K^L_{S^*}\,(\cdot, W^*)\,G\right)_{[0]}\right) Z \end{pmatrix} \\
&= \begin{pmatrix} (S - S^{\triangle}(W))\,(I - Z^*W)^{-1} G - S\left(I - S^*_{[0]} S^{\triangle}(W)\right) G \\ \left((I - S^* S^{\triangle}(W))\,(I - Z^*W)^{-1}\,G - \left(I - S^*_{[0]} S^{\triangle}(W)\right) G\right) Z \end{pmatrix} \\
&= \begin{pmatrix} (S - S^{\triangle}(W))\,(I - Z^*W)^{-1} Z^*WG \\ (I - S^* S^{\triangle}(W))\,(I - Z^*W)^{-1}\,Z^*WG \end{pmatrix} \\
&\quad + \begin{pmatrix} -S\left(I - S^*_{[0]} S^{\triangle}(W)\right) G \\ \left((I - S^* S^{\triangle}(W)) - \left(I - S^*_{[0]} S^{\triangle}(W)\right) G\right) Z \end{pmatrix} \\
&= \begin{pmatrix} (S - S^{\triangle}(W))\,(Z - W)^{-1} WG \\ K^L_{S^*}\,(\cdot, W^*)\,(WG)^{(1)} \end{pmatrix} \\
&\quad - \begin{pmatrix} K^L_S\,(\cdot, 0)\,(S^{\triangle}(W)G) \\ \left(S^* - S^*_{[0]}\right) Z\,(S^{\triangle}(W)G)^{(1)} \end{pmatrix} \\
&= D^L_S(\cdot, W) \begin{pmatrix} 0 \\ (WG)^{(1)} \end{pmatrix} - D^L_S(\cdot, 0) \begin{pmatrix} S^{\triangle}(W)G \\ 0 \end{pmatrix}.
\end{aligned}
$$

We conclude that

$$
\tilde{\alpha}\begin{pmatrix} F \\ H \end{pmatrix} = D^L_S(\cdot, W) \begin{pmatrix} E \\ (WG)^{(1)} \end{pmatrix} - D^L_S(\cdot, 0) \begin{pmatrix} E + S^{\triangle}(W)G \\ 0 \end{pmatrix}, \tag{7.7}
$$

and therefore $\tilde{\alpha}$ is well defined.

Let $E \in \mathcal{D}_2$. Then,

$$\tilde{\gamma}(E) = \begin{pmatrix} \left(I - SS_{[0]}^*\right) E \\ \left(S^* - S_{[0]}^*\right) EZ \end{pmatrix} = D_S^L(\cdot, 0) \begin{pmatrix} E \\ 0 \end{pmatrix}.$$

Therefore $\tilde{\gamma}$ is well defined. Obviously $\tilde{\beta}$ and $\tilde{\delta}$ are well defined, and therefore the colligation $\tilde{\mathcal{U}}$ is well defined.

We now show that $\mathcal{U}_L^* = \tilde{\mathcal{U}}$, and first check that $\alpha_L^* = \tilde{\alpha}$. One can show that

$$\begin{aligned}
&\left\langle D_S^L(\cdot, V) \begin{pmatrix} E_1 \\ (VG_1)^{(1)} \end{pmatrix}, \alpha_L^* \left(D_S^L(\cdot, W) \begin{pmatrix} (W^*E_2)^{(-1)} \\ G_2 \end{pmatrix} \right) \right\rangle_{\mathcal{D}_L(S)} \\
&= \left\langle \alpha_L \left(D_S^L(\cdot, V) \begin{pmatrix} E_1 \\ (VG_1)^{(1)} \end{pmatrix} \right), D_S^L(\cdot, W) \begin{pmatrix} (W^*E_2)^{(-1)} \\ G_2 \end{pmatrix} \right\rangle_{\mathcal{D}_L(S)} \\
&= \left\langle D_S^L(\cdot, V) \begin{pmatrix} (V^*E_1)^{(-1)} \\ G_1 \end{pmatrix}, D_S^L(\cdot, W) \begin{pmatrix} (W^*E_2)^{(-1)} \\ G_2 \end{pmatrix} \right\rangle_{\mathcal{D}_L(S)} \\
&\quad - \left\langle D_S^L(\cdot, 0) \begin{pmatrix} 0 \\ G_1 + S^{\wedge}(V)^*E_1 \end{pmatrix}, D_S^L(\cdot, W) \begin{pmatrix} (W^*E_2)^{(-1)} \\ G_2 \end{pmatrix} \right\rangle_{\mathcal{D}_L(S)} \\
&= \left\langle D_S^L(\cdot, V) \begin{pmatrix} E_1 \\ (VG_1)^{(1)} \end{pmatrix}, D_S^L(\cdot, W) \begin{pmatrix} E_2 \\ (WG_2)^{(1)} \end{pmatrix} \right\rangle_{\mathcal{D}_L(S)} \\
&\quad - \left\langle D_S^L(\cdot, V) \begin{pmatrix} E_1 \\ (VG_1)^{(1)} \end{pmatrix}, D_S^L(\cdot, 0) \begin{pmatrix} E_2 + S^{\triangle}(W)G_2 \\ 0 \end{pmatrix} \right\rangle_{\mathcal{D}_L(S)} \\
&= \left\langle D_S^L(\cdot, V) \begin{pmatrix} E_1 \\ (VG_1)^{(1)} \end{pmatrix}, \tilde{\alpha} \left(D_S^L(\cdot, W) \begin{pmatrix} (W^*E_2)^{(-1)} \\ G_2 \end{pmatrix} \right) \right\rangle_{\mathcal{D}_L(S)}
\end{aligned}$$

and therefore $\alpha_L^* = \tilde{\alpha}$.

Let $E \in \mathcal{D}_2$ and $\begin{pmatrix} F \\ H \end{pmatrix} \in \mathring{\mathcal{D}}_L(S)$. Then

$$\begin{aligned}
\left\langle E, \beta_L^* \begin{pmatrix} F \\ H \end{pmatrix} \right\rangle_{\mathcal{D}_2} &= \left\langle \beta_L(E), \begin{pmatrix} F \\ H \end{pmatrix} \right\rangle_{\mathcal{D}_L(S)} \\
&= \left\langle D_S^L(\cdot, 0) \begin{pmatrix} 0 \\ E \end{pmatrix}, \begin{pmatrix} F \\ H \end{pmatrix} \right\rangle_{\mathcal{D}_L(S)} \\
&= \left\{ \operatorname{Tr} E^* H^{\nabla}(0) \right\}^* \\
&= \left\{ \operatorname{Tr} E^* H_{[0]} \right\}^* \\
&= \operatorname{Tr} H_{[0]}^* E \\
&= \langle E, H_{[0]} \rangle_{\mathcal{D}_2},
\end{aligned}$$

and thus

$$\beta_L^*\begin{pmatrix} F \\ H \end{pmatrix} = H_{[0]} = \tilde{\beta}\begin{pmatrix} F \\ H \end{pmatrix}.$$

In a similar way,

$$\begin{aligned}
\left\langle \begin{pmatrix} F \\ H \end{pmatrix}, \gamma_L^*(E) \right\rangle_{\mathcal{D}_L(S)} &= \left\langle \gamma_L \begin{pmatrix} F \\ H \end{pmatrix}, E \right\rangle_{\mathcal{D}_2} \\
&= \langle F_{[0]}, E\rangle_{\mathcal{D}_2} \\
&= \mathrm{Tr} E^* F_{[0]} \\
&= \mathrm{Tr} E^* F^\wedge(0) \\
&= \left\langle \begin{pmatrix} F \\ H \end{pmatrix}, D_S^L(\cdot, 0)\begin{pmatrix} E \\ 0 \end{pmatrix} \right\rangle_{\mathcal{D}_L(S)},
\end{aligned}$$

and therefore

$$\gamma_L^*(E) = D_S^L(\cdot, 0)\begin{pmatrix} E \\ 0 \end{pmatrix} = \tilde{\gamma}(E).$$

Obviously $\delta_L^*(E) = S_{[0]}^* E = \tilde{\delta}(E)$.

Step 3. *The colligation $\mathcal{U}_L$ is unitary.*

Proof of Step 3. First we show that $\mathcal{U}_L$ is coisometric. Let $\begin{pmatrix} F \\ H \end{pmatrix} \in \mathring{\mathcal{D}}_L(S)$ then

$$\begin{aligned}
(\alpha_L\alpha_L^* + \beta_L\beta_L^*)\begin{pmatrix} F \\ H \end{pmatrix} &= \alpha_L \begin{pmatrix} FZ - SH_{[0]} \\ (H - H_{[0]})Z \end{pmatrix} + \beta_L\left(H_{[0]}\right) \\
&= \begin{pmatrix} \left((FZ - SH_{[0]}) - (FZ - SH_{[0]})_{[0]}\right)Z^{-1} \\ (H - H_{[0]})ZZ^* - S^*(FZ - SH_{[0]})_{[0]} \end{pmatrix} \\
&\quad + \begin{pmatrix} (S - S_{[0]})H_{[0]}Z^{-1} \\ (I - S^*S_{[0]})H_{[0]} \end{pmatrix} \\
&= \begin{pmatrix} F - (S - S_{[0]})H_{[0]}Z^{-1} \\ (H - H_{[0]}) + S^*S_{[0]}H_{[0]} \end{pmatrix} \\
&\quad + \begin{pmatrix} (S - S_{[0]})H_{[0]}Z^{-1} \\ (I - S^*S_{[0]})H_{[0]} \end{pmatrix} \\
&= \begin{pmatrix} F \\ H \end{pmatrix},
\end{aligned}$$

therefore

$$\alpha_L\alpha_L^* + \beta_L\beta_L^* = I_{\mathcal{D}_L(S)}.$$

Let $E \in \mathcal{D}_2$:

$$(\alpha_L\gamma_L^* + \beta_L\delta_L^*)(E) = \alpha_L \begin{pmatrix} \left(I - SS_{[0]}^*\right)E \\ \left(S^* - S_{[0]}^*\right)EZ \end{pmatrix} + \beta_L\left(S_{[0]}^*E\right)$$

$$
\begin{aligned}
&= \begin{pmatrix} \left(\left(I - SS^*_{[0]}\right)E - \left(\left(I - SS^*_{[0]}\right)E\right)_{[0]}\right)Z^{-1} \\ \left(S^* - S^*_{[0]}\right)EZZ^* - S^*\left(\left(I - SS^*_{[0]}\right)E\right)_{[0]} \end{pmatrix} \\
&\quad + \begin{pmatrix} \left(S - S_{[0]}\right)S^*_{[0]}EZ^{-1} \\ \left(I - S^*S_{[0]}\right)S^*_{[0]}E \end{pmatrix} \\
&= \begin{pmatrix} \left(I - SS^*_{[0]}\right)EZ^{-1} - \left(I - S_{[0]}S^*_{[0]}\right)EZ^{-1} \\ \left(S^* - S^*_{[0]}\right)EZZ^* - S^*\left(I - S_{[0]}S^*_{[0]}\right)E \end{pmatrix} \\
&\quad + \begin{pmatrix} \left(S - S_{[0]}\right)S^*_{[0]}EZ^{-1} \\ \left(I - S^*S_{[0]}\right)S^*_{[0]}E \end{pmatrix} \\
&= \begin{pmatrix} 0 \\ 0 \end{pmatrix},
\end{aligned}
$$

so that $\alpha_L\gamma_L^* + \beta_L\delta_L^* = 0$. Still for $E \in \mathcal{D}_2$ we have

$$
\begin{aligned}
\left(\gamma_L\gamma_L^* + \delta_L\delta_L^*\right)(E) &= \gamma_L \begin{pmatrix} \left(I - SS^*_{[0]}\right)E \\ \left(S^* - S^*_{[0]}\right)E \end{pmatrix} + \delta_L\left(S^*_{[0]}E\right) \\
&= \left(\left(I - SS^*_{[0]}\right)E\right)_{[0]} + S_{[0]}S^*_{[0]}E = E,
\end{aligned}
$$

and therefore $\gamma_L\gamma_L^* + \delta_L\delta_L^* = I_{\mathcal{D}_2}$. We conclude that the colligation $\mathcal{U}_L$ is coisometric. Next we show that $\mathcal{U}_L$ is isometric.

Let $\begin{pmatrix} F \\ H \end{pmatrix} \in \mathring{\mathcal{D}}_L(S)$ then

$$
\begin{aligned}
\left(\alpha_L^*\alpha_L + \gamma_L^*\gamma_L\right)\begin{pmatrix} F \\ H \end{pmatrix} &= \alpha_L^* \begin{pmatrix} \left(F - F_{[0]}\right)Z^{-1} \\ HZ^* - S^*F_{[0]} \end{pmatrix} + \gamma_L^*\left(F_{[0]}\right) \\
&= \begin{pmatrix} \left(F - F_{[0]}\right)Z^{-1}Z - S\left(HZ^* - S^*F_{[0]}\right)_{[0]} \\ \left(\left(HZ^* - S^*F_{[0]}\right) - \left(HZ^* - S^*F_{[0]}\right)_{[0]}\right)Z \end{pmatrix} \\
&\quad + \begin{pmatrix} \left(I - SS^*_{[0]}\right)F_{[0]} \\ \left(S^* - S_{[0]}\right)F_{[0]}Z \end{pmatrix} \\
&= \begin{pmatrix} \left(F - F_{[0]}\right) + SS^*_{[0]}F_{[0]} \\ H - S^*F_{[0]}Z + S^*_{[0]}F_{[0]}Z \end{pmatrix} \\
&\quad + \begin{pmatrix} \left(I - SS^*_{[0]}\right)F_{[0]} \\ \left(S^* - S_{[0]}\right)F_{[0]}Z \end{pmatrix} \\
&= \begin{pmatrix} F \\ H \end{pmatrix},
\end{aligned}
$$

and so $\alpha_L^*\alpha_L + \gamma_L^*\gamma_L = I_{\mathcal{D}_L(S)}$.

Similarly,

$$\begin{aligned}(\alpha_L^*\beta_L+\gamma_L^*\delta_L)(E) &= \alpha_L^*\begin{pmatrix}(S-S_{[0]})EZ^{-1}\\ (I-S^*S_{[0]})E\end{pmatrix}+\gamma_L^*(S_{[0]}E)\\ &= \begin{pmatrix}\left(S-S_{[0]}^*\right)EZ^{-1}Z-S\left(\left(I-S^*S_{[0]}\right)E\right)_{[0]}\\ \left(\left(I-S^*S_{[0]}\right)E-\left(\left(I-S^*S_{[0]}\right)E\right)_{[0]}\right)Z\end{pmatrix}\\ &\quad+\begin{pmatrix}\left(I-SS_{[0]}^*\right)S_{[0]}E\\ \left(S^*-S_{[0]}^*\right)S_{[0]}EZ\end{pmatrix}\\ &= \begin{pmatrix}\left(S-S_{[0]}\right)E-S\left(I-S_{[0]}^*S_{[0]}\right)E\\ \left(I-S^*S_{[0]}\right)EZ-\left(I-S_{[0]}^*S_{[0]}\right)EZ\end{pmatrix}\\ &\quad+\begin{pmatrix}\left(I-SS_{[0]}^*\right)S_{[0]}E\\ \left(S^*-S_{[0]}^*\right)S_{[0]}EZ\end{pmatrix}\\ &= \begin{pmatrix}0\\0\end{pmatrix},\end{aligned}$$

so that $\alpha_L^*\beta_L+\gamma_L^*\delta_L=0$.

Finally,

$$\begin{aligned}(\beta_L^*\beta_L+\delta_L^*\delta_L)(E) &= \beta_L^*\begin{pmatrix}(S-S_{[0]})EZ^{-1}\\ (I-S^*S_{[0]})E\end{pmatrix}+\delta_L^*(S_{[0]}E)\\ &= \left(\left(I-S^*S_{[0]}\right)E\right)_{[0]}+S_{[0]}^*S_{[0]}E=E.\end{aligned}$$

Thus $\beta_L^*\beta_L+\delta_L^*\delta_L=I_{\mathcal{D}_2}$. Therefore the colligation $\mathcal{U}_L$ is isometric. We conclude that the colligation $\mathcal{U}_L$ is unitary. □

In the next theorem we introduce the unitary realization of S.

Theorem 7.2. *Let $W\in\Omega$ with $\|W\|<1$. For every $E\in\mathcal{D}_2$ we have*

$$(SE)^{\triangle}(W)=\left(\delta_L+M_W^R\gamma_L\left(I_{\mathcal{D}_L(S)}-\alpha_LM^R_{\text{diag }(W,W^{(1)})}\right)^{-1}\beta_L\right)(E),$$

(where $M^R_{\text{diag }(W,W^{(1)})}$ is the operator (6.4)*). This realization is closely connected and is unique up to a unitary equivalence.*

Proof. We first note that (with the already mentioned abuse of notation $M_W^R=M^R_{\text{diag }(W,W^{(1)})}$) $M_W^R\alpha_L=\alpha_LM^R_{W^{(1)}}$. Indeed, let $\begin{pmatrix}F\\H\end{pmatrix}\in\mathcal{D}_L(S)$. Then,

$$M_W^R\alpha_L\begin{pmatrix}F\\H\end{pmatrix}=M_W^R\begin{pmatrix}(F-F_{[0]})Z^{-1}\\ HZ^*-S^*F_{[0]}\end{pmatrix}$$

$$= \begin{pmatrix} \left(F - F_{[0]}\right) Z^{-1} W \\ \left(HZ^* - S^* F_{[0]}\right) W^{(1)} \end{pmatrix}$$
$$= \begin{pmatrix} \left(FW^{(1)} - F_{[0]} W^{(1)}\right) Z^{-1} \\ HW^{(2)} Z^* - S^* F_{[0]} W^{(1)} \end{pmatrix}$$
$$= \alpha_L M^R_{W^{(1)}} \begin{pmatrix} F \\ H \end{pmatrix}.$$

Recall that for any $F \in \mathcal{U}$ and any $W \in \Omega$

$$(FZ)^{\triangle} = F^{\triangle}\left(W^{(-1)}\right) W.$$

Straightforward computations show that

$$\gamma_L M^R_W \alpha_L \begin{pmatrix} F \\ H \end{pmatrix} = Z F_{[1]} Z^* W.$$

In the same manner one can show that for every $n \geq 1$

$$\gamma_L \left(M^R_W \alpha_L\right)^n \begin{pmatrix} F \\ H \end{pmatrix} = Z^n F_{[n]} \left(Z^* W\right)^n.$$

Therefore

$$\gamma_L \left(I_{\mathcal{D}_L(S)} - M^R_W \alpha_L\right)^{-1} \begin{pmatrix} F \\ H \end{pmatrix} = F^{\triangle}(W).$$

Note that $\alpha_L \alpha_L^* + \beta_L \beta_L^* = I_{\mathcal{D}_L(S)}$ and therefore α_L is contraction, but M^R_W is a strict contraction, therefore the indicated inverse exists.

Let $E \in \mathcal{D}_2$. Then,

$$\left(\gamma_L \left(I_{\mathcal{D}_L(S)} - M^R_W \alpha_L\right)^{-1} \beta_L\right)(E) = \left(\left(S - S_{[0]}\right) E Z^{-1}\right)^{\triangle}(W),$$

and therefore

$$\left(M^R_W \gamma_L \left(I_{\mathcal{D}_L(S)} - M^R_{W^{(-1)}} \alpha_L\right)^{-1} \beta_L\right)(E) = \left(\left(S - S_{[0]}\right) E\right)^{\triangle}(W).$$

The realization formula

$$\left(\delta_L + M^R_W \gamma_L \left(I_{\mathcal{D}_L(S)} - \alpha_L M^R_W\right)^{-1} \beta_L\right)(E) = (SE)^{\triangle}(W).$$

follows. Moreover, one easily shows that

$$(\alpha_L)^n \begin{pmatrix} F \\ H \end{pmatrix} = \begin{pmatrix} \left(F - \sum_{k=0}^{n-1} Z^k F_{[k]}\right) Z^{-n} \\ HZ^{*n} - S^* \left(\sum_{k=0}^{n-1} Z^k F_{[k]}\right) Z^{*(n-1)} \end{pmatrix}$$

and that

$$(\alpha_L^*)^m \begin{pmatrix} F \\ H \end{pmatrix} = \begin{pmatrix} FZ^m - S\left(\sum_{k=0}^{m-1} H_{[k]} Z^{*k}\right) Z^{m-1} \\ \left(H - \sum_{k=0}^{m-1} H_{[k]} Z^{*k}\right) Z^m \end{pmatrix},$$

for any $n, m \geq 1$. Therefore

$$\gamma_L (\alpha_L)^n \begin{pmatrix} F \\ H \end{pmatrix} = Z^n F_{[n]} Z^{*n}$$

for $n \geq 0$, and we conclude that $F = 0$ whenever $\begin{pmatrix} F \\ H \end{pmatrix}$ belongs to

$$\cap_{n=0}^{\infty} \mathrm{Ker} \gamma_L (\alpha_L)^n .$$

Similarly, for $m \geq 0$

$$\beta_L^* (\alpha_L^*)^m \begin{pmatrix} F \\ H \end{pmatrix} = H_{[m]} Z^{*m} Z^m = H_{[m]}$$

and we conclude that $H = 0$ whenever $\begin{pmatrix} F \\ H \end{pmatrix}$ belongs to $\cap_{m=0}^{\infty} \mathrm{Ker} \beta_L^* (\alpha_L^*)^m$. Therefore the realization is closely connected.

Let $W = \lambda I$ with $|\lambda| < 1$, then

$$\left(\delta_L + \lambda \gamma_L \left(I_{\mathcal{D}_L(S)} - \lambda \alpha_L\right)^{-1} \beta_L\right)(E) = (SE)^{\triangle}(\lambda I),$$

where

$$\theta_L(\lambda) = \delta_L + \lambda \gamma_L \left(I_{\mathcal{D}_L(S)} - \lambda \alpha_L\right)^{-1} \beta_L,$$

is the characteristic function of the colligation $\mathcal{U}_L$ which is unitary and closely connected and therefore $\mathcal{U}_L$ is unique up to a unitary equivalence. □

Acknowledgements

It is a pleasure to thank Professor Joe Ball for his very useful comments at the MTNS96 in Saint-Louis, where the results of this paper were reported on. We wish also to thank the referee for his remarks on the paper.

References

[1] N.I. Akhiezer and I.M. Glazman. *Theory of linear operators. (Vol. I)*. Pitman Advanced Publishing Program, 1981.

[2] D. Alpay. Algorithme de Schur, espaces à noyau reproduisant et théorie des systèmes. To appear in the series *Panoramas et Synthèses, Société Mathématique de France, 1998.*

[3] ______ A theorem on reproducing kernel Hilbert spaces of pairs. *Rocky Mountain J. Math.*, 22:1243–1257, 1992.

[4] D. Alpay, J. Ball, and Y. Perets. Model theory and nonstationary de Branges Rovnyak spaces. In preparation.

[5] D. Alpay, V. Bolotnikov, A. Dijksma, and H.S.V. de Snoo. *On some operator colligations and associated reproducing kernel Hilbert spaces*, volume 61 of *Operator theory: Advances and Applications*, pages 89–159. Birkhäuser Verlag, Basel, 1993.

[6] ______ Interpolation problems, extensions of symmetric operators and reproducing kernel spaces II. *Integral Equations Operator Theory*, 14:465–500, 1991.

[7] D. Alpay and P. Dewilde. Time varying signal approximation and estimation. In M. Kaashoek, J.H. van Schuppen, and A.C.M. Ran, editors, *Proceedings of the international symposium MTNS 89*, volume 5 of *Progress in systems and control theory*, pages 1–22. Birkhäuser Verlag, Basel, 1990.

[8] D. Alpay, P. Dewilde, and H. Dym. *Lossless inverse scattering and reproducing kernels for upper triangular operators*, volume 47 of *Operator Theory: Advances and Applications*, pages 61–133. Birkhäuser Verlag, Basel, 1990.

[9] D. Alpay, A. Dijksma, and Y. Perets. Nonstationary analogue of the Herglotz representation theorem: the discrete case. In preparation.

[10] D. Alpay, A. Dijksma, J. Rovnyak, and H. de Snoo. Fonctions de Schur, colligations d'opérateurs, et espaces de Pontryagin à noyaux reproduisants. *C. R. Acad. Sci. Paris Sér. I Math.*, 322:15–20, 1996.

[11] ______ *Schur functions, operator colligations, and reproducing kernel Pontryagin spaces*, volume 96 of *Operator theory: Advances and Applications*. Birkhäuser Verlag, Basel, 1997.

[12] D. Alpay and Y. Peretz. Special realizations for Schur upper triangular operators centered at an arbitrary point. Preprint, 1997.

[13] T. Ando. de Branges spaces and analytic operator functions. Lecture notes, Hokkaido University, Sapporo, 1990.

[14] N. Aronszjan. Theory of reproducing kernels. *Trans. Amer. Math. Soc.*, 68:337–404, 1950.

[15] D. Arov, M. Kaashoek, and D. Pik. Minimal and optimal linear discrete time–varying dissipative scattering systems. Preprint, Vrije Universiteit Amsterdam, 1997.

[16] Gr. Arsene, Z. Ceauşescu, and T. Constantinescu. Schur analysis of some completion problems. *Linear Algebra Appl.*, 109:1–35, 1988.

[17] J. Ball. Factorization and model theory for contraction operators with unitary parts. *Memoirs of the American Mathematical Society*, 13, 1978.

[18] J. Ball, I. Gohberg, and M.A. Kaashoek. Nevanlinna–Pick interpolation for time–varying input–output maps: the discrete case. In I. Gohberg, editor, *Time–variant systems and interpolation*, volume 56 of *Operator Theory: Advances and Applications*, pages 1–51. Birkhäuser Verlag, Basel, 1992.

[19] ——— Nevanlinna–Pick interpolation for time–varying input–output maps: the continuous case. In I. Gohberg, editor, *Time–variant systems and interpolation*, volume 56 of *Operator Theory: Advances and Applications*, pages 52–89. Birkhäuser Verlag, Basel, 1992.

[20] ——— Bitangential interpolation for input–output maps of time–varying systems: the continuous time case. *Integral Equations Operator Theory*, 20:1–43, 1994.

[21] ——— Two sided Nudelman interpolation for input–output operators of discrete time–varying systems. *Integral Equations Operator Theory*, 21:174–211, 1995.

[22] J. Ball and T.Trent. Unitary colligations, reproducing kernel Hilbert spaces and Nevanlinna–Pick interpolation in several variables. Preprint, 1996.

[23] H. Bart, I. Gohberg, and M. Kaashoek. *Minimal factorization of matrix and operator functions*, volume 1 of *Operator Theory: Advances and Applications*. Birkhäuser Verlag, Basel, 1979.

[24] L. de Branges. Complementation in Kreĭn spaces. *Trans. Amer. Math. Soc.*, 305:277–291, 1988.

[25] L. de Branges and J. Rovnyak. Canonical models in quantum scattering theory. In C. Wilcox, editor, *Perturbation theory and its applications in quantum mechanics*, pages 295–392. Wiley, New–York, 1966.

[26] ——— *Square summable power series*. Holt, Rinehart and Winston, New–york, 1966.

[27] T. Constantinescu. *Schur parameters, factorization and dilation problems*, volume 82 of *Operator Theory: Advances and Applications*. Birkhäuser Verlag, Basel, 1996.

[28] B. Curgus, A. Dijksma, H. Langer, and H. de Snoo. *characteristic functions of unitary colligations and of bounded operators in Kreĭn spaces*, volume 41 of *Operator theory: Advances and Applications* , pages 125–152. Birkhäuser Verlag, Basel, 1989.

[29] E. Deprettere and P. Dewilde. *The generalized Schur algorithm*, volume 29 of *Operator Theory: Advances and Applications*, pages 97–115. Birkhäuser Verlag, Basel, 1988.

[30] P. Dewilde and H. Dym. Interpolation for upper triangular operators. In I. Gohberg, editor, *Time–variant systems and interpolation*, volume 56 of *Operator Theory: Advances and Applications*, pages 153–260. Birkhäuser Verlag, Basel, 1992.

[31] P. Dewilde, M.A. Kaashoek, and M. Verhaegen, editors. *Challenges of a generalized system theory*. Koninklijke Nederlandse Akademie van Wetenschappen, Verhandelingen Afd. Natuurkunde, Eerste reeks, deel 40. North–Holland, 1993.

[32] A. Dijksma, H. Langer, and H.S.V. de Snoo. *Characteristic functions of unitary operator colligations in Π_k spaces*, volume 19 of *Operator theory: Advances and Applications*, pages 125–194. Birkhäuser Verlag, Basel, 1986.

[33] ——— Unitary colligations in Π_k spaces, characteristic functions and Štraus extensions. *Pacific J. Math.*, 125:347–362, 1986.

[34] H. Dym. *J contractive matrix functions, reproducing kernel spaces and interpolation*, volume 71 of *CBMS Lecture Notes*. Amer. Math. Soc., Rhodes Island, 1989.

[35] P.A. Fillmore and J.P. Williams. On operator ranges. *Adv. in Math.*, 7:254–281, 1971.

[36] C. Foias, A.E. Frazho, I. Gohberg, and M.A. Kaashoek. Discrete time–variant interpolation as classical interpolation with an operator argument. *Integral Equations Operator Theory*, 26:371–403, 1996.

[37] P.A. Fuhrmann. *Linear systems and operators in Hilbert space*. McGraw-Hill international book company, 1981.

[38] I. Gohberg, editor. *I. Schur methods in operator theory and signal processing*, volume 18 of *Operator theory: Advances and Applications*. Birkhäuser Verlag, Basel, 1986.

[39] ——— *Time variant systems and interpolation*, volume 56 of *Operator theory: Advances and Applications*. Birkhäuser Verlag, Basel, 1992.

[40] I. Gohberg and M.G. Kreĭn. *Introduction to the theory of linear nonselfadjoint operators*, volume 18 of *Translations of mathematical monographs*. American Mathematical Society, Rhode Island, 1969.

[41] S. Itoh. Reproducing kernels in modules over C^*–algebras and their applications. *Bull. Kyushu Inst. Tech.*, 37:1–20, 1990.

[42] T. Kailath. *A theorem of I. Schur and its impact on modern signal processing*, volume 18 of *Operator theory: Advances and Applications*, pages 9–30. Birkhäuser Verlag, Basel, 1986.

[43] P. D. Lax and R. S. Phillips. *Scattering theory (revised edition)*, volume 26 of *Pure and Applied Mathematics*. Academic Press, New–York, 1989.

[44] N.K. Nikolskii and V.I. Vasyunin. Notes on two function models. In *Proceedings of the symposium on the occasion of the proof of the Bieberbach conjecture,* math. survey monographs 21, pages 113–141. American Mathematical Society, Providence, R.I., 1986.

[45] W. Paschke. Inner product spaces over B^*–algebras. *Trans. Amer. Math. Soc.*, 1982:443–468, 1973.

[46] W. Rudin. *Analyse réelle et complexe.* Masson, Paris, 1980.

[47] S. Saitoh. *Theory of reproducing kernels and its applications*, volume 189. Longman scientific and technical, 1988.

[48] D. Sarason. Shift–invariant spaces from the brangesian point of view. In *Proceedings of the symposium on the occasion of the proof of the Bieberbach conjecture,* math. survey monographs 21. American Mathematical Society, Providence, R.I., 1986.

[49] ——— *Function theory and de Branges spaces*, volume 51 of *Proceedings of symposia in pure mathematics*, pages 495–501. American Mathematical Society, 1990.

[50] L. Schwartz. Sous espaces hilbertiens d'espaces vectoriels topologiques et noyaux associés (noyaux reproduisants). *J. Analyse Math.*, 13:115–256, 1964.

[51] Yu.L. Shmul'yan. Division in the class of j–expansive operators. *Mat. Sb. (N.S)*, 74, 1967. English translation: Math. USSR–Sbornik, **3** (1967), 471–479.

Department of Mathematics
Ben–Gurion University of the Negev
POB 653. 84105 Beer-Sheva
Israel

1991 Mathematics Subject Classification. Primary 46E22; Secondary 47A48

Operator Theory:
Advances and Applications, Vol. 106

On the defect of noncontractive operators in Kreĭn spaces: a new formula and some applications

TOMAS YA. AZIZOV, AAD DIJKSMA, AND VLADIMIR L. KHATSKEVICH

To Heinz Langer on the occasion of his 60th birthday

We prove a formula for the defect of a noncontractive operator from one Kreĭn space to another and give some applications of this notion to the study of the spectral properties of nondecreasing continuous or real analytic functions whose values are selfadjoint operators on a Hilbert space.

1. Introduction

Let $\mathcal{H}$ be a Hilbert space and let A_0 and $A_1 \in \mathbf{L}(\mathcal{H})$ be selfadjoint operators such that $A_0 \leq A_1$. By $\mathcal{H}_i^{\pm}$ we denote the spectral subspace of the operator A_i related to the set $\sigma(A_i) \cap \mathbb{R}^{\pm}$, $i = 0, 1$; here, $\mathbb{R}^+ = [0, \infty)$ and $\mathbb{R}^- = (-\infty, 0]$. In [Pe], [S] and [AKh2] the next result is proved:

Theorem 1.1. *Assume* $0 \in \rho(A_0) \cap \rho(A_1)$. *Then the following statements are equivalent:*

1. $A_0^{-1} \geq A_1^{-1}$.
2. $\mathcal{H}_1^+ \cap \mathcal{H}_0^- = \{0\}$.
3. *If* $A = A^* \in \mathbf{L}(\mathcal{H})$ *and* $A_0 \leq A \leq A_1$ *then* $0 \in \rho(A)$.
4. *There exists a continuous function A on* $[0, 1]$ *whose values are selfadjoint operators in* $\mathbf{L}(\mathcal{H})$ *with* $A_0 = A(0) \leq A(t) \leq A(1) = A_1$ *and* $0 \in \rho(A(t)), t \in [0, 1]$.
5. *If A is a continuous function on* $[0, 1]$ *whose values are selfadjoint operators in* $\mathbf{L}(\mathcal{H})$ *and* $A_0 = A(0) \leq A(t) \leq A(1) = A_1$, *then* $0 \in \rho(A(t)), t \in [0, 1]$.

The case where the selfadjoint operators are unbounded is considered in [AKh1] and [AKh2]. As corollaries sufficient conditions are obtained to ensure the existence of a unique solution of nonlinear periodic Hamiltonian systems, wave equations and Schrödinger equations. In this note these papers will be generalized (see Section 3) and these generalizations are applied to give an upper bound for the dimension of the space of solutions of certain elliptic problems (see Section 4). A lower bound in a similar situation is given in [BBF] and [R]. In our proofs of these generalizations we use the theory of noncontractive operators on Kreĭn spaces, and in particular apply the notion of the defect of such operators.

In Section 2 we derive a new formula for the defect of a noncontractive operator from one Kreĭn space to another. The particular case where the operator is bi-contractive or bi-noncontractive is considered in [Po], [Gi] and [KSh1]. For another formula we refer the reader to [Gh].

2. The defect of noncontractive operators

Let $T \in \mathbf{L}(\mathcal{K}_0, \mathcal{K}_1)$ be a noncontractive operator from the Kreĭn space $\mathcal{K}_0$ to the Kreĭn space $\mathcal{K}_1$, that is,

$$[Tx, Tx]_1 \geq [x, x]_0, \quad x \in \mathcal{K}_0.$$

If $\mathcal{L}_0^+$ is a maximal nonnegative subspace of $\mathcal{K}_0$ then the subspace $T\mathcal{L}_0^+$ is nonnegative in $\mathcal{K}_1$ and admits an extension to a maximal nonnegative subspace $\mathcal{L}_1^+$ of $\mathcal{K}_1$: $T\mathcal{L}_0^+ \subseteq \mathcal{L}_1^+$. The dimension of the quotient space $\mathcal{L}_1^+/T\mathcal{L}_0^+$ is called the defect of T and is denoted by $\operatorname{def} T$. The number is independent of the choice of $\mathcal{L}_0^+$ and of $\mathcal{L}_1^+$ containing $T\mathcal{L}_0^+$; see [KSh1, Theorem 2.3] (or [AI, Chapter 2, Theorem 4.15]), the definition given there coincides with one given here. Since T is bi–noncontractive (that is, T and T^* are both noncontractive) if and only if T maps some (and hence every) maximal nonnegative subspace into a maximal nonnegative subspace (see [KSh1, Theorem 3.2], or [AI, Chapter 2, Theorem 4.17]), we have that T is bi–noncontractive if and only if $\operatorname{def} T = 0$. If T is an isometry then

$$\operatorname{def} T = \dim (\ker T^*)^+. \tag{2.1}$$

In the formula we use that any subspace $\mathcal{L}$ in $\mathcal{K}_0$, say, admits the orthogonal direct sum (fundamental) decomposition of the form

$$\mathcal{L} = \mathcal{L}^+ \dot{+} \mathcal{L}^0 \dot{+} \mathcal{L}^-,$$

where $\mathcal{L}^0 = \mathcal{L} \cap \mathcal{L}^\perp$, and $\mathcal{L}^+$ and $\mathcal{L}^-$ are positive and negative subspaces of $\mathcal{L}$, respectively; the triple $(\dim \mathcal{L}^+, \dim \mathcal{L}^0, \dim \mathcal{L}^-)$ is called the signature of $\mathcal{L}$, and is independent of the chosen decomposition.

To see (2.1), let $\mathcal{K}_0 = \mathcal{K}_0^+ \oplus \mathcal{K}_0^-$ be a fundamental decomposition of $\mathcal{K}_0$. Then $T\mathcal{K}_0^+$ and $T\mathcal{K}_0^-$ are uniformly positive and negative subspaces, respectively, and there exists a fundamental decomposition $\mathcal{K}_1 = \mathcal{K}_1^+ \oplus \mathcal{K}_1^-$ of $\mathcal{K}_1$ such that $T\mathcal{K}_0^\pm \subset \mathcal{K}_1^\pm$. The equality (2.1) follows from the decomposition

$$(\ker T^*)^+ = (\mathcal{K}_1 \ominus T\mathcal{K}_0)^+ = \mathcal{K}_1^+ \ominus T\mathcal{K}_0^+.$$

Theorem 2.1. *Let $\mathcal{K}_0$ and $\mathcal{K}_1$ be Kreĭn spaces and let $T \in \mathbf{L}(\mathcal{K}_0, \mathcal{K}_1)$ be a noncontractive operator. If $\operatorname{def} T < \infty$, then*

$$\operatorname{def} T = \sum_{\lambda<0} \dim \ker (TT^* - \lambda) \; + \; \dim \left((\ker TT^*)^+ \dot{+} (\ker TT^*)^0 \right). \tag{2.2}$$

We note that if T is a noncontractive operator, then $\ker T^* = \ker TT^*$: If x belongs to the space on the righthand side, then T^*x belongs to $(\ker T)^0$, which is equal to $\{0\}$, since $\ker T$ is a uniformly negative subspace (due to Yu.P. Ginzburg, see [KSh1] or [AI, Chapter 2, Proposition 4.14]). Hence $\ker TT^* \subseteq \ker T^*$; the converse inclusion is obvious. Thus in (2.2) we may replace $\ker TT^*$ by $\ker T^*$, and also, since $\lambda \neq 0$, $\dim\ker(TT^* - \lambda)$ by $\dim\ker(T^*T - \lambda)$. The definition of the defect of a nonexpansive operator is the same as for a noncontractive operator, except that the word nonnegative has to be replaced everywhere by the word nonpositive. Then for a nonexpansive operator T with finite defect, the above formula holds if we replace $(\ker TT^*)^+$ by $(\ker TT^*)^-$.

To prove the theorem we first prove two lemmas. The first one gives a sufficient condition for the existence of the square root of an operator and the second one concerns a special case of Theorem 2.1.

Lemma 2.2. *If $A \in \mathbf{L}(\mathcal{K})$ is a selfadjoint operator in the Kreĭn space $\mathcal{K}$ such that*

$$[Ax, x] \geq [x, x], \quad x \in \mathcal{K},$$

and $\sigma(A) \subseteq [0, \infty)$, then there exists a unique operator $B = B^* \in \mathbf{L}(\mathcal{K})$ with $\sigma(B) \subseteq [0, \infty)$ such that $B^2 = A$.

The lemma can be proved using the functional calculus for definitizable operators, see H. Langer [L] or the generalization of P.Jonas [J]. We give a direct proof following [KSh2, Theorem 4.2]; see also [AI, Chapter 3, Theorem 1.12].

Proof of Lemma 2.2. We denote by E the spectral function for the nonnegative operator $A - 1$. Then for $0 < \varepsilon < 1$, the space $\mathcal{L}_\varepsilon = E(-\infty, \varepsilon - 1)\mathcal{K}$ is uniformly negative. We fix an $\varepsilon \in (0, 1)$ and write $\mathcal{K}$ as $\mathcal{K} = \mathcal{L}_\varepsilon \oplus \mathcal{L}_\varepsilon^\perp$. The matrix representation of A with respect to this decomposition is of the form

$$A = \begin{pmatrix} A_\varepsilon & 0 \\ 0 & A'_\varepsilon \end{pmatrix} : \begin{pmatrix} \mathcal{L}_\varepsilon \\ \mathcal{L}_\varepsilon^\perp \end{pmatrix} \to \begin{pmatrix} \mathcal{L}_\varepsilon \\ \mathcal{L}_\varepsilon^\perp \end{pmatrix}.$$

Here, A_ε is a selfadjoint operator in the Hilbert space $(\mathcal{L}_\varepsilon, -[\cdot,\cdot])$ and $\sigma(A_\varepsilon) \subseteq [0, \varepsilon)$. Hence the operator $B_\varepsilon = A_\varepsilon^{1/2}$ is well defined and uniquely determined by the properties: B_ε is selfadjoint, $B_\varepsilon^2 = A_\varepsilon$, and $\sigma(B_\varepsilon) \subseteq [0, \infty)$. As to A'_ε we have that $\sigma(A'_\varepsilon) \subseteq [\varepsilon, \infty)$ and A'_ε hence admits the integral representation

$$A'_\varepsilon = -\frac{1}{2\pi i}\int_{\Gamma_\varepsilon} \lambda (A'_\varepsilon - \lambda)^{-1} d\lambda,$$

where Γ_ε is a Jordan contour in the open right half plane which is symmetric with respect to the real axis and contains $\sigma(A'_\varepsilon)$ in its interior. The operator

$$B'_\varepsilon = -\frac{1}{2\pi i}\int_{\Gamma_\varepsilon} \lambda^{1/2} (A'_\varepsilon - \lambda)^{-1} d\lambda,$$

is selfadjoint and ${B'_\varepsilon}^2 = A'_\varepsilon$ and $\sigma(B'_\varepsilon) \subseteq [0,\infty)$. B'_ε is the only operator with these three properties. The operator

$$B = \begin{pmatrix} B_\varepsilon & 0 \\ 0 & B'_\varepsilon \end{pmatrix} : \begin{pmatrix} \mathcal{L}_\varepsilon \\ \mathcal{L}_\varepsilon^\perp \end{pmatrix} \to \begin{pmatrix} \mathcal{L}_\varepsilon \\ \mathcal{L}_\varepsilon^\perp \end{pmatrix}$$

has the desired property. To show that it is unique, let B_1 be a selfadjoint operator such that $B_1^2 = A$ and $\sigma(B_1) \subseteq [0,\infty)$. Then B_1 commutes with A and hence with $E(-\infty, \varepsilon - 1)$. Therefore $\mathcal{L}_\varepsilon$ and $\mathcal{L}_\varepsilon^\perp$ are B_1-invariant subspaces, and so

$$B_1 = \begin{pmatrix} B_{1\varepsilon} & 0 \\ 0 & B'_{1\varepsilon} \end{pmatrix} : \begin{pmatrix} \mathcal{L}_\varepsilon \\ \mathcal{L}_\varepsilon^\perp \end{pmatrix} \to \begin{pmatrix} \mathcal{L}_\varepsilon \\ \mathcal{L}_\varepsilon^\perp \end{pmatrix}.$$

From $B_{1,\varepsilon}^2 = A_\varepsilon$ and $\sigma(B_{1,\varepsilon}) \subseteq [0,\infty)$, ${B'_{1,\varepsilon}}^2 = A'_\varepsilon$ and $\sigma(B'_{1,\varepsilon}) \subseteq [0,\infty)$ it follows that $B_{1,\varepsilon} = B_\varepsilon$, $B'_{1,\varepsilon} = B'_\varepsilon$, and hence $B = B_1$. □

Lemma 2.3. *Let $\mathcal{K}_0$ and $\mathcal{K}_1$ be Kreĭn spaces and let $T \in \mathbf{L}(\mathcal{K}_0, \mathcal{K}_1)$ be a noncontractive operator such that $\sigma(T^*T) \subseteq [0,\infty)$. Then*

$$\operatorname{def} T = \dim \left((\ker TT^*)^+ \dot{+} (\ker TT^*)^0 \right).$$

The equality means that the two numbers are both infinite or both finite and equal.

Proof of Lemma 2.3. By Lemma 2.2, the operator T^*T on the Kreĭn space $\mathcal{K}_0$ has a unique square root $(T^*T)^{1/2}$ with $(T^*T)^{1/2} \subseteq [0,\infty)$. Since $(T^*T)^{1/2}$ is noncontractive and selfadjoint it is bi–noncontractive. Let $\mathcal{K}_0 = \mathcal{K}_0^+ \oplus \mathcal{K}_0^-$ be a fundamental decomposition of $\mathcal{K}_0$. Then $(T^*T)^{1/2}\mathcal{K}_0^+$ is a maximal uniformly positive subspace of $\mathcal{K}_0$. We claim that

$$\mathcal{L} = (T\mathcal{K}_0^+)^\perp \cap \overline{\operatorname{ran}} T$$

is a nonpositive subspace of $\mathcal{K}_1$. To prove this we first note that, since $T\mathcal{K}_0^+$ is uniformly positive,

$$T\mathcal{K}_0^+ \oplus \overline{(T\mathcal{K}_0^+)^\perp \cap \operatorname{ran} T} = \overline{\operatorname{ran}} T = T\mathcal{K}_0^+ \oplus \mathcal{L},$$

so that $(T\mathcal{K}_0^+)^\perp \cap \operatorname{ran} T$ is dense in $\mathcal{L}$. If the claim is not true there is a $y \in (T\mathcal{K}_0^+)^\perp \cap \operatorname{ran} T$ such that $[y,y]_1 > 0$. We choose $x \in \mathcal{K}_0$ such that $y = Tx$. Then

$$[(T^*T)^{1/2}x, (T^*T)^{1/2}\mathcal{K}_0^+]_0 = [y, T\mathcal{K}_0^+]_1 = \{0\},$$

and hence span $\{(T^*T)^{1/2}\mathcal{K}_0^+, (T^*T)^{1/2}x\}$ is uniformly positive. Since $(T^*T)^{1/2}\mathcal{K}_0^+$ is maximal with respect to this property, we have that $(T^*T)^{1/2}x \in (T^*T)^{1/2}\mathcal{K}_0^+$, which readily yields the contradiction

$$0 = [(T^*T)^{1/2}x, (T^*T)^{1/2}x]_0 = [y,y]_1 > 0.$$

Hence the claim is true. We now use that $\mathcal{K} = (T\mathcal{K}_0^+)^\perp$ is a Kreĭn space with positive index $\operatorname{def} T$ and that

$$\mathcal{L}^\perp \cap \mathcal{K} = \ker T^* = (\ker T^*)^+ \dot{+} (\ker T^*)^0 \dot{+} (\ker T^*)^-.$$

Since $\mathcal{L}$ is nonpositive in $\mathcal{K}$, $\operatorname{def} T$ is equal to the dimension of a maximal nonnegative subspace of $\mathcal{L}^\perp \cap \mathcal{K}$, and this number is equal to

$$\dim \left((\ker TT^*)^+ \dot{+} (\ker TT^*)^0 \right).$$

This proves the lemma. □

Proof of Theorem 2.1. Let E be the spectral function for the nonnegative operator $T^*T - 1$ in $\mathcal{K}_0$. Then for $\mu < -1$, the spectral subspace $\mathcal{L}_\mu = E(-\infty, \mu)\mathcal{K}_0$ is T^*T–invariant and uniformly negative in $\mathcal{K}_0$. Moreover, $\sigma((T^*T-1)|_{\mathcal{L}_\mu}) \subseteq (-\infty, \mu]$ and hence $\sigma(T^*T|_{\mathcal{L}_\mu}) \subseteq (-\infty, 1+\mu]$. The operator $T^*T|_{\mathcal{L}_\mu}$ is selfadjoint in the Hilbert space $(\mathcal{L}_\mu, -[\cdot,\cdot]_0)$, and therefore

$$-[T^*Tx, x]_0 \leq -(1+\mu)[x,x]_0, \quad x \in \mathcal{L}_\mu.$$

We denote by $\|\cdot\|_j$ the norm associated with the Hilbert space inner product on $\mathcal{K}_j$ induced by a fundamental decomposition of $\mathcal{K}_j$, $j = 0, 1$. Then, since $\mathcal{L}_\mu$ is uniformly negative, there is a $\delta > 0$ such that

$$[x,x]_0 < -\delta \|x\|_0^2, \quad x \in \mathcal{L}_\mu.$$

Hence, with $\delta' = -\delta(1+\mu)(>0)$,

$$[Tx, Tx]_1 \geq \delta' \|x\|_0^2 \geq \frac{\delta'}{\|T\|^2} \|Tx\|_1^2, \quad x \in \mathcal{L}_\mu,$$

where $\|\cdot\|$ is the operator norm associated with the norms $\|\cdot\|_j$ on $\mathcal{K}_j$, $j = 0, 1$. It follows that $T\mathcal{L}_\mu$ is a uniformly positive subspace in $\mathcal{K}_1$. The matrix representation of T with respect to the decompositions

$$\mathcal{K}_0 = \mathcal{L}_\mu \oplus \mathcal{L}_\mu^\perp, \quad \mathcal{K}_1 = T\mathcal{L}_\mu \oplus (T\mathcal{L}_\mu)^\perp \tag{2.3}$$

is given by

$$T = \begin{pmatrix} T_0 & 0 \\ 0 & T_1 \end{pmatrix} : \begin{pmatrix} \mathcal{L}_\mu \\ \mathcal{L}_\mu^\perp \end{pmatrix} \to \begin{pmatrix} T\mathcal{L}_\mu \\ (T\mathcal{L}_\mu)^\perp \end{pmatrix}. \tag{2.4}$$

To verify this diagonal matrix form we only have to check that $T(\mathcal{L}_\mu^\perp) \subseteq (T\mathcal{L}_\mu)^\perp$; but this follows from the T^*T–invariance of $\mathcal{L}_\mu$:

$$[T(\mathcal{L}_\mu^\perp), T\mathcal{L}_\mu]_1 = [\mathcal{L}_\mu^\perp, T^*T\mathcal{L}_\mu]_0 = \{0\}.$$

Since $0 \in \rho(T^*T|_{\mathcal{L}_\mu})$, we have that $\ker T|_{\mathcal{L}_\mu} = \{0\}$, so that $\dim T\mathcal{L}_\mu = \dim \mathcal{L}_\mu$. The diagonal matrix form of T implies that

$$\operatorname{def} T = \operatorname{def} T_0 + \operatorname{def} T_1 = \dim T\mathcal{L}_\mu + \operatorname{def} T_1 = \dim \mathcal{L}_\mu + \operatorname{def} T_1. \tag{2.5}$$

We conclude that for every $\mu < -1$,

$$\dim E(-\infty, \mu)\mathcal{K}_0 = \dim \mathcal{L}_\mu \le \operatorname{def} T.$$

Hence the set $\sigma(T^*T - 1) \cap (-\infty, -1)$ consists of finitely many points $\mu_1, \dots, \mu_m$, say, and

$$\sum_{\mu<-1} \dim \ker (T^*T - (1+\mu)) = \sum_{j=1}^{m} \dim \ker (T^*T - (1+\mu_j)) \le \operatorname{def} T.$$

In the following we choose μ such that $\max_{j=1,\dots,m} \mu_j < \mu < -1$. Then

$$\sum_{\lambda<0} \dim \ker (T^*T - \lambda) = \dim \mathcal{L}_\mu. \tag{2.6}$$

We may apply Lemma 2.3 to the operator $T_1 \in \mathbf{L}(\mathcal{L}_\mu^\perp, (T\mathcal{L}_\mu)^\perp)$, and we obtain

$$\operatorname{def} T_1 = \dim \left((\ker T_1 T_1^*)^+ \dot{+} (\ker T_1 T_1^*)^0 \right). \tag{2.7}$$

The formula (2.2) for the defect of T now follows from (2.5), (2.6) and (2.7) and the observation that $\ker T_1 T_1^* = \ker TT^*$. □

Remark 2.4. Let $T : \mathcal{K}_0 \to \mathcal{K}_1$ be a noncontractive operator with arbitrary defect, and assume there exists an $\varepsilon > 0$ such that $(-\varepsilon, 0) \subset \rho(T^*T)$. If also

$$(\ker T^*)^+ + (\ker T^*)^0 = \{0\},$$

then for $\mu \in (-1-\varepsilon, -1)$, the subspaces $\mathcal{L}_\mu$ and $T\mathcal{L}_\mu$ in the decompositions (2.3) are uniformly negative and uniformly positive, respectively, and T_1 in (2.4) is a bi–noncontractive operator, and hence $\dim \mathcal{L}_\mu = \operatorname{def} T$. We note that if $T : \mathcal{K}_0 \to \mathcal{K}_1$ is a noncontractive operator and $\operatorname{ran} T = \mathcal{K}_1$, then there exists an $\varepsilon > 0$ such that $(-\varepsilon, 0) \cup (0, \varepsilon) \subset \rho(T^*T)$. Indeed, from $\operatorname{ran} T = \mathcal{K}_1$, it follows that $\operatorname{ran} T^*$ is closed and $\ker T^* = \{0\}$. Since $\ker T$ is uniformly negative, $\mathcal{K}_1 = \operatorname{ran} T^* + \ker T$, hence $\operatorname{ran} TT^* = \mathcal{K}_1$. The injectivity of TT^* implies that $0 \in \rho(TT^*)$. The result now follows from the fact that outside the point 0 the resolvent sets of TT^* and of T^*T coincide.

3. The defect between two operators

We first recall that a point $\lambda \in \mathbb{C}$ is a normal eigenvalue of an operator A, if the root subspace of this operator in λ is finite dimensional, λ is an isolated point of the spectrum $\sigma(A)$ of A and $\operatorname{ran}(A-\lambda) = \overline{\operatorname{ran}}(A-\lambda)$. If A is a selfadjoint operator in a Hilbert space then λ is an isolated point of $\sigma(A)$ if and only if $\operatorname{ran}(A-\lambda) = \overline{\operatorname{ran}}(A-\lambda)$. The set of all normal eigenvalues of A will be denoted by $\sigma_{np}(A)$; the points in the set $\tilde{\rho}(A) = \sigma_{np}(A) \cup \rho(A)$ are called the normal points of A. In this section we consider a Hilbert space $(\mathcal{H}, (\cdot,\cdot))$ and two selfadjoint operators A_0 and A_1 in $\mathbf{L}(\mathcal{H})$ with $A_0 \leq A_1$ and $0 \in \tilde{\rho}(A_0) \cap \tilde{\rho}(A_1)$. We want to define the defect $\operatorname{def} A_1/A_0$ of A_0 in A_1, and for this we first consider a special case.

(I) *Assume* $0 \in \rho(A_0) \cap \rho(A_1)$.
Then for $i = 0, 1$, the space $\mathcal{K}_i = (\mathcal{H}, [\cdot,\cdot]_i = (A_i\cdot,\cdot))$ is a Kreĭn space, and the decomposition $\mathcal{K}_i = \mathcal{K}_i^+ \oplus \mathcal{K}_i^-$, where $\mathcal{K}_i^{\pm}$ is the spectral subspace of the operator A_i related to the set $\sigma(A_i) \cap \mathbb{R}^{\pm}$, $i = 0, 1$, is fundamental. We define the defect of A_0 in A_1 by

$$\operatorname{def} A_1/A_0 = \dim \mathcal{K}_1^+ \cap \mathcal{K}_0^-.$$

Associated with the operators A_0 and A_1 is the operator $T : \mathcal{K}_0 \to \mathcal{K}_1$ with $Tx = x$. Its adjoint is given by $T^* = A_0^{-1}A_1$, and T is noncontractive operator. We have

$$\operatorname{def} A_1/A_0 = \operatorname{def} T. \tag{3.1}$$

To see (3.1), consider the subspace $\mathcal{K}_0^+ \dot{+} \mathcal{K}_1^-$ in $\mathcal{K}_1$. Since $\mathcal{K}_1^-$ is a maximal uniformly negative subspace and $\mathcal{K}_0^+$ is a nonnegative subspace,

$$\operatorname{def} T = \dim \mathcal{K}_1/(\mathcal{K}_0^+ \dot{+} \mathcal{K}_1^-) = \dim (\mathcal{K}_0^+ \dot{+} \mathcal{K}_1^-)^{\perp} = \dim (\mathcal{K}_0^+)^{\perp} \cap (\mathcal{K}_1^-)^{\perp} = \operatorname{def} A_1/A_0.$$

Remark 3.1. In the proof of the equality (3.1) we have used only that $\mathcal{K}_0^+$ is a maximal nonnegative subspace of $\mathcal{K}_0$ and $\mathcal{K}_1^-$ is a maximal uniformly negative subspace of $\mathcal{K}_1$. Therefore

$$\operatorname{def} A_1/A_0 = \dim \mathcal{M}_1^+ \cap \mathcal{M}_0^-, \tag{3.2}$$

where $\mathcal{M}_1^+$ is an arbitrary maximal uniformly positive subspace of $\mathcal{K}_1$ and $\mathcal{M}_0^-$ is an arbitrary maximal nonpositive subspace of $\mathcal{K}_0$.
(II) *Assume* $0 \in \tilde{\rho}(A_0) \cap \tilde{\rho}(A_1)$ *(the general case).*
For $i = 0, 1$, the spaces $\mathcal{K}_i = (\mathcal{H}, [\cdot,\cdot]_i = (A_i\cdot,\cdot))$ may be degenerate in which case they can be decomposed as the direct sum

$$\mathcal{K}_i = \mathcal{K}_i^0 \dot{+} \mathcal{K}_i^1,$$

where $\mathcal{K}_i^0 = \mathcal{K}_i \cap \mathcal{K}_i^{\perp} = \ker A_i$ is the finite dimensional isotropic part of $\mathcal{K}_i$ and $\mathcal{K}_i^1$ is a Kreĭn space, orthogonal to $\mathcal{K}_i^0$ in the inner products of $\mathcal{H}$ and $\mathcal{K}_i$. By P_i we denote the projection in $\mathcal{K}_i$ onto $\mathcal{K}_i^1$ parallel to $\mathcal{K}_i^0$. The operator $T : \mathcal{K}_0 \to \mathcal{K}_1$ with $Tx = x$ is noncontractive, and so is the operator $V = P_1TP_0 \in \mathbf{L}(\mathcal{K}_0^1, \mathcal{K}_1^1)$. If $\mathcal{L}_0$ and $\mathcal{L}_1$ are maximal nonnegative subspaces of $\mathcal{K}_0$ and $\mathcal{K}_1$, respectively (hence $\mathcal{K}_i^0 \subset \mathcal{L}_i$), and if $\mathcal{L}_0 \subseteq \mathcal{L}_1$, then

$$\dim \mathcal{L}_1/\mathcal{L}_0 = \operatorname{def} V + \dim \mathcal{K}_1^0 - \dim \mathcal{K}_0^0.$$

This shows that the number on the lefthand side is independent of the chosen maximal subspaces, and so we may define the defect of the operator A_0 in the operator A_1 and the defect of the operator T by

$$\operatorname{def} A_1/A_0 = \operatorname{def} T = \dim \mathcal{L}_1/\mathcal{L}_0. \tag{3.3}$$

If $G \in \mathbf{L}(\mathcal{H})$ is a uniformly positive operator then the inner product $(G\cdot,\cdot)$ is called the G–inner product on $\mathcal{H}$; if B_0 and B_1 are selfadjoint operators and $B_0 \le B_1$ in the G–inner product, then we denote by $\operatorname{def}_G B_1/B_0$ the defect of B_0 in B_1 with respect to the G–inner product.

Lemma 3.2. *Assume that $0 \in \rho(A_0) \cap \rho(A_1)$. There exists a uniformly positive operator $G \in \mathbf{L}(\mathcal{H})$ such that the space $\mathcal{H}$ admits a G–orthogonal decomposition $\mathcal{H} = \mathcal{H}_0 \oplus \mathcal{H}_1$ in which the subspaces $\mathcal{H}_0$ and $\mathcal{H}_1$ are $G^{-1}A_i$–invariant, $i = 0, 1$. If $B_i^j = G^{-1}A_i|_{\mathcal{H}_j}$, $i, j = 0, 1$, then B_0^0 is G–uniformly negative, B_1^0 is G–uniformly positive and $\operatorname{def}_G B_1^1/B_0^1 = 0$.*

Proof. We use that $T^* = T^*T = TT^* = A_0^{-1}A_1$. Since $0 \in \rho(A_0^{-1}A_1)$, there exists an $\varepsilon > 0$ such that $(-\varepsilon, 0) \subset \rho(A_0^{-1}A_1)$. Consider the first decomposition in (2.3):

$$\mathcal{H} = \mathcal{L}_\mu \oplus \mathcal{L}_\mu^\perp \tag{3.4}$$

with $\mu \in (-1-\varepsilon, -1)$ and introduce in $\mathcal{H}$ a new Hilbert inner product:

$$(x, y)_1 = (x_1, y_1) + (x_2, y_2),\ x = x_1 + x_2,\ y = y_1 + y_2,\ x_1, y_1 \in \mathcal{L}_\mu,\ x_2, y_2 \in \mathcal{L}_\mu^\perp.$$

It is generated by a uniformly positive operator $G \in \mathbf{L}(\mathcal{H})$: $(x, y)_1 = (Gx, y)$. Hence $[x, y]_i = (A_i x, y) = (G^{-1}A_i x, y)_1$, $i = 0, 1$. Now the decomposition (3.4) is $(\cdot,\cdot)_1$– as well as $[\cdot,\cdot]_i$–orthogonal, $i = 0, 1$. It follows that the subspaces $\mathcal{H}_0 = \mathcal{L}_\mu$ and $\mathcal{H}_1 = \mathcal{L}_\mu^\perp$ are $G^{-1}A_i$–invariant, $i = 0, 1$. The subspace $\mathcal{H}_0$ is $A_0^{-1}A_1$–invariant, uniformly $[\cdot,\cdot]_0$-negative and uniformly $[\cdot,\cdot]_1$–positive (recall that $\mathcal{L}_\mu = T\mathcal{L}_\mu$), and by Remark 2.4, the operator $T|_{\mathcal{H}_1}$ is bi-noncontractive. Thus B_0^0 and B_1^0 are uniformly negative and uniformly positive operators, respectively, and $\operatorname{def}_G B_1^1/B_0^1 = 0$. □

Remark 3.3. By Lemma 3.2, we can assume below without loss of generality that $G = I$, that is, $G^{-1}A_i = A_i$, $i = 0, 1$.

Theorem 3.4. *Assume that $0 \in \rho(A_0) \cap \rho(A_1)$ and let $\kappa \in \{0, 1, \ldots\}$. The following statements are equivalent:*
(a) $\operatorname{def} A_1/A_0 = \kappa$.
(b) *For every $A = A^* \in \mathbf{L}(\mathcal{H})$ with $A_0 \le A \le A_1$, the point $\lambda = 0$ belongs to $\tilde\rho(A)$ and $\dim\ker A \le \kappa$, and there exists an operator $A' = A'^* \in \mathbf{L}(\mathcal{H})$ with $A_0 \le A' \le A_1$ such that $\dim\ker A' = \kappa$.*
(c) *For every nondecreasing continuous function A on $[0, 1]$ whose values are selfadjoint operators in $\mathbf{L}(\mathcal{H})$ with $A_0 = A(0) \le A(t) \le A(1) = A_1$, the equality*

$$\dim \operatorname{span} \{\ker A(t) \mid 0 \le t \le 1\} = \kappa \tag{3.5}$$

holds and $0 \in \tilde\rho(A(t))$ for all $t \in [0, 1]$.
(d) *There exists a nondecreasing continuous function A on $[0, 1]$ whose values are selfadjoint operators in $\mathbf{L}(\mathcal{H})$ with $A_0 = A(0) \le A(t) \le A(1) = A_1$, such that the equality (3.5) holds and $0 \in \tilde\rho(A(t))$ for all $t \in [0, 1]$.*

If (a)–(d) *are valid, then (3.5) can be replaced by*

$$\sum_{t\in[0,1]} \dim\ker A(t) = \kappa \tag{3.6}$$

if and only if there exists at most a finite set $\{t_1, t_2, \ldots, t_m\} \subset [0,1]$ *with* $m \leq \kappa$ *such that* $0 \in \sigma_{np}(A(t_i)), i = 1, 2, \ldots, m$.

Proof (a) $\Leftrightarrow$ (b). Assume $A \in \mathbf{L}(\mathcal{H})$ satisfies $A_0 \leq A = A^* \leq A_1$. According to Lemma 3.2 and Remark 3.3, $\mathcal{H}$ can be written as the orthogonal sum $\mathcal{H} = \mathcal{H}_0 \oplus \mathcal{H}_1$, which is also $[\cdot,\cdot]_i$–orthogonal, $i = 0, 1$, such that for $i = 0, 1$, A_i has the matrix representation

$$A_i = \begin{pmatrix} A_i^0 & 0 \\ 0 & A_i^1 \end{pmatrix} : \begin{pmatrix} \mathcal{H}_0 \\ \mathcal{H}_1 \end{pmatrix} \longrightarrow \begin{pmatrix} \mathcal{H}_0 \\ \mathcal{H}_1 \end{pmatrix},$$

in which A_0^0 is uniformly negative, A_1^0 is uniformly positive, and $\operatorname{def} A_1^1/A_0^1 = 0$. Hence $\operatorname{def} A_1/A_0 = \dim \mathcal{H}_0$. If we write

$$A = \begin{pmatrix} A^0 & A_{01} \\ A_{01}^* & A^1 \end{pmatrix} : \begin{pmatrix} \mathcal{H}_0 \\ \mathcal{H}_1 \end{pmatrix} \longrightarrow \begin{pmatrix} \mathcal{H}_0 \\ \mathcal{H}_1 \end{pmatrix},$$

then $A_0^1 \leq A^1 \leq A_1^1$ and, on account of Theorem 1.1, $0 \in \rho(A^1)$. If (a) holds then $\dim \mathcal{H}_0 = \kappa < \infty$ and $0 \in \tilde{\rho}(A)$. From

$$\ker A = \left\{ \begin{pmatrix} x \\ -(A^1)^{-1} A_{01}^* x \end{pmatrix} \mid x \in \ker\left(A^0 - A_{01}(A^1)^{-1}A_{01}^*\right) \right\}.$$

and (a) it follows that $\dim\ker A \leq \dim \mathcal{H}_0 = \kappa$. The operator

$$A' = \begin{pmatrix} 0 & 0 \\ 0 & A^1 \end{pmatrix} : \begin{pmatrix} \mathcal{H}_0 \\ \mathcal{H}_1 \end{pmatrix} \longrightarrow \begin{pmatrix} \mathcal{H}_0 \\ \mathcal{H}_1 \end{pmatrix}$$

has the properties mentioned in (b). This proves the implication (a) $\Rightarrow$ (b).
To prove (b) $\Rightarrow$ (a), it is sufficient to show that $\operatorname{def} A_1/A_0 < \infty$ for then we can use the implication (a) $\Rightarrow$ (b). Assume that $\operatorname{def} A_1/A_0 = \infty$. Then $\dim \mathcal{H}_0 = \infty$ and if we take

$$A' = \begin{pmatrix} 0 & 0 \\ 0 & A_1^1 \end{pmatrix}, \tag{3.7}$$

then $A_0 \leq A' = A'^* \leq A_1$, $\dim\ker A' = \infty$, and hence $0 \notin \tilde{\rho}(A')$. This contradicts the assumption (b). So $\operatorname{def} A_1/A_0 < \infty$.
(a) $\Rightarrow$ (c). Let the operator valued function $A(t)$ be as in (c). From the equivalence between (a) and (b) we have $0 \in \tilde{\rho}(A(t))$. Hence the space $\mathcal{K}_t = (\mathcal{H}, [\cdot,\cdot]_t = (A(t)\cdot,\cdot))$ can be degenerate, and the defect of the identity operator $T_{ts} : \mathcal{K}_s \to \mathcal{K}_t$, $0 \leq s \leq t \leq 1$, is defined in (II) directly after Remark 3.1:

$$\operatorname{def} T_{ts} = \dim \mathcal{L}_t / \mathcal{L}_s,$$

where $\mathcal{L}_s$ and $\mathcal{L}_t$ are maximal nonnegative subspaces of $\mathcal{K}_s$ and $\mathcal{K}_t$, respectively with $\mathcal{L}_s \subset \mathcal{L}_t$. If $0 \le r \le s \le t \le 1$, then

$$\operatorname{def} T_{tr} = \operatorname{def} T_{ts} + \operatorname{def} T_{sr}. \tag{3.8}$$

Later on we use the following implications:

$$0 \le r \le s \le t \le 1 \Rightarrow \ker A(r) \cap \ker A(t) \subseteq \ker A(s). \tag{3.9}$$

In particular,

$$\left.\begin{array}{l} 0 \le r \le s \le t \le 1 \\ \{0\} \ne \ker A(t) \subseteq \ker A(r) \end{array}\right\} \Rightarrow \ker A(t) \subseteq \ker A(s). \tag{3.10}$$

To prove (3.9), let x belong to the intersection. Then, as $A(r) \le A(s) \le A(t)$, $(A(s)x, x) = 0$. From $A(s) - A(r) \ge 0$ and $((A(s) - A(r))x, x) = 0$, we get $x \in \ker (A(s) - A(r))$ and hence $x \in \ker A(s)$.
(A direct proof of (3.10) runs as follows: With respect to the decomposition $\mathcal{H} = \ker A(t) \oplus \overline{\operatorname{ran}} A(t)$ we have

$$A(r) = \begin{pmatrix} 0 & 0 \\ 0 & A_{11}(r) \end{pmatrix}$$
$$\le A(s) = \begin{pmatrix} A_{00}(s) & A_{01}(s) \\ A_{01}(s)^* & A_{11}(s) \end{pmatrix} \le A(t) = \begin{pmatrix} 0 & 0 \\ 0 & A_{11}(t) \end{pmatrix}.$$

These inequalities imply that $A_{00}(s) = 0$ and $A_{01}(s) = 0$. Hence (3.10) holds.)

We claim that for every $t \in (0,1)$, there is an $\varepsilon_t > 0$ such that $(t - \varepsilon_t, t + \varepsilon_t) \subset (0,1)$ and

$$\ker A(s) \cap \ker A(t) \text{ is independent of } s \in (t - \varepsilon_t, t), \tag{3.11}$$

$$\operatorname{def} T_{ts} = \dim \ker A(t) - \dim (\ker A(s) \cap \ker A(t)), \quad s \in (t - \varepsilon_t, t), \tag{3.12}$$

$$\ker A(s) \subseteq \ker A(t), \quad s \in (t, t + \varepsilon_t), \tag{3.13}$$

$$\operatorname{def} T_{st} = 0, \quad s \in (t, t + \varepsilon_t). \tag{3.14}$$

For the points $t = 0$ and $t = 1$ we have that there is an $\varepsilon_0 \in (0,1)$ such that for $s \in (0, \varepsilon_0)$,

$$\ker A(s) = \{0\}, \quad \operatorname{def} T_{s0} = 0,$$

and also there is an $\varepsilon_1 \in (0,1)$ such that for $s \in (1 - \varepsilon_1, 1)$,

$$\ker A(s) = \{0\}, \quad \operatorname{def} T_{1s} = 0.$$

The last two statements follow from Theorem 1.1 in the Introduction, but can also be proved using the arguments below. Before proving the claim, we first show how it is used to prove the implication (a) $\Rightarrow$(c). From (3.8) and (3.12) it follows that there are at most κ points $t \in (0,1)$ for which the number $\operatorname{def} T_{ts}$ in (3.12) is not zero: If there are more than κ points t_i with this property then for any $s_i \in (t_i - \varepsilon_{t_i}, t_i)$ sufficiently close to t_i,

$$\kappa < \sum_i \operatorname{def} T_{t_i s_i} \leq \operatorname{def} T_{10} = \kappa.$$

Let

(3.15) $$0 < t_1 < t_2 < \ldots < t_k < 1$$

be the unique points for which the number $\operatorname{def} T_{t_i s}$ in (3.12) is not zero. Then, as we shall show in a moment, for $i = 1, \ldots k+1$,

(3.16) $$\ker A(s) \subseteq \ker A(t_{i-1}), \quad s \in [t_{i-1}, t_i),$$

(3.17) $$\ker A(s) \cap \ker A(t_i) = \ker A(t_{i-1}) \cap \ker A(t_i), \quad s \in [t_{i-1}, t_i),$$

(3.18) $$\operatorname{def} T_{st_{i-1}} = 0, \quad s \in [t_{i-1}, t_i).$$

Here we have set $t_0 = 0$ and $t_{k+1} = 1$. We select points $t_i' \in (t_i - \varepsilon_{t_i}, t_i), i = 1, \ldots, k+1$, such that

$$0 < t_1' < t_1 < t_2' < t_2 < \ldots < t_k' < t_k < t_{k+1}' < 1.$$

From (3.8), (3.12), (3.17) and (3.18) with $r = t_{i-1}$, $s = t_i'$, $t = t_i$, we see that for $i = 1, \ldots, k+1$,

$$\operatorname{def} T_{t_i t_{i-1}} = \dim \ker A(t_i) - \dim\,(\ker A(t_{i-1}) \cap \ker A(t_i))$$

and hence

$$\begin{aligned}
\kappa &= \operatorname{def} T_{10} = \textstyle\sum_{i=1}^{k} \operatorname{def} T_{t_i t_{i-1}} \\
&= \textstyle\sum_{i=1}^{k} \dim \ker A(t_i) - \dim\,(\ker A(t_{i-1}) \cap \ker A(t_i)) \\
&\overset{(1)}{=} \dim \operatorname{span}\,\{\ker A(t_i) \ominus (\ker A(t_{i-1}) \cap \ker A(t_i)) \mid i = 1, 2 \ldots, k\} \\
&= \dim \operatorname{span}\,\{\ker A(t_i) \mid i = 1, 2, \ldots, k\} \\
&\overset{(2)}{=} \dim \operatorname{span}\,\{\ker A(t) \mid t \in [0,1]\}.
\end{aligned}$$

The equality $\overset{(1)}{=}$ follows from (3.9), as it implies that the spaces

$$\ker A(t_i) \ominus (\ker A(t_{i-1}) \cap \ker A(t_i)), \quad i = 1, 2 \ldots, k,$$

on the righthand side are disjoint. To prove the equality $\overset{(2)}{=}$, it suffices to show that if $\ker A(s) \neq \{0\}$, then

$$\ker A(s) \subseteq \operatorname{span}\,\{\ker A(t_i) \mid i = 1, \ldots, k\}.$$

If this inclusion does not hold then for all $i = 1, \ldots, k+1$,

$$\ker A(s) \cap \ker A(t_{i-1}) \neq \ker A(s).$$

But for i such that $s \in (t_{i-1}, t_i)$ this inequality is in contradiction with (3.16). This proves $\overset{(2)}{=}$.
It remains to prove the claim and (3.16)–(3.18).
We fix a $t \in (0,1)$. For $s \in (0,1)$ we consider the direct sum decomposition

$$\mathcal{K}_s = \mathcal{K}_t^+ \dot{+} \mathcal{K}_t^- \dot{+} \mathcal{K}_t^0. \tag{3.19}$$

There is a $k_t > 0$ such that for all $x \in \mathcal{K}_t^+$,

$$(A(s)x, x) = (A(t)x, x) + ((A(s) - A(t))x, x) \geq k_t\|x\|^2 - \|A(s) - A(t)\|\|x\|^2.$$

By the continuity of A, for s sufficiently close to t, the space $\mathcal{K}_t^+$ is uniformly positive in $\mathcal{K}_s$, and, as can be shown in a similar way, the space $\mathcal{K}_t^-$ is uniformly negative in $\mathcal{K}_s$. We only consider such s.
Assume $s < t$, then $A(s) \leq A(t)$ implies that $\mathcal{K}_t^- \dot{+} \mathcal{K}_t^0$ is a nonpositive subspace in $\mathcal{K}_s$. Hence in the decomposition (3.19) the spaces $\mathcal{K}_t^+$ and $\mathcal{K}_t^- \dot{+} \mathcal{K}_t^0$ are maximal uniformly positive and maximal nonpositive, respectively. We note that in the decomposition

$$\mathcal{K}_s = \left(\mathcal{K}_t^+ \dot{+} (\mathcal{K}_t^0 \cap \mathcal{K}_s^0)\right) \dot{+} \left(\mathcal{K}_t^- \dot{+} \mathcal{K}_t^0 \ominus (\mathcal{K}_t^0 \cap \mathcal{K}_s^0)\right).$$

the first summand $\mathcal{K}_t^+ \dot{+} (\mathcal{K}_t^0 \cap \mathcal{K}_s^0)$ is a maximal nonnegative subspace of $\mathcal{K}_s$. If this is not the case then there exists a maximal nonnegative subspace $\mathcal{L}$ of $\mathcal{K}_s$ which contains $\mathcal{K}_t^+ \dot{+} (\mathcal{K}_t^0 \cap \mathcal{K}_s^0)$. Choose $x \in \mathcal{L} \setminus \left(\mathcal{K}_t^+ \dot{+} (\mathcal{K}_t^0 \cap \mathcal{K}_s^0)\right)$ such that

$$x = x^- + x^0, \quad x^- \in \mathcal{K}_t^-, \quad x^0 \in \mathcal{K}_t^0 \ominus (\mathcal{K}_t^0 \cap \mathcal{K}_s^0).$$

Then

$$0 \leq [x, x]_s \leq [x, x]_t = [x^-, x^-]_t \leq 0.$$

So $x^- = 0$, $x = x^0$, and $[x, x]_s = 0$. Since $x \in \mathcal{L}$ and $\mathcal{L}$ is nonnegative, $x \in \mathcal{L}^{\perp_s}$, the orthogonal complement of $\mathcal{L}$ in $\mathcal{K}_s$. Also, since $x \in \mathcal{K}_t^0 \ominus (\mathcal{K}_t^0 \cap \mathcal{K}_s^0)$ and $\mathcal{K}_t^- \dot{+} \mathcal{K}_t^0 \ominus (\mathcal{K}_t^0 \cap \mathcal{K}_s^0)$ is nonpositive, $x \in (\mathcal{K}_t^- \dot{+} \mathcal{K}_t^0 \ominus (\mathcal{K}_t^0 \cap \mathcal{K}_s^0))^{\perp_s}$. Hence $x \in \mathcal{K}_s^0$, that is, $x = 0$. This contradiction implies that $\mathcal{K}_t^+ \dot{+} (\mathcal{K}_t^0 \cap \mathcal{K}_s^0)$ is a maximal nonnegative subspace of $\mathcal{K}_s$ and hence

$$\operatorname{def} T_{ts} = \dim \mathcal{K}_t^0 - \dim (\mathcal{K}_t^0 \cap \mathcal{K}_s^0)$$

from which (3.12) follows. By (3.9), $\ker A(s) \cap \ker A(t)$ is increasing as $s \nearrow t$ and contained in $\ker A(t)$, and therefore we can choose ε_t so small that $\ker A(s) \cap \ker A(t)$ is independent of $s \in (t - \varepsilon_t, t)$. This proves (3.11).
Assume $t < s$ and consider (3.19). Then for s sufficiently close to t, $\mathcal{K}_t^+ \dot{+} \mathcal{K}_t^0$ is maximal nonnegative (and $\mathcal{K}_t^-$ is maximal uniformly negative) in $\mathcal{K}_s$. Therefore (3.14) holds and

$$\mathcal{K}_s^0 = (\mathcal{K}_t^+ \dot{+} \mathcal{K}_t^0) \cap (\mathcal{K}_t^+ \dot{+} \mathcal{K}_t^0)^{\perp_s}.$$

This equality implies that $\mathcal{K}_s^0 \subseteq \mathcal{K}_t^0$: If $x \in \mathcal{K}_s^0$ and $x = x^+ + x^0$, $x^+ \in \mathcal{K}_t^+$, $x^0 \in \mathcal{K}_t^0$, then

$$0 \leq [x^+, x^+]_t = [x, x]_t \leq [x, x]_s = 0,$$

which shows that $x^+ = 0$ and $x = x^0 \in \mathcal{K}_t^0$. Hence $\ker A(s) \subseteq \ker A(t)$. This proves (3.13) and completes the proof of the claim.
To prove (3.16), we define

$$\tilde{t} = \sup\{t \in [t_{i-1}, t_i] \mid \ker A(s) \subseteq \ker A(t_{i-1}), s \in [t_{i-1}, t)\}.$$

By (3.13), $\tilde{t} \in (t_{i-1}, t_i]$. We claim $\tilde{t} = t_i$. Assume $\tilde{t} < t_i$. On account of (3.10) with $r = t_{i-1}$,

$$\ker A(t) \subseteq \ker A(s) \subseteq \ker A(t_{i-1}), \quad t_{i-1} \leq s \leq t < \tilde{t}.$$

It follows that for t sufficiently close to $\tilde{t}$, $\ker A(t) = \mathcal{N}$, say, is independent of t. The continuity of A implies that $\mathcal{N} \subseteq \ker A(\tilde{t})$. If equality prevails, then by (3.13), there is an $\tilde{\varepsilon} > 0$ such that

$$\ker A(s) \subseteq \ker A(\tilde{t}) = \mathcal{N} \subseteq \ker A(t_{i-1}), \quad s \in [\tilde{t}, \tilde{t} + \tilde{\varepsilon}).$$

But this contradicts the definition of $\tilde{t}$ as a supremum. Hence $\mathcal{N}$ is properly contained in $\ker A(\tilde{t})$, and in a small left neighborhood of $\tilde{t}$,

$$\dim \ker A(\tilde{t}) - \dim \left(\ker A(\tilde{t}) \cap \ker A(t)\right) \neq 0.$$

This implies $\tilde{t} = t_i$.
(3.17) follows from (3.16) and (3.9).
We prove (3.18). Define

$$\tilde{s} = \sup\{t \in [t_{i-1}, t_i] \mid \operatorname{def} T_{tt_{i-1}} = 0\}.$$

From (3.14) we have $\tilde{s} > t_{i-1}$ and $\operatorname{def} T_{\tilde{s}t_{i-1}} \neq 0$. On account of (3.8),

$$\operatorname{def} T_{tt_{i-1}} = 0, \quad t_{i-1} \leq t \leq \tilde{s},$$

which implies that for s in a small left neighborhood of $\tilde{s}$, $\operatorname{def} T_{\tilde{s}s} \neq 0$. Now (3.12) implies that $\tilde{s} = t_i$, that is, (3.18) holds.
(c) $\Rightarrow$ (d). The function $A(t) = (1-t)A_0 + tA_1$ has the desired properties.
(d) $\Rightarrow$ (a). This implication and the last statement in the theorem can be proved using arguments as in the proof of (a) $\Rightarrow$ (c). The details are left to the reader. □

Remark 3.5. Assume that $0 \in \rho(A_0) \cap \rho(A_1)$. If there exists a nondecreasing continuous function A on $[0,1]$ whose values are selfadjoint operators in $\mathbf{L}(\mathcal{H})$ with $A_0 = A(0) \leq A(t) \leq A(1) = A_1$, such that for all $t \in [0,1]$, $0 \in \tilde{\rho}(A(t))$, then $\operatorname{def} A_1/A_0$ is finite. Indeed, from the proof of Theorem 3.4 it follows that for every t there is an ε-neighborhood such that $\operatorname{def} A(\tau)/A(t) = 0$ for $\tau \in (t, t+\varepsilon)$ and

$\operatorname{def} A(t)/A(\tau)$ is constant for $\tau \in (t-\varepsilon, t)$. Therefore the function $\operatorname{def} A(t)/A(0)$ has at most finitely many jumps. Hence $\operatorname{def} A(1)/A(0) < \infty$.

Corollary 3.6. *Let A be a nondecreasing continuous function on $[0,1]$ whose values are selfadjoint operators in $\mathbf{L}(\mathcal{H})$ with $0 \in \tilde{\rho}(A(t))$ for all $t \in [0,1]$ and let $T : (\mathcal{H}, (A(0)\cdot,\cdot)) \to (\mathcal{H}, (A(1)\cdot,\cdot))$ be the identity operator. Then*

$$\dim \operatorname{span}\{\ker A(t) \mid 0 \le t \le 1\} = \operatorname{def} T + \dim\ker A(0). \tag{3.20}$$

Moreover,

$$\sum_{t\in[0,1]} \dim\ker A(t) = \operatorname{def} T + \dim\ker A(0) \tag{3.21}$$

if and only if there exists at most a finite set $\{t_1, t_2, \ldots, t_m\} \subset [0,1]$ with $m \le 1 + \operatorname{def} T$ such that $0 \in \sigma_{np}(A(t_i)), i = 1, 2, \ldots, m$.

Proof. Since $0 \in \tilde{\rho}(A(t))$ there is an $\varepsilon > 0$ such that $(0, \varepsilon] \subset \rho(A(0))$ and $[-\varepsilon, 0) \subset \rho(A(1))$. Consider the operator function $\widetilde{A}$:

$$\widetilde{A}(t) = \begin{cases} A(0) + t, & t \in [-\varepsilon, 0), \\ A(t), & t \in [0,1], \\ A(1) + t - 1, & t \in (1, 1+\varepsilon]. \end{cases}$$

$\widetilde{A}$ is a nondecreasing function on $[-\varepsilon, 1+\varepsilon]$ and

$$\dim \operatorname{span}\{\ker \widetilde{A}(t) \mid -\varepsilon \le t \le 1+\varepsilon\} = \dim \operatorname{span}\{\ker A(t) \mid 0 \le t \le 1\}.$$

From (3.9) it follows that the subspaces $\mathcal{K}^0 = \ker A(0) \cap \ker A(1)$ and $\mathcal{K}^1 = \mathcal{K}^{0\perp}$ are $\widetilde{A}(t)$–invariant for all $t \in [-\varepsilon, 1+\varepsilon]$. We write $\widetilde{A}(t)^i = \widetilde{A}(t)|_{\mathcal{K}^i}$, $i = 0, 1$. Then

$$\operatorname{def} \widetilde{A}(1+\varepsilon)/\widetilde{A}(-\varepsilon) = \operatorname{def} \widetilde{A}(1+\varepsilon)^0/\widetilde{A}(-\varepsilon)^0 + \operatorname{def} \widetilde{A}(1+\varepsilon)^1/\widetilde{A}(-\varepsilon)^1.$$

By definition of $\mathcal{K}^0$ we have

$$\operatorname{def} \widetilde{A}(1+\varepsilon)^0/\widetilde{A}(-\varepsilon)^0 = \dim \mathcal{K}^0$$

and

$$\operatorname{def} \widetilde{A}(1+\varepsilon)^1/\widetilde{A}(-\varepsilon)^1 = \operatorname{def} T + \dim(\ker A(0) \ominus \mathcal{K}^0).$$

The first equality follows from the operator equalities $\widetilde{A}(-\varepsilon)^0 = -\varepsilon < 0$ and $\widetilde{A}(1+\varepsilon)^0 = \varepsilon > 0$. To prove the second equality, we use the formula (3.8):

$$\begin{aligned} &\operatorname{def} \widetilde{A}(1+\varepsilon)^1/\widetilde{A}(-\varepsilon)^1 = \\ &\quad = \operatorname{def} \widetilde{A}(1+\varepsilon)^1/\widetilde{A}(1)^1 + \operatorname{def} \widetilde{A}(1)^1/\widetilde{A}(0)^1 + \operatorname{def} \widetilde{A}(0)^1/\widetilde{A}(-\varepsilon)^1 \\ &\quad = 0 + \operatorname{def} T|_{\mathcal{K}^1} + \dim(\ker A(0) \ominus \mathcal{K}^0) \\ &\quad = \operatorname{def} T + \dim(\ker A(0) \ominus \mathcal{K}^0). \end{aligned}$$

The equalities (3.20) and (3.21) now follow from (3.5) and (3.6) if we replace there A and $[0,1]$ by $\widetilde{A}$ and $[-\varepsilon, 1+\varepsilon]$, respectively. □

Corollary 3.7. *If A is a nondecreasing real analytic function on $(-\varepsilon, 1+\varepsilon), \varepsilon > 0$, whose values are selfadjoint operators in $\mathbf{L}(\mathcal{H})$ with $0 \in \rho(A(0)) \cap \rho(A(1))$ and* $\operatorname{def} A(1)/A(0) = \kappa < \infty$, *then*

$$\sum_{t \in [0,1]} \dim \ker A(t) = \kappa.$$

Proof . In the proof of the implications (a) $\Rightarrow$ (c) of Theorem 3.4 we have shown that there is a unique finite set $\{t_i\}_1^k$ ordered as in 3.15 such that the numbers $\operatorname{def} T_{t_i s}$ in (3.12) is not zero. Since $A(t)$ is an analytic function, $0 \in \rho(A(t))$ for all $t \in [0,1] \setminus \{t_i\}_1^k$. Indeed, if this is not the case, then there is $t \in (t_i, t_{i+1})$ such that $0 \in \sigma_{np}(A(t))$ and $\ker A(t) \subseteq \ker A(t_i)$. By (3.16) and (3.10),

$$\ker A(t) \subseteq \ker A(s) \subseteq \ker A(t_i), \quad s \in [t_i, t].$$

Since A is an analytic function, $A(s)x = 0$ for all $x \in \ker A(t)$ and $s \in [0,1]$, that is,

$$\ker A(t) \subseteq \cap\{\ker A(t) \mid t \in [0,1]\}.$$

This is impossible since $0 \in \rho(A_0) \cap \rho(A_1)$. The corollary now follows from the formula (3.6). □

Corollary 3.8. *Let A be a nondecreasing analytic function on $(-\varepsilon, 1+\varepsilon)$, $\varepsilon > 0$, whose values are Hermitian matrices acting on a finite dimensional Euclidean space $\mathcal{E}$. If $\mathcal{E}_0 = \cap\{\ker A(t) \mid t \in [0,1]\}$* and *$\mathcal{E}_1 = \mathcal{E}_0^{\perp}$ then for all $t \in [0,1]$,* $\mathcal{E}_0$ and $\mathcal{E}_1$ *are $A(t)$-invariant and*

$$\sum_{t \in [0,1]} \dim \ker A_1(t) = \frac{1}{2} \left(\operatorname{sign} A(1) - \operatorname{sign} A(0) + \dim \ker A(1) + \dim \ker A(0) \right),$$

where $A_1(t) = A(t)|_{\mathcal{E}_1}$ and $\operatorname{sign} A(t)$ *denotes the difference between the number of positive and the number of negative eigenvalues of the operator $A(t)$.*
In particular,

$$A(0) < 0 \text{ and } A(1) > 0 \Rightarrow \sum_{t \in [0,1]} \dim \ker A(t) = \dim \mathcal{E}.$$

The subspace $\mathcal{E}_0$ in this corollary coincides with the subspace $\ker A(0) \cap \ker A(1)$. This follows immediately from (3.9).

Proof of Corollary 3.8. The same arguments we have used in the proof of Corollary 3.7 yield a finite set $\{t_i\}_1^k \subset [0,1]$ such that $0 \in \sigma_p(A(t_i))$, $i = 1, 2, ..., k$, and $0 \in \rho(A(t))$ for $t \in [0,1] \setminus \{t_i\}_1^k$. The first equality of the corollary now follows from (3.21) and the formula

$$\operatorname{def} T = \frac{1}{2} \left(\operatorname{sign} A(1) - \operatorname{sign} A(0) + \dim \ker A(1) - \dim \ker A(0) \right).$$

If we write

$$\mathcal{E} = \ker A(0) + \mathcal{E}_0^+ + \mathcal{E}_0^- = \ker A(1) + \mathcal{E}_1^+ + \mathcal{E}_1^-.$$

then the formula for $\operatorname{def} T$ follows from the equalities $\operatorname{sign} A(i) = \dim \mathcal{E}_i^+ - \dim \mathcal{E}_i^-$, $i = 0, 1$, and (see (3.3))

$$\operatorname{def} T = (\dim \ker A(1) + \dim \mathcal{E}_1^+) - (\dim \ker A(0) + \dim \mathcal{E}_0^+).$$

The second equality of the corollary follows from the first one, because $\operatorname{sign} A(0) = -\dim \mathcal{E}$, $\operatorname{sign} A(1) = \dim \mathcal{E}$ and $\dim \ker A(0) = \dim \ker A(1) = 0$. □

Corollary 3.9. *Let A be a nondecreasing continuous function on $[0,1]$, whose values are Hermitian matrices acting on a finite dimensional Euclidean space $\mathcal{E}$. Assume $0 \in \rho(A(t))$, except for finitely many points $\{t_i\}_{i=1}^k \subset [0,1]$. Then $\operatorname{sign} A(t)$ is nondecreasing on $[0,1]$ and constant on (t_{i-1}, t_i). If $n(A)$ is the number of zeros of $\det A(t)$, then*

$$n(A) = \frac{1}{2}\left(\operatorname{sign} A(1) - \operatorname{sign} A(0) + \dim \ker A(1) + \dim \ker A(0)\right).$$

In particular, these results hold if A is analytic and nondecreasing on $(-\varepsilon, 1+\varepsilon)$, $\varepsilon > 0$, and $\det A(t) \not\equiv 0$.

Proof. Set $\mathcal{E} = \ker A(s) + \mathcal{E}_s^+ + \mathcal{E}_s^-$, where $\mathcal{E}_s^+$ and $\mathcal{E}_s^-$ are the subspaces of $\mathcal{E}$ on which $A(s)$ is positive and negative, respectively. Then

$$\operatorname{sign} A(s) = 2\dim \mathcal{E}_s^+ + \ker A(s) - \dim \mathcal{E}$$

which is nondecreasing because the space $\ker A(s) + \mathcal{E}_s^+$ is nonnegative in the inner product $(A(t)\cdot, \cdot)$ on $\mathcal{E}$ for $t > s$. On account of (3.3) and (3.18), we have for $t_{i-1} < s \leq t < t_i$,

$$0 = \operatorname{def} T_{ts} = \frac{1}{2}(\operatorname{sign} A(t) - \operatorname{sign} A(s)),$$

which proves that $\operatorname{sign} A(s)$ is constant on (t_{i-1}, t_i). The formula for $n(A)$ follows from Corollary (3.8). □

For related results we refer to [DM, Theorem 4]. In particular we mention formula (4.29) of [DM]: If $A(t)$, $t \in (a,b)$, is continuous, monotonically decreasing and satisfies certain additional assumptions, then this formula states that

$$\sum_{t \in (a,b)} \dim \ker A(t) = \dim \operatorname{ran} E_{A(b)}(-\infty, 0) - \dim \operatorname{ran} E_{A(a)}(-\infty, 0),$$

where $E_{A(t)}$ is the spectral measure for $A(t)$.

Example 3.10. We give an example of a continuous Hermitian *nonmonotonic* 3×3–matrix function A on $[0,1]$ such that $A(0) \leq A(t) \leq A(1)$, $0 \in \rho(A(0)) \cap \rho(A(1))$, def $A(1)/A(0) = 1$ but dim span $\{\ker A(t) \mid t \in [0,1]\} = 2$ (compare with equality (3.5)):

Let

$$A_0 = \begin{pmatrix} -1 & 0 & 0 \\ 0 & 1 & 0 \\ 0 & 0 & 1 \end{pmatrix} \quad \text{and} \quad A_1 = 2\begin{pmatrix} 1 & 0 & 0 \\ 0 & 1 & 0 \\ 0 & 0 & 1 \end{pmatrix}.$$

Then $A_0 \leq A_1$, $0 \in \rho(A_0) \cap \rho(A_1)$ and def $A_1/A_0 = 1$. If

$$B = \begin{pmatrix} 1/2 & \sqrt{3}/2 & 0 \\ \sqrt{3}/2 & 3/2 & 0 \\ 0 & 0 & 3/2 \end{pmatrix} \quad \text{and} \quad C = \begin{pmatrix} 1/2 & \sqrt{6}/4 & 0 \\ \sqrt{6}/4 & 3/2 & 3/\sqrt{8} \\ 0 & 3/\sqrt{8} & 3/2 \end{pmatrix},$$

then $A_0 \leq B \leq A_1$ and $A_0 \leq C \leq A_1$. Consider the matrix function A:

$$A(t) = \begin{cases} (1-4t)A_0 + 4tB, & 0 \leq t \leq 1/4, \\ (3/2 - 2t)B + (2t - 1/2)C, & 1/4 \leq t \leq 3/4, \\ (4-4t)C + (4t-3)A_1, & 3/4 \leq t \leq 1. \end{cases}$$

It is continuous, $A(0) = A_0, A(1) = A_1$ and $A_0 \leq A(t) \leq A_1$. Moreover, $0 \in \rho(A(t))$ for $t \in [0,1] \setminus \{1/4, 3/4\}$, $A(1/4) = B$, $A(3/4) = C$ and

$$\ker B = \operatorname{span}\left\{\begin{pmatrix} -\sqrt{3}/2 \\ 1/2 \\ 0 \end{pmatrix}\right\}, \quad \ker C = \operatorname{span}\left\{\begin{pmatrix} -\sqrt{6}/4 \\ 1/2 \\ -1/\sqrt{8} \end{pmatrix}\right\}.$$

Therefore dim span $\{\ker B, \ker C\} = 2$. □

Example 3.11. Here is an example of a continuous function A on $[0,1]$ whose values are selfadjoint operators in $\mathbf{L}(\mathcal{H})$, such that $A(0) \leq A(1)$, def $A(1)/A(0) = 1$ and $0 \in \rho(A(t)), t \in [0,1]$. In this example the condition $A(0) \leq A(t) \leq A(1), t \in [0,1]$, is not valid.

Let $\{e_i\}_{-\infty}^{\infty}$ be an orthonormal basis in the Hilbert space $\mathcal{H}$, let S be the shift operator: $Se_i = e_{i+1}, i = 0, \pm 1, \pm 2, \ldots$. Let A_0, $A_1 \in \mathbf{L}(\mathcal{H})$ be defined by

$$A_0 e_i = \begin{cases} e_i, & i = 1, 2, \ldots, \\ -e_i, & i = 0, -1, -2, \ldots, \end{cases}$$

and $A_1 = S^* A_0 S$. Then $0 \in \rho(A_0) \cap \rho(A_1)$, $A_0 \leq A_1$ and def $A_1/A_0 = 1$. By Kuiper's theorem (see [K]), there is a function U on $[0,1]$ whose values are unitary operators in $\mathbf{L}(\mathcal{H})$ such that $U(0) = I$, $U(1) = S$. If $A(t) = U(t)^* A_0 U(t)$ then $A(t) = A(t)^* \in \mathbf{L}(\mathcal{H})$, $t \in [0,1]$, $A(0) = A_0 \leq A_1 = A(1)$, def $A(1)/A(0) = 1$ and $0 \in \rho(A(t))$, $t \in [0,1]$. □

4. Unbounded selfadjoint operators

Let $\mathcal{H}$ be a Hilbert space and let $A : \mathcal{H} \to \mathcal{H}$ be a selfadjoint operator. Denote by $\mathcal{K}_A$ the space $\operatorname{dom}|A|^{1/2}$ equipped with the scalar product

$$(x,y)_A = (x_0, y_0) + (|A|^{1/2}x_1, |A|^{1/2}y_1),$$
$$x, y \in \operatorname{dom}|A|^{1/2},\ x = x_0 + x_1,\ y = y_0 + y_1,\ x_0, y_0 \in \ker A,\ x_1, y_1 \in \overline{\operatorname{ran}} A,$$

and by $[x,y]_A$ the indefinite form which is the continuous extension under $(x,y)_A$ of $[x,y]_A = (Ax,y), x,y \in \operatorname{dom} A$. The space $\mathcal{K}_A$ admits a decomposition

$$\mathcal{K}_A = \mathcal{K}_A^0 + \mathcal{K}_A^1,$$

where $\mathcal{K}_A^0$ is the dimensional isotropic part of $\mathcal{K}_A$ and $\mathcal{K}_A^1$ is a Kreĭn space if and only if $0 \in \tilde{\rho}(A)$; $\mathcal{K}_A$ is a Kreĭn space if and only if $0 \in \rho(A)$.
Let A_0 and A_1 be selfadjoint operators. We say $A_0 \preceq A_1$ if $[x,y]_{A_0} \le [x,y]_{A_1}$ for all $x \in \operatorname{dom}|A_0|^{1/2} \cap \operatorname{dom}|A_1|^{1/2}$. If $\operatorname{dom}|A_0|^{1/2} \subseteq \operatorname{dom}|A_1|^{1/2}$ we consider the operator $\tilde{T} : \mathcal{K}_{A_0} \to \mathcal{K}_{A_1}$ with $\tilde{T}x = x$, $x \in \mathcal{K}_{A_0}$.

Theorem 4.1. *Let A be a selfadjoint operator on a Hilbert space $\mathcal{H}$ and let C be a nondecreasing continuous function on $[0,1]$ whose values are selfadjoint operators in $\mathbf{L}(\mathcal{H})$, such that for $t \in [0,1]$, $0 \in \tilde{\rho}(A + C(t))$, then*

$$\dim \operatorname{span}\{\ker(A + C(t)) \mid t \in [0,1]\} = \operatorname{def}\tilde{T} + \dim\ker(A + C(0)), \tag{4.1}$$

where $\tilde{T} : \mathcal{K}_{A+C(0)} \to \mathcal{K}_{A+C(1)}$ with $\tilde{T}x = x$. Moreover if $B \in \mathbf{L}(\mathcal{H})$ is a selfadjoint operator such that $C(0) \le B \le C(1)$, then $0 \in \tilde{\rho}(A + B)$ and

$$\dim\ker(A + B) \le \operatorname{def}\tilde{T} + \dim\ker(A + C(0)). \tag{4.2}$$

Proof . We denote by P_0 the orthogonal projection from $\mathcal{H}$ on $\ker(A + C(0))$. Then $D = P_0 + |A + C(0)|^{1/2}$ is a boundedly invertible operator on $\mathcal{H}$. We now use the following well known result: If X and Y are nonnegative selfadjoint operators on a Hilbert space with $\operatorname{dom} X \subset \operatorname{dom} Y$, then $\operatorname{dom} X^{1/2} \subset \operatorname{dom} Y^{1/2}$. It implies that $\operatorname{dom}|A + C(t)|^{1/2} = \operatorname{dom}|A|^{1/2}$ is independent of $t \in [0,1]$, and hence the operator function $D^{-1}(A + C(t))D^{-1}$, $t \in [0,1]$, can be extended by continuity to a selfadjoint operator $F(t)$ in $\mathbf{L}(\mathcal{H})$: $F(t) = (D^{-1}AD^{-1})^c + D^{-1}C(t)D^{-1}$, where X^c stands for closure of the operator X. Moreover, the function F is nondecreasing and continuous on $[0,1]$, and $0 \in \tilde{\rho}(F(t))$. From Remark 3.5 it follows that $\operatorname{def} F(1)/F(0) < \infty$. We claim that $\operatorname{def} F(1)/F(0) = \operatorname{def}\tilde{T}$. To see this consider on $\mathcal{H}$ the indefinite inner product $[u,v]_t = (F(t)u,v)$. Since for $x, y \in \operatorname{dom} A$,

$$[Dx, Dy]_t = (F(t)Dx, Dy) = ((A + C(t))x, y) = [x,y]_{A+C(t)},$$

the operator D considered as an operator from $\mathcal{K}_{A+C(t)}$ to $(\mathcal{H}, [\cdot,\cdot]_t)$ is isometric. Therefore, if $\mathcal{L}_i$ is a maximal nonnegative subspace of $\mathcal{K}_{A+C(i)}$, $D\mathcal{L}_i$ is a maximal nonnegative subspace of $(\mathcal{H}, [\cdot,\cdot]_i)$, $i = 0, 1$. If $\mathcal{L}_0 \subset \mathcal{L}_1$, then

$$\operatorname{def} \tilde{T} = \dim \mathcal{L}_1/\mathcal{L}_0 = \dim D\mathcal{L}_1/D\mathcal{L}_0 = \operatorname{def} F(1)/F(0).$$

This proves the claim.

The formula (4.1) now follows from the equality $\dim\ker F(t) = \dim\ker(A+C(t))$, $t \in [0,1]$, and Corollary 3.6 applied to $F(t)$. Note that $\operatorname{def} T = \operatorname{def} F(1)/F(0) = \operatorname{def} \tilde{T}$, according to the claim just proved.

Consider the operator $F_B = (D^{-1}(A+B)D^{-1})^c$. Since $C(0) \leq B \leq C(1)$, we have $A + C(0) \preceq A + B \preceq A + C(1)$ and therefore $F(0) \leq F_B \leq F(1)$. Again we apply Corollary 3.6, but now to the operator function

$$A(t) = \begin{cases} (1-2t)F(0) + 2tF_B, & 0 \leq t \leq 1/2, \\ (2-2t)F_B + (2t-1)F(1), & 1/2 \leq t \leq 1. \end{cases}$$

It follows that $\dim\ker F_B \leq \operatorname{def} T + \dim\ker F(0)$, where $\operatorname{def} T = \operatorname{def} A(1)/A(0) = \operatorname{def} F(1)/F(0) = \operatorname{def} \tilde{T}$. The inequality (4.2) now follows from the equalities $\dim\ker F_B = \dim\ker(A+B)$ and $\dim\ker F(0) = \dim\ker(A + C(0))$. □

Corollary 4.2. *Let A be a selfadjoint operator with compact resolvent and let B be a bounded selfadjoint operator with*

$$\lambda_{\min} = \min\{\lambda \in \sigma(B)\}, \quad \lambda_{\max} = \max\{\lambda \in \sigma(B)\}.$$

Then

(4.3) $$\dim\ker(A+B) \leq \sum_{\lambda_{\min} \leq \lambda \leq \lambda_{\max}} \dim\ker(A+\lambda).$$

Proof. Apply (4.1) and (4.2) to $C(t) = (1-t)\lambda_{\min} + t\lambda_{\max}$. □

Professor A. Markus, Ben-Gurion University of the Negev, Beer-Sheva, Israel, sent us a direct and simple proof of this corollary. We thank him for allowing us to include it in this note.

Proof of Corollary 4.2 (A. Markus). Let $\mathcal{M}$ be the span of all eigenvectors of A corresponding to the eigenvalues in $[-\lambda_{\max}, -\lambda_{\min}]$, and let $\mathcal{L} = \mathcal{H} \ominus \mathcal{M}$. The spectrum of A is discrete, and hence there exists a number $\delta > 0$ such that

(4.4) $$\sigma(A|_{\mathcal{L}}) \subset (-\infty, -\lambda_{\max} - \delta] \cup [-\lambda_{\min} + \delta, +\infty)$$

Suppose that (4.3) does not hold. This means that

$$\dim\ker(A+B) > \dim \mathcal{M}.$$

Then there exists a vector $f \in \ker(A+B)$ such that $\|f\| = 1$ and $f \perp \mathcal{M}$, that is, $f \in \mathcal{L}$. We have

$$f = \sum_j \alpha_j e_j,$$

where $\{e_j\}$ is an orthonormal sequence of eigenvectors of A corresponding to the eigenvalues $\{\lambda_j\}$ which belong to the set on the righthand side of (4.4). Set $\alpha = -\frac{1}{2}(\lambda_{\min} + \lambda_{\max})$. Then

$$(4.5)\quad \|Af - \alpha f\| = \|\sum_j (\lambda_j - \alpha)e_j\| \geq \min_j |\lambda_j - \alpha| \geq \frac{1}{2}(\lambda_{\max} - \lambda_{\min}) + \delta.$$

On the other hand, $\sigma(B+\alpha) \subset [-\frac{1}{2}(\lambda_{\max} - \lambda_{\min}), \frac{1}{2}(\lambda_{\max} - \lambda_{\min})]$ and

$$(4.6)\qquad \|Bf + \alpha f\| \leq \frac{1}{2}(\lambda_{\max} - \lambda_{\min}).$$

But $f \in \ker(A+B)$ implies $Af - \alpha f = -(Bf + \alpha f)$. Hence $\|Af - \alpha f\| = \|Bf + \alpha f\|$, which contradicts (4.5) and (4.6). □

Example 4.3. Let $\iota = [0, 2\pi]$, $\iota^2 = [0, 2\pi] \times [0, 2\pi]$, $f \in L^2(\iota^2)$, and $g = \overline{g} \in L^\infty(\iota^2)$. Let $\Delta = \partial^2/\partial x^2 + \partial^2/\partial y^2$ be the Laplacian in two variables. A function $u \in L^2(\iota^2)$ is called a generalized solution of the "periodic" problem

$$(4.7)\qquad -\Delta u + gu = f \text{ on } \iota^2,$$

if

$$(4.8)\qquad -(u, \Delta v) + (gu, v) = (f, v).$$

for all functions $v \in C^2(\iota^2)$ with the "periodic" properties

$$v(0,y) = v(2\pi, y), \quad \tfrac{\partial}{\partial x} v(0,y) = \tfrac{\partial}{\partial x} v(2\pi, y),$$

$$v(x,0) = v(x, 2\pi), \quad \tfrac{\partial}{\partial y} v(x,0) = \tfrac{\partial}{\partial y} v(x, 2\pi).$$

Let

$$\lambda_{\min} = \inf\left\{ \int_{\iota^2} g|u|^2 dxdy \mid \int_{\iota^2} |u|^2 dxdy = 1 \right\},$$

and

$$\lambda_{\max} = \sup\left\{ \int_{\iota^2} g|u|^2 dxdy \mid \int_{\iota^2} |u|^2 dxdy = 1 \right\}.$$

Then

$$\text{ess.inf}\,\{g(x,y) \mid x, y \in \iota\} = \lambda_{\min} \leq \lambda_{\max} = \text{ess.sup}\,\{g(x,y) \mid x, y \in \iota\}.$$

We claim that the number of linear independent generalized solutions of the homogeneous problem (4.7), (4.8) is less than or equal to the number of solutions $(m,p) \in \mathbb{Z}^2$ of the system

$$\begin{cases} m^2 + p^2 + \lambda_{\min} \leq 0, \\ m^2 + p^2 + \lambda_{\max} \geq 0. \end{cases}$$

Indeed, this follows from Corollary 4.2 with $A = -\overline{\Delta}$ and B defined by $Bu = gu, u \in L^2(\iota^2)$: The resolvent of A is compact and

$$\sigma(A) = \left\{m^2 + p^2 \mid m, p \in \mathbb{Z}\right\}.$$

In particular, if

$$[-\lambda_{\max}, -\lambda_{\min}] \cap \left\{m^2 + p^2 \mid m, p \in \mathbb{Z}\right\} = \emptyset$$

then $0 \in \rho(A+B)$. □

Acknowledgements

The research of T. Ya. Azizov was supported by the Netherlands Organization for Scientific Research NWO, and by INTAS (project 93–0249). The research of V.L. Khatskevich was supported by the International Science Foundation (grant NZP000).

References

[AI] T.Ya. Azizov, I.S. Iokhvidov, Foundation of the theory of linear operators in spaces with an indefinite metric (Russian), "Nauka", Moscow, 1986; English transl.: Linear operators in spaces with an indefinite metric, Wiley, New York, 1989.

[AKh1] T.Ya. Azizov, V.L. Khatskevich, On some applications of Kreĭ n space operator theory to the solution of nonlinear Hamiltonian systems (Russian), Math. Notes **50** (1991), 4, 1-9.

[AKh2] ———, On selfadjoint operators connected with inequalities and on applications to problems in mathematical physics (Russian), Math. Notes **55** (1994), 6, 3-12.

[AL] T.Ya. Azizov, H. Langer, Some spectral properties of contractive and expansive operators in indefinite inner product spaces, Math. Nachr. **162** (1993), 247–259.

[BBF] P. Bartolo, V. Benci, D. Fortunato, Abstract critical point theorems and applications to some nonlinear problems with "strong" resonance at infinity, Nonlinear Analysis, Theory, Methods and Applications **7** (1983), 9, 981-1012.

[DM] V.A. Derkach, M.M. Malamud, Generalized resolvents and the boundary value problems for Hermitian operators with gaps, J. Functional Analysis **95** (1991), 1, 1–95.

[Gh] A. GHEONDEA, Quasi-contractions on Kreĭn spaces, Operator Theory: Advances and Applications **61**, Birkhäuser Verlag, Basel, 1993, 123–148.

[Gi] YU.P. GINZBURG, On J-nonexpansive operators in a Hilbert space (Russian), Research Notes Department of Mathematical Physics, Odessa State Pedagogical Institute **22** (1958), 1.

[IKL] I.S. IOKHVIDOV, M.G. KREĬN, H. LANGER, Introduction to the spectral theory of operators in spaces with an indefinite metric, Akademie Verlag, Berlin, 1982.

[J] P. JONAS, On the functional calculus and the spectral function for definitizable operators in Krein space, Beiträge Anal. **16** (1981), 121–135.

[KSh1] M.G. KREĬN, YU.L. SHMUL'YAN, On plus-operators in an indefinite metric space (Russian), Math. Research, Kishinev, **1** (1966), 1, 131-161; English transl.: Amer. Math. Soc. Transl. **2** (1969), 8, 93-113.

[KSh2] ———, The J-polar representation of plus-operators (Russian), Math. Research, Kishinev, **1** (1966), 2, 172-210; English transl.: Amer. Math. Soc. Transl. **2** (1969), 85, 115-153.

[K] N.H. KUIPER, The homotopy type of the unitary group of Hilbert space, Topology **3** (1965), 1, 19-30.

[L] H. LANGER, Spektraltheorie linearer Operatoren in J-Räumen und einige Anwendungen auf die Schar $L(\lambda) = \lambda^2 I + \lambda B + C$, Habilitationsschrift, Technische Unversität Dresden, 1965.

[Pe] A.I. PEROV, Variational methods on the nonlinear vibrations theory (Russian), Voronezh State University, Voronezh, 1981.

[Po] V.P. POTAPOV, The multiplicative structure of J-nonexpansive matrix function (Russian), Works of Moscow Math. Soc. **4** (1955), 125-236.

[R] P. RABINOWITZ, Periodic solutions of Hamiltonian systems, SIAM J. Math. Anal. **13** (1982), 343-352.

[S] YU.L. SHMUL'YAN, Concerning a problem on inequalities between Hermitian operators (Russian), Math. Notes **49** (1991), 4, 138-141.

Department of Mathematics
Voronezh State University
Universitetskaja pl.,1
394693 Voronezh
Russia

Department of Mathematics
University of Groningen
P.O. Box 800
Groningen
The Netherlands

Department of Mathematics
Voronezh State University
Universitetskaja pl.,1
394693 Voronezh
Russia

1991 Mathematics Subject Classification. Primary 46C20, 46N20; Secondary 47F05

Operator Theory:
Advances and Applications, Vol. 106

Positive differential operators in the Krein space $L^2(\mathbb{R}^n)$

BRANKO ĆURGUS AND BRANKO NAJMAN†

To Heinz Langer on the occasion of his 60th birthday.

We characterize a class of indefinite partial differential operators which are similar to selfadjoint operators in the Hilbert space $L^2(\mathbb{R}^n)$.

1. Introduction

In this paper we consider the weighted eigenvalue problem

$$Lu = \lambda\,(\operatorname{sgn} x_n)u, \tag{1.1}$$

on the whole space $\mathbb{R}^n$ where $L = p(D)$ is a positive symmetric partial differential operator with constant coefficients. Our goal is to characterize a class of nonnegative polynomials p for which the operator associated with the problem (1.1) in the Hilbert space $L^2(\mathbb{R}^n)$ is similar to a selfadjoint operator. For example, our results imply that the operator $(\operatorname{sgn} x_n)\Delta$ defined on $H^2(\mathbb{R}^n)$ is similar to a selfadjoint operator in $L^2(\mathbb{R}^n)$.

The natural setting to study the problem (1.1) is the space $L^2(\mathbb{R}^n)$ with the indefinite inner product $[u,v] = \int u(x)\overline{v(x)}\operatorname{sgn} x_n dx$. The space $L^2(\mathbb{R}^n)$ with this inner product is a Krein space. The operator $A = (\operatorname{sgn} x_n)L$ is positive in this Krein space. In order to apply H. Langer's spectral theory of definitizable operators in Krein spaces we need to prove that the resolvent set $\rho(A)$ is not empty. In the setting of this paper, a useful tool for this is a simple result stated in Lemma 2.1. The spectral theory of definitizable operators is a generalization of the spectral theory of selfadjoint operators in Hilbert spaces. In particular, a definitizable operator in a Krein space has a spectral function. With exception of finitely many critical points this spectral function has properties analogous to the properties of the spectral function of a selfadjoint operator in a Hilbert space. Definitizable operators in this paper are of the simplest kind: positive operators in a Krein space with nonempty resolvent set. For such operators only 0 and ∞ may be critical points. The projector valued spectral function G of a positive operator A with nonempty resolvent set is defined on open intervals in $\mathbb{R}$ with the endpoints different from 0 and ∞. The ranges of projectors corresponding to intervals with positive endpoints are Hilbert subspaces and the ranges of projectors corresponding to intervals with

---

†Branko Najman died unexpectedly in August 1996.

negative endpoints are anti-Hilbert subspaces of the Krein space $L^2(\mathbb{R}^n)$. In general, for a definitizable operator T with the spectral function E in a Krein space $\mathcal{K}$ a spectral point λ is of *positive type* (*negative type*) if there exists an open interval $\imath$ such that $\lambda \in \imath$ and the range $E(\imath)\mathcal{K}$ is a Hilbert (anti-Hilbert) subspace of $\mathcal{K}$. A spectral point of T is *critical* if it is neither of positive nor of negative type. A critical point λ is *regular* if the spectral function is bounded near λ. A critical point is *singular* if it is not regular. For a positive operator A the points 0 and ∞ are the only possible critical points of A.

We are primarily interested in the case when neither 0 nor ∞ is a singular critical point of A. In this case A is similar to a selfadjoint operator in $L^2(\mathbb{R}^n)$. When $n = 1$ and $p(t) = t^2$ we proved in [5] that A is similar to a selfadjoint operator in $L^2(\mathbb{R})$. In [13] this result was extended to more general weight functions (see also Example 3.6 below) and in [6] the result was extended to more general polynomials p (see also Corollary 3.5 below). In this paper we characterize a class of polynomials p in n variables for which the corresponding operator $A = (\operatorname{sgn} x_n)p(D)$ is similar to a selfadjoint operator in the Hilbert space $L^2(\mathbb{R}^n)$. The problem with a definite discontinuous weight has recently been considered in [19].

The question of regularity of the critical point ∞ of definitizable operators in Krein spaces has attracted considerable interest, see for example [2, 14, 15, 23]. Corresponding questions for the Sturm-Liouville problem and the elliptic eigenvalue problem with indefinite weight were also studied extensively, see the references in [3, 4, 10, 11, 12, 24]. One of the reasons for this is the following: if a definitizable operator T in a Krein space $\mathcal{K}$ has a discrete spectrum, only ∞ may be an accumulation point of spectral points of both positive and negative type. In this case regularity of the critical point ∞ is equivalent to the existence of a Riesz basis of $\mathcal{K}$ which consists of eigenvectors and generalized eigenvectors of T (see [4, Proposition 2.3]). The regularity of the critical point ∞ of a definitizable operator was characterized in [2] in terms of the operator domain. This was used in [3] (case $n = 1$) and in [4] (case $n > 1$) to prove regularity of the critical point ∞ for differential operators with more general weight functions and more general differential expressions L.

Our main interest in this paper is the case when the operator A is positive (not uniformly positive as in [4]) and this is why the critical point 0 may appear as a critical point. If the spectrum of A accumulates at 0 from both sides, then 0 is a critical point of A. To determine whether it is singular or regular we need to investigate the range of A. This question is harder than the investigation of the domain.

For the readers convenience in Section 2 we prove several simple lemmas which we use later on in the paper. We give a sufficient condition for $\operatorname{ran}(B+V) = \operatorname{ran}(B)$ for a closed operator B. For further results related to the stability of the range under additive perturbations see [7]. From [6] we recall a necessary and sufficient condition for $\operatorname{ran}(B) = \operatorname{ran}(C)$ for multiplication operators $B,\ C$ in $L^2(\mathbb{R}^n)$.

In Section 3 we prove several stability theorems for the regularity of the critical points 0 and ∞ of positive definitizable operators in a Krein space. As a consequence we get a stability theorem for the similarity to a selfadjoint operator in a Hilbert space. These results are improvements of the corresponding results in [6] since they do not require a priori knowledge of nonemptyness of the resolvent sets of the resulting operators. For related results in this direction see [14].

In Section 4 we consider partial differential operators with constant coefficients. For polynomials p of the form $p(\hat{x}, x_n) = q(\hat{x}) + r(x_n)$ we establish the formula (4.5) expressing the spectral function of A in terms of the spectral functions of the operators

$$(\operatorname{sgn} x_n)\left(r\left(\frac{1}{i}\frac{d}{dx_n}\right) + q(\hat{x})I\right) .$$

For such polynomials we give a detailed analysis of the spectrum and the critical points. We show that ∞ is a regular critical point and give sufficient conditions for 0 to be a regular critical point. These results about critical points are extended to more general polynomials p using the perturbation results from Section 2. These perturbation results are used in Section 5 to treat a variable coefficient operator.

The study of spectral properties of indefinite eigenvalue problems for differential operators has been motivated by the investigation of the half-range completeness property, see [1]. It follows from the general operator theory in Krein spaces (see [3, 6]) that an operator which is positive in the Krein space $(L^2(\mathbb{R}^n), [\cdot\,,\cdot])$ and similar to a selfadjoint operator in the Hilbert space $L^2(\mathbb{R}^n)$ has the half-range completeness property. Therefore our results in Sections 4 and 5 give sufficient conditions for the half-range completeness property for the problem (1.1).

For definitions and basic results of the theory of definitizable operators see [8, 17].

2. Preliminaries

We start with a simple lemma that assures preservation of nonemptyness of resolvent sets under bounded additive perturbations. For a closed operator T in a Hilbert space $\mathcal{H}$, $\rho(T)$ denotes the resolvent set of T.

Lemma 2.1. *Let A be an operator in a Hilbert space $\mathcal{H}$ which is similar to a selfadjoint operator and let B be a bounded operator in $\mathcal{H}$. There exists $K > 0$ such that $\lambda \in \rho(A + B)$ whenever $|\operatorname{Im}\lambda| > K$.*

Proof. Since A is similar to a selfadjoint operator there exists a constant $C > 0$ such that $\|(A - \lambda I)^{-1}\| < C|\operatorname{Im}\lambda|^{-1}$ for all $\lambda \in \mathbb{C}\setminus\mathbb{R}$. Therefore, $B(A - \lambda I)^{-1}$ is a bounded operator with norm < 1 whenever $|\operatorname{Im}\lambda| > C\|B\|$. Thus, $I + B(A - \lambda I)^{-1}$ has a bounded inverse for all $\lambda \in \mathbb{C}$ such that $|\operatorname{Im}\lambda| > C\|B\|$. Since $A + B - \lambda I = (I + B(A - \lambda I)^{-1})(A - \lambda I)$, it follows that $\lambda \in \rho(A + B)$ whenever $|\operatorname{Im}\lambda| > C\|B\|$. □

Lemma 2.2. *Let A and B be definitizable operators in the Krein space $(\mathcal{K}, [\cdot, \cdot])$ such that 0 is neither an eigenvalue of A nor of B. Assume that* $\operatorname{ran}(A) = \operatorname{ran}(B)$. *Then 0 is not a singular critical point of A if and only if 0 is not a singular critical point of B.*

Proof. Both operators A^{-1} and B^{-1} are definitizable and 0 is not a singular critical point of A if and only if ∞ is not a singular critical point of A^{-1}. Since $\operatorname{dom}(A^{-1}) = \operatorname{dom}(B^{-1})$, [2, Corollary 3.3] implies that ∞ is not a singular critical point of A^{-1} if and only if ∞ is not a singular critical point of B^{-1}. Since ∞ is not a singular critical point of B^{-1} if and only if 0 is not a singular critical point of B, the lemma is proved. □

Motivated by Lemma 2.2 we prove a result on the preservation of ranges under additive perturbations. The following is a restatement of [16, Lemma VI.2.30].

Lemma 2.3. *Let A and V be closed densely defined operators in the Hilbert space $\mathcal{H}$. Let A be injective. Assume that* $\operatorname{dom}(A^*) \subseteq \operatorname{dom}(V^*)$ *and that there exists $\beta \geq 0$ such that*

$$\|V^*x\| \leq \beta\|A^*x\| \quad \textit{for all} \quad x \in \operatorname{dom}(A^*) \ . \tag{2.1}$$

Then $\operatorname{ran}(V) \subseteq \operatorname{ran}(A)$ *and* $\|A^{-1}Vy\| \leq \beta\|y\|$ *for all* $y \in \operatorname{dom}(V)$.

Corollary 2.4. *In addition to the assumptions of Lemma 2.3 assume that (2.1) holds with $\beta < 1$. Then $A + V$ is injective and*

$$\operatorname{ran}(A + V) = \operatorname{ran}(A) \ .$$

Proof. Lemma 2.3 implies that $\operatorname{ran}(A+V) \subseteq \operatorname{ran}(A)$. Next we prove the opposite inclusion. We have $\operatorname{dom}((A + V)^*) \subseteq \operatorname{dom}(V^*)$. Further it follows from (2.1) that

$$\|V^*u\| \leq \beta\|A^*u\| \leq \beta\|(A^* + V^*)u\| + \beta\|V^*u\|,$$

implying

$$\|V^*u\| \leq \frac{\beta}{1-\beta}\|(A^* + V^*)u\| \ \text{ for all } \ u \in \operatorname{dom}((A + V)^*) \ .$$

Applying Lemma 2.3 to the operators $A + V$ and $-V$ we conclude that $\operatorname{ran}(A) = \operatorname{ran}((A + V) - V) \subseteq \operatorname{ran}(A + V)$.

From Lemma 2.3 it also follows that the operator $A^{-1}V$ is defined on $\operatorname{dom}(V)$ and bounded and with the norm is less than or equal to β. If $x \in \operatorname{dom}(V)$ satisfies $(A + V)x = 0$, then $x = -A^{-1}Vx$. Therefore $x = 0$. □

Corollary 2.5. *Let A be selfadjoint and V a closed symmetric operator in the Hilbert space $\mathcal{H}$ and* $\mathrm{dom}(A) \subseteq \mathrm{dom}(V)$. *Assume that (2.1) holds with $\beta < 1$. Then*

$$\mathrm{ran}(A+V) = \mathrm{ran}(A) \quad \text{and} \quad \mathrm{dom}(A+V) = \mathrm{dom}(A)\ .$$

Proof. It follows from (2.1) that $\ker(A) \subseteq \ker(V)$. Denote the closure of $\mathrm{ran}(A)$ by $\mathcal{L}$. Then $\mathcal{L}$ is invariant under A and V and the restriction A_r of A to $\mathcal{L}$ is injective and it satisfies all the assumptions of Corollary 2.4. □

Let μ be a Borel measure on $\mathbb{R}^n$. A μ-measurable function $f : \mathbb{R}^n \to \mathbb{C}$ is *nonnegative* if $f(x) \geq 0$ for μ-almost all $x \in \mathbb{R}^n$. Denote by M_f the operator of multiplication by f in the Hilbert space $L^2(\mathbb{R}^n, \mu)$.

Lemma 2.6. *Let g and h be nonnegative μ-measurable functions on $\mathbb{R}^n$.*

(1) *The following statements are equivalent.*

  (a) $\mathrm{dom}(M_g) = \mathrm{dom}(M_h)$

  (b) *There exists $c > 0$ such that the functions $\frac{h}{c+g}$ and $\frac{g}{c+h}$ are μ-essentially bounded.*

(2) *The following statements are equivalent.*

  (a) $\mathrm{ran}(M_g) = \mathrm{ran}(M_h)$.

  (b) *There exists a constant $C > 0$ such that*

$$g \leq Ch(1+g) \ \ \mu - a.e. \quad \text{and} \quad h \leq Cg(1+h) \ \ \mu - a.e.\ . \tag{2.2}$$

Proof. The statement (1) is evident. To prove (2), for a μ-measurable function $f : \mathbb{R}^n \to \mathbb{C}$ denote by N_f the set $\{x \in \mathbb{R}^n | f(x) = 0\}$. Note that the conditions (2.2) imply that the symmetric difference of the sets N_g and N_h has μ-measure zero. Therefore $\ker(M_g) = \ker(M_h)$. Let

$$G(x) = \begin{cases} 0 & \text{if } g(x) = 0, \\ \frac{1}{g(x)} & \text{if } g(x) \neq 0 \end{cases} \quad \text{and} \quad H(x) = \begin{cases} 0 & \text{if } h(x) = 0, \\ \frac{1}{h(x)} & \text{if } h(x) \neq 0. \end{cases}$$

It follows from (1) that the condition (2.2) is equivalent to $\mathrm{dom}(M_G) = \mathrm{dom}(M_H)$. Since $\mathrm{dom}(M_G) = \mathrm{ran}(M_g) \oplus \ker(M_g)$, (2a) and (2b) are equivalent. □

We need the following simple lemma in Section 4.

Lemma 2.7. *Let A be a uniformly positive operator in the Krein space $(\mathcal{K}, [\cdot\,,\cdot])$. Let $\gamma > 0$ be a lower bound of the uniformly positive operator $B = JA$ in the Hilbert space $(\mathcal{K}, \langle\cdot\,,\cdot\rangle)$. Then the interval $(-\gamma, \gamma)$ is contained in the resolvent set of A.*

Proof. Clearly $\gamma^{-1} = \|B^{-1}\| = \|A^{-1}\|$. Let $|\lambda| < \gamma$. Then $\lambda^{-1} \in \rho(A^{-1})$, hence $\lambda \in \rho(A)$. □

3. Similarity to selfadjoint operators

In this section we reformulate and improve some results from [2] and [6]. Let $(\mathcal{K}, [\cdot\,,\cdot])$ be a Krein space, let J be a fundamental symmetry in $\mathcal{K}$ and let $\langle\cdot\,,\cdot\rangle = [J\cdot,\cdot]$ be the corresponding Hilbert space inner product.

Lemma 3.1. *Let $\eta > 0$. The following statements are equivalent.*

(a) *The operator JP is positive in $(\mathcal{K}, [\cdot\,,\cdot])$, $\rho(JP) \neq \emptyset$ and ∞ is not a singular critical point of JP.*

(b) *The operator JP^{η} is positive in $(\mathcal{K}, [\cdot\,,\cdot])$, $\rho(JP^{\eta}) \neq \emptyset$ and ∞ is not a singular critical point of JP^{η}.*

Proof. Assume (a). Then $J(P+I)$ is a uniformly positive operator in $(\mathcal{K}, [\cdot\,,\cdot])$. Since $\mathrm{dom}(J(P+I)) = \mathrm{dom}(JP)$, [2, Corollary 3.3] (see also [8, Theorem 1.6]) implies that ∞ is not a singular critical point of $J(P+I)$. [2, Theorem 2.9] implies that ∞ is not a singular critical point of $J(P+I)^{\eta}$. Since $\mathrm{dom}(J(P+I)^{\eta}) = \mathrm{dom}(J(P^{\eta}+I))$, and since both operators $J(P+I)^{\eta}$ and $J(P^{\eta}+I)$ are uniformly positive, [2, Corollary 3.3] implies that ∞ is not a singular critical point of $J(P^{\eta}+I)$. By [2, Theorem 2.5] (or [8, Theorem 1.6]) the operator $J(P^{\eta}+I)$ is similar to a selfadjoint operator in $(\mathcal{K}, \langle\cdot\,,\cdot\rangle)$. Lemma 2.1 implies that $\rho(JP^{\eta}) \neq \emptyset$, so JP^{η} is a definitizable operator. As $\mathrm{dom}(JP^{\eta}) = \mathrm{dom}(J(P^{\eta}+I))$, the statement (b) follows from [2, Corollary 3.3]. The implication (b) $\Rightarrow$ (a) follows by applying (a) $\Rightarrow$ (b) to the operator JP^{η} and the positive number $1/\eta$. □

Corollary 3.2. *Let $\eta > 0$. The following statements are equivalent.*

(a) *The operator JP is positive in $(\mathcal{K}, [\cdot\,,\cdot])$, 0 is not an eigenvalue of P, $\rho(JP) \neq \emptyset$ and 0 is not a singular critical point of the operator JP.*

(b) *The operator JP^{η} is positive in $(\mathcal{K}, [\cdot\,,\cdot])$, 0 is not an eigenvalue of P^{η}, $\rho(JP^{\eta}) \neq \emptyset$ and 0 is not a singular critical point of the operator JP^{η}.*

Corollary 3.3. *Let $\eta \neq 0$. The following statements are equivalent:*

(a) *The operator JP is positive in $(\mathcal{K}, [\cdot\,,\cdot])$, 0 is not an eigenvalue of P and JP is similar to a selfadjoint operator in $(\mathcal{K}, \langle\cdot\,,\cdot\rangle)$.*

(b) *The operator JP^{η} is positive in $(\mathcal{K}, [\cdot\,,\cdot])$, 0 is not an eigenvalue of P^{η} and JP^{η} is similar to a selfadjoint operator in $(\mathcal{K}, \langle\cdot\,,\cdot\rangle)$.*

The following theorem is an improvement of [6, Theorem 1.4] since it does not require a priori knowledge of nonemptyness of the resolvent set of the operator $Jh(S)$. It also can be considered as an abstract version of [6, Theorem 2.3].

Theorem 3.4. *Let S be a selfadjoint operator in the Hilbert space $(\mathcal{K}, \langle\cdot\,,\cdot\rangle)$ and let $h : \mathbb{R} \to \mathbb{R}$ be a nonnegative continuous function.*

(1) *Assume that there exists $\eta > 0$ such that the functions $g(t) = |t|^\eta$ and h satisfy the conditions (1b) of Lemma 2.6. The following statements are equivalent.*

   (a) *∞ is not a singular critical point of $J(S^2 + I)$.*

   (b) *$\rho(Jh(S)) \neq \emptyset$ and ∞ is not a singular critical point of $Jh(S)$.*

(2) *Assume that 0 is not an eigenvalue of S and that there exists $\eta > 0$ such that the functions $g(t) = |t|^\eta$ and h satisfy the condition (2.2). Then the following statements are equivalent.*

   (a) *$\rho(JS^2) \neq \emptyset$ and 0 is not a singular critical point of $J(S^2)$.*

   (b) *$\rho(Jh(S)) \neq \emptyset$ and 0 is not a singular critical point of $Jh(S)$.*

Proof. The proof combines ideas used in the proofs of Lemma 3.1 and [6, Theorem 1.4]. We prove (2). The proof of (1) is similar. Note that Lemma 2.6 (2), with $n = 1$, implies that for any Borel measure μ the multiplication operators M_g and M_h in $L^2(\mathbb{R}, \mu)$ have the same range. The Spectral Theorem, see [25, Theorem 7.18], implies $\operatorname{ran}(|S|^\eta) = \operatorname{ran}(h(S))$. Therefore, $\operatorname{ran}(J|S|^\eta) = \operatorname{ran}(Jh(S))$. Assume (2a). Corollary 3.2 implies that 0 is not an eigenvalue of $J|S|^\eta$, $\rho(J|S|^\eta) \neq \emptyset$ and 0 is not a singular critical point of $J|S|^\eta$. Therefore ∞ is not a singular critical point of $(J|S|^\eta)^{-1}$. Since $(Jh(S))^{-1} + J$ is uniformly positive and since its domain coincides with the domain of $(J|S|^\eta)^{-1}$ we conclude that ∞ is not a singular critical point of $(Jh(S))^{-1} + J$, that is $(Jh(S))^{-1} + J$ is similar to a selfadjoint operator in $(\mathcal{K}, \langle\cdot\,,\cdot\rangle)$. Lemma 2.1 implies that $\rho((Jh(S))^{-1}) \neq \emptyset$. Consequently, $\rho((Jh(S))) \neq \emptyset$. The equality $\operatorname{ran}(J|S|^\eta) = \operatorname{ran}(Jh(S))$ implies that 0 is not an eigenvalue of $Jh(S)$ and 0 is not a singular critical point of $Jh(S)$. This proves (2b). The proof of the converse is similar. □

The combination of parts (1) and (2) of Theorem 3.4 gives sufficient conditions under which the similarity to a selfadjoint operator of JS^2 is equivalent to the similarity to a selfadjoint operator of $Jh(S)$. If the function h is a polynomial this takes a particularly simple form which we state in the following corollary.

Corollary 3.5. *Let S be a selfadjoint operator in the Hilbert space $(\mathcal{K}, \langle\cdot\,,\cdot\rangle)$ and let p be a nonnegative polynomial on $\mathbb{R}$ with 0 being its only root. The following statements are equivalent.*

(a) *JS^2 is similar to a selfadjoint operator in the Hilbert space $(\mathcal{K}, \langle\cdot\,,\cdot\rangle)$.*

(b) *$Jp(S)$ is similar to a selfadjoint operator in the Hilbert space $(\mathcal{K}, \langle\cdot\,,\cdot\rangle)$.*

Proof. Let $2k, k > 0$, be the degree of p and let $2j, j > 0$, be the multiplicity of the root 0 of p. Let $g_1(t) = t^{2k}$ and $g_2(t) = t^{2j}$. Then g_1 and p satisfy the conditions (1b) in Lemma 2.6 and g_2 and p satisfy the conditions (2.2) in Lemma 2.6. Therefore the equivalence of (a) and (b) follows from Theorem 3.4. □

Example 3.6. Let $w(t) = |t|^\tau \operatorname{sgn} t$, $\tau > -1$, and $S = -i|t|^{-\tau/2}\frac{d}{dt}$. Let $\mathcal{K} = L^2(\mathbb{R}, |w|)$ be a Krein space with the indefinite inner product $[f,g] = \int_{\mathbb{R}} f(t)\overline{g(t)}w(t)dt$. The operator $(Jf)(t) = (\operatorname{sgn} t)f(t)$ is a fundamental symmetry on $\mathcal{K}$. By [13, Theorem 2.7] the operator JS^2 is similar to a selfadjoint operator in the Hilbert space $L^2(\mathbb{R}, |w|)$. Let p be a nonnegative polynomial on $\mathbb{R}$ with 0 being its only root. Corollary 3.5 implies that the operator $Jp(S)$ is similar to a selfadjoint operator in $L^2(\mathbb{R}, |w|)$. Using [6, Proposition 2.4] we can extend this result to nonnegative polynomials with exactly one real root.

4. Partial differential operators with constant coefficients

In this section $\mathcal{K}$ denotes the Krein space $L^2(\mathbb{R}^n)$ with the inner product $[f,g] = \int_{\mathbb{R}^n} f(x)\overline{g(x)} \operatorname{sgn} x_n \, dx$, where $x = (x_1, \ldots, x_n)$. The multiplication operator

$$(Jy)(x) = (\operatorname{sgn} x_n)y(x)$$

is a fundamental symmetry on $(L^2(\mathbb{R}^n), [\,\cdot\,,\,\cdot\,])$ and the corresponding Hilbert space inner product is $\langle f,g\rangle = \int_{\mathbb{R}^n} f(x)\overline{g(x)}\, dx$. The points $x \in \mathbb{R}^n$ are denoted by $x = (\hat{x}, t)$, where $\hat{x} = (x_1, \ldots, x_{n-1}), t = x_n$. The partial Fourier transform with respect to $\hat{x}$ is denoted by F. It is a unitary operator in $L^2(\mathbb{R}^n)$.

We study partial differential operators with constant coefficients. Let p be a nonconstant polynomial of degree m in n variables,

$$p(x) = \sum_{|\alpha| \le m} c_\alpha x_1^{\alpha_1} \cdots x_n^{\alpha_n},$$

where $(x_1, \ldots, x_n) \in \mathbb{R}^n$, $\alpha = (\alpha_1, \ldots, \alpha_n)$ is a multiindex, $c_\alpha \in \mathbb{R}$ and $|\alpha| = \sum \alpha_j$. Denote by D^α the partial differential expression

$$\left(\frac{1}{i}\right)^{|\alpha|} \frac{\partial^{\alpha_1}}{\partial x_1^{\alpha_1}} \cdots \frac{\partial^{\alpha_n}}{\partial x_n^{\alpha_n}}$$

and let B be the closed operator associated with the differential expression

$$p(D) = \sum_{|\alpha| \le m} c_\alpha D^\alpha$$

in the Hilbert space $(L^2(\mathbb{R}^n), \langle\,\cdot\,,\,\cdot\,\rangle)$. Instead of B we will often write $p(D)$ to emphasize its dependence on p. The operator B is selfadjoint in the Hilbert space $(L^2(\mathbb{R}^n), \langle\,\cdot\,,\,\cdot\,\rangle)$. The operator $A = JB$ is selfadjoint in the Krein space $(L^2(\mathbb{R}^n), [\,\cdot\,,\,\cdot\,])$. We will prove that, under certain assumptions on the polynomial p, the operator A is similar to a selfadjoint operator in $(L^2(\mathbb{R}^n), \langle\,\cdot\,,\,\cdot\,\rangle)$.

Definition 4.1. Let p be a nonnegative polynomial in n variables, let $q(\hat{x}) = p(\hat{x}, 0)$, let $a_0 t^{2k}$, $a_0 \geq 0$, be the leading term of the polynomial $p(0,t) - p(0,0)$ and put

$$p = p_1 + p_2 \quad \text{with} \quad p_1(x) = a_0 t^{2k} + q(\hat{x}) \quad \text{and} \quad p_2(x) = p(x) - p_1(x). \tag{4.1}$$

The polynomial p is *weakly separated* if there exist $\gamma_1, \gamma_2, \beta \geq 0$, $\gamma_1 < 1$ such that

$$-\gamma_1 p_1(x) - \beta \leq p_2(x) \leq \gamma_2 p_1(x) + \beta \tag{4.2}$$

The polynomial p is *strongly separated* if (4.2) holds with $\beta = 0$.

Lemma 4.2. *Let $p(y,t) = ay^2 + byt + ct^2 + \mu t + \nu$, with $a, c > 0, \delta := \frac{|b|}{2\sqrt{ac}} < 1$ and not both μ and ν equal 0. Then*

(i) *p is weakly separated.*

(ii) *If $4c\nu \leq \mu^2$, then p is not strongly separated.*

(iii) *If $4(1-\delta)^2 c\nu > \mu^2$, then p is strongly separated.*

Proof. By Definition 4.1 $p_1(y,t) = ct^2 + ay^2 + \nu$ and $p_2(y,t) = byt + \mu t$. To prove (i) note that $|byt| < \frac{|b|}{2\sqrt{ac}}(ct^2 + ay^2)$. Since $\frac{|b|}{2\sqrt{ac}} < 1$, there exists $\epsilon > 0$ such that $\frac{|b|}{2\sqrt{ac}} + \epsilon < 1$. Choosing $\tau \geq \frac{\mu^2}{4\epsilon c}$, we get that $|\mu t| \leq \epsilon c t^2 + \tau$. Therefore

$$|p_2(y,t)| \leq \left(\frac{|b|}{2\sqrt{ac}} + \epsilon\right)(ct^2 + ay^2) + \tau = \left(\frac{|b|}{2\sqrt{ac}} + \epsilon\right) p_1(y,t) + \beta$$

for some real number β. Thus p is weakly separated.

To prove (ii) assume that p is strongly separated. Then for some $0 \leq \gamma_1 < 1$ we have $-\gamma_1(ct^2 + ay^2 + \nu) \leq byt + \mu t$. With $y = 0$, this inequality implies $\mu^2 - 4\gamma_1^2 c\nu \leq 0$, and therefore $\mu^2 < 4c\nu$.

To prove (iii) assume that $4(1-\delta)^2 c\nu > \mu^2$. Then $\nu > 0$ and there exists $\epsilon > 0$ such that $\mu^2 - 4(1-\delta-\epsilon)^2 c\nu < 0$. Consequently $|\mu t| \leq (1-\delta-\epsilon)(ct^2+\nu)$. Together with the first inequality used in the proof of (i), this yields

$$|byt + \mu t| \leq \delta(ct^2 + ay^2) + (1-\delta-\epsilon)(ct^2 + ay^2 + \nu) \leq (1-\epsilon)p_1(y,t)\ .$$

Thus p is strongly separated. □

Lemma 4.3. *Let p be a nonnegative polynomial in n variables and let p_1 be the polynomial introduced in Definition 4.1.*

(a) *Assume that p is weakly separated. Then p does not depend on t if and only if p_1 does not depend on t.*

(b) *If p is weakly separated, then the multiplication operators M_p and M_{p_1} have the same domain in $\mathcal{K}$.*

(c) *If p is strongly separated, then $p(x) = 0$ if and only if $p_1(x) = 0$.*

(d) *If p is strongly separated, then the multiplication operators M_p and M_{p_1} have the same range in $\mathcal{K}$.*

Proof. The statements in (a), (c) follow directly from Definition 4.1. Note that p_1 does not depend on t if and only if $a_0 = 0$. Assume that p is weakly separated. Then p and p_1 satisfy the conditions in (1b) of Lemma 2.6 with $c = \beta + 1 > 0$. Indeed, the condition (4.2) yields

$$\frac{p_1}{\beta + 1 + p} \leq \frac{1}{1 - \alpha_1} \quad \text{and} \quad \frac{p}{\beta + 1 + p_1} \leq 1 + \alpha_2 + \beta\,.$$

Therefore (b) follows from Lemma 2.6. If p is strongly separated, then (4.2), with $\beta = 0$, implies

$$\frac{p_1}{p(1 + p_1)} \leq \frac{1}{1 - \alpha_1} \quad \text{and} \quad \frac{p}{p_1(1 + p)} \leq 1 + \alpha_2\,,$$

and (d) is a consequence of Lemma 2.6. □

Denote by P the operator $-\frac{d^2}{dt^2}$ in $L^2(\mathbb{R})$ on $H^2(\mathbb{R})$. By [6, Theorem 2.5] (see also Example 3.6) for any $b \geq 0$ and k a natural number the operator $(\operatorname{sgn} t)(P^k + bI)$ is similar to a selfadjoint operator in $L^2(\mathbb{R})$.

Lemma 4.4. *Let p be a nonnegative polynomial such that $p_2 = 0$. Assume that $\rho(A) \neq \emptyset$. Then $A = Jp(D)$ is similar to a selfadjoint operator in the Hilbert space $(L^2(\mathbb{R}^n), \langle\cdot\,,\cdot\rangle)$.*

Proof. Let $p(x) = q(\hat{x}) + a_0 t^{2k}$, $a_0 \geq 0$. If $a_0 = 0$, the operator A commutes with the fundamental symmetry J and consequently A is similar to a selfadjoint operator in the Hilbert space $(L^2(\mathbb{R}^n), \langle\cdot\,,\cdot\rangle)$.

If $a_0 > 0$ without loss of generality we can assume that $a_0 = 1$. Since we assume that $\rho(A) \neq \emptyset$, we only have to prove that the points 0 and ∞ are not singular critical points of A. Let $y \in \operatorname{dom}(A)$ and $\lambda \in \mathbb{C} \setminus \mathbb{R}$. It follows from the basic properties of the partial Fourier transform F that

$$((A - \lambda I)y)(x) = (F^{-1}(JP^k + q(\hat{x})J - \lambda I)Fy)(x). \tag{4.3}$$

Denote by E the spectral function of A and by G_α the spectral function of the operator $J(P^k + \alpha I)$. Consider an interval $\imath = (a, b)$ with $0 < a < b$. It follows from the definition of the spectral function and (4.3) that

$$(E(\imath)y)(x) = (F^{-1}G_{q(\hat{x})}(\imath)Fy)(x)\,. \tag{4.4}$$

Let $\alpha > 0$. The operator $J(P^k + \alpha I)$ is uniformly positive in the Krein space $(L^2(\mathbb{R}^n), [\cdot\,,\cdot])$ and the lower bound of $P^k + \alpha I$ is α. Lemma 2.7 implies that the interval $(-\alpha,\alpha)$, belongs to the resolvent set of $J(P^k+\alpha I)$ and consequently $G_\alpha(\imath)=0$ for $b<\alpha$. Thus, it follows from (4.4) that

$$\|E(\imath)y\|^2=\int_{q(\hat{x})\le b}\|(G_{q(\hat{x})}(\imath)Fy)(\hat{x},\cdot)\|^2d\hat{x}\ . \tag{4.5}$$

Denote by $U(\delta)$, $\delta\in\mathbb{R}\setminus\{0\}$ the dilation operator: $(U(\delta)f)(x)=f(\delta x)$, $x\in\mathbb{R}^n$. Then $U(\delta)$ is a bounded operator with the bounded inverse $U(1/\delta)$. We have

$$\langle U(\delta)f,U(\delta)f\rangle=|\delta|^{-n}\langle f,f\rangle \tag{4.6}$$

and

$$U(\delta)^{-1}P^kU(\delta)=a^{2k}P^k\ . \tag{4.7}$$

From

$$J(P^k+\alpha I)=\alpha U\Big(\alpha^{\frac{1}{2k}}\Big)\,J(P^k+I)U\Big(\alpha^{-\frac{1}{2k}}\Big)\ ,$$

it follows that

$$G_\alpha(\imath)=U\Big(\alpha^{\frac{1}{2k}}\Big)\,G_1(\imath_\alpha)U\Big(\alpha^{-\frac{1}{2k}}\Big)\ , \tag{4.8}$$

where $\imath_\alpha=\left(\frac{a}{\alpha},\frac{b}{\alpha}\right)$. From (4.5) and (4.8) we conclude

$$\|E(\imath)y\|^2=\int_{q(\hat{x})\le b}\Big\|\Big(U\Big(q(\hat{x})^{\frac{1}{2k}}\Big)\,G_1(\imath_{q(\hat{x})})U\Big(q(\hat{x})^{-\frac{1}{2k}}\Big)\,Fy\Big)\,(\hat{x},\cdot)\Big\|^2\,d\hat{x}\ . \tag{4.9}$$

Since $U(t)$ is a multiple of an isometry, it follows from the Plancherel theorem that

$$\|E(a,b)\|\le\sup_{q(\hat{x})\le b}\left\|G_1\left(\frac{a}{q(\hat{x})},\frac{b}{q(\hat{x})}\right)\right\|\le\sup_{0<\alpha\le b}\left\|G_1\left(\frac{a}{\alpha},\frac{b}{\alpha}\right)\right\|\ .$$

A similar formula holds for $a<b<0$. Since $J(P^k+I)$ is similar to a selfadjoint operator in the Hilbert space $L^2(\mathbb{R})$, it follows that both 0 and ∞ are not singular critical points of A. □

Corollary 4.5. *Let p be a nonnegative polynomial and assume that $p_2=0$. Then $A=Jp(D)$ is similar to a selfadjoint operator in the Hilbert space $(L^2(\mathbb{R}^n),\langle\cdot,\cdot\rangle)$.*

Proof. The polynomial $p+1$ is strictly positive. The operator $p(D)+I$ is uniformly positive in $(L^2(\mathbb{R}^n),\langle\cdot,\cdot\rangle)$. Therefore the operator $J(p(D)+I)$ is uniformly positive in $(L^2(\mathbb{R}^n),[\cdot,\cdot])$ and consequently $0\in\rho(J(p(D)+I))$. Lemma 4.4 implies that $J(p(D)+I)$ is similar to a selfadjoint operator in $(L^2(\mathbb{R}^n),\langle\cdot,\cdot\rangle)$. By Lemma 2.1 the operator $Jp(D)$ has a nonempty resolvent set. Applying Lemma 4.4 again yields that $Jp(D)$ is similar to a selfadjoint operator in $(L^2(\mathbb{R}^n),\langle\cdot,\cdot\rangle)$. □

Theorem 4.6. *Let p be a nonnegative polynomial and $A = Jp(D)$.*

(a) *If p is a weakly separated polynomial, then A is a positive operator in the Krein space $L^2(\mathbb{R}^n)$, $\rho(A) \neq \emptyset$ and ∞ is not a singular critical point of A.*

(b) *If p is a strongly separated polynomial, then 0 is not a singular critical point of A. The operator A is similar to a selfadjoint operator in $(L^2(\mathbb{R}^n), \langle \cdot\,, \cdot \rangle)$.*

Proof. If p does not depend on t the operator A commutes with the fundamental symmetry J and consequently A is similar to a selfadjoint operator in $(L^2(\mathbb{R}^n), \langle \cdot\,, \cdot \rangle)$. By Lemma 4.3 (a) p does not depend on t if and only if $a_0 = 0$. Thus, in the rest of the proof we can assume that $p_1(x) = a_0 t^{2k} + q(\hat{x})$, where $a_0 > 0$. Put $A_1 = Jp_1(D)$. By Corollary 4.5 the operator A_1 is similar to a selfadjoint operator in $(L^2(\mathbb{R}^n), \langle \cdot\,, \cdot \rangle)$. The operator $A = Jp(D)$ is positive in $(L^2(\mathbb{R}^n), [\cdot\,, \cdot])$. Lemma 4.3 implies that $\mathrm{dom}(M_p) = \mathrm{dom}(M_{p_1})$. Applying the inverse Fourier transform we conclude that $\mathrm{dom}(A) = \mathrm{dom}(A_1)$. Clearly the operator $J(p(D) + I) = A + J$ is uniformly positive in $(L^2(\mathbb{R}^n), [\cdot\,, \cdot])$ and $\mathrm{dom}(A+J) = \mathrm{dom}(A) = \mathrm{dom}(A_1)$. Since ∞ is not a singular critical point of A_1, [2, Corollary 3.3] implies that ∞ is not a singular critical point of $A+J$. Therefore $A + J = J(B + I)$ is similar to a selfadjoint operator in $(L^2(\mathbb{R}^n), \langle \cdot\,, \cdot \rangle)$. Lemma 2.1 implies that $\rho(A) \neq \emptyset$ and consequently A is a definitizable operator. Since $\mathrm{dom}(A+J) = \mathrm{dom}(A)$, [2, Corollary 3.3] implies that ∞ is not a singular critical point of A. This proves part (a).

We prove part (b) for a strongly separated polynomial p. It remains to prove that 0 is not a singular critical point of A. By Lemma 4.3 the ranges of the multiplication operators M_p and M_{p_1} coincide. Applying the inverse Fourier transform we conclude that $\mathrm{ran}(A) = \mathrm{ran}(A_1)$. Note that 0 is not an eigenvalue neither of A nor of A_1. Since 0 is not a singular critical point of A_1, we conclude that 0 is not a singular critical point of A. This proves the theorem. □

Proposition 4.7. *Let q be a nonnegative polynomial in $n-1$ variables, r a nonnegative and nonconstant polynomial in one variable and $p(x) = q(\hat{x}) + r(t)$. Let $A = Jp(D)$. Then:*

(a) *The operator A has no eigenvalues.*

(b) *The spectrum of A is given by*

$$\sigma(A) = (-\infty, -m_p] \cup [m_p, +\infty) \, , \tag{4.10}$$

where $m_p = \min\{p(x) : x \in \mathbb{R}^n\}$.

Proof. (a) The operator A is definitizable by Theorem 4.6. Let $\lambda \in \mathbb{R}$ and $y \in \mathrm{dom}(A)$ satisfy $Jp(D)y = \lambda y$. Let $z = Fy$ be the partial Fourier transform of y. Then

$$J\left(r\left(\frac{1}{i}\frac{d}{dt}\right) + q(\hat{x})I\right) z(\hat{x}, t) = \lambda z(\hat{x}, t) \, .$$

[6, Theorem 2.2 (b)] implies that $z(\hat{x}, \cdot) = 0$ for all $\hat{x} \in \mathbb{R}^{n-1}$. Thus $y = 0$.

To prove (b) we extend the argument of Lemma 4.4. Denote by E the spectral function of A and by G_α the spectral function of the operator $J(r(-i\frac{d}{dt}) + \alpha I)$. The equalities (4.4) and (4.5) hold true for newly defined G_α.

We prove that for all positive a, b such that $b > m_p$ and $a < b$ we have $E(a,b) \neq 0$. Note that $m_p = m_r + m_q$. Let $\hat{x}_0$ be such that $m_r + q(\hat{x}_0) < b$. By [6, Theorem 2.2] the spectrum of the operator $J(r(-i\frac{d}{dt}) + q(\hat{x}_0)I)$ is $(-\infty, -m_r - q(\hat{x}_0)] \cup [m_r + q(\hat{x}_0), +\infty)$. Therefore there exists $h \in L^2(\mathbb{R})$ such that $G_{q(\hat{x}_0)}(a,b)h \neq 0$. The function $\alpha \mapsto \|G_\alpha(a,b)h\|$ is continuous on $\mathbb{R}_+$ by [18, Theorem 3.1. part 3)]. Therefore the function

$$\hat{x} \mapsto \|G_{q(\hat{x})}(a,b))h\|$$

is continuous on $\mathbb{R}^{n-1}$. Hence the set

$$\mathcal{O} = \{\hat{x} \in \mathbb{R}^{n-1} : \|G_{q(\hat{x})}(a,b)h\| > 0\}$$

is open. This set is nonempty since $\hat{x}_0 \in \mathcal{O}$. The set $\mathcal{O}$ is contained in $\{\hat{x} \in \mathbb{R}^{n-1} : q(\hat{x}) \leq b\}$. Choose $z \in L^2(\mathbb{R}^{n-1})$ such that $z \neq 0$ almost everywhere. Let $y(x) = h(t)(F^{-1}z)(\hat{x})$. From (4.5) it follows

$$\begin{aligned} \|E(a,b)y\|^2 &= \int_{q(\hat{x})\leq b} |z(\hat{x})|^2 \|G_{q(\hat{x})}(a,b)h\|^2 d\hat{x} \\ &\geq \int_{\mathcal{O}} |z(\hat{x})|^2 \|G_{q(\hat{x}}(a,b)h\|^2 d\hat{x} > 0 . \end{aligned}$$

We have proved that for arbitrary $b > m_p$ and $0 < a < b$ we have $E(a,b) \neq 0$.

This implies that the spectrum of A in $\mathbb{R}_+$ contains $[m_p, +\infty)$. If $m_p > 0$ and $0 < \lambda < m_p$, then (4.5) implies that $\lambda \in \rho(A)$. In this case $0 \in \rho(A)$ since A is a uniformly positive operator. Therefore the spectrum of A in $\mathbb{R}_+$ coincides with $[m_p, +\infty)$. Similarly one proves that the spectrum of A in $\mathbb{R}_-$ coincides with $(-\infty, -m_p]$. □

Corollary 4.8. *Let q be a nonnegative polynomial in $n-1$ variables, r a nonnegative and nonconstant polynomial in one variable and $p(x) = q(\hat{x}) + r(t)$. Let $A = Jp(D)$.*

(a) *The point ∞ <u>is</u> a regular critical point of $A = Jp(D)$.*

(b) *Assume that the polynomial r has at most one root. The following statements are equivalent:*

 (i) *p has a zero.*

 (ii) $0 \in \sigma(A)$.

 (iii) *0 <u>is</u> a regular critical point of A.*

5. Variable coefficients[1]

In this section we use Corollary 2.5 to extend results from Section 4. To illustrate the method, we consider the Schrödinger operator with indefinite weight $(\operatorname{sgn} x_n)(-\Delta + q)$ on $\mathbb{R}^n$.

Let $H = -\Delta$ be defined on its natural domain in $L^2(\mathbb{R}^n)$. Its inverse is an unbounded integral operator.

Proposition 5.1. *Let* $5 \leq n \leq 8$ *and* $q \in L^{n/2}(\mathbb{R}^n)$. *There exists* $\kappa_0 > 0$ *such that for all real* κ *with* $|\kappa| < \kappa_0$ *the operator* $(\operatorname{sgn} x_n)(-\Delta + \kappa q)$ *is similar to a selfadjoint operator in* $L^2(\mathbb{R}^n)$.

Proof. Since $n \leq 8$, it follows from the Sobolev embedding theorem that

$$\operatorname{dom}(H) = H^2(\mathbb{R}^n) \subseteq \operatorname{dom}(q) .$$

We show that the operator qH^{-1} is bounded by a constant multiple of $\|q\|_{n/2}$.

Note that $H^{-1} = h(-i\nabla)$ with $h(x) = |x|^{-2}$. Therefore $h \in L_w^{n/2}(\mathbb{R}^n)$, see [21, Example IX.4.2] . By [22, Theorem 4.2] $q(x)h(-i\nabla) \in L_w^{n/2}(\mathbb{R}^n)$, and moreover

$$\|q(x)h(-i\nabla)\|_{n/2,w} \leq C\|q\|_{n/2}\|h\|_{n/2,w} ,$$

where $\|\cdot\|_{p,w}$ are the functions defined in [22, p. 13] and [21, Definition IX.4]. Hence (see [22, p. 13]) $\|q(x)h(-i\nabla)\|_{n/2,w} \leq C_1\|q\|_{n/2}$. Next we can use the inequalities on p. 13 of [22] to conclude

$$\|q(x)h(-i\nabla)\| = \|q(x)h(-i\nabla)\|_{\infty} \leq C_2\|q(x)h(-i\nabla)\|_{n/2,w} \leq C_3\|q\|_{n/2} .$$

It follows from [16, Theorems IV.1.1, IV.2.14, IV.3.1 and VI.3.1] that for $|\kappa|$ sufficiently small we have that the operator $J(H+\kappa q)$ is positive in $(L^2(\mathbb{R}^n), [\cdot\,,\cdot])$; the resolvent set $\rho(J(H+\kappa q))$ is nonempty and $\operatorname{dom}(J(H+\kappa q)) = \operatorname{dom}(H)$. The conclusion of the proposition follows from Theorem 4.6, Lemma 2.2, Corollary 2.5 and [2, Corollary 3.3]. □

Note that we needed $n \leq 8$ only to prove that the operator qH^{-1} is densely defined. However, the Gagliardo-Nirenberg inequality implies that $\operatorname{dom}(H) \subseteq \operatorname{dom}(q)$ (and also that qH^{-1} is bounded) as soon as $n \geq 5$. This shows that the assumption $n \leq 8$ is in fact redundant.

We prove a strengthening of Proposition 5.1 .

Theorem 5.2. *Let* $n \geq 5$ *and*

$$B = \sum_{0 \leq i+j \leq 2} b_{ij} D^{ij}$$

[1]For the case of a more general elliptic operator with indefinite weight $\frac{1}{r}(L+q)$ we can use the results from [9].

be a partial differential operator with the coefficients b_{ij} satisfying

$$b_{ij} \in L^\infty \quad \text{if} \quad i+j=2\ ,\ b_{ij} \in L^n(\mathbb{R}^n) \ \text{if} \ i+j=1\ , \ \text{and}\ b_{00} \in L^{n/2}(\mathbb{R}^n)\ .$$

Further assume that B is symmetric in $(L^2(\mathbb{R}^n), \langle\cdot\,,\,\cdot\rangle)$.

Then the operator B, defined on $\operatorname{dom}(B) = H^2(\mathbb{R}^n)$ is a closed operator in $L^2(\mathbb{R}^n)$. There exists κ_0 such that if

$$\sum_{1\le i\le n} \|b_{ii} - \operatorname{sgn} x_n\|_\infty + \sum_{i+j=2, j\neq i} \|b_{ij}\|_\infty + \sum_{i+j=1} \|b_{ij}\|_n + \|b_{00}\|_{n/2} \le \kappa_0\ ,$$

then $(\operatorname{sgn} x_n)B$ is similar to a selfadjoint operator in $(L^2(\mathbb{R}^n), \langle\cdot\,,\,\cdot\rangle)$.

Proof. The first statement easily follows from the Sobolev embedding theorem. Let $A_0 = (\operatorname{sgn} x_n)(-\Delta) = JH$, defined on $\operatorname{dom}(A_0) = \operatorname{dom}(B)$, $V = JB - A_0$. By Theorem 4.6 the operator A_0 is similar to a selfadjoint operator in $L^2(\mathbb{R}^n)$. Note that

$$JV = \sum_{0\le i+j\le 2} v_{ij} D^{ij}$$

with $v_{ii} = b_{ii} - \operatorname{sgn} x_n, 1 \le i \le n$, $v_{ij} = b_{ij}$ for all other i, j. It is sufficient to show that JVA_0^{-1} or equivalently VH^{-1} is a bounded densely defined operator. To this end, we show that $v_{ij}D^{ij}H^{-1}$ is bounded and densely defined. In fact, it is sufficient to show that

$$\|v_{ij}D^{ij}u\| \le C_{ij}\|\Delta u\| \quad u \in H^2(\mathbb{R}^n) \tag{5.1}$$

for all i,j with $i+j \le 2$. If $i+j=2$, the estimate (5.1) is evident. If $i=j=0$, let $\frac{1}{p} = \frac{1}{2} - \frac{2}{n}$. Hölder's inequality yields

$$\|qu\|_2 \le \|q\|_{n/2}\|u\|_p \quad (u \in L^p(\mathbb{R}^n)),$$

From the Gagliardo-Nirenberg inequality, see [20, p. 125], it follows that

$$\|u\|_p \le C\|\Delta u\|_2 \quad (u \in H^2(\mathbb{R}^n))\ .$$

This implies (5.1) if $i+j=0$. Finally, if $i+j=1$, then $v_{ij}D^{ij} = b_k D^k$ for some $k \in \{1, \ldots, n\}$, where $b_k = b_{k0}$ or b_{0k}. Let $p = \frac{2n}{n-2}$. From Hölder's and Gagliardo-Nirenberg inequality we again find

$$\|v_{ij}D^{ij}u\| = \|b_k D^k u\| \le \|b_k\|_n \|D^k u\|_p \le C\|b_k\|_n\|\Delta u\|\ ,$$

and this proves (5.1). □

References

[1] Beals, R., Indefinite Sturm-Liouville problems and half-range completeness. J. Differential Equations 56 (1985), 391–407.

[2] Ćurgus, B., On the regularity of the critical point infinity of definitizable operators. Integral Equations Operator Theory 8 (1985), 462–488.

[3] Ćurgus, B., Langer, H., A Krein space approach to symmetric ordinary differential operators with an indefinite weight function. J. Differential Equations 79 (1989), 31–61.

[4] Ćurgus, B., Najman, B., A Krein space approach to elliptic eigenvalue problems with indefinite weights. Differential and Integral Equations 7 (1994), 1241–1252.

[5] Ćurgus, B., Najman, B., The operator $(\operatorname{sgn} x)\frac{d^2}{dx^2}$ is similar to a selfadjoint operator in $L^2(\mathbb{R})$. Proc. Amer. Math. Soc. 123 (1995), 1125–1128.

[6] Ćurgus, B., Najman, B., Positive differential operators in Krein space $L^2(\mathbb{R})$. Recent developments in operator theory and its applications (Winnipeg, MB, 1994), 95–104, Oper. Theory Adv. Appl., 87, Birkhäuser, Basel, 1996.

[7] Ćurgus, B., Najman, B., Preservation of the range under perturbations of an operator. Proc. Amer. Math. Soc. 125 (1997), 2627–2631.

[8] Dijksma, A., Langer, H., Operator theory and ordinary differential operators. Lectures on operator theory and its applications (Waterloo, ON, 1994), 73–139, Fields Inst. Monogr., 3, Amer. Math. Soc., Providence, RI, 1996.

[9] Edmunds, D. E., Triebel, H., Eigenvalue distributions of some degenerate elliptic operators, an approach via entropy numbers. Math. Ann. 299 (1994), 311–340.

[10] Faierman, M., Langer, H., Elliptic problems involving an indefinite weight function. Recent developments in operator theory and its applications (Winnipeg, MB, 1994), 105–124, Oper. Theory Adv. Appl., 87, Birkhäuser, Basel, 1996.

[11] Fleige, A., A spectral theory of indefinite Krein-Feller differential operators, Mathematical Research 98, Akademie Verlag, Berlin 1996.

[12] Fleige, A., A counterexample to completeness properties for indefinite Sturm-Liouville problems. Math. Nach. 190 (1998), 123–128.

[13] Fleige, A., Najman, B., Nonsingularity of critical points of some differential and difference operators, Differential and Integral Operators (Regensburg, 1995), 85–95, Oper. Theory: Adv. Appl., 102, Birkhäuser, Basel, 1998.

[14] Jonas, P., Compact perturbations of definitizable operators. II. J. Operator Theory 8 (1982), 3–18.

[15] Jonas, P., On a problem of the perturbation theory of selfadjoint operators in Krein spaces. J. Operator Theory 25(1991), 183–211.

[16] Kato, T., Perturbation Theory of Linear Operators. Springer-Verlag, Berlin, 1966.

[17] Langer, H., Spectral function of definitizable operators in Krein spaces. Functional Analysis, Proceedings, Dubrovnik 1981. Lecture Notes in Mathematics 948, Springer-Verlag, Berlin, 1982, 1–46.

[18] LANGER, H., NAJMAN, B., Perturbation theory for definitizable operators in Krein spaces. J. Operator Theory 9 (1983), 297–317.

[19] MEISTER, E., LATZ, N., SCHEURER, J., Spectral analysis of a transmission problem for the Helmholtz equation on the half-space, Rendiconti di Matematica, Ser. VII, 13(1993), 751–772.

[20] NIRENBERG, L., On elliptic partial differential equations, Ann. Scuola Norm. Pisa 13 (1959), 115–162.

[21] REED, M., SIMON, B., Methods of Modern Mathematical Physics II: Fourier Analysis, Self-adjointness. Academic Press, New York, 1975.

[22] SIMON, B., Trace Ideals and Their Applications, Cambridge University Press, Cambridge, 1979.

[23] VESELIĆ, K., On spectral properties of a class of J-selfadjoint operators. I. Glasnik Mat. Ser. III 7(27) (1972), 229–248.

[24] Volkmer, H., Sturm-Liouville problems with indefinite weights and Everitt's inequality, Proc. Roy. Soc. Edinburgh Sect. A 126 (1966), 1097–1112.

[25] WEIDMANN, J., Linear Operators in Hilbert Spaces. Springer-Verlag, Berlin, 1980.

Department of Mathematics
Western Washington University
Bellingham, WA 98225, USA

1991 Mathematics Subject Classification. Primary 47B50, 47E05; Secondary 47B25, 34L05

Operator Theory:
Advances and Applications, Vol. 106

Singular values of positive pencils and applications

ROBERT L. ELLIS, ISRAEL GOHBERG, AND DAVID C. LAY

Dedicated to Professor Heinz Langer on the occasion of his 60-th birthday

Singular values are introduced and studied for pencils $A - \lambda G$ of selfadjoint matrices which for some values of λ are positive definite. These singular values describe the widths of certain unbounded sets.

1. Introduction

In this paper, which is related to [BG], we investigate a generalization of the singular values of a complex matrix. This generalization applies to selfadjoint pencils $A - \lambda G$ that have positive definite values for some choices of the parameter λ. The generalized singular values are the square roots of the absolute values of some of the eigenvalues of the pencil. For the case in which $G = I$ and $A = T^*T$, they coincide with the usual singular values of T. We obtain a generalized singular value decomposition and a canonical form of the pencil. Some of the new singular values may be interpreted geometrically as widths (in the sense of Gelfand) of the possibly unbounded images $T(B_+)$ and $T(B_-)$ of the indefinite unit balls $B_+ = \{x \mid \langle Gx, x\rangle \leq 1\}$ and $B_- = \{x \mid \langle Gx, x\rangle \geq -1\}$.

There are four sections in this paper. In Section 2, we give necessary and sufficient conditions for a pencil of the form $T^*T - \lambda G$ to assume a positive definite value, define the new singular values, and give several examples. The third section is devoted to the geometric interpretation of singular values as widths. For the examples of Section 2, this interpretation yields formulas that generalize well-known formulas for the usual singular values. In the final section we study the set of values of the parameter λ for which $T^*T - \lambda G$ is positive definite.

2. Positive pencils and singular values

Let A and G be selfadjoint $r \times r$ matrices. We say that the pencil $\{A - \lambda G \mid \lambda \in \mathbf{R}\}$ is **positive** if it contains a positive definite matrix. This means that there exists a real number ν such that

$$A - \nu G > 0.$$

Of special interest is the case in which $A = T^*T$ for some $r \times r$ matrix T. The following theorem describes all positive pencils of the form $T^*T - \lambda G$. For any integer n, I_n will denote the $n \times n$ identity matrix.

Theorem 2.1. *Let G be a nonzero $r \times r$ selfadjoint matrix with p positive, q negative and s zero eigenvalues and let T be an $r \times r$ matrix. Then the pencil $T^*T - \lambda G$ is positive if and only if there exist an invertible $r \times r$ matrix S and a nonnegative diagonal matrix $D = diag(d_1, \ldots, d_r)$ such that*

$$S^*GS = G_1 \tag{2.1}$$

where

$$G_1 = \begin{pmatrix} I_p & 0 & 0 \\ 0 & 0 & 0 \\ 0 & 0 & -I_q \end{pmatrix} \tag{2.2}$$

and

$$S^*T^*TS = D \tag{2.3}$$

with

$$d_1 \geq \cdots \geq d_p, \; d_{p+1} \geq \cdots \geq d_{p+s} > 0, \; d_{p+s+1} \geq \cdots \geq d_r \tag{2.4}$$

and with

$$d_p \neq 0 \text{ or } d_r \neq 0.$$

Proof. Suppose the pencil $T^*T - \lambda G$ is positive, and let ν be any real number for which $T^*T - \nu G > 0$. Then it follows [LT, Theorem 2, page 185] that $T^*T - \nu G$ and G are simultaneously diagonalizable by congruence. This implies that T^*T and G are simultaneously diagonalizable by congruence. Thus there is an invertible $r \times r$ matrix S such that S^*T^*TS and S^*GS are diagonal. Since G has p positive, q negative and s zero eigenvalues, we may multiply S on the right by an invertible matrix that interchanges and scales columns in order to obtain (2.1) and (2.2). This will not change the fact that S^*T^*TS is diagonal. Then we may multiply S on the right by a permutation matrix that interchanges some of the first p columns, some of the next s columns, and some of the last q columns to obtain (2.3) with

$$d_1 \geq \cdots \geq d_p, \; d_{p+1} \geq \cdots \geq d_{p+s}, d_{p+s+1} \geq \cdots \geq d_r \tag{2.5}$$

This will not destroy (2.1) and (2.2). Since

$$T^*T - \nu G > 0 \tag{2.6}$$

it follows from (2.1) and (2.3) that

$$D - \nu G_1 > 0. \tag{2.7}$$

This implies that

$$d_j - \nu > 0 \qquad (1 \leq j \leq p) \tag{2.8}$$

$$d_j > 0 \qquad (p + 1 \leq j \leq p + s) \tag{2.9}$$

$$d_j + \nu > 0 \qquad (p + s + 1 \leq j \leq r). \tag{2.10}$$

Then (2.4) follows from (2.5) and (2.9). The inequalities (2.8) and (2.10) imply that

$$\text{(2.11)}\quad \max\{-d_j \mid p+s+1 \le j \le r\} = -d_r < \nu < d_p = \min\{d_j \mid 1 \le j \le p\}$$

so that either $d_r \neq 0$ or $d_p \neq 0$. Conversely, suppose there exists an invertible S and a nonnegative diagonal matrix D such that (2.1)–(2.4) hold. Let ν be any real number satisfying (2.11). Then (2.8)–(2.10) and hence (2.7) are true. This along with (2.1), (2.3) and the invertibility of S implies (2.6), which means that the pencil $T^*T - \lambda G$ is positive. □

Example 2.2. Let $G = I_r$. Then $G_1 = I_r$ in (2.2), so that S is unitary by (2.1). Thus (2.3) implies that the diagonal entries of D are the eigenvalues of T^*T. Therefore $d_1^{1/2}, \dots, d_r^{1/2}$ are the singular values of T.

Example 2.3. Let G be positive definite. Then the pencil $T^*T - \lambda G$ is positive for any $r \times r$ matrix T. For any real number ν, $T^*T - \nu G > 0$ if and only if $(TG^{-1/2})^*(TG^{-1/2}) - \nu I > 0$. Furthermore, an invertible matrix S and a nonnegative diagonal matrix D satisfy (2.1) and (2.3) if and only if

$$(G^{1/2}S)^* I (G^{1/2}S) = G_1$$

and

$$(G^{1/2}S)^*(TG^{-1/2})^*(TG^{-1/2})(G^{1/2}S) = D.$$

From Example 2.2 we conclude that $d_1^{1/2}, \dots, d_r^{1/2}$ are the singular values of $TG^{-1/2}$.

Example 2.4. Let $T = I_r$. Then the pencil $T^*T - \lambda G = I_r - \lambda G$ is positive for every selfadjoint $r \times r$ marix G, and (2.3) becomes

$$S^*S = D.$$

Therefore D is invertible and $SD^{-1/2}$ is unitary. By (2.1),

$$\begin{aligned}(SD^{-1/2})^*G(SD^{-1/2}) &= D^{-1/2}G_1D^{-1/2}\\ &= \operatorname{diag}(d_1^{-1}, \dots, d_p^{-1}, 0, \dots, 0, -d_{p+s+1}^{-1}, \dots, -d_r^{-1}).\end{aligned}$$

Since $SD^{-1/2}$ is unitary, it follows that the eigenvalues of G_1, which we will denote by $\lambda_1, \dots, \lambda_r$, are the diagonal entries of the preceding diagonal matrix. Therefore

$$d_j = \begin{cases} \lambda_j^{-1} & (1 \le j \le p) \\ -\lambda_j^{-1} & (p+s+1 \le j \le r) \end{cases}$$

provided the eigenvalues $\lambda_1, \ldots, \lambda_r$ are ordered so that

$$0 < \lambda_1 \leq \cdots \leq \lambda_p, \ \lambda_{p+1} = \cdots = \lambda_{p+s} = 0, \ 0 > \lambda_{p+s+1} \geq \cdots \geq \lambda_r.$$

In Examples 2.2–2.4, the numbers $d_1, \ldots, d_p, \ d_{p+s+1}, \ldots, d_r$ are uniquely determined by G and T and hence do not depend on S. In the next section we will prove that that is true in general. In view of Example 2.2, we will call the numbers $d_1^{1/2}, \ldots, d_p^{1/2}, \ d_{p+s+1}^{1/2}, \ldots, d_r^{1/2}$ the **G-singular values** of T. However, if $s \neq 0$, then the numbers $d_{p+1}, \ldots, d_{p+s}$ do depend on S. In fact, S can always be chosen so that $d_{p+1}, \ldots, d_{p+s}$ are any preassigned s positive numbers satisfying $d_{p+1} \geq \cdots \geq d_{p+s}$. To prove this, we let D_s be any $s \times s$ positive definite diagonal matrix whose diagonal entries are nondecreasing with index, and define

$$S' = S \begin{pmatrix} I_p & 0 & 0 \\ 0 & D_s & 0 \\ 0 & 0 & I_q \end{pmatrix}$$

and

$$D' = D \begin{pmatrix} I_p & 0 & 0 \\ 0 & D_s^2 & 0 \\ 0 & 0 & I_q \end{pmatrix}.$$

Then (2.1)–(2.3) remain true when S is replaced by S' and D is replaced by D'. In that case, the numbers $d_1, \ldots, d_p, \ d_{p+s+1}, \ldots, d_r$ remain the same in (2.4), but the numbers $d_{p+1}, \ldots, d_{p+s}$ are multiplied by the diagonal entries of D_s^2.

Finally, we remark that by (2.1) and (2.3),

$$\det G_1 \mid \det T \mid^2 = \det G \prod_{j=1}^{r} d_j$$

If G is invertible, it follows that

$$|\det T|^2 = |\det G| \prod_{j=1}^{r} d_j.$$

3. Widths

In this section we will use the notion of the Gelfand width of a set [P, page 7] to interpret the G-singular values of a matrix T.

For any centrally symmetric subset E of $\mathbb{C}^r$ and for $1 \leq k \leq r$, the **kth width** of E is defined by

(3.1) $$\omega_k(E) = \inf_{L \in \mathcal{L}_k} \sup\{\|x\| \mid x \in L \cap E\}$$

where $\mathcal{L}_k$ is the set of k-dimensional subspaces of $\mathbb{C}^r$. Let G be a nonzero $r \times r$ selfadjoint matrix and let

$$B_+ = \{x \in \mathbf{C}^r \mid \langle Gx, x\rangle \leq 1\}$$

and

$$B_- = \{x \in \mathbf{C}^r \mid \langle Gx, x\rangle \geq -1\}$$

where $\langle\cdot,\cdot\rangle$ denotes the usual scalar product on $\mathbf{C}^r$. The next theorem shows the connection between the G-singular values of an $r\times r$ matrix T for which the pencil $T^*T - \lambda G$ is positive, and the widths of $T(B_+)$ and $T(B_-)$.

Theorem 3.1. *Let G be a nonzero $r\times r$ selfadjoint matrix with p positive, q negative and s zero eigenvalues. Let T be an $r \times r$ matrix such that the pencil $T^*T - \lambda G$ is positive, and let $d_1^{1/2}, \ldots, d_p^{1/2}$, $d_{p+s+1}^{1/2}, \ldots, d_r^{1/2}$ be the G-singular values of T.*

a. If T is invertible, then

$$\omega_k(T(B_+)) = \begin{cases} d_{p-k+1}^{1/2} & , \quad 1 \leq k \leq p \\ \infty & , \quad p < k \leq r \end{cases} \tag{3.2}$$

and

$$\omega_k(T(B_-)) = \begin{cases} d_{r-k+1}^{1/2} & , \quad 1 \leq k \leq q \\ \infty & , \quad q < k \leq r \end{cases} \tag{3.3}$$

b. If T is not invertible and $d_r \neq 0$, then

$$\omega_k(T(B+)) = \begin{cases} d_{p-k+1}^{1/2} & , \quad 1 \leq k \leq p \\ \infty & , \quad p < k \leq r \end{cases} \tag{3.4}$$

and

$$\omega_k(T(B_-)) = \begin{cases} 0 & , \quad 1 \leq k \leq p-m \\ \infty & , \quad p-m < k \leq r \end{cases} \tag{3.5}$$

where $m = \max\{j \mid 1 \leq j \leq p$ and $d_j \neq 0\}$.

c. If T is not invertible and $d_p \neq 0$, then

$$\omega_k(T(B_+)) = \begin{cases} 0 & , \quad 1 \leq k \leq q-n \\ \infty & , \quad q-n < k \leq r \end{cases} \tag{3.6}$$

and

$$\omega_k(T(B_-)) = \begin{cases} d_{r-k+1}^{1/2} & , \quad 1 \leq k \leq q \\ \infty & , \quad q < k \leq r \end{cases} \tag{3.7}$$

where $n = \max\{j \mid 1 \leq j \leq q$ and $d_{p+s+j} \neq 0\}$.

Proof. Before beginning the proof, we observe from (3.1) that

$$(3.8)\quad \omega_k(T(B_+)) = \inf_{L\in\mathcal{L}_k} \{\sup \|Tz\| \mid Tz \in L,\ \langle Gz, z\rangle \leq 1\} \qquad (1 \leq k \leq r)$$

and

$$\omega_k(T(B_-)) = \inf_{L\in\mathcal{L}_k} \{\sup \|Tz\| \mid Tz \in L,\ \langle Gz, z\rangle \geq -1\} \qquad (1 \leq k \leq r).$$

We make the change of variable

$$y = S^{-1}z$$

where S is an invertible matrix such that (2.1)–(2.4) hold. Using (2.1), we find that

$$(3.9a)\quad \omega_k(T(B_+)) = \inf_{L\in\mathcal{L}_k} \sup \{\|TSy\| \mid TSy \in L,\ \langle G_1y, y\rangle \leq 1\} \quad (1 \leq k \leq r)$$

and

$$(3.9b)\quad \omega_k(T(B_-)) = \inf_{L\in\mathcal{L}_k} \sup \{\|TSy\| \mid TSy \in L,\ \langle G_1y, y\rangle \geq -1\} \quad (1 \leq k \leq r).$$

We will first prove (3.2), (3.4) and (3.5). Then we will derive (3.3), (3.6) and (3.7) from (3.2), (3.4) and (3.5).

For the case in which $G \leq 0$ we have $p = 0$, $s + q = r$ and $B_+ = \mathbf{C}^r$. Thus (3.2) is obviously true. If $G \leq 0$ and $d_r \neq 0$, then T is invertible, so part (b) does not arise if $G \leq 0$. Thus we may assume in the rest of the proof of (3.2), (3.4) and (3.5) that $p \geq 1$. Because (3.2), (3.4) and (3.5) appear in parts (a) and (b), we also assume that either T is invertible, or T is not invertible and $d_r \neq 0$. Since

$$\dim\ker T = \dim\ker TS = \dim\ker S^*T^*TS = \dim\ker D = p - m$$

we have

$$\operatorname{codim} \mathcal{R}(T) = \operatorname{codim} \mathcal{R}(TS) = p - m$$

where $\mathcal{R}(\cdot)$ denotes the range of an operator. If T is not invertible and $1 \leq k \leq p - m$, then there is a k-dimensional subspace L of $\mathbf{C}^r$ such that $L \cap \mathcal{R}(T) = \{0\}$. Thus

$$\sup\{\|Tz\| \mid Tz \in L,\ \langle Gz, z\rangle \leq 1\} = \{0\}$$

so it follows from (3.8) that

$$(3.10)\qquad \omega_k(T(B_+)) = 0 \qquad (1 \leq k \leq p - m).$$

Since $d_{m+1}, \ldots, d_p = 0$ in case T is not invertible and $d_r \neq 0$, this proves (3.4) for $1 \leq k \leq p - m$. Now we will assume that $r \geq k > p - m$, which is no restriction if T is invertible. Let L be any k-dimensional subspace of $\mathbf{C}^r$. Since $\dim L > p - m = \operatorname{codim} \mathcal{R}(TS)$, it follows that

$$\dim(L \cap \mathcal{R}(TS)) \geq k - p + m > 0.$$

Let $s' = k-p+m$. Then there are vectors $w_1, \ldots, w_{s'}$ such that $TSw_1, \ldots, TSw_{s'}$ are linearly independent vectors in L. Write

$$w_j = w_j^+ + w_j^0 + w_j^{00} + w_j^- \qquad (1 \leq j \leq s') \tag{3.11}$$

where $w_j^+ \in \operatorname{Span}\{e_1, \ldots, e_m\}$, $w_j^0 \in \operatorname{Span}\{e_{m+1}, \ldots, e_p\} = \operatorname{Ker} TS$, $w_j^{00} \in \operatorname{Span}\{e_{p+1}, \ldots, e_{p+s}\}$, and $w_j^- \in \operatorname{Span}\{e_{p+s+1}, \ldots, e_r\}$. Then

$$TS(w_j) = TS(w_j^+ + w_j^{00} + w_j^-). \tag{3.12}$$

Furthermore, since $s' + (p-k+1) = m+1$ and $w_1^+, \ldots, w_{s'}^+$, $e_1, \ldots, e_{p-k+1}$ are all in $\operatorname{Span}\{e_1, \ldots, e_m\}$, there are scalars $\alpha_1, \ldots, \alpha_{s'}$ (not all 0) and scalars $c_1, \ldots, c_{p-k+1}$ such that

$$\sum_{j=1}^{s'} \alpha_j w_j^+ = \sum_{j=1}^{p-k+1} c_j e_j \tag{3.13}$$

We will show that a scalar γ exists so that the vector

$$y' = \gamma \sum_{j=1}^{s'} \alpha_j (w_j^+ + w_j^{00} + w_j^-) \tag{3.14}$$

satisfies

$$\langle G_1 y', y' \rangle \leq 1 \text{ and } \|TSy'\| \geq d_{p-k+1}^{1/2}.$$

Then by (3.12), $TSy' \in L$ so that

$$\sup\{\|TSy\| \mid TSy \in L,\ \langle G_1 y, y \rangle \leq 1\} \geq d_{p-k+1}^{1/2}.$$

By (3.9a) this implies that

$$\omega_k(T(B_+)) \geq d_{p-k+1}^{1/2} \qquad (p-m+1 \leq k \leq p). \tag{3.15}$$

Case 1 $\|\sum_{j=1}^{s'} \alpha_j w_j^+\| \leq \|\sum_{j=1}^{s'} \alpha_j w_j^-\|$

For any γ it follows from (2.2) and (3.14) that

$$\langle G_1 y', y' \rangle = |\gamma|^2 \left(\Big\| \sum_{j=1}^{s'} \alpha_j w_j^+ \Big\|^2 - \Big\| \sum_{j=1}^{s'} \alpha_j w_j^- \Big\|^2 \right) \leq 0.$$

Moreover,

$$\|TSy'\| = |\gamma|\ \|\sum_{j=1}^{s'} \alpha_j TS(w_j^+ + w_j^{00} + w_j^-)\|$$

$$= |\gamma|\ \|\sum_{j=1}^{s'} \alpha_j TSw_j\|.$$

Since $TSw_1, \ldots, TSw_{s'}$ are linearly independent and $\alpha_1, \ldots, \alpha_{s'}$ are not all 0, it follows that $\sum_{j=1}^{s'} \alpha_j TSw_j$ is not 0. Thus γ may be chosen so that $\|TSy'\| \geq d_{p-k+1}^{1/2}$.

Case 2 $\|\sum_{j=1}^{s'} \alpha_j w_j^+\| > \|\sum_{j=1}^{s'} \alpha_j w_j^-\|$

As in case 1,

$$\langle G_1 y', y'\rangle = |\gamma|^2 (\|\sum_{j=1}^{s'} \alpha_j w_j^+\|^2 - \|\sum_{j=1}^{s'} \alpha_j w_j^-\|^2).$$

In this case we may choose γ so that

$$\langle G_1 y', y'\rangle = 1$$

By (3.14) and (3.13), y' has the form

$$y' = \sum_{j=1}^{p-k+1} y_j e_j + \sum_{j=p+1}^{r} y_j e_j$$

and hence

$$\sum_{j=1}^{p-k+1} |y_j|^2 - \sum_{j=p+s+1}^{r} |y_j|^2 = \langle G_1 y', y'\rangle = 1.$$

It follows that

$$\begin{aligned}
\|TSy'\|^2 &= \langle TSy', TSy'\rangle = \langle S^*T^*TSy', y'\rangle \\
&= \langle Dy', y'\rangle = \sum_{j=1}^{p-k+1} d_j |y_j|^2 + \sum_{j=p+1}^{r} d_j |y_j|^2 \\
&\geq \sum_{j=1}^{p-k+1} d_j |y_j|^2 \geq d_{p-k+1} \sum_{j=1}^{p-k+1} |y_j|^2 \\
&\geq d_{p-k+1} \left(\sum_{j=1}^{p-k+1} |y_j|^2 - \sum_{j=p+s+1}^{r} |y_j|^2 \right) \\
&\geq d_{p-k+1}.
\end{aligned}$$

Thus we have constructed the vector y' in (3.14) in both cases, so (3.15) holds. Now let

$$L = \mathrm{Span}\{TSe_{p-k+1}, \ldots, TSe_m\} \oplus \mathcal{R}(TS)^{\perp},$$

a k-dimensional subspace of $\mathbf{C}^r$. Consider any vector y such that $\langle G_1 y, y\rangle \leq 1$ and $TSy \in L$. Then

$$TSy = TS\left(\sum_{j=p-k+1}^{m} \alpha_j e_j\right)$$

for suitable scalars α_j $(p-k+1 \leq j \leq m)$, so that

$$y = \sum_{j=p-k+1}^{p} \alpha_j e_j$$

for suitable scalars α_j $(m+1 \leq j \leq p)$. Therefore

$$1 \geq \langle G_1 y, y\rangle = \sum_{j=p-k+1}^{p} |\alpha_j|^2$$

and hence

$$\sum_{j=p-k+1}^{m} |\alpha_j|^2 \leq 1.$$

Furthermore

$$\begin{aligned} \|TSy\|^2 &= \sum_{j=p-k+1}^{m} |\alpha_j|^2 \langle TSe_j, TSe_j\rangle \\ &= \sum_{j=p-k+1}^{m} |\alpha_j|^2 d_j \\ &\leq d_{p-k+1} \sum_{j=p-k+1}^{m} |\alpha_j|^2 \\ &\leq d_{p-k+1}. \end{aligned}$$

This and (3.9a) imply that

$$\omega_k(T(B_+)) \leq d_{p-k+1}^{1/2}.$$

Combining this with (3.15) yields

$$\omega_k(T(B_+)) = d_{p-k+1}^{1/2}.$$

This completes the proof of (3.2) and (3.4) for $1 \leq k \leq p$. Next we suppose that $p+1 \leq k \leq r$. Let L be any k-dimensional subspace of $\mathbf{C}^r$. Since $\mathrm{codim}\mathcal{R}(TS) = p-m$, it follows that

$$L \cap \mathrm{Span}\{TSe_{p+1}, \ldots, TSe_r\} \neq \{0\}.$$

Thus there is a nonzero vector

$$z = \sum_{j=p+1}^{r} c_j e_j$$

such that TSz is a nonzero vector in L. For any scalar γ, consider the vector

$$y' = \gamma \left(\sum_{j=m+1}^{p} c_j e_j + \sum_{j=p+1}^{r} c_j e_j \right)$$

where γ and $c_{m+1}, \ldots, c_p$ are to be chosen. Then

$$\langle G_1 y', y' \rangle = |\gamma|^2 \left(\sum_{j=m+1}^{p} |c_j|^2 - \sum_{j=p+s+1}^{r} |c_j|^2 \right). \tag{3.16}$$

We may choose $c_{m+1}, \ldots, c_p$ so that

$$\sum_{j=m+1}^{p} |c_j|^2 = \sum_{j=p+s+1}^{r} |c_j|^2.$$

Then by (3.16),

$$\langle G_1 y', y' \rangle = 0$$

for any choice of γ. Since

$$\|TSy\| = |\gamma| \, \|TSz\|$$

and since $TSz \neq 0$, it follows that

$$\sup\{\|TSy\| \mid TSy \in L, \ \langle G_1 y, y \rangle \leq 1\} = \infty.$$

Therefore

$$\omega_k(T(B_+)) = \infty \qquad (p+1 \leq k \leq r).$$

This concludes the proof of (3.2) and (3.4).

Now we turn to the proof of (3.5). Assume that T is not invertible and $d_r \neq 0$. Then $d_p = 0$ and hence $m < p$. Consider any integer k with $1 \leq k \leq p - m$. Since $\operatorname{codim} \mathcal{R}(T) = p - m$, there is a k-dimensional subspace of $\mathbf{C}^r$ such that

$$L \cap \mathcal{R}(TS) = \{0\}.$$

As in the proof of (3.10), it follows that

$$\omega_k(T(B_-)) = 0 \qquad (1 \leq k \leq p - m).$$

Thus (3.5) holds for $1 \leq k \leq p - m$. Suppose $p - m < k \leq r$, and let L be a k-dimensional subspace of $\mathbf{C}^r$. Then $L \cap \mathcal{R}(TS) \neq \{0\}$. Let w be a vector such that TSw is a nonzero vector in L. Write

$$w = w^+ + w^0 + w^{00} + w^-$$

as in (3.11). Then $TS(w^+ + w^{00} + w^-) = TS(w)$.

Case 1 $\|w^+\| \geq \|w^-\|$. For any scalar γ, let

$$y' = \gamma(w^+ + w^{00} + w^-).$$

Then

$$\langle G_1 y', y' \rangle = |\gamma|^2(\|w^+\|^2 - \|w^-\|^2) \geq -1$$

and

$$\|TSy'\| = |\gamma| \ \|TSw\|.$$

Since $TSw \neq 0$, it follows that $\|TSy'\|$ can be made arbitrarily large by choosing $|\gamma|$ sufficiently large. Thus

$$\sup\{\|TSy\| \mid TSy \in L, \ \langle G_1 y, y \rangle \geq -1\} = \infty. \tag{3.17}$$

Case 2 $\|w^+\| < \|w^-\|$. For any scalars γ and μ, let

$$y' = \mu e_p + \gamma(w^+ + w^{00} + w^-).$$

Then

$$\langle G_1 y', y' \rangle = |\mu|^2 + |\gamma|^2(\|w^+\|^2 - \|w^-\|^2)$$

and

$$\|TSy'\| = |\gamma| \ \|TSw\|.$$

Since $TSw \neq 0$, $\|TSy'\|$ can be made arbitrarily large by choosing $|\gamma|$ sufficiently large. Then μ can be chosen to make

$$\langle G_1 y', y' \rangle \geq -1.$$

Thus (3.17) holds in this case also. We conclude from (3.9b) that

$$\omega_k(T(B_-)) = \infty \qquad (p - m \leq k \leq r).$$

This completes the proof of (3.5).

Finally, we will prove that (3.2), (3.4) and (3.5) imply (3.3), (3.6) and (3.7). Let

$$G' = -G$$

and

$$S' = S\Sigma$$

where

$$\Sigma = \begin{pmatrix} 0 & 0 & I_p \\ 0 & I_s & 0 \\ I_q & 0 & 0 \end{pmatrix}.$$

Then

$$(S')^* G' S' = \begin{pmatrix} I_q & 0 & 0 \\ 0 & 0 & 0 \\ 0 & 0 & -I_p \end{pmatrix} \tag{3.18}$$

and

$$(S')^* T^* T S' = \operatorname{diag}(d_{p+s+1}, \ldots, d_r, d_{p+1}, \ldots, d_{p+s}, d_1, \ldots, d_p) \tag{3.19}$$

where

$$d_{p+s+1} \geq \cdots \geq d_r, d_{p+1} \geq \cdots \geq d_{p+s} > 0 \text{ and } d_1 \geq \cdots \geq d_p. \tag{3.20}$$

Observe that (3.18), (3.19) and (3.20) have the same form as (2.1), (2.3) and (2.4), with G replaced by G', S replaced by S', p and q interchanged, and $\{d_1, \ldots, d_p\}$ and $\{d_{p+s+1}, \ldots, d_r\}$ interchanged. Notice also that the unit balls B'_+ and B'_- for G' are the same as the unit balls B_- and B_+, respectively, for G. Applying (3.2), (3.4) and (3.5) for G', we obtain (3.3), (3.7) and (3.6), respectively, for G. □

Example 3.2. Let $G = I_r$. By Example 2.2, the numbers $d_1^{1/2}, \ldots, d_r^{1/2}$ are the singular values $s_1(T), \ldots, s_r(T)$ of T, arranged in nonincreasing order. Therefore Theorem 3.1 yields the well-known result that

$$\begin{aligned} s_k(T) &= \omega_{r-k+1}(T(B_+)) \\ &= \inf_{L \in \mathcal{L}_{r-k+1}} \sup\{\|Tx\| \,\big|\, \|x\| \leq 1 \text{ and } Tx \in L\} \qquad (1 \leq k \leq r). \end{aligned}$$

Example 3.3. Let G be positive definite. By Example 2.3, the numbers $d_1^{1/2}, \ldots, d_r^{1/2}$ are the singular values of $TG^{-1/2}$, arranged in nonincreasing order. Therefore Theorem 3.1 implies

$$\begin{aligned} s_k(TG^{-1/2}) &= \omega_{r-k+1}(T(B_+)) \\ &= \inf_{L \in \mathcal{L}_{r-k+1}} \sup\{\|Tx\| \mid \langle Gx, x\rangle \leq 1 \text{ and } Tx \in L\}. \end{aligned}$$

Example 3.4. Let $T = I_r$. For any nonzero $r \times r$ selfadjoint matrix G, let $\lambda_1, \ldots, \lambda_r$ be the eigenvalues of G, arranged so that

$$0 < \lambda_1 \leq \cdots \leq \lambda_p, \ \lambda_{p+1} = \cdots = \lambda_{p+s} = 0, \ 0 > \lambda_{p+s+1} \geq \cdots \geq \lambda_r.$$

Then Theorem 3.1 along with Example 2.4 implies that

$$\begin{aligned}\lambda_k^{-1/2} &= \omega_{p-k+1}(B_+)\\ &= \inf_{L\in\mathcal{L}_{p-k+1}} \sup\{\|x\| \mid \langle Gx, x\rangle \leq 1 \text{ and } x \in L\} \qquad (1 \leq k \leq p)\end{aligned}$$

and

$$\begin{aligned}\lambda_k^{-1/2} &= -\omega_{r-k+1}(B_-)\\ &= -\inf_{L\in\mathcal{L}_{r-k+1}} \sup\{\|x\| \mid \langle Gx, x\rangle \geq -1 \text{ and } x \in L\}(p+s+1 \leq k \leq r).\end{aligned}$$

4. The interval of positivity

In this section we will investigate the set $\{\nu \mid T^*T - \nu G > 0\}$, assuming that the pencil $T^*T - \lambda G$ is positive, as in Theorem 2.1. We will let S and D be matrices for which (2.1)–(2.4) hold.

Proposition 4.1. *Let T and G be as in Theorem 2.1, and assume that the pencil $T^*T - \lambda G$ is positive. Then the set $\mathcal{S} = \{\nu \mid T^*T - \nu G > 0\}$ is a (possibly unbounded) interval (ν_+, ν_-). Furthermore, $\nu_+ = -\infty$ if and only if $G \geq 0$; $\nu_- = \infty$ if and only if $G \leq 0$; and $(\nu_+, \nu_-) = (-d_r, d_p)$ otherwise.*

Proof. Since the set of positive definite $r \times r$ matrices is open and $T^*T - \nu G$ is a continuous function of $\nu, \mathcal{S}$ is open. Suppose ν_1 and ν_2 are in $\mathcal{S}$ and let λ be any number with $0 < \lambda < 1$. Then

$$T^*T - \nu_1 G > 0$$

and

$$T^*T - \nu_2 G > 0.$$

Multiplying the first of these by λ and the second by $1 - \lambda$ and adding yields

$$T^*T - [\lambda\nu_1 + (1-\lambda)\nu_2]G > 0.$$

Thus $\lambda\nu_1 + (1-\lambda)\nu_2 \in \mathcal{S}$. This implies that $\mathcal{S}$ is an open interval (ν_+, ν_-). If $G \geq 0$, then $q = 0$, so the condition in (2.10) is absent, and (2.11) becomes $\nu < d_p$. Thus $\nu_+ = -\infty$ and $\nu_- = d_p$. Conversely, if $\nu_+ = -\infty$, then for any negative number ν in $\mathcal{S}$, (2.6)implies

$$G > \frac{1}{\nu} T^*T.$$

Letting ν tend to $-\infty$ proves that G has no negative eigenvalues, so $G \geq 0$. A similar argument proves that $\nu_- = \infty$ if and only if $G \leq 0$, in which case $\nu_+ = -d_r$.

If neither $G \geq 0$ nor $G \leq 0$, then $p \neq 0$ and $p + s < r$, so that neither (2.8) nor (2.10) is vacuous. Thus (2.6) holds if and only if (2.8)–(2.10) are satisfied, which means that $\nu \in (-d_r, d_p)$. Thus $(\nu_+, \nu_-) = (-d_r, d_p)$ if neither $G \geq 0$ nor $G \leq 0$.

We call the interval (ν_+, ν_-) in the proposition the **interval of positivity** of T with respect to G. When we wish to indicate their dependence on T and G, we will denote ν_+ and ν_- by $\nu_+^G(T)$ and $\nu_-^G(T)$, respectively.

In the next theorem we will investigate the effect on $\nu_+^G(T)$ and $\nu_-^G(T)$ of a perturbation of T by an $r \times r$ matrix K. For convenience, we denote by $M(G,T)$ the set of all $r \times r$ matrices K such that the pencil

$$(T - K)^*(T - K) - \lambda G$$

is positive. □

Theorem 4.2. *Let T and G be as in Theorem 2.1 and assume that the pencil $T^*T - \lambda G$ is positive.*

a. Suppose $p > 0$. Then for $0 \leq j \leq p - 1$ and for any $K \in M(G,T)$ with rank $K = j$,

$$\nu_-^G(T - K) \leq d_{p-j}.$$

If $q = 0$ or $d_r \neq 0$, then for $1 \leq j \leq p - 1$, there exists some $K \in M(G,T)$ with rank $\leq j$ such that

$$\nu_-^G(T - K) = 0.$$

b. Suppose $q > 0$. Then for $0 \leq j \leq q - 1$ and for any $K \in M(G,T)$ with rank $K = j$,

$$\nu_+^G(T - K) \geq -d_{r-j}.$$

If $p = 0$ or $d_p \neq 0$, then for $1 \leq j \leq q - 1$ there exists some $K \in M(G,T)$ such that

$$\nu_+^G(T - K) = 0.$$

Proof. a. Since $p \neq 0$, it follows from Proposition 4.1 that $\nu_-^G(T) = d_p$. If $j = 0$ and rank $K = 0$, then $\nu_-^G(T - K) = \nu_-^G(T) = d_p = d_{p-j}$. Assume that $1 \leq j \leq p - 1$ and $K \in M(G,T)$ with rank $K = j$. Then rank $KS = j$. Therefore there exists some nonzero $v \in \mathrm{Span}\{e_{p-j}, \ldots, e_p\}$ such that

$$KSv = 0.$$

For any $\nu \in (\nu_+^G(T - K), \nu_-^G(t - K))$,

$$(TS - KS)^*(TS - KS) - \nu G_1 = S^*[(T - k)^*(T - K) - \nu G]S > 0.$$

It follows that

$$\langle (TS - KS)^*(TS - KS)v, v\rangle - \nu\langle G_1 v, v\rangle > 0.$$

Since $KSv = 0$, this implies that

$$\langle S^*T^*TSv, v\rangle - \nu\langle G_1 v, v\rangle > 0. \tag{4.1}$$

But since $v \in \operatorname{Span}\{e_{p-j}, \dots, e_p\}$,

$$\langle G_1 v, v\rangle = \|v\|^2$$

and

$$\begin{aligned}\langle S^*T^*TSv, v\rangle &= \langle Dv, v\rangle \\ &= \sum_{k=p-j}^{p} d_k |v_k|^2 \\ &\leq d_{p-j}\|v\|^2.\end{aligned}$$

Therefore (4.1) implies that

$$d_{p-j}\|v\|^2 - \nu\|v\|^2 > 0$$

so that

$$\nu < d_{p-j}.$$

It follows that

$$\nu_-^G(T-K) \leq d_{p-j}.$$

If $q = 0$ or $d_r \neq 0$, then for any j with $1 \leq j \leq p-1$, let $K = TSES^{-1}$, where

$$E = \sum_{k=p-j+1}^{p} e_k e_k^*.$$

Then rank $E = j$ and rank $K \leq j$. Furthermore

$$\begin{aligned}S^*(T-K)^*(T-K)S &= (I-E)^*S^*T^*TS(I-E) \\ &= (I-E)^*D(I-E) \\ &= \operatorname{diag}(d_1, \dots, d_{p-j}, 0, \dots, 0, d_{p+1}, \dots, d_r).\end{aligned}$$

From Theorem 2.1 we conclude that the pencil $(T-K)^*(T-K) - \lambda G$ is positive. By Proposition 4.1,

$$\nu_-^G(T-K) = 0.$$

b. Let

$$G' = -G \text{ and } S' = S\Sigma$$

where

$$\sum = \begin{pmatrix} 0 & 0 & I_p \\ 0 & I_s & 0 \\ I_q & 0 & 0 \end{pmatrix}$$

so that (3.18)– (3.20) + hold. Then

$$\nu_+^G(T-K) = -\nu_-^{G'}(T-K)$$

so that (b) follows from (a). □

References

[BG] Asher Ben-Artzi and Israel Gohberg, Singular Numbers of Contractions in Spaces with an Indefinite Metric, to appear.

[GK] Israel Gohberg and Mark Krein, Introduction to the Theory of Linear Non-selfadjoint Operators, Translations of Mathematicsal Monographs, Vol. 18, American Mathematical Society, Providence, Rhode Island, 1969.

[LT] Peter Lancaster and Miron Tismenetsky, *The Theory of Matrices, Second Edition with Applications* (Computer Science and Applied Mathematics), Academic Press, Inc., Orlando, 1985.

[P] Allan Pinkus, *n-Widths in Approximation Theory* (Ergebnisse der Mathematik und ihrer Grenzgebiete; 3. Folge, Band 7), Springer-Verlag, Berlin Heidelberg, 1985.

Department of Mathematics
University of Maryland
College Park
Maryland 20742, USA

Faculty of Exact Sciences
Tel Aviv University
Ramat Aviv 69978
Israel

Department of Mathematics
University of Maryland
College Park
Maryland 20742, USA

1991 Mathematics Subject Classification. Primary 15A18, 15A22; Secondary 15A21, 15A48, 47A56

Operator Theory:
Advances and Applications, Vol. 106
© 1998 Birkhäuser Verlag Basel/Switzerland

Perturbations of Krein spaces preserving the nonsingularity of the critical point infinity

ANDREAS FLEIGE AND BRANKO NAJMAN[1]

Dedicated to Heinz Langer on the occasion of his 60th birthday

We consider a nonnegative selfadjoint operator A in a Krein space such that $\rho(A) \neq \emptyset$ and $\infty \notin c_s(A)$ (i.e. ∞ is not a singular critical point of A). Then we show that these properties remain true for a certain perturbation of the operator, acting in a slightly perturbed Krein space. This result is applied to elliptic differential operators with indefinite weights and to certain difference operators.

1. Introduction

Let $(\mathcal{K}, [\cdot|\cdot])$ be a Krein space, A a nonnegative selfadjoint operator in $\mathcal{K}$ with nonempty resolvent set. Then, according to the theory of definitizable operators (see [L]), A has a spectral function with the set $c(A)$ of critical points and $c(A) \subset \{0, \infty\}$. We assume that ∞ does not belong to the set $c_s(A)$ of the singular critical points of A. In recent years, a number of sufficient conditions have been found in order that an additive perturbation $A + V$ of A also does not have a singular critical point at ∞.

In this note we consider perturbations of both the Krein space $\mathcal{K}$ and the operator A. More precisely, we consider a second inner product $[\cdot|\cdot]^\sim$ in $\mathcal{K}$ (which induces the same topology as $[\cdot|\cdot]$) and the operator $\tilde{A}$ with the property $[\tilde{A}u|v]^\sim = [Au|v]$ ($u \in \mathcal{D}(\tilde{A}) = \mathcal{D}(A), v \in \mathcal{K}$). Then we give sufficient conditions on A and on $[\cdot|\cdot]^\sim$ so that the perturbation $\tilde{A}$ is a positive operator in $(\mathcal{K}, [\cdot|\cdot]^\sim)$ with a nonempty resolvent set and ∞ is not a singular critical point of $\tilde{A}$. We apply the obtained abstract result to certain difference operators in Krein spaces of complex sequences and to the operators $A = -\dfrac{1}{\rho}\triangle$ in L^2_ρ and $\tilde{A} = -\dfrac{1}{\tilde{\rho}}\triangle$ in $L^2_{\tilde{\rho}}$, where the weight functions ρ and $\tilde{\rho}$ have common sign changes. These applications motivated the present research.

---

[1]The first author wants to express his deep sorrow about the tragic death of Branko Najman who died unexpectedly in August 1996

2. The results

Let $(\mathcal{K}, [\cdot|\cdot])$ be a Krein space, J a fundamental symmetry in $\mathcal{K}$, $(u|v) = [Ju|v]$ the corresponding positive definite inner product and $\|u\| = (u|u)^{1/2}$ the corresponding norm on $\mathcal{K}$. We assume that a selfadjoint bounded and boundedly invertible operator M in $\mathcal{K}$ is given, i.e. [2]

(M) $$M^+ = M \in \mathcal{L}(\mathcal{K}), \quad 0 \in \rho(M)$$

where M^+ denotes the indefinite adjoint of M. Then the inner product

$$[u|v]^\sim = [Mu|v] \quad (u, v \in \mathcal{K})$$

is a nondegenerate inner product on $\mathcal{K}$ and the space $\tilde{\mathcal{K}} := (\mathcal{K}, [\cdot|\cdot]^\sim)$ is a Krein space with the same topology as $\mathcal{K}$.

Now let A be a nonnegative selfadjoint operator in $\mathcal{K}$, with $\rho(A) \neq \emptyset$, and let a be the associated hermitian sesquilinear form (see [K, Chapter VI]): a is the closure of the form $[Au|v]$ $(u, v \in \mathcal{D}(A))$. Then the domain $\mathcal{D}(a)$ of a is the domain of the nonnegative square root $(JA)^{1/2}$ of the nonnegative operator JA in the Hilbert space $(\mathcal{K}, (\cdot|\cdot))$. Let $(\cdot|\cdot)_+$ be the positive definite inner product

$$(u|v)_+ = a(u, v) + (u|v) \quad (u, v \in \mathcal{D}(a))$$

on $\mathcal{D}(a)$. Denote by $\mathcal{K}_+$ the Hilbert space $(\mathcal{D}(a), (\cdot|\cdot)_+)$; its norm can be expressed as $\|u\|_+ = \|(JA+I)^{1/2}u\|$ $(u \in \mathcal{K}_+)$. In addition to condition (M) we assume that

(AM) $$M\mathcal{K}_+ \subset \mathcal{K}_+.$$

Let $\tilde{A}$ be the nonnegative selfadjoint operator in $\tilde{\mathcal{K}}$ corresponding to the closed nonnegative form a in $\tilde{\mathcal{K}}$ (the existence of $\tilde{A}$ follows from [K, Theorem VI.2.1]):

$$[\tilde{A}u|v]^\sim = a(u, v) \quad (u \in \mathcal{D}(\tilde{A}), v \in \mathcal{K}_+).$$

Then $M\tilde{A} = A$, hence

(2.1) $$\tilde{A} = M^{-1}A, \quad \mathcal{D}(\tilde{A}) = \mathcal{D}(A).$$

Lemma 2.1. *Assume that* $\infty \notin c_s(A)$ *and*

(2.2) $$\gamma = \|M^{-1}(I - M)\| < 1.$$

Then the resolvent set of $\tilde{A}$ *is nonempty.*

Proof. From (2.2) it follows that $\|(\tilde{A} - A)u\| = \|M^{-1}(I - M)Au\| \leq \gamma\|Au\|$, hence $\tilde{A} - A$ is relatively A-bounded with the A-bound less than 1. Since $\|(i\mu - A)^{-1}\| \leq \frac{C}{|\mu|}$ $(\mu \in \mathbb{R}, |\mu| \geq \mu_0)$, it follows that $i\mu \in \rho(\tilde{A})$ for $|\mu|$ sufficiently large. □

---

[2] We thank Professor Peter Jonas for a helpful hint by which we could weaken our original conditions (M) and (AM)

Under the conditions of Lemma 2.1 $\tilde{A}$ is a nonnegative selfadjoint operator in $\tilde{\mathcal{K}}$ with nonempty resolvent set, hence it has a spectral function and $c(\tilde{A}) \subset \{0, \infty\}$. Next we state the main result of this note.

Theorem 2.2. *Let $(\mathcal{K}, [\cdot|\cdot])$ be a Krein space, A and M selfadjoint operators in $\mathcal{K}$ such that A satisfies*

$$\text{(A)} \qquad A \textit{ is nonnegative}, \quad \rho(A) \neq \emptyset, \quad \infty \notin c_s(A)$$

and let $\mathcal{K}_+$ be the domain of the closure of the form $[A \cdot|\cdot]$. Further assume that M satisfies

$$\text{(M)} \qquad M^+ = M \in \mathcal{L}(\mathcal{K}), \quad 0 \in \rho(M),$$

$$\|M^{-1}(I - M)\| < 1, \tag{2.2}$$

and

$$\text{(AM)} \qquad M\mathcal{K}_+ \subset \mathcal{K}_+ .$$

Then the operator $\tilde{A} = M^{-1}A$ also satisfies (A) *in $\tilde{\mathcal{K}}$, that is, $\tilde{A}$ is a nonnegative selfadjoint operator in $\tilde{\mathcal{K}}$ with nonempty resolvent set and ∞ is not a singular critical point of $\tilde{A}$.*

Proof. The nonemptiness of the resolvent set of $\tilde{A}$ has been proved in Lemma 2.1. Put $C := JA + I$ and $B := JC = A + J$. Then B and $JC^{\frac{1}{2}}$ are nonnegative, selfadjoint and boundedly invertible in $\mathcal{K}$. Moreover we have $D((JA)^{\frac{1}{2}}) = D(C^{\frac{1}{2}}) = D((JB)^{\frac{1}{2}})$ (see e.g. [C, Corollary 1.3]). Therefore by [C, Corollary 3.6] $\infty \notin c_s(A)$ implies $\infty \notin c_s(B)$. By [C, Theorem 2.9] this is equivalent to $\infty \notin c_s(JC^{\frac{1}{2}})$. Consequently $JC^{\frac{1}{2}}$ is fundamentally reducible (see e.g. [C, Theorem 2.5]). Then there exists a fundamental symmetry J_0 of $\mathcal{K}$ such that $J_0(D(JC^{\frac{1}{2}})) \subset D(JC^{\frac{1}{2}})$. Since $D(JC^{\frac{1}{2}}) = D(C^{\frac{1}{2}}) = \mathcal{K}_+$ we obtain $J_0M\mathcal{K}_+ \subset J_0\mathcal{K}_+ \subset \mathcal{K}_+$, using condition (AM). The operator J_0M is positive, bounded and boundedly invertible in $\tilde{\mathcal{K}}$ and $\mathcal{K}_+$ is the domain of the closure of the form $[\tilde{A}u|v]^{\sim}$ $(u, v \in \mathcal{D}(\tilde{A}))$ in $\tilde{\mathcal{K}}$. Thererfore $\infty \notin c_s(\tilde{A})$ follows from [C, Proposition 3.5, Remark 1.5]. □

3. Application to partial differential equations

Let ρ be a real measurable function on $\mathbb{R}^n$ such that $\rho \neq 0$ (a.e). Consider $\mathcal{K} = L^2(\mathbb{R}^n, |\rho|)$ with the indefinite form

$$[u|v] = \int_{\mathbb{R}^n} u\bar{v}\rho dx \quad (u, v \in L^2(\mathbb{R}^n, |\rho|)).$$

Then $(\mathcal{K}, [\cdot|\cdot])$ is a Krein space. A natural fundamental symmetry on $\mathcal{K}$ is the operator J given by $Ju(x) = \text{sgn}\rho(x)\, u(x)$. Then

$$(u|v) = \int u\bar{v}|\rho|dx$$

and the Hilbert space $(\mathcal{K}, (\cdot|\cdot))$ is the space $L^2(\mathbb{R}^n, |\rho|)$ with its natural norm. Our basic assumption on ρ is an implicit one:

(C1) $A = -\frac{1}{\rho}\triangle$ defined on
$\mathcal{D}(A) = \{u \in L^2(\mathbb{R}^n; |\rho|) : \frac{1}{|\rho|^{1/2}}\triangle u \in L^2(\mathbb{R}^n)\}$
satisfies **(A)**,

that is, $\rho(A) \neq \emptyset$ and $\infty \notin c_s(A)$. Note that

$$a(u,u) = \int_{\mathbb{R}^n} |\nabla u|^2 dx.$$

In [CN] it has been shown that if $\rho(x) = \text{sgn} x_n$ then (C1) is satisfied. More sufficient conditions for (C1) can be found in [FL] (and for a similar differential operator, definded on a bounded region $G \subset \mathbb{R}^n$, in [P]).

Let $\tilde{\rho}$ be another real measurable function such that there exists $\gamma \in (0,1)$ with the property

(C2) $$\frac{1}{1+\gamma}|\rho(x)| < |\tilde{\rho}(x)| < \frac{1}{1-\gamma}|\rho(x)| \,, \quad \rho(x)\tilde{\rho}(x) > 0 \ \ (a.e.).$$

Then $\tilde{\rho} \neq 0$ (a.e.) and the operator M of multiplication by $\dfrac{\tilde{\rho}}{\rho}$ is well defined on $\mathcal{K}$; it follows from (C2) that M and M^{-1} are bounded operators and that the assumption (M) is satisfied. Note that $\tilde{\mathcal{K}} = L^2(\mathbb{R}^n, |\tilde{\rho}|) = \mathcal{K}$ with the inner product

$$[u|v]^{\sim} = \int u\bar{v}\tilde{\rho}dx.$$

Then $\tilde{A} = -\frac{1}{\tilde{\rho}}\triangle$. Moreover, it follows from (C2) that $-\gamma < \dfrac{\rho(x) - \tilde{\rho}(x)}{\tilde{\rho}(x)} < \gamma$ (a.e.), hence $\|\dfrac{\rho-\tilde{\rho}}{\tilde{\rho}}\|_{L^\infty(R^n)} \leq \gamma$. Consequently $\|M^{-1}(I-M)\|_{\mathcal{L}(L^2(R^n,|\rho|))} \leq \|\dfrac{\rho-\tilde{\rho}}{\tilde{\rho}}\| \leq \gamma < 1$, so (2.2) is also satisfied. It remains to verify (AM). We first find a sufficient condition.

Lemma 3.1. *If there exists $C \geq 0$ such that*

$$\|u\frac{\nabla\rho}{\rho}\|_{L^2(R^n)} + \|u\frac{\nabla\tilde{\rho}}{\rho}\|_{L^2(R^n)} \leq C(\|\nabla u\|_{L^2(R^n)} + \|u\|_{L^2(R^n,|\rho|)}) \tag{3.1}$$

for all u with the finite right-hand side, then (AM) *is satisfied (and hence $\tilde{A}$ also satisfies* (A) *in $\tilde{\mathcal{K}}$).*

Proof. Condition (C2) implies $\frac{\tilde{\rho}}{\rho}, \frac{\rho}{\tilde{\rho}} \in L^\infty(\mathbb{R}^n)$. Note that $\nabla(Mu) = \frac{\tilde{\rho}}{\rho}\nabla u + u\nabla(\frac{\tilde{\rho}}{\rho}) = \frac{\tilde{\rho}}{\rho}\nabla u + u(\frac{\nabla\tilde{\rho}}{\rho} - \frac{\tilde{\rho}}{\rho}\cdot\frac{\nabla\rho}{\rho})$. Hence $\|\nabla(Mu)\|_{L^2} \leq C_1(\|\nabla u\|_{L^2} + \|u\frac{\nabla\tilde{\rho}}{\rho}\|_{L^2} + \|u\frac{\nabla\rho}{\rho}\|_{L^2})$. From (3.1) it follows that $\|\nabla(Mu)\|_{L^2} \leq C_2(\|\nabla u\|_{L^2} + \|u\|_{L^2(|\rho|)}) \leq C_3\|u\|_+$. Therefore $u \in \mathcal{K}_+$ satisfies $Mu \in \mathcal{K}_+$. □

We give two instances when the assumption (3.1) can be made more explicit:

a) $\rho \in L^\infty(\mathbb{R}^n)$, $\frac{1}{\rho} \in L^\infty(\mathbb{R}^n)$.

Note that (C2) implies $\tilde{\rho}, \frac{1}{\tilde{\rho}} \in L^\infty(\mathbb{R}^n)$. In this case (3.1) can be replaced by

$$\|u\nabla\rho\|_{L^2} + \|u\nabla\tilde{\rho}\|_{L^2} \leq C\|u\|_{H^1}. \tag{3.2}$$

Since $H^1 \subset L^p$ for all p with $\frac{1}{2} \geq \frac{1}{p} \geq \frac{1}{2} - \frac{1}{n} = \frac{n-2}{2n}$, it follows that

$$\|u\nabla\rho\|_{L^2} \leq \|\nabla\rho\|_{L^q}\|u\|_{L^p} \leq \|\nabla\rho\|_{L^q}\|u\|_{H^1}$$

where $\frac{1}{q} = \frac{1}{2} - \frac{1}{p} \leq \frac{1}{n}$. Hence (3.2) is satisfied if

$$\nabla\rho, \nabla\tilde{\rho} \in L^q(\mathbb{R}^n) \tag{3.3}$$

Therefore it is sufficient that (3.3) holds for some q with $n < q < \infty$.

b) Assume that $n \geq 3$ and there exists s, $1 < s < 2$, such that

$$\frac{\nabla\rho}{\rho^{1+s}} \in L^{\frac{n}{1-2s}}(\mathbb{R}^n). \tag{3.4}$$

We shall use the inequality (see [BCLSC])

$$\|u\|_{2n/(n-2)} \leq C\|\nabla u\|_2. \tag{3.5}$$

It follows from the inequality

$$\|u\frac{\nabla\rho}{\rho}\|_2 = \|\rho^s u^{2s} u^{1-2s}\frac{\nabla\rho}{\rho^{1+s}}\|_2 \leq$$

$$\|\rho^s u^{2s}\|_{1/s}\|u^{1-2s}\|_b\|\frac{\nabla\rho}{\rho^{1+s}}\|_{n/(1-2s)} \leq \|\frac{\nabla\rho}{\rho^{1+s}}\|_{n/(1-2s)}\|\rho^{1/2}u\|_2^{2s}\|u\|_{b(1-2s)}^{1-2s}$$

where $\frac{1}{b} = \frac{1}{2} - s - \frac{1-2s}{n}$. From (3.4) and (3.5) we obtain

$$\|u\frac{\nabla\rho}{\rho}\|_2 \leq C(\||\rho|^{1/2}u\|_2 + \|\nabla u\|_2).$$

Similarly, if there exists $\tilde{s} \in (1,2)$ such that

$$\frac{\nabla\tilde{\rho}}{\rho^{1+\tilde{s}}} \in L^{\frac{n}{1-2\tilde{s}}}(\mathbb{R}^n), \tag{3.6}$$

then

$$\|u\frac{\nabla\tilde{\rho}}{\rho}\|_{L^2} \leq C(\||\rho|^{1/2}u\|_2 + \|\nabla u\|_2).$$

Therefore (3.4) and (3.6) imply (3.1).

4. Application to difference operators

Let $t_{-1} < t_{-2} < \ldots < 0 < \ldots t_2 < t_1$ and $m_n > 0 \quad (n \in \mathbb{N})$, $m_n < 0 \quad (n \in -\mathbb{N})$ such that $t_n \nearrow 0 \quad (n \to -\infty)$, $t_n \searrow 0 \quad (n \to \infty)$, $\sum_{n\in\mathbb{Z}_0} |m_n| < \infty$, where $\mathbb{Z}_0 := \mathbb{Z}\backslash\{0\}$. Let further

$$\mathcal{K} := l^2(m_n)_{n\in\mathbb{Z}_0} := \{(f_n)_{n\in\mathbb{Z}_0} : \sum_{n\in\mathbb{Z}_0} |f_n|^2|m_n| < \infty\}$$

and

$$[f|g] := \sum_{n\in\mathbb{Z}_0} f_n\bar{g}_n m_n, \quad (f,g) := \sum_{n\in\mathbb{Z}_0} f_n\bar{g}_n|m_n| \quad (f,g \in l^2(m_n)_{n\in\mathbb{Z}_0}).$$

Then $(\mathcal{K}, [\cdot|\cdot])$ is a Krein space,

$$Jf := (\operatorname{sgn}(m_n)\cdot f_n)_{n\in\mathbb{Z}_0} \quad (f \in \mathcal{K})$$

is a fundamental symmetry, and $(\cdot|\cdot)$ is the corresponding positive definite scalar product. Now we define a difference operator on the subspace

$$\begin{aligned} D(A) := \{f \in l^2(m_n)_{n\in\mathbb{Z}_0} : &\lim_{n\to\infty} f_n = \lim_{n\to-\infty} f_n, \\ &\lim_{n\to\infty}\frac{f_n - f_{n+1}}{t_n - t_{n+1}} = \lim_{n\to-\infty}\frac{f_n - f_{n+1}}{t_n - t_{n+1}}, \\ \sum_{n\in\mathbb{Z}\backslash\{-1,0,1\}} &\frac{|(t_{n+1}-t_n)f_{n-1} - (t_{n+1}-t_{n-1})f_n + (t_n - t_{n-1})f_{n+1}|^2}{|m_n|(t_{n-1}-t_n)^2(t_n - t_{n+1})^2} < \infty\}. \end{aligned}$$

Let $\alpha \in (0, \frac{\pi}{2}]$ and $\beta \in [0, \frac{\pi}{2})$. Then for $f \in D(A)$, $g \in \mathcal{K}$ we put $Af = g$, if and only if

$$\frac{(t_{n+1}-t_n)f_{n-1} - (t_{n+1}-t_{n-1})f_n + (t_n - t_{n-1})f_{n+1}}{(t_{n-1}-t_n)(t_n - t_{n+1})} = m_n \cdot g_n \qquad (n \in \mathbb{Z}\backslash\{-1,0,1\}),$$

$$\cot\alpha \; f_{-1} - \frac{f_{-2}-f_{-1}}{t_{-2}-t_{-1}} = m_{-1}\cdot g_{-1}, \quad \tan\beta \; f_1 + \frac{f_1 - f_2}{t_1 - t_2} = m_1 \cdot g_1.$$

By [F, Theorem 2.13, Example 2.16] the operator A is selfadjoint and nonnegative in $\mathcal{K}$ with $\rho(A) \neq \emptyset$. Moreover $\sigma(A)(\subset \mathbb{R})$ is discrete and has no finite accumulation point. From [F, Example 2.40] it follows that

$$\mathcal{D}(a) = \{f \in l^2(m_n)_{n\in\mathbb{Z}_0} : \sum_{n\in\mathbb{Z}\setminus\{-1,0\}} \frac{|f_n - f_{n+1}|^2}{t_n - t_{n+1}} < \infty, \lim_{n\to\infty} f_n = \lim_{n\to-\infty} f_n\},$$

$$a(f,g) = \sum_{n\in\mathbb{Z}\setminus\{-1,0\}} \frac{(f_n - f_{n+1})(\bar{g}_n - \bar{g}_{n+1})}{t_n - t_{n+1}} + \cot\alpha\, f_{-1}\bar{g}_{-1} + \tan\beta\, f_1\bar{g}_1.$$

In the following we assume that

$$\infty \notin c_s(A)$$

(which implies (A)). Note that by [F, Theorem 3.6] this condition is satisfied if

$$\lim_{n\to\infty} \frac{m_n}{m_{n-1}} \neq 1, \tag{4.1}$$

$$C_1 \leq \frac{t_{n-2} - t_{n-1}}{t_{n-1} - t_n} \leq C_2, \quad \left|\frac{m_{n-1}^2 - m_n m_{n-2}}{(t_{n-1} - t_n)m_{n-2}m_{n-1}}\right| \leq C_3 \quad (n \geq 4) \tag{4.2}$$

with some constants $C_1, C_2, C_3 > 0$. Now we consider another sequence $(\tilde{m}_n)_{n\in\mathbb{Z}_0}$ such that

$$c_n := \frac{\tilde{m}_n}{m_n} \in [\gamma_1, \gamma_2] \quad (n \in \mathbb{Z}_0)$$

with some constants $\frac{1}{2} < \gamma_1 < 1 < \gamma_2$. For $f \in \mathcal{K}$ define $Mf := (c_n \cdot f_n)_{n\in\mathbb{Z}_0}$. Then we have

$$[Mf|f] = \sum_{n\in\mathbb{Z}_0} |f_n|^2 \tilde{m}_n, \quad (Mf|f) = \sum_{n\in\mathbb{Z}_0} |f_n|^2 |\tilde{m}_n|.$$

Consequently M is an operator in $\mathcal{K}$, satisfying condition (M). We have $\tilde{\mathcal{K}} = l^2(\tilde{m}_n)_{n\in\mathbb{Z}_0}$ and the operator $\tilde{A}$ in $\tilde{\mathcal{K}}$ has the same form as A with m_n replaced by $\tilde{m}_n$. Moreover for $f \in \mathcal{K}$ it holds

$$\|M^{-1}f - f\|^2 = \sum_{n\in\mathbb{Z}_0} |c_n^{-1} - 1|^2 |f_n|^2 |m_n| \leq \gamma \|f\|^2$$

with some $0 \leq \gamma < 1$. This implies (2.2). Again it remains to verify (AM).

Lemma 4.1. *Assume that*

$$\lim_{n\to\infty} c_n = \lim_{n\to-\infty} c_n = 1, \qquad \sum_{n\in\mathbb{Z}\setminus\{-1,0\}} \frac{(c_n - c_{n+1})^2}{t_n - t_{n+1}} < \infty.$$

Then (AM) *is satisfied (and hence* $\tilde{A}$ *satisfies condition* (A) *in* $\tilde{\mathcal{K}}$*).*

Proof. Let $f \in \mathcal{K}_+(= \mathcal{D}(a))$ and $g := Mf$. Then we have

$$\lim_{n\to\pm\infty} g_n = \lim_{n\to\pm\infty} c_n \cdot f_n = \lim_{n\to\pm\infty} f_n,$$

$$\begin{aligned}
&\sum_{n\in\mathbb{Z}\setminus\{-1,0\}} \frac{|g_n - g_{n+1}|^2}{t_n - t_{n+1}} \\
\leq\ &\sum_{n\in\mathbb{Z}\setminus\{-1,0\}} (t_n - t_{n+1})^{-1}(2|c_n f_n - c_n f_{n+1}|^2 + 2|c_n f_{n+1} - c_{n+1} f_{n+1}|^2) \\
\leq\ & 2\gamma_2^2 \sum_{n\in\mathbb{Z}\setminus\{-1,0\}} \frac{|f_n - f_{n+1}|^2}{t_n - t_{n+1}} + 2 \sum_{n\in\mathbb{Z}\setminus\{-1,0\}} \frac{(c_n - c_{n+1})^2}{t_n - t_{n+1}} |f_{n+1}|^2.
\end{aligned}$$

The first series is convergent and, since $(f_n)_{n\in\mathbb{Z}_0} \in \mathcal{D}(a)$ is bounded, the second series too. This implies $M\mathcal{K}_+ \subset \mathcal{K}_+$. □

In order to show that Lemma 4.1 improves the regularity criterion (4.1),(4.2) we consider the following example:

$$m_n := 2^{-n}, \quad t_n := 2^{1-n}, \quad \tilde{m}_n := 2^{-n} - 2^{-\frac{7}{4}n - \frac{3}{4}},$$
$$m_{-n} := -m_n, \quad t_{-n} := -t_n, \quad \tilde{m}_{-n} := -\tilde{m}_n \quad (n \in \mathbb{N}).$$

Then $m_n, t_n (n \in \mathbb{Z}_0)$ satisfy the conditions (4.1), (4.2) and

$$c_n = \frac{\tilde{m}_n}{m_n} (= 1 - 2^{-\frac{3}{4}|n| - \frac{3}{4}} > \frac{1}{2},\ n \in \mathbb{Z}_0)$$

satisfy the assumptions of Lemma 4.1. However, (4.1),(4.2) are not valid with m_n replaced by $\tilde{m}_n$.

References

[BCLSC] Bakry, D., Coulhon, T., Ledoux, M., Saloff-Coste, L.,: Sobolev inequalities in disguise, Indiana Univ. Math. J., 44(1995), 1033–1074.

[C] Ćurgus, B.: On the regularity of the critical point infinity of definitizable operators. J. Integral Equations Operator Theory **8** (1985), 462–488.

[CN] Ćurgus, B., Najman, B.: Positive differential operators in Krein space $L^2(\mathbf{R}^n)$. In: Gohberg, I.; Lancaster, P.; Shivakumar, P.N. (eds.): Recent developments in operator theory and its applications. International conference in Winnipeg, October 2–6, 1994. Operator Theory: Advances and Applications, Vol. 87, Birkhäuser Verlag, Basel, Boston, 1996.

[FL] Faierman, M., Langer, H.: Elliptic problems involving an indefinite weight function. In: Gohberg, I.; Lancaster, P.; Shivakumar, P.N. (eds.): Recent developments in operator theory and its applications. International conference in Winnipeg, October 2–6, 1994. Operator Theory: Advances and Applications, Vol. 87, Birkhäuser Verlag, Basel, Boston, 1996, 105–127.

[F] Fleige, A.: Spectral theory of indefinite Krein-Feller differential operators. Mathematical Research, Vol. 98, Akademie Verlag, Berlin, 1996.

[K] Kato, T.: Perturbation Theory for Linear Operators. 2nd ed. Springer-Verlag, Berlin, 1976.

[L] Langer, H.: Spectral function of definitizable operators in Krein spaces. Functional Analysis, Proceedings, Dubrovnik 1981. Lecture Notes in Mathematics 948, Springer-Verlag, Berlin, 1982, 1–46.

[P] Pyatkov, S.G.: Elliptic eigenvalue problems with an indefinite weight function. Siberian Advances in Mathematics, V.4, N2 (1994), 87–121.

Universität GH Essen
FB6 – Mathematik und Informatik
D-45117 Essen
Germany

1991 Mathematics Subject Classification. Primary 47B50; Secondary 47B39, 39A70, 47F05

Operator Theory:
Advances and Applications, Vol. 106

An analysis of the block structure of j_{qq}-inner functions

BERND FRITZSCHE, BERND KIRSTEIN, AND KARSTEN MÜLLER

To Heinz Langer, in honour and admiration, on his sixtieth birthday

This paper is aimed at analyzing of the canonical block structure of j_{qq}-inner functions. Inspired by papers of Arov [1] and Dewilde/Dym [10], [11] a concept of parametrization of j_{qq}–inner functions is developed.

0. Introduction

The class of j_{pq}–inner functions turned out to play an important role in the framework of matricial generalizations of classical interpolation problems of Schur-Nevanlinna-Pick type. Namely, the set of solutions of such an interpolation problem can be parametrized with the aid of linear fractional transformations the generating matrix-valued functions of which are j_{pq}–inner functions appropriately constructed from the given data (see, e.g., [4], [9], [12], [15], [22], [24], [19]). The inverse question of constructing interpolation problems such that their solution sets can be parametrized by a given function in the above described way, was studied in [4] and [8]. Inverse problems for j_{pq}–inner functions with prescribed block information are the content of the papers [1], [5], [6] and [7].

In this paper we are looking for appropriate parametrizations of j_{qq}–inner functions. Particular cases of rational j_{pq}–inner functions this problem were treated by Dubovoj [12], [13], Galstjan [22], Kovalishina [24] and in [16], [18]. Our concept of parametrizing j_{qq}–inner functions is based on methods used by Arov [2] and Dewilde/Dym [10], [11], where the special case of j_{qq}–inner functions belonging to the Smirnov class was treated. For this reason, we will call the parametrization worked out in Section 5 the ADD-parametrization of j_{qq}–inner functions. A remarkable feature of this parametrization is the fact that a j_{qq}–inner function can be described by three $q \times q$ matrix-valued functions which belong to the Hardy class $[H^2(\mathbb{D})]^{q\times q}$ and the Carathéodory class. Moreover, we will present a procedure of constructing j_{qq}–inner functions with prescribed ADD-parameters.

1. Some preliminaries and notations

In the first section we will summarize some facts on several classes of meromorphic functions. For a detailed treatment we refer the reader to the monograph of R. Nevanlinna [26] and P. L. Duren [14]. We will start with some notations.

Throughout this paper, let p and q be positive integers. We will use $\mathbb{C}$, $\mathbb{D}$, $\mathbb{T}$, $\mathbb{C}_0$ and $\mathbb{E}$ to denote the set of complex numbers, the open unit disc, the unit circle, the extended complex plane and the exterior of the closed unit disc, respectively:

$$\mathbb{D} := \{z \in \mathbb{C} : |z| < 1\}\,,\ \mathbb{T} := \{z \in \mathbb{C} : |z| = 1\}\,,\ \mathbb{C}_0 := \mathbb{C} \cup \{\infty\}\,,\ \mathbb{E} := \mathbb{C}_0 \setminus (\mathbb{D} \cup \mathbb{T})\,.$$

If $\mathfrak{X}$ is a nonempty set, then $\mathfrak{X}^{p\times q}$ stands for the set of $p \times q$ matrices each entry of which belongs to $\mathfrak{X}$. The null matrix which belongs to $\mathbb{C}^{p\times q}$ will be denoted by $0_{p\times q}$. The identity matrix which belongs to $\mathbb{C}^{q\times q}$ will be designated by I_q. If the size of a null matrix or an identity matrix is clear then we will omit the indexes. The set of all $q \times q$ nonnegative Hermitian matrices will be denoted by $\mathbb{C}^{q\times q}_{\geq}$. A matrix $A \in \mathbb{C}^{p\times q}$ is called *contractive* (respectively, *strictly contractive*) if $I - A^*A$ is nonnegative Hermitian (respectively, positive Hermitian). We will use the notation tr A to denote the trace of a square matrix A. If A belongs to $\mathbb{C}^{q\times q}$, then let Re A and Im A be the real part of A and imaginary part of A, respectively. The linear Lebesgue-Borel measure on $\mathbb{T}$ will be designated by $\underline{\lambda}$ whereas $\mathfrak{B}_{\mathbb{T}}$ stands for the σ-algebra of all Borelian subsets of $\mathbb{T}$. If $t \in (0, \infty)$, then let $\mathcal{L}^t(\mathbb{T})$ denote the set of all Borel measurable functions $g : \mathbb{T} \longrightarrow \mathbb{C}$ for which $|g|^t$ is integrable which respect to $\underline{\lambda}$ on $\mathbb{T}$, whereas $\mathcal{L}^\infty(\mathbb{T})$ stands for the set of all functions $g : \mathbb{T} \longrightarrow \mathbb{C}$ which are bounded $\underline{\lambda}$-almost everywhere on $\mathbb{T}$.

Assume that G is a simply connected domain of $\mathbb{C}_0$. Then let $\mathcal{NM}(G)$ be the *Nevanlinna class* of all functions which are meromorphic in G and which can be represented as a quotient of two bounded holomorphic functions in G. If $g \in [\mathcal{NM}(\mathbb{D})]^{p\times q}$ (respectively, $g \in [\mathcal{NM}(\mathbb{E})]^{p\times q}$), then a well-known theorem due to Fatou implies that there exist a Borelian subset B_0 of the unit circle $\mathbb{T}$ with $\underline{\lambda}(B_0) = 0$ and a Borel measurable function $\underline{g} : \mathbb{T} \longrightarrow \mathbb{C}^{p\times q}$ such that

$$\lim_{r\to 1-0} g(rz) = \underline{g}(z) \qquad \text{(respectively.} \ \lim_{r\to 1+0} g(rz) = \underline{g}(z)\ )$$

for all $z \in \mathbb{T} \setminus B_0$. In the following, we will continue to use the symbol $\underline{g}$ to denote a radial boundary function of a function g which belongs to $[\mathcal{NM}(\mathbb{D})]^{p\times q}$ or $[\mathcal{NM}(\mathbb{E})]^{p\times q}$.

Let $g \in [\mathcal{NM}(\mathbb{D})]^{p\times q}$. Then one says that g *admits a pseudocontinuation (into* $\mathbb{E}$*)* if there exists a function $g^\# \in [\mathcal{NM}(\mathbb{E})]^{p\times q}$ such that the radial boundary values $\underline{g}$ and $\underline{g^\#}$ of g and $g^\#$, respectively, coincide $\underline{\lambda}$-almost everywhere on $\mathbb{T}$. It is obvious that a function $g \in [\mathcal{NM}(\mathbb{D})]^{p\times q}$ admits at most one pseudocontinuation. Note that if $g \in [\mathcal{NM}(\mathbb{D})]^{p\times q}$ admits a pseudocontinuation $g^\#$ and if, additionally, g is analytically continuable through some open arc of $\mathbb{T}$, then the analytic continuation coincides with the pseudocontinuation. In the sequel, we will continue to write $g^\#$ for the pseudocontinuation of g.

Let $\mathfrak{X}$ be a nonempty subset of the extended complex plane $\mathbb{C}_0$, and let $f : \mathfrak{X} \longrightarrow \mathbb{C}^{p\times q}$ be a matrix-valued function. Then we will use the symbol $\widehat{f}$ to denote the function $\widehat{f} : \widehat{\mathfrak{X}} \longrightarrow \mathbb{C}^{q\times p}$ which is given by $\widehat{\mathfrak{X}} := \{z \in \mathbb{C}_0 : 1/\overline{z} \in \mathfrak{X}\}$ and $\widehat{f}(z) := [f(1/\overline{z})]^*$. The following result, which can be easily checked, will play an essential role in our further considerations.

Remark 1.1. If f belongs to $[\mathcal{NM}(\mathbb{D})]^{p\times q}$ (respectively, $[\mathcal{NM}(\mathbb{E})]^{p\times q}$), then $\widehat{f}$ belongs to $[\mathcal{NM}(\mathbb{E})]^{q\times p}$ (respectively, $[\mathcal{NM}(\mathbb{D})]^{q\times p}$) and $\underline{f}^*$ is a radial boundary function of $\widehat{f}$.

The set of all $g \in \mathcal{NM}(\mathbb{D})$ which are holomorphic in $\mathbb{D}$ will be denoted by $\mathcal{N}(\mathbb{D})$. The class $\mathcal{N}(\mathbb{D})$ can be described as the set of all functions g which are holomorphic in $\mathbb{D}$ and which fulfill

$$\sup_{r\in[0,1)} \int_{\mathbb{T}} \log^+ |g(rz)| \underline{\lambda}(\mathrm{d}z) < +\infty$$

where $\log^+ x := \max(\log x, 0)$ for each $x \in [0,\infty)$. If $g : \mathbb{D} \longrightarrow \mathbb{C}$ admits a representation

$$g(w) = \alpha \cdot \exp\left\{\frac{1}{2\pi}\int_{\mathbb{T}} \frac{z+w}{z-w} \log k(z) \underline{\lambda}(\mathrm{d}z)\right\}, \qquad w \in \mathbb{D},$$

with some $\alpha \in \mathbb{T}$ and some Borel measurable function $k : \mathbb{T} \longrightarrow [0,\infty)$ which satisfies $(1/2\pi)\int_{\mathbb{T}} |\log k| \mathrm{d}\underline{\lambda} < \infty$, then g belongs to $\mathcal{N}(\mathbb{D})$. Such functions g are called *outer*. For all $g \in \mathcal{N}(\mathbb{D})$, the inequality

$$\frac{1}{2\pi}\int_{\mathbb{T}} \log^+ |\underline{g}(z)| \underline{\lambda}(\mathrm{d}z) \leq \lim_{r\to 1-0} \frac{1}{2\pi}\int_{\mathbb{T}} \log^+ |g(rz)| \underline{\lambda}(\mathrm{d}z) \tag{1.1}$$

holds true. By the *Smirnov class* $\mathcal{N}_+(\mathbb{D})$ we will mean the set of all $g \in \mathcal{N}(\mathbb{D})$ for which equality holds true in (1.1). The class $\mathcal{N}_+(\mathbb{D})$ proves to be a subalgebra of $\mathcal{N}(\mathbb{D})$. If g is outer in $\mathcal{N}(\mathbb{D})$, then g necessarily belongs to $\mathcal{N}_+(\mathbb{D})$. Note that the Hardy classes $H^t(\mathbb{D})$, $t \in (0,\infty]$, are subsets of $\mathcal{N}_+(\mathbb{D})$.

A function $\Phi \in [\mathcal{N}_+(\mathbb{D})]^{q\times q}$ is called *outer* (in $[\mathcal{N}_+(\mathbb{D})]^{q\times q}$) if $\det\Phi$ is outer in $\mathcal{N}(\mathbb{D})$. Basic facts on outer functions in $[\mathcal{N}_+(\mathbb{D})]^{q\times q}$ can be found in [4]. In particular, if Φ is an outer function in $[\mathcal{N}_+(\mathbb{D})]^{q\times q}$, then $\det\Phi(w) \neq 0$ for all $w \in \mathbb{D}$ and Φ^{-1} is also an outer function in $[\mathcal{N}_+(\mathbb{D})]^{q\times q}$. Conversely, if $\Phi \in [\mathcal{N}_+(\mathbb{D})]^{q\times q}$ satisfies $\det\Phi(w) \neq 0$ for all $w \in \mathbb{D}$ and if $\Phi^{-1} \in [\mathcal{N}_+(\mathbb{D})]^{q\times q}$, then Φ and Φ^{-1} are necessarily outer functions in $[\mathcal{N}_+(\mathbb{D})]^{q\times q}$. If $\Phi \in [\mathcal{N}_+(\mathbb{D})]^{q\times q}$ and $\Psi \in [\mathcal{N}_+(\mathbb{D})]^{q\times q}$ are outer functions then the product $\Phi\Psi$ is also an outer function in $[\mathcal{N}_+(\mathbb{D})]^{q\times q}$. An outer function $\Phi \in [\mathcal{N}_+(\mathbb{D})]^{q\times q}$ is called *normalized* if $\Phi(0)$ is nonnegative Hermitian.

A function $f : \mathbb{D} \longrightarrow \mathbb{C}^{p\times q}$ is said to be a $p\times q$ *Schur function* if f is both holomorphic and contractive in $\mathbb{D}$. The set $\mathcal{S}_{p\times q}(\mathbb{D})$ of all $p\times q$ Schur functions is obviously a subset of the Hardy class $[H^\infty(\mathbb{D})]^{p\times q}$. A function $f \in \mathcal{S}_{q\times q}(\mathbb{D})$ is called an *inner function* if f has unitary radial boundary values $\underline{\lambda}$-almost everywhere on $\mathbb{T}$. If $f \in \mathcal{S}_{p\times q}(\mathbb{D})$ has even strictly contractive values $f(z)$ for all $z \in \mathbb{D}$, then f is said to be a *strictly contractive* $p\times q$ *Schur function.*

Let $f \in [\mathcal{NM}(\mathbb{D})]^{p\times q}$. Then an inner function B that belongs to $\mathcal{S}_{p\times p}(\mathbb{D})$ (respectively, $\mathcal{S}_{q\times q}(\mathbb{D})$) is called a *left* (respectively, *right*) *denominator of* f if Bf

(respectively, fB) belongs to $[\mathcal{N}_+(\mathbb{D})]^{p\times q}$. It is readily checked that every function $g \in [\mathcal{NM}(\mathbb{D})]^{p\times q}$ has left and right denominators. The concept of left and right denominators was created by Arov [1] during his investigations on Darlington synthesis.

2. Left and right connected pairs of $[H^2(\mathbb{D})]^{q\times q}$-functions

In this section, we will present a concept of some association between matrix-valued functions which belong to the Hardy class $[H^2(\mathbb{D})]^{q\times q}$. These considerations will help us to find convenient parametrizations of $j_{qq}-$inner functions.

Definition 2.1. An ordered pair $[\Phi, \Psi]$ of functions which belong to $[H^2(\mathbb{D})]^{q\times q}$ is called *left* (respectively, *right*) *connected pair of* $[H^2(\mathbb{D})]^{q\times q}$*-functions* if there is an inner $q \times q$ Schur function V such that

$$\underline{\Psi} = \underline{V}\,\underline{\Phi}^* \qquad \text{(respectively, } \underline{\Psi} = \underline{\Phi}^*\underline{V}\text{)} \tag{2.1}$$

holds true $\underline{\lambda}$-a. e. on $\mathbb{T}$. Every such function V is said to be an *inner function which realizes this left* (respectively, *right*) *connection of* $[\Phi, \Psi]$.

Remark 2.2. $[\Phi, \Psi]$ is a left connected pair of $[H^2(\mathbb{D})]^{q\times q}$-functions if and only if $[\Psi, \Phi]$ is a right connected pair of $[H^2(\mathbb{D})]^{q\times q}$-functions.

The following result indication the close interrelation between left (respectively, right) connected pairs of $[H^2(\mathbb{D})]^{q\times q}-$functions and pseudocontinuability.

Proposition 2.3. *Let $[\Phi, \Psi]$ be a left (respectively, right) connected pair of $[H^2(\mathbb{D})]^{q\times q}$-functions, and let $V \in \mathcal{S}_{q\times q}(\mathbb{D})$ be an inner function which realizes this left (respectively, right) connection of $[\Phi, \Psi]$. Then both functions Φ and Ψ admit pseudocontinuations $\Phi^\#$ and $\Psi^\#$ which satisfy*

$$\Psi = V\widehat{\Phi^\#} \quad \textit{and} \quad \Phi = \widehat{\Psi^\#}V$$

$$\left(\textit{respectively, } \Psi = \widehat{\Phi^\#}V \quad \textit{and} \quad \Phi = V\widehat{\Psi^\#}\right). \tag{2.2}$$

In particular, V is a left denominator of $\widehat{\Phi^\#}$ and a right denominator of $\widehat{\Psi^\#}$ (respectively, a right denominator of $\widehat{\Phi^\#}$ and a left denominator of $\widehat{\Psi^\#}$).

Proof. From Remark 1.1 and (2.1) we see that $\Phi^\# = \widehat{\Psi}\widehat{V}^{-1}$ (respectively, $\Phi^\# = \widehat{V}^{-1}\widehat{\Psi}$) is a pseudocontinuation of Φ, and that $\Psi^\# = \widehat{V}^{-1}\widehat{\Phi}$ (respectively, $\Psi^\# = \widehat{\Phi}\widehat{V}^{-1}$) is a pseudocontinuation of Ψ. In view of $H^2(\mathbb{D}) \subseteq \mathcal{N}_+(\mathbb{D})$ the proof is complete. □

Remark 2.4. Let $[\Phi, \Psi]$ be a left or right connected pair of $[H^2(\mathbb{D})]^{q\times q}$-functions. Then $\det \Phi$ does not identically vanish if and only if $\det \Psi$ does not identically vanish.

Lemma 2.5. *Let $[\Phi, \Psi]$ be a left (respectively, right) connected pair of $[H^2(\mathbb{D})]^{q\times q}$-functions. Suppose that the function $\det \Phi$ does not identically vanish. Then $\det \Psi$ does not identically vanish, and there is a unique inner function $V \in \mathcal{S}_{q\times q}(\mathbb{D})$ which realizes the left (respectively, right) connection of $[\Phi, \Psi]$. This function V admits the representations*

$$V = \Psi\left(\widehat{\Phi\#}\right)^{-1} \quad \text{and} \quad V = \left(\widehat{\Psi\#}\right)^{-1}\Phi$$

$$\left(\text{respectively, } V = \left(\widehat{\Phi\#}\right)^{-1}\Psi \quad \text{and} \quad V = \Phi\left(\widehat{\Psi\#}\right)^{-1}\right). \tag{2.3}$$

Proof. From Proposition 2.3 we obtain that (2.2) holds. Thus the assumption that $\det \Phi$ does not identically vanish and Remark 2.4 yield the asserted statements. □

Proposition 2.6. *Let Φ be a function which belongs to $[H^2(\mathbb{D})]^{q\times q}$ and which admits a pseudocontinuation $\Phi^{\#}$. Let V be a left (respectively, right) denominator of $\widehat{\Phi\#}$, and let $\Psi := V\widehat{\Phi\#}$ (respectively, $\Psi := \widehat{\Phi\#}V$). Then Ψ belongs to $[H^2(\mathbb{D})]^{q\times q}$, and $[\Phi, \Psi]$ is a left (respectively, right) connected pair of $[H^2(\mathbb{D})]^{q\times q}$-functions where V is an inner function which realizes this left (respectively, right) connection of $[\Phi, \Psi]$.*

Proof. Since V is a left (respectively, right) denominator of $\widehat{\Phi\#}$ it follows that Ψ and $\underline{V}$ belong to $[\mathcal{N}_+(\mathbb{D})]^{q\times q}$ and $[\mathcal{L}^\infty(\mathbb{T})]^{q\times q}$ respectively. On the other hand, we have $\underline{\Phi} \in [\mathcal{L}^2(\mathbb{T})]^{q\times q}$. In view of Remark 1.1, $\underline{\Psi} = \underline{V}\,\underline{\Phi}^*$ holds $\underline{\lambda}$-almost everywhere on $\mathbb{T}$. Hence $\underline{\Psi} \in [\mathcal{L}^2(\mathbb{T})]^{q\times q}$. Thus the maximum modulus principle for the Smirnov class (see, e. g., [14, Theorem 2.11]) provides that Ψ even belongs to $[H^2(\mathbb{D})]^{q\times q}$. The rest of the assertion follows immediately. □

Proposition 2.7. *Let $\Phi \in [H^2(\mathbb{D})]^{q\times q}$ and $\Psi \in [H^2(\mathbb{D})]^{q\times q}$ be such that $\det \Phi$ and $\det \Psi$ do not identically vanish in $\mathbb{D}$. Then the following statements are equivalent:*

(i) *$[\Phi, \Psi]$ is a left connected pair of $[H^2(\mathbb{D})]^{q\times q}$-functions.*
(ii) *Φ admits a pseudocontinuation $\Phi^{\#}$ and $V := \Psi(\widehat{\Phi\#})^{-1}$ is an inner $q \times q$ Schur function.*
(iii) *Ψ admits a pseudocontinuation $\Psi^{\#}$ and $W := (\widehat{\Psi\#})^{-1}\Phi$ is an inner $q \times q$ Schur function.*

Proof. (i)⇒(ii), (i)⇒(iii): Use Proposition 2.3 and Lemma 2.5.
(ii)⇒(i): Because of (ii) we see $V\widehat{\Phi\#} = \Psi \in [H^2(\mathbb{D})]^{q\times q} \subseteq [\mathcal{N}_+(\mathbb{D})]^{q\times q}$. Therefore V is a left denominator of $\widehat{\Phi\#}$. From Proposition 2.6 it follows (i).
(iii)⇒(i): From (iii) we get $\widehat{\Psi\#}W = \Phi \in [\mathcal{N}_+(\mathbb{D})]^{q\times q}$. Hence W is a right denominator of $\widehat{\Psi\#}$. Thus Proposition 2.6 shows that $[\Psi, \Phi]$ is a right connected pair of $[H^2(\mathbb{D})]^{q\times q}$-functions. Remark 2.2 then yields (i). □

Using Remark 2.2 the following analogous result can be immediately derived from Proposition 2.7.

Proposition 2.8. *Let* $\Phi \in [H^2(\mathbb{D})]^{q\times q}$ *and* $\Psi \in [H^2(\mathbb{D})]^{q\times q}$ *be such that* $\det \Phi$ *and* $\det \Psi$ *do not identically vanish in* $\mathbb{D}$. *Then the following statements are equivalent:*

(i) $[\Phi, \Psi]$ *is a right connected pair of* $[H^2(\mathbb{D})]^{q\times q}$*-functions.*
(ii) Φ *admits a pseudocontinuation* $\Phi^\#$ *and* $V := (\widehat{\Phi\#})^{-1}\Psi$ *is an inner* $q \times q$ *Schur function.*
(iii) Ψ *admits a pseudocontinuation* $\Psi^\#$ *and* $W := \Phi(\widehat{\Psi\#})^{-1}$ *is an inner* $q \times q$ *Schur function.*

3. Some considerations on particular matrix-valued Carathéodory functions

This section is aimed to study a particular subclass of matrix-valued Carathéodory functions.

A function $\Omega : \mathbb{D} \longrightarrow \mathbb{C}^{q\times q}$ is said to be a $q \times q$ *Carathéodory function (on* $\mathbb{D}$*)* if Ω is holomorphic in $\mathbb{D}$ and if Re $\Omega(z)$ is nonnegative Hermitian for all $z \in \mathbb{D}$. The set $\mathcal{C}_q(\mathbb{D})$ of all $q \times q$ Carathéodory functions is a subset of $[\mathcal{N}_+(\mathbb{D})]^{q\times q}$ (see, e. g., [20, Corollary 2]). In particular, every $q \times q$ Carathéodory function Ω has radial boundary values $\underline{\Omega}$ with nonnegative Hermitian real part Re $\underline{\Omega}$ $\underline{\lambda}$-almost everywhere on $\mathbb{T}$. If $\Omega \in \mathcal{C}_q(\mathbb{D})$, then the matricial version of a famous theorem due to F. Riesz and Herglotz (see, e. g., [13, Theorem 2.2.2]) provides that there is a unique nonnegative Hermitian-valued Borel measure F on the unit circle $\mathbb{T}$ such that

$$\Omega(w) = \int_{\mathbb{T}} \frac{z+w}{z-w} F(dz) + i \operatorname{Im} [\Omega(0)] \tag{3.1}$$

is satisfied for all $w \in \mathbb{D}$. This nonnegative Hermitian-valued measure F is called the *F. Riesz-Herglotz measure associated with* Ω. A $q \times q$ Carathéodory function Ω (on $\mathbb{D}$) is said to be *absolutely continuous* (respectively, *singular*) if the F. Riesz-Herglotz measure associated with Ω is absolutely continuous (respectively, singular) with respect to the linear Lebesgue-Borel measure $\underline{\lambda}$ on $\mathbb{T}$. In the following, we will use $\mathcal{C}_q^{(a)}(\mathbb{D})$ (respectively, $\mathcal{C}_q^{(s)}(\mathbb{D})$) in order to denote the set of all absolutely continuous (respectively, singular) $q \times q$ Carathéodory functions. A function $\Omega \in \mathcal{C}_q(\mathbb{D})$ is called *normalized* if $\operatorname{Im}[\Omega(0)] = 0_{q\times q}$. We will write $\mathcal{C}_q^{\square}(\mathbb{D})$ for the set of all normalized $q \times q$ Carathéodory functions (on $\mathbb{D}$).

Lemma 3.1. *Every singular $q \times q$ Carathéodory function Ω_s admits a pseudocontinuation $\Omega_s^{\#}$, namely $\Omega_s^{\#} = -\widehat{\Omega_s}$, and fulfills* $\operatorname{Re} \underline{\Omega_s} = 0_{q\times q}$ *$\underline{\lambda}$-almost everywhere on $\mathbb{T}$.*

Proof. By virtue of [20, Lemma 4], the function $\operatorname{Re} \underline{\Omega_s}$ is a version of the Radon-Nikodym derivative of F_a with respect to $(1/2\pi)\underline{\lambda}$ where F_a is the absolutely continuous part in the Lebesgue decomposition of the F. Riesz-Herglotz measure associated with Ω_s with respect to $(1/2\pi)\underline{\lambda}$. Since Ω_s is singular thus it follows $\operatorname{Re} \underline{\Omega_s} = 0_{q\times q}$ $\underline{\lambda}$-almost everywhere on $\mathbb{T}$. Because of $\mathcal{C}_q(\mathbb{D}) \subseteq [\mathcal{NM}(\mathbb{D})]^{q\times q}$ we see that $g := -\widehat{\Omega_s}$ belongs to $[\mathcal{NM}(\mathbb{E})]^{q\times q}$. From Remark 1.1 we get finally $\underline{g} = -\underline{\Omega_s^*} = -2(\operatorname{Re} \underline{\Omega_s}) + \underline{\Omega_s} = \underline{\Omega_s}$ $\underline{\lambda}$-almost everywhere on $\mathbb{T}$. □

Lemma 3.2. *Let $\Sigma \in [\mathcal{L}^1(\mathbb{T})]^{q\times q}$ be such that $\Sigma(z)$ is nonnegative Hermitian for $\underline{\lambda}$-almost all $z \in \mathbb{T}$. Then $\Omega_\Sigma : \mathbb{D} \longrightarrow \mathbb{C}^{q\times q}$ given by*

$$\Omega_\Sigma(w) := \frac{1}{2\pi}\int_{\mathbb{T}} \frac{z+w}{z-w}\Sigma(z)\underline{\lambda}(\mathrm{d}z) \tag{3.2}$$

belongs to $\mathcal{C}_q^{(a)}(\mathbb{D}) \cap \mathcal{C}_q^{\square}(\mathbb{D})$ and satisfies $\operatorname{Re} \underline{\Omega_\Sigma} = \Sigma$ *$\underline{\lambda}$-almost everywhere on $\mathbb{T}$. The F. Riesz-Herglotz measure F associated with Ω_Σ admits the representation*

$$F(B) = \frac{1}{2\pi}\int_B \Sigma \,\mathrm{d}\underline{\lambda} \tag{3.3}$$

for every Borel subset B of $\mathbb{T}$.

Proof. Let $F : \mathfrak{B}_{\mathbb{T}} \longrightarrow \mathbb{C}^{q\times q}$ be given by (3.3). Then it is readily checked that F is a Hermitian-valued Borel measure on $\mathbb{T}$ which satisfies

$$\Omega_\Sigma(w) = \frac{1}{2\pi}\int_{\mathbb{T}} \frac{z+w}{z-w}F(\mathrm{d}z)$$

for all $w \in \mathbb{D}$. Thus one can easily see that Ω_Σ belongs to $\mathcal{C}_q^{(a)}(\mathbb{D}) \cap \mathcal{C}_q^{\square}(\mathbb{D})$. Lemma 4 in [20] yields finally that $\operatorname{Re} \underline{\Omega_\Sigma} = \Sigma$ holds $\underline{\lambda}$-almost everywhere on $\mathbb{T}$. □

If Σ and Ξ are functions which belong to $[\mathcal{L}^1(\mathbb{T})]^{q\times q}$, which have nonnegative Hermitian values $\underline{\lambda}$-almost everywhere on $\mathbb{T}$ and which satisfy $\Sigma = \Xi$ $\underline{\lambda}$-almost everywhere on $\mathbb{T}$, then the functions Ω_Σ and Ω_Ξ given by (3.2) coincide, i. e., the function Ω_Σ depends only on the equivalence class $\langle\Sigma\rangle$ of all functions $\Xi : \mathbb{T} \longrightarrow \mathbb{C}^{q\times q}$ which fulfill $\Xi = \Sigma$ $\underline{\lambda}$-almost everywhere on $\mathbb{T}$. In the following, we will continue to use this notation $\langle\Sigma\rangle$. Moreover, we will write $\Omega_{\langle\Sigma\rangle}$ for the function which is given by $\Omega_{\langle\Sigma\rangle} := \Omega_\Sigma$ and (3.2).

Lemma 3.3. *Let $\Sigma \in [\mathcal{L}^1(\mathbb{T})]^{q\times q}$ be such that $\Sigma(z)$ is nonnegative Hermitian for $\underline{\lambda}$-almost all $z \in \mathbb{T}$. Then the set*

$$\mathcal{C}_{q,\langle\Sigma\rangle} := \{\Omega \in \mathcal{C}_q(\mathbb{D}) : \langle \operatorname{Re} \underline{\Omega}\rangle = \langle\Sigma\rangle\} \tag{3.4}$$

admits the representation

$$\mathcal{C}_{q,\langle\Sigma\rangle} = \{\Omega_{\langle\Sigma\rangle} + \Omega_s : \Omega_s \in \mathcal{C}_q^{(s)}(\mathbb{D})\} . \tag{3.5}$$

Proof. For each $\Omega \in \mathcal{C}_q(\mathbb{D})$, let F_Ω be the F. Riesz-Herglotz measure associated with Ω, and let $F_{\Omega,a}$ (respectively, $F_{\Omega,s}$) be the absolutely continuous part (respectively, the singular part) of F_Ω in the Lebesgue decomposition of F_Ω with respect to $(1/2\pi)\underline{\lambda}$. According to Lemma 3.1, we have $\operatorname{Re} \underline{\Omega_s} = 0_{q\times q}$ $\underline{\lambda}$-almost everywhere on $\mathbb{T}$ for every singular $q \times q$ Carathéodory function Ω_s. Applying Lemma 3.2 we then see that, for each $\Omega_s \in \mathcal{C}_q^{(s)}(\mathbb{D})$, the function $\Omega_\blacksquare := \Omega_{\langle\Sigma\rangle} + \Omega_s$ belongs to $\mathcal{C}_{q,\langle\Sigma\rangle}$. Conversely, now assume that Ω is an arbitrary function which belongs to $\mathcal{C}_{q,\langle\Sigma\rangle}$. Using the arguments mentioned above then we see that $\Omega_\square := \Omega - \Omega_{\langle\Sigma\rangle}$ is a function which is holomorphic in $\mathbb{D}$ and which fulfills

$$\begin{aligned}\Omega_\square(w) &= \frac{1}{2\pi}\int_{\mathbb{T}} \frac{z+w}{z-w} \operatorname{Re} \underline{\Omega}(z)\underline{\lambda}(\mathrm{d}z) \\ &\quad + \int_{\mathbb{T}} \frac{z+w}{z-w} F_{\Omega,s}(\mathrm{d}z) - \frac{1}{2\pi}\int_{\mathbb{T}} \frac{z+w}{z-w}\Sigma(z)\underline{\lambda}(\mathrm{d}z) \\ &= \int_{\mathbb{T}} \frac{z+w}{z-w} F_{\Omega,s}(\mathrm{d}z)\end{aligned}$$

for all $w \in \mathbb{D}$. Thus $\Omega_\square$ is a singular $q \times q$ Carathéodory function satisfying $\Omega = \Omega_{\langle\Sigma\rangle} + \Omega_s$. □

If Φ belongs to $[H^2(\mathbb{D})]^{q\times q}$, then both functions $\underline{\Phi}\,\underline{\Phi}^*$ and $\underline{\Phi}^*\underline{\Phi}$ belong to $[\mathcal{L}^1(\mathbb{T})]^{q\times q}$. In view of Lemma 3.3 we then introduce the following notion.

Definition 3.4. If $\Phi \in [H^2(\mathbb{D})]^{q\times q}$, then the set $\mathcal{C}_{q,\langle\underline{\Phi}\,\underline{\Phi}^*\rangle}$ (respectively, $\mathcal{C}_{q,\langle\underline{\Phi}^*\underline{\Phi}\rangle}$) defined by (3.4) is called the *subclass of $\mathcal{C}_q(\mathbb{D})$ which is left* (respectively, *right*) *generated by Φ*.

Remark 3.5. (a) Let $[\Phi, \Psi]$ be a left connected pair of $[H^2(\mathbb{D})]^{q\times q}$-functions. Then the subclass of $\mathcal{C}_q(\mathbb{D})$ which is left generated by Φ coincides with the subclass of $\mathcal{C}_q(\mathbb{D})$ which is right generated by Ψ.
(b) Let $[\Phi, \Psi]$ be a right connected pair of $[H^2(\mathbb{D})]^{q\times q}$-functions. Then the subclass of $\mathcal{C}_q(\mathbb{D})$ which is right generated by Φ coincides with the subclass of $\mathcal{C}_q(\mathbb{D})$ which is left generated by Ψ.

Lemma 3.6. *Let Φ be a function which belongs to $[H^2(\mathbb{D})]^{q\times q}$ and which admit a pseudocontinuation $\Phi^\#$. Then every function Ω that belongs to $\mathcal{C}_{q,\langle\underline{\Phi}\,\underline{\Phi}^*\rangle}$ admits a pseudocontinuation $\Omega^\#$, namely $\Omega^\# = 2\Phi^\#\widehat{\Phi} - \widehat{\Omega}$.*

Proof. Let $\Omega \in \mathcal{C}_{q,\langle \underline{\Phi}\,\underline{\Phi}^* \rangle}$, i. e., Ω belongs to $\mathcal{C}_q(\mathbb{D})$ and satisfies $\underline{\Omega}+\underline{\Omega}^* = 2\underline{\Phi}\,\underline{\Phi}^*$ $\underline{\lambda}$-almost everywhere on $\mathbb{T}$. Since $[H^2(\mathbb{D})]^{q\times q}$ and $\mathcal{C}_q(\mathbb{D})$ are subsets of $[\mathcal{NM}(\mathbb{D})]^{q\times q}$, the function $g := 2\Phi^{\#}\widehat{\Phi} - \widehat{\Omega}$ belongs to $[\mathcal{NM}(\mathbb{E})]^{q\times q}$. From Remark 1.1 we then see that $\underline{g} = 2\underline{\Phi}\,\underline{\Phi}^* - \underline{\Omega}^* = \underline{\Omega}$ $\underline{\lambda}$-almost everywhere on $\mathbb{T}$. Hence g is a pseudocontinuation of Ω. □

Analogously, the following result can be proved.

Lemma 3.7. *Let Φ be a function which belongs to $[H^2(\mathbb{D})]^{q\times q}$ and which admit a pseudocontinuation $\Phi^{\#}$. Then every function Ω that belongs to $\mathcal{C}_{q,\langle \underline{\Phi}^*\underline{\Phi} \rangle}$ admits a pseudocontinuation $\Omega^{\#}$, namely $\Omega^{\#} = 2\widehat{\Phi}\Phi^{\#} - \widehat{\Omega}$.*

4. Some remarks on J-inner functions

Throughout this section, let m be a positive integer, and let J be an $m \times m$ signature matrix, i. e., J belongs to $\mathbb{C}^{m\times m}$ and satisfies as well $J = J^*$ as $J^2 = I$. A matrix $A \in \mathbb{C}^{m\times m}$ is called *J-contractive* if $B := J - A^*JA$ is nonnegative Hermitian. If $A \in \mathbb{C}^{m\times m}$ even satisfies $A^*JA = J$, then A is said to be *J-unitary.* The *Potapov class* $\mathfrak{P}_J(\mathbb{D})$ consists of all $m \times m$ matrix-valued functions W which satisfy the following three conditions:

(i) W is meromorphic in $\mathbb{D}$.
(ii) The function $\det W$ does not identically vanish in $\mathbb{D}$.
(iii) For each z which belongs to the set $\mathbb{H}_W$ of all points of analyticity of W, the matrix $W(z)$ is J-contractive.

The Potapov class $\mathfrak{P}_J(\mathbb{D})$ is a subclass of $[\mathcal{NM}(\mathbb{D})]^{m\times m}$ (see, e. g., [15, Corollary 2]). In particular, every function W which belongs to $\mathfrak{P}_J(\mathbb{D})$ has radial boundary values $\underline{W}$ $\underline{\lambda}$-almost everywhere on $\mathbb{T}$. If $W \in \mathfrak{P}_J(\mathbb{D})$ satisfies $\underline{W}^*J\underline{W} = J$ $\underline{\lambda}$-almost everywhere on $\mathbb{T}$, then W is said to be a *J-inner function.*

Remark 4.1. Every J-inner function W admits a pseudocontinuation $W^{\#}$ (into $\mathbb{E}$). For each $z \in \mathbb{E}$ which fulfills $1/\overline{z} \in \mathbb{H}_W$ and $\det W(1/\overline{z}) \neq 0$ this pseudocontinuation $W^{\#}$ admits the representation $W^{\#}(z) = J[W(1/\overline{z})]^{-*}J$.

Now we will focus our attention to the special $2q \times 2q$ signature matrix

$$j_{qq} := \operatorname{diag}\left(I_q, -I_q\right). \tag{4.1}$$

In the sequel, when we will consider a $2q \times 2q$ complex matrix W or a $2q \times 2q$ matrix-valued function W, then we will often work with the $q \times q$ block partition

$$W = \begin{bmatrix} W_{11} & W_{12} \\ W_{21} & W_{22} \end{bmatrix} \tag{4.2}$$

of W. Furthermore, if a $2q \times 2q$ matrix-valued function W is given, then we will use $\mathbb{H}_W$ to denote the set of all points of analyticity of W.

A useful tool to treat problems which are formulated for functions which belong to $\mathfrak{P}_{j_{qq}}(\mathbb{D})$ is the transformation into the Schur class. The following result gives a summary of facts which are useful to do this.

Proposition 4.2. *Let $W \in \mathfrak{P}_{j_{qq}}(\mathbb{D})$. Then the following statements are fulfilled:*

(a) *For each $z \in \mathbb{H}_W$, the inequalities* $\det W_{22}(z) \neq 0$, $\det[W_{22}(z) + W_{21}(z)] \neq 0$ *and* $\det[W_{22}(z) + W_{12}(z)] \neq 0$ *hold true. Moreover, the functions* $\det(\widehat{W_{11}^{\#}} + \widehat{W_{12}^{\#}})$ *and* $\det(\widehat{W_{11}^{\#}} + \widehat{W_{21}^{\#}})$ *do not identically vanish.*

(b) *The function*

$$S := \begin{bmatrix} W_{11} - W_{12}W_{22}^{-1}W_{21} & W_{12}W_{22}^{-1} \\ -W_{22}^{-1}W_{21} & W_{22}^{-1} \end{bmatrix} \tag{4.3}$$

belongs to the Schur class $\mathcal{S}_{2q \times 2q}(\mathbb{D})$. In particular, $S_{11} := W_{11} - W_{12}W_{22}^{-1}W_{21}$, $S_{12} := W_{12}W_{22}^{-1}$, $S_{21} := -W_{22}^{-1}W_{21}$ and $S_{22} := W_{22}^{-1}$ are matrix-valued Schur functions, whereby S_{12} and S_{21} are even strictly contractive.

(c) *The functions* $\det S_{11}$ *and* $\det S_{22}$ *do not identically vanish.*

(d) $\mathbb{H}_W = \{z \in \mathbb{D} : \det S_{22}(z) \neq 0\}$.

(e) *If W is a j_{qq}-inner function, then S is an inner function and the following identities are valid:*

$$S_{12} = (\widehat{W_{11}^{\#}})^{-1}\widehat{W_{21}^{\#}}\,, \quad S_{21} = -\widehat{W_{12}^{\#}}(\widehat{W_{11}^{\#}})^{-1} \quad \text{and} \quad S_{11} = (\widehat{W_{11}^{\#}})^{-1}. \tag{4.4}$$

A proof of the results stated in Proposition 4.2 is given in [1], [23], [10] and [11]. Observe that the function S defined by (4.3) is called the *Potapov-Ginzburg transform of W (with respect to j_{qq})*. In some sense, the following result, which can be verified by straightforward calculation, is a converse statement to part (b) of Proposition 4.2.

Proposition 4.3. *Let W be an inner $2q \times 2q$ Schur function, and let (4.2) be the $q \times q$ block partition of W. Suppose that* $\det W_{22}$ *does not identically vanish. Then S given by (4.3) is a j_{qq}-inner function.*

5. ADD-parametrization of j_{qq}-inner functions

Our first goal of this section consists of some analysis of the $q \times q$ block structure of j_{qq}–inner functions.

Proposition 5.1. *Let W be a j_{qq}-inner function. Then:*

(a) *The pair $[\Phi_{W,l}, \Psi_{W,l}]$ given by*

$$\Phi_{W,l} := (W_{22} + W_{21})^{-1} \quad \text{and} \quad \Psi_{W,l} := (\widehat{W_{11}^{\#}} + \widehat{W_{12}^{\#}})^{-1} \tag{5.1}$$

is a left connected pair of $[H^2(\mathbb{D})]^{q\times q}$-functions, where

$$V_{W,l} := (W_{11} + W_{12})(W_{22} + W_{21})^{-1} \tag{5.2}$$

is the unique inner $q \times q$ Schur function which realizes this left connection.

(b) *Both functions $\Phi_{W,l}$ and $\Psi_{W,l}$ admit pseudocontinuations $\Phi_{W,l}^{\#}$ and $\Psi_{W,l}^{\#}$, respectively, and satisfy the identities*

$$V_{W,l}\widehat{\Phi_{W,l}^{\#}} = \Psi_{W,l} \quad \text{and} \quad \widehat{\Psi_{W,l}^{\#}}V_{W,l} = \Phi_{W,l} \,. \tag{5.3}$$

(c) *The function*

$$\Omega_{W,l} := (W_{22} + W_{21})^{-1}(W_{22} - W_{21}) \tag{5.4}$$

belongs to the subclass $\mathcal{C}_{q,\langle \underline{\Phi_{W,l}} \,\underline{\Phi_{W,l}^{*}}\rangle}$ of $\mathcal{C}_q(\mathbb{D})$ which is left generated by $\Phi_{W,l}$ and which admits the representation

$$\Omega_{W,l} = (I + W_{22}^{-1}W_{21})^{-1}(I - W_{22}^{-1}W_{21}) \,. \tag{5.5}$$

Proof. From Proposition 4.2 we see that $S_{11} := (\widehat{W_{11}^{\#}})^{-1}$, $S_{22} := W_{22}^{-1}$, $S_{12} := (\widehat{W_{11}^{\#}})^{-1}\widehat{W_{21}^{\#}}$ and $S_{21} := -W_{22}^{-1}W_{21}$ are (well-defined) $q \times q$ Schur functions where both functions S_{12} and S_{21} are even strictly contractive. In view of a result due to Arov [1] we can conclude that $T_{21} := I - S_{21}$ and $T_{12} := I + S_{12}$ are outer functions in $[H^\infty(\mathbb{D})]^{q\times q}$. Thus $(I - S_{21})^{-1}$ and $(I + S_{12})^{-1}$ are outer functions in $[\mathcal{N}_+(\mathbb{D})]^{q\times q}$. Since $\mathcal{N}_+(\mathbb{D})$ is an algebra over $\mathbb{D}$, we get from the identities

$$\Phi_{W,l} = (I - S_{21})^{-1}S_{22} \quad \text{and} \quad \Psi_{W,l} = (I + S_{12})^{-1}S_{11} \tag{5.6}$$

that $\Phi_{W,l}$ and $\Psi_{W,l}$ belong to $[\mathcal{N}_+(\mathbb{D})]^{q\times q}$. Thus, according to the maximum modulus principle for the Smirnov class (see, e. g., [14, Theorem 2.11]), it is sufficient to verify that the radial boundary functions $\underline{\Phi_{W,l}}$ and $\underline{\Psi_{W,l}}$ of $\Phi_{W,l}$ and $\Psi_{W,l}$, respectively, belong to $[\mathcal{L}^2(\mathbb{T})]^{q\times q}$ in order to prove that $\Phi_{W,l}$ and $\Psi_{W,l}$ belong to $[H^2(\mathbb{D})]^{q\times q}$. Since $\Omega_{W,l}$ admits the representation $\Omega_{W,l} = (I - S_{21})^{-1}(I + S_{21})$ the function $\Omega_{W,l}$ belongs to $\mathcal{C}_q(\mathbb{D})$ (see, e. g., [13, Propositions 2.1.2 and 2.1.3]). Since W has j_{qq}-unitary radial boundary values $\underline{\lambda}$-almost everywhere on $\mathbb{T}$, we obtain that $\underline{W}^*$ has j_{qq}-unitary values $\underline{\lambda}$-almost everywhere on $\mathbb{T}$. Consequently, we have

$$(5.7) \qquad \underline{W_{22}}\,\underline{W_{22}^*} - \underline{W_{21}}\,\underline{W_{21}^*} = I \qquad \underline{\lambda}\text{-a. e. on } \mathbb{T}$$

and hence

$$\begin{aligned} \operatorname{Re} \underline{\Omega_{W,l}} &= \frac{1}{2}(\underline{\Omega_{W,l}} + \underline{\Omega_{W,l}^*}) \\ &= (\underline{W_{22}} + \underline{W_{21}})^{-1}(\underline{W_{22}}\,\underline{W_{22}^*} - \underline{W_{21}}\,\underline{W_{21}^*})(\underline{W_{22}} + \underline{W_{21}})^{-*} \\ &= \underline{\Phi_{W,l}}\,\underline{\Phi_{W,l}^*} \qquad \underline{\lambda}\text{-a. e. on } \mathbb{T}\,. \end{aligned} \tag{5.8}$$

From [20, Lemma 4] we know that Re $\underline{\Omega_{W,l}}$ belongs to $[\mathcal{L}^1(\mathbb{T})]^{q\times q}$. Thus the identity (5.8) provides $\underline{\Phi_{W,l}} \in [\mathcal{L}^2(\mathbb{T})]^{q\times q}$. Therefore, $\Phi_{W,l} \in [H^2(\mathbb{D})]^{q\times q}$ is verified. Let $e :=$ (I_q, I_q). Since W is a j_{qq}-inner function, the function $g := e(j_{qq} - W^* j_{qq} W)e^*$ has nonnegative Hermitian values the radial boundary values of which fulfill $\underline{g} = 0_{q\times q}$ $\underline{\lambda}$-almost everywhere on $\mathbb{T}$. Using the block form of this inequality and this equality, one can easily verify that $V_{W,l}$ is an inner $q \times q$ Schur function. Since W has j_{qq}-unitary radial boundary values $\underline{\lambda}$-almost everywhere on $\mathbb{T}$ we get from Remark 1.1 that $\underline{V_{W,l}}\,\underline{\Phi_{W,l}^{-1}} = \underline{\Psi_{W,l}^{-*}}$ and hence $\underline{\Psi_{W,l}} = \underline{V_{W,l}}\,\underline{\Phi_{W,l}^*}$ $\underline{\lambda}$-almost everywhere on $\mathbb{T}$. Taking into account that $V_{W,l}$ is an inner $q \times q$ Schur function and that $\underline{\Phi_{W,l}}$ belongs to $[\mathcal{L}^2(\mathbb{T})]^{q\times q}$ we thus see that $\underline{\Psi_{W,l}}$ also belongs to $[\mathcal{L}^2(\mathbb{T})]^{q\times q}$. Hence $\Psi_{W,l}$ belongs to $[H^2(\mathbb{D})]^{q\times q}$. Consequently, we get that $[\Phi_{W,l}, \Psi_{W,l}]$ is a left connected pair of $[H^2(\mathbb{D})]^{q\times q}$-functions. In view of Lemma 2.5 it follows that $V_{W,l}$ is the unique inner function which realizes this left connection. Proposition 2.3 yields that $\Phi_{W,l}$ and $\Psi_{W,l}$ admit pseudocontinuations and that (5.3) holds true. Finally, since the $q \times q$ Carathéodory function $\Omega_{W,l}$ satisfies (5.8) we see that $\Omega_{W,l}$ belongs to $\mathcal{C}_{q,\langle \underline{\Phi_{W,l}}\,\underline{\Phi_{W,l}^*}\rangle}$. □

Analogously to Proposition 5.1 the following result can be proved.

Proposition 5.2. *Let W be a j_{qq}-inner function. Then:*

(a) *The pair $[\Phi_{W,r}, \Psi_{W,r}]$ given by*

$$(5.9) \qquad \Phi_{W,r} := (W_{22} + W_{12})^{-1} \quad \textit{and} \quad \Psi_{W,r} := (\widehat{W_{11}^{\#}} + \widehat{W_{21}^{\#}})^{-1}$$

is a right connected pair of $[H^2(\mathbb{D})]^{q\times q}$-functions, where

$$(5.10) \qquad V_{W,r} := (W_{22} + W_{12})^{-1}(W_{11} + W_{21})$$

is the unique inner $q \times q$ Schur function which realizes this right connection.

(b) *Both functions* $\Phi_{W,r}$ *and* $\Psi_{W,r}$ *admit pseudocontinuations* $\Phi^{\#}_{W,r}$ *and* $\Psi^{\#}_{W,r}$, *respectively, and satisfy the identities*

$$\widehat{\Phi^{\#}_{W,r}} V_{W,r} = \Psi_{W,r} \quad \text{and} \quad V_{W,r}\widehat{\Psi^{\#}_{W,r}} = \Phi_{W,r} \ . \tag{5.11}$$

(c) *The function*

$$\Omega_{W,r} := (W_{22} - W_{12})(W_{22} + W_{12})^{-1} \tag{5.12}$$

belongs to the subclass $\mathcal{C}_{q,\langle \underline{\Phi^{*}_{W,r}}, \underline{\Phi_{W,r}} \rangle}$ *of* $\mathcal{C}_q(\mathbb{D})$ *which is right generated by* $\Phi_{W,r}$ *and which admits the representation*

$$\Omega_{W,r} = (I - W_{12}W_{22}^{-1})(I + W_{12}W_{22}^{-1})^{-1} \ . \tag{5.13}$$

Propositions 5.1 and 5.2 lead us to the following notions.

Definition 5.3. Let W be a j_{qq}-inner function.

(a) The pair $[\Phi_{W,l}, \Psi_{W,l}]$ given by (5.1) (respectively, $[\Phi_{W,r}, \Psi_{W,r}]$ given by (5.9)) is called the *left* (respectively, *right*) *connected pair of* $[H^2(\mathbb{D})]^{q\times q}$*-functions generated by* W.
(b) The function $\Omega_{W,l}$ given by (5.4) (respectively, $\Omega_{W,r}$ given by (5.12)) is said to be the *left* (respectively, *right*) $q \times q$ *Carathéodory function generated by* W.

Now we are going to investigate the Carathéodory functions generated by a j_{qq}−inner function.

Proposition 5.4. *Let* W *be a* j_{qq}*-inner function, and let* $[\Phi_{W,l}, \Psi_{W,l}]$ *be the left connected pair of* $[H^2(\mathbb{D})]^{q\times q}$*-functions generated by* W. *Then:*

(a) *The left* $q \times q$ *Carathéodory function* $\Omega_{W,l}$ *generated by* W *can be represented via*

$$\Omega_{W,l} = (\widehat{W^{\#}_{11}} - \widehat{W^{\#}_{12}})(\widehat{W^{\#}_{11}} + \widehat{W^{\#}_{12}})^{-1} \ . \tag{5.14}$$

Moreover, $\Omega_{W,l}$ *admits a pseudocontinuation* $\Omega^{\#}_{W,l}$ *which satisfies the identities*

$$\Omega_{W,l} + \widehat{\Omega^{\#}_{W,l}} = 2\Phi_{W,l}\widehat{\Phi^{\#}_{W,l}} \tag{5.15}$$

and

$$\widehat{\Omega^{\#}_{W,l}} = (W_{11} + W_{12})^{-1}(W_{11} - W_{12}) \ . \tag{5.16}$$

(b) *The function*

$$\Omega_{W,l,s} := \Omega_{W,l} - \Omega_{\langle \underline{\Phi_{W,l}}, \underline{\Phi^{*}_{W,l}} \rangle} \tag{5.17}$$

is a singular $q \times q$ *Carathéodory function.*

Proof. From Remark 4.1 we know that all the $q \times q$ matrix-valued functions W_{11}, W_{12}, W_{21} and W_{22} admit pseudocontinuations. Hence $V := W_{22} - W_{21}$ admits a pseudocontinuation. On the other hand, we know from Proposition 2.3, that $\Phi_{W,l}$ admits a pseudocontinuation. As the product of the pseudocontinuable functions $\Phi_{W,l}$ and V the function $\Omega_{W,l}$ admits a pseudocontinuation $\Omega^{\#}_{W,l}$ as well. From Proposition 5.1 we see that $\Omega_{W,l}$ belongs to $\mathcal{C}_{q,\langle \underline{\Phi_{W,l}}\, \underline{\Phi^*_{W,l}}\rangle}$. In particular, we have $\operatorname{Re}\, \underline{\Omega_{W,l}} = \underline{\Phi_{W,l}}\; \underline{\Phi^*_{W,l}}$ λ-almost everywhere on $\mathbb{T}$. Thus Remark 1.1 implies (5.15). Moreover, since $\underline{W}$ is j_{qq}-unitary λ-almost everywhere on $\mathbb{T}$, we get

$$\underline{W_{21}}\,\underline{W^*_{11}} - \underline{W_{22}}\,\underline{W^*_{12}} = 0 \qquad \lambda\text{-a. e. on } \mathbb{T} \tag{5.18}$$

and hence

$$(\underline{W_{22}} - \underline{W_{21}})(\underline{W_{11}} + \underline{W_{12}})^* = (\underline{W_{22}} + \underline{W_{21}})(\underline{W_{11}} - \underline{W_{12}})^* \quad \lambda\text{-a. e. on } \mathbb{T}. \tag{5.19}$$

Using Remark 1.1 we thus obtain

$$(W_{22} - W_{21})(\widehat{W^{\#}_{11}} + \widehat{W^{\#}_{12}}) = (W_{22} + W_{21})(\widehat{W^{\#}_{11}} - \widehat{W^{\#}_{12}})\,. \tag{5.20}$$

From Proposition 4.2 then it follows (5.14). Remark 1.1 implies

$$\underline{\Omega^*_{W,l}} = [(\underline{W^*_{11}} - \underline{W^*_{12}})(\underline{W^*_{11}} + \underline{W^*_{12}})^{-1}]^* \tag{5.21}$$

and consequently (5.16). Finally, the assertion stated in part (b) is an immediate consequence of $\Omega_{W,l} \in \mathcal{C}_{q,\langle \underline{\Phi_{W,l}}\, \underline{\Phi^*_{W,l}}\rangle}$ and Lemma 3.3. □

Similarly to Proposition 5.4, an analogous result for right connected pairs of $[H^2(\mathbb{D})]^{q\times q}$-functions can be proved.

Proposition 5.5. *Let W be a j_{qq}-inner function, and let $[\Phi_{W,r}, \Psi_{W,r}]$ be the right connected pair of $[H^2(\mathbb{D})]^{q\times q}$-functions generated by W. Then:*

(a) *The right $q\times q$ Carathéodory function $\Omega_{W,r}$ generated by W can be represented via*

$$\Omega_{W,r} = (\widehat{W^{\#}_{11}} + \widehat{W^{\#}_{21}})^{-1}(\widehat{W^{\#}_{11}} - \widehat{W^{\#}_{21}})\,. \tag{5.22}$$

Moreover, $\Omega_{W,r}$ admits a pseudocontinuation $\Omega^{\#}_{W,r}$ which satisfies the identities

$$\Omega_{W,r} + \widehat{\Omega^{\#}_{W,r}} = 2\widehat{\Phi^{\#}_{W,r}}\Phi_{W,r} \tag{5.23}$$

and

$$\widehat{\Omega^{\#}_{W,r}} = (W_{11} - W_{21})(W_{11} + W_{21})^{-1}\,. \tag{5.24}$$

(b) *The function*

$$\Omega_{W,r,s} := \Omega_{W,r} - \Omega_{\langle \underline{\Phi^*_{W,r}}\, \underline{\Phi_{W,r}}\rangle} \tag{5.25}$$

is a singular $q\times q$ Carathéodory function.

Propositions 5.4 and 5.5 lead us to the following notions.

Definition 5.6. Let W be a j_{qq}-inner function. Then $\Omega_{W,l,s}$ defined by (5.17) (respectively, $\Omega_{W,r,s}$ defined by (5.25)) is called the *left* (respectively, *right*) *singular* $q \times q$ *Carathéodory function generated by* W.

Now we are able to formulate and prove the first main result of the paper.

Theorem 5.7. *Let W be a j_{qq}-inner function, let $[\Phi_{W,l}, \Psi_{W,l}]$ and $\Omega_{W,l}$ be the left connected pair of $[H^2(\mathbb{D})]^{q\times q}$-functions and the left $q \times q$ Carathéodory function, respectively, generated by W. Then W admits the representation*

$$W = \frac{1}{2} \cdot \operatorname{diag}\left[(\widehat{\Psi^{\#}_{W,l}})^{-1}, \Phi^{-1}_{W,l}\right] \cdot \begin{bmatrix} I + \widehat{\Omega^{\#}_{W,l}} & I - \widehat{\Omega^{\#}_{W,l}} \\ I - \Omega_{W,l} & I + \Omega_{W,l} \end{bmatrix}. \tag{5.26}$$

If $[\Phi, \Psi]$ is a left connected pair of $[H^2(\mathbb{D})]^{q\times q}$-functions such that the function $\det \Phi$ *does not identically vanish in $\mathbb{D}$ and if Ω is a $q \times q$ Carathéodory function which admits a pseudocontinuation $\Omega^{\#}$ such that the representation*

$$W = \frac{1}{2} \cdot \operatorname{diag}\left[(\widehat{\Psi^{\#}})^{-1}, \Phi^{-1}\right] \cdot \begin{bmatrix} I + \widehat{\Omega^{\#}} & I - \widehat{\Omega^{\#}} \\ I - \Omega & I + \Omega \end{bmatrix} \tag{5.27}$$

of W is satisfied, then $\Phi = \Phi_{W,l}$, $\Psi = \Psi_{W,l}$ and $\Omega = \Omega_{W,l}$.

Proof. In view of Remark 1.1, it is sufficient to verify

$$\underline{W} = \frac{1}{2} \cdot \operatorname{diag}\left[\underline{\Psi^{-*}_{W,l}}, \underline{\Phi^{-1}_{W,l}}\right] \cdot \begin{bmatrix} I + \underline{\Omega^{*}_{W,l}} & I - \underline{\Omega^{*}_{W,l}} \\ I - \underline{\Omega_{W,l}} & I + \underline{\Omega_{W,l}} \end{bmatrix} \tag{5.28}$$

$\underline{\lambda}$-almost everywhere on $\mathbb{T}$ in order to prove (5.26). However, according to Remark 1.1 and equation (5.16) we get

$$\begin{aligned} \frac{1}{2}\underline{\Psi^{-*}_{W,l}}(I + \underline{\Omega^{*}_{W,l}}) &= \frac{1}{2}(\underline{W_{11}} + \underline{W_{12}})[I + (\underline{W_{11}} + \underline{W_{12}})^{-1}(\underline{W_{11}} - \underline{W_{12}})] \\ &= \underline{W_{11}} \end{aligned}$$

and analogously

$$\frac{1}{2}\underline{\Psi^{-*}_{W,l}}(I - \underline{\Omega^{*}_{W,l}}) = \underline{W_{12}}\,, \quad \frac{1}{2}\underline{\Phi^{-1}_{W,l}}(I - \underline{\Omega_{W,l}}) = \underline{W_{21}}\,, \quad \frac{1}{2}\underline{\Phi^{-1}_{W,l}}(I + \underline{\Omega_{W,l}}) = \underline{W_{22}}$$

$\underline{\lambda}$-almost everywhere on $\mathbb{T}$. Thus (5.28) and hence (5.26) are checked. Now assume that $[\Phi, \Psi]$ is a left connected pair of $[H^2(\mathbb{D})]^{q\times q}$-functions such that $\det \Phi$ does not identically vanish in $\mathbb{D}$. Then Proposition 2.3 shows that $\det \widehat{\Psi^{\#}}$ does not identically vanish in $\mathbb{D}$. Further assume that Ω is a pseudocontinuable $q \times q$ Carathéodory function such that (5.27) is fulfilled. Then

$$W_{21} + W_{22} = \frac{1}{2}\Phi^{-1}[(I - \Omega) + (I + \Omega)] = \Phi^{-1} \tag{5.29}$$

and therefore $\Phi = \Phi_{W,l}$. Using (5.29) and $W_{22} - W_{21} = \frac{1}{2}\Phi^{-1}[(I+\Omega)-(I-\Omega)] = \Phi^{-1}\Omega$, we can conclude that $\Omega = (W_{22}+W_{21})^{-1}W_{22} - W_{21} = \Omega_{W,l}$. Moreover, we see from (5.27) that

$$W_{11} + W_{12} = \frac{1}{2}(\widehat{\Psi\#})^{-1}[(I+\widehat{\Omega\#}) + (I-\widehat{\Omega\#})] = (\widehat{\Psi\#})^{-1} \tag{5.30}$$

and consequently $\widehat{\Psi\#} = (W_{11}+W_{12})^{-1}$. This implies finally $\Psi = \Psi_{W,l}$. □

Corollary 5.8. *Let W be a j_{qq}-inner function. Then there exist a unique left connected pair $[\Phi, \Psi]$ of $[H^2(\mathbb{D})]^{q\times q}$-functions such that the functions* $\det\Phi$ *and* $\det\Psi$ *do not identically vanish in $\mathbb{D}$ and a unique singular $q \times q$ Carathéodory function Ω_s such that*

$$W = \frac{1}{2}\cdot \operatorname{diag}\left[(\widehat{\Psi\#})^{-1}, \Phi^{-1}\right] \cdot \begin{bmatrix} I + \widehat{\Omega^{\#}_{\langle\underline{\Phi}\,\underline{\Phi}^*\rangle}} + \widehat{\Omega^{\#}_s} & I - \widehat{\Omega^{\#}_{\langle\underline{\Phi}\,\underline{\Phi}^*\rangle}} - \widehat{\Omega^{\#}_s} \\ I - \Omega_{\langle\underline{\Phi}\,\underline{\Phi}^*\rangle} - \Omega_s & I + \Omega_{\langle\underline{\Phi}\,\underline{\Phi}^*\rangle} + \Omega_s \end{bmatrix}, \tag{5.31}$$

namely the left connected pair of $[H^2(\mathbb{D})]^{q\times q}$-functions and the left singular $q \times q$ Carathéodory function generated by W.

Proof. From Lemma 3.1 we know that every singular $q \times q$ Carathéodory function admits a pseudocontinuation. Further we can conclude from Proposition 2.3 and Lemma 3.6 that $\Omega_{\langle\underline{\Phi}\,\underline{\Phi}^*\rangle}$ also admits a pseudocontinuation. Thus the application of Theorem 5.7 and Lemma 3.3 provides the assertion. □

Analogously to Theorem 5.7 the following result can be verified.

Theorem 5.9. *Let W be a j_{qq}-inner function, let $[\Phi_{W,r}, \Psi_{W,r}]$ and $\Omega_{W,r}$ be the right connected pair of $[H^2(\mathbb{D})]^{q\times q}$-functions and the right $q \times q$ Carathéodory function, respectively, generated by W. Then W admits the representation*

$$W = \frac{1}{2}\cdot \begin{bmatrix} I + \widehat{\Omega^{\#}_{W,r}} & I - \Omega_{W,r} \\ I - \widehat{\Omega^{\#}_{W,r}} & I + \Omega_{W,r} \end{bmatrix} \cdot \operatorname{diag}\left[(\widehat{\Psi^{\#}_{W,r}})^{-1}, \Phi^{-1}_{W,r}\right]. \tag{5.32}$$

If $[\Phi, \Psi]$ is a right connected pair of $[H^2(\mathbb{D})]^{q\times q}$-functions such that the function $\det\Phi$ *does not identically vanish in $\mathbb{D}$ and if Ω is a $q \times q$ Carathéodory function which admits a pseudocontinuation $\Omega^{\#}$ such that the representation*

$$W = \frac{1}{2}\cdot \begin{bmatrix} I + \widehat{\Omega\#} & I - \Omega \\ I - \widehat{\Omega\#} & I + \Omega \end{bmatrix} \cdot \operatorname{diag}\left[(\widehat{\Psi\#})^{-1}, \Phi^{-1}\right] \tag{5.33}$$

of W is satisfied, then $\Phi = \Phi_{W,r}$, $\Psi = \Psi_{W,r}$ and $\Omega = \Omega_{W,r}$.

Corollary 5.10. *Let W be a j_{qq}-inner function. Then there exist a unique right connected pair $[\Phi, \Psi]$ of $[H^2(\mathbb{D})]^{q\times q}$-functions such that the functions* $\det \Phi$ *and* $\det \Psi$ *do not identically vanish in $\mathbb{D}$ and a unique singular $q \times q$ Carathéodory function Ω_s such that*

$$(5.34)\quad W = \frac{1}{2} \cdot \begin{bmatrix} I + \widehat{\Omega^{\#}_{\langle \underline{\Phi}^* \underline{\Phi} \rangle}} + \widehat{\Omega^{\#}_s} & I - \Omega_{\langle \underline{\Phi}^* \underline{\Phi} \rangle} - \Omega_s \\ I - \widehat{\Omega^{\#}_{\langle \underline{\Phi}^* \underline{\Phi} \rangle}} - \widehat{\Omega^{\#}_s} & I + \Omega_{\langle \underline{\Phi}^* \underline{\Phi} \rangle} + \Omega_s \end{bmatrix} \cdot \operatorname{diag}\left[(\widehat{\Psi^{\#}})^{-1}, \Phi^{-1} \right],$$

namely the right connected pair of $[H^2(\mathbb{D})]^{q\times q}$-functions and the right singular $q\times q$ Carathéodory function generated by W.

Proof. In view of Lemmas 3.1 and 3.7 and Proposition 2.3, the assertion follows easily from Theorem 5.9 and Lemma 3.3. We omit the details. □

Corollaries 5.8 and 5.10 lead us to the desired parametrization of j_{qq}–inner functions.

Definition 5.11. Let W be a j_{qq}-inner function.

(a) The triple $[\Phi_l, \Psi_l, \Omega_{l,s}]$ where $[\Phi_l, \Psi_l]$ is the left connected pair of $[H^2(\mathbb{D})]^{q\times q}$-functions generated by W and where $\Omega_{l,s}$ is the left singular $q \times q$ Carathéodory function generated by W is called the *left ADD-parametrization of W*.

(b) The triple $[\Phi_r, \Psi_r, \Omega_{r,s}]$ where $[\Phi_r, \Psi_r]$ is the right connected pair of $[H^2(\mathbb{D})]^{q\times q}$-functions generated by W and where $\Omega_{r,s}$ is the right singular $q \times q$ Carathéodory function generated by W is said to be the *right ADD-parametrization of W*.

Now we turn our attention to the inverse question, namely to construct j_{qq}–inner functions with prescribed ADD-parametrizations.

Theorem 5.12. *Let $[\Phi, \Psi]$ be a left connected pair of $[H^2(\mathbb{D})]^{q\times q}$-functions such that the function* $\det \Phi$ *does not identically vanish. Further, let $\Omega \in \mathcal{C}_{q,\langle \underline{\Phi}\, \underline{\Phi}^* \rangle}$. Then Ω admits a pseudocontinuation $\Omega^{\#}$ and*

$$(5.35)\qquad W := \frac{1}{2} \cdot \operatorname{diag}\left[(\widehat{\Psi^{\#}})^{-1}, \Phi^{-1} \right] \cdot \begin{bmatrix} I + \widehat{\Omega^{\#}} & I - \widehat{\Omega^{\#}} \\ I - \Omega & I + \Omega \end{bmatrix}$$

is a j_{qq}-inner function. Moreover, $[\Phi, \Psi]$ and Ω are the left connected pair of $[H^2(\mathbb{D})]^{q\times q}$-functions and the left $q \times q$ Carathéodory function, respectively, generated by W. If V is the (unique) inner $q \times q$ Schur function which realizes the left connection of $[\Phi, \Psi]$, then the Potapov-Ginzburg transform S of W admits the representation

$$(5.36)\qquad S = \begin{bmatrix} 2\Psi(I + \Omega)^{-1} & V - 2\Psi(I + \Omega)^{-1}\Phi \\ -(I - \Omega)(I + \Omega)^{-1} & 2(I + \Omega)^{-1}\Phi \end{bmatrix}.$$

Proof. Since Ω belongs to $\mathcal{C}_{q,\langle\Phi\,\Phi^*\rangle}$ and the function $\det\Phi$ does not identically vanish in $\mathbb{D}$, the functions $\det(\Omega+\Omega^*)$ and $\det\Psi$ do not identically vanish in $\mathbb{D}$. Using Remark 3.5 and the identities $\underline{\Omega}(\underline{\Omega}+\underline{\Omega}^*)^{-1}+\underline{\Omega}^*(\underline{\Omega}+\underline{\Omega}^*)^{-1}=I$ and

$$\begin{aligned}&\underline{\Omega}(\underline{\Omega}+\underline{\Omega}^*)^{-1}\underline{\Omega}^*-\underline{\Omega}^*(\underline{\Omega}+\underline{\Omega}^*)^{-1}\underline{\Omega}\\ &\quad= [\underline{\Omega}(\underline{\Omega}+\underline{\Omega}^*)^{-1}\underline{\Omega}^*+\underline{\Omega}^*(\underline{\Omega}+\underline{\Omega}^*)^{-1}\underline{\Omega}^*]-[\underline{\Omega}^*(\underline{\Omega}+\underline{\Omega}^*)^{-1}\underline{\Omega}^*+\underline{\Omega}^*(\underline{\Omega}+\underline{\Omega}^*)^{-1}\underline{\Omega}]\\ &\quad= \underline{\Omega}^*-\underline{\Omega}^* \;=\; 0\,,\end{aligned}$$

which hold true $\underline{\lambda}$-almost everywhere on $\mathbb{T}$, we obtain that

$$(5.37)\qquad \begin{aligned}&\frac{1}{4}(I+\underline{\Omega})(\underline{\Psi}^*\underline{\Psi})^{-1}(I+\underline{\Omega}^*)-(I-\underline{\Omega}^*)(\underline{\Phi}\,\underline{\Phi}^*)^{-1}(I-\underline{\Omega}) =\\ &=\frac{1}{2}[(I+\underline{\Omega})(\underline{\Omega}+\underline{\Omega}^*)^{-1}(I+\underline{\Omega}^*)-(I-\underline{\Omega}^*)(\underline{\Omega}+\underline{\Omega}^*)^{-1}(I-\underline{\Omega})]=I\end{aligned}$$

and

$$(5.38)\qquad \frac{1}{4}(I-\underline{\Omega})(\underline{\Psi}^*\underline{\Psi})^{-1}(I-\underline{\Omega}^*)-(I+\underline{\Omega}^*)(\underline{\Phi}\,\underline{\Phi}^*)^{-1}(I+\underline{\Omega}) \;=\; -I$$

are valid $\underline{\lambda}$-almost everywhere on $\mathbb{T}$. Applying the same arguments we also get the equations

$$(5.39)\qquad (I+\underline{\Omega})(\underline{\Psi}^*\underline{\Psi})^{-1}(I-\underline{\Omega}^*)-(I-\underline{\Omega}^*)(\underline{\Phi}\,\underline{\Phi}^*)^{-1}(I+\underline{\Omega}) \;=\; 0$$

and

$$(5.40)\qquad (I-\underline{\Omega})(\underline{\Psi}^*\underline{\Psi})^{-1}(I+\underline{\Omega}^*)-(I+\underline{\Omega}^*)(\underline{\Phi}\,\underline{\Phi}^*)^{-1}(I-\underline{\Omega}) \;=\; 0$$

are satisfied $\underline{\lambda}$-almost everywhere on $\mathbb{T}$. In view of (5.37), (5.38), (5.39), (5.40) and Remark 1.1, then it follows

(5.41)

$$\begin{aligned}\underline{W}^*j_{qq}\underline{W} \;&=\; \frac{1}{4}\begin{bmatrix} I+\underline{\Omega} & I-\underline{\Omega}^* \\ I-\underline{\Omega} & I+\underline{\Omega}^*\end{bmatrix}\operatorname{diag}\left[\underline{\Psi}^{-1}\underline{\Psi}^{-*},-\underline{\Phi}^{-*}\underline{\Phi}^{-1}\right]\begin{bmatrix} I+\underline{\Omega}^* & I-\underline{\Omega}^* \\ I-\underline{\Omega} & I+\underline{\Omega}\end{bmatrix}\\ &=\; j_{qq}\qquad \underline{\lambda}\text{-a. e. on } \mathbb{T}\,.\end{aligned}$$

Let W be partitioned into $q\times q$ blocks via (4.2). Our following considerations are aimed to show that

$$(5.42)\qquad S_{11}:=W_{11}-W_{12}W_{22}^{-1}W_{21}\quad,\qquad S_{12}:=W_{12}W_{22}^{-1}$$

and

$$(5.43)\qquad S_{21}:=-W_{22}^{-1}W_{21}\quad,\qquad S_{22}:=W_{22}^{-1}$$

are well-defined functions which belong to $[\mathcal{N}_+(\mathbb{D})]^{q\times q}$. Obviously, $W_{22}=\frac{1}{2}\Phi^{-1}(I+\Omega)$. Since Ω belongs to $\mathcal{C}_q(\mathbb{D})$, we can conclude that $\det W_{22}$ does not identically vanish in $\mathbb{D}$ (see, e. g., [13, Part (a) of Proposition 2.1.3]) and that S_{22} is a well-defined function which admits the representation

$$(5.44)\qquad S_{22}=2(I+\Omega)^{-1}\Phi\,.$$

From [20, Proposition 3] we know that $(I+\Omega)^{-1}$ is an outer function in $\mathcal{S}_{q\times q}(\mathbb{D})$. Since $\mathcal{S}_{q\times q}(\mathbb{D})$ and $[H^2(\mathbb{D})]^{q\times q}$ are subsets of $[\mathcal{N}_+(\mathbb{D})]^{q\times q}$ and because $\mathcal{N}_+(\mathbb{D})$ is an algebra over $\mathbb{C}$, we see then that S_{22} belongs to $[\mathcal{N}_+(\mathbb{D})]^{q\times q}$. An easy calculation shows

$$S_{21} = -(I-\Omega)(I+\Omega)^{-1}\ . \tag{5.45}$$

Since $\Omega \in \mathcal{C}_q(\mathbb{D})$ thus S_{21} belongs to $\mathcal{S}_{q\times q}(\mathbb{D})$ and therefore to $[\mathcal{N}_+(\mathbb{D})]^{q\times q}$ (see, e. g., [13, Part (b) of Proposition 2.1.3]). From Proposition 2.3 we know that

$$(\widehat{\Psi\#})^{-1}\Phi = V\ . \tag{5.46}$$

Because of $\Omega \in \mathcal{C}_{q,\langle \underline{\Phi}\,\underline{\Phi}^*\rangle}$ and Remark 3.5 we have $\Omega \in \mathcal{C}_{q,\langle \underline{\Psi}^*\underline{\Psi}\rangle}$. Hence the identity $\underline{\Psi}^* = \frac{1}{2}(\underline{\Omega}+\underline{\Omega}^*)\underline{\Psi}^{-1}$ $\underline{\lambda}$-almost everywhere on $\mathbb{T}$. Thus Lemma 3.6 and Remark 1.1 imply that Ω admits a pseudocontinuation $\Omega^\#$ for which the identity $\widehat{\Psi\#} = \frac{1}{2}(\Omega+\widehat{\Omega\#})\Psi^{-1}$ and consequently

$$(\widehat{\Psi\#})^{-1} = 2\Psi(\Omega+\widehat{\Omega\#})^{-1} \tag{5.47}$$

are valid. Using (5.35), (5.46) and (5.47) we obtain

$$\begin{aligned} S_{12} &= (\widehat{\Psi\#})^{-1}(I-\widehat{\Omega\#})(I+\Omega)^{-1}\Phi \\ &= (\widehat{\Psi\#})^{-1}[(I+\Omega)-(\Omega+\widehat{\Omega\#})](I+\Omega)^{-1}\Phi \\ &= (\widehat{\Psi\#})^{-1}\Phi-(\widehat{\Psi\#})^{-1}(\Omega+\widehat{\Omega\#})(I+\Omega)^{-1}\Phi \\ &= V-2\Psi(I+\Omega)^{-1}\Phi\ . \end{aligned} \tag{5.48}$$

Since V, Ψ, $(I+\Omega)^{-1}$ and Φ belong to $[\mathcal{N}_+(\mathbb{D})]^{q\times q}$ we thus see that S_{12} belongs to $[\mathcal{N}_+(\mathbb{D})]^{q\times q}$ as well. Since W has j_{qq}-unitary radial boundary values $\underline{\lambda}$-almost everywhere on $\mathbb{T}$, we obtain that $\det \underline{W_{11}} \neq 0$, $\det \underline{W_{22}} \neq 0$ and $\underline{W_{11}}^{-*} = \underline{W_{11}} - \underline{W_{12}W_{22}^{-1}W_{21}} = \underline{S_{11}}$ hold $\underline{\lambda}$-almost everywhere on $\mathbb{T}$. In view of Remark 1.1 this implies $(\widehat{W_{11}^\#})^{-1} = S_{11}$. On the other hand, formula (5.35) provides

$$\widehat{W_{11}^\#} = \frac{1}{2}(I+\Omega)\Psi^{-1}\ . \tag{5.49}$$

Consequently

$$S_{11} = 2\Psi(I+\Omega)^{-1}\ . \tag{5.50}$$

The same arguments as above yield then $S_{11} \in [\mathcal{N}_+(\mathbb{D})]^{q\times q}$. Therefore all the functions S_{11}, S_{12}, S_{21} and S_{22} belong to $[\mathcal{N}_+(\mathbb{D})]^{q\times q}$. In view of (5.41) then from a result due to Arov (see [1]) it follows that W is a j_{qq}-inner function. According to (5.35) we have $W_{12} = \frac{1}{2}(\widehat{\Psi\#})^{-1}(I-\widehat{\Omega\#})$ and hence $\widehat{W_{12}^\#} = \frac{1}{2}(I-\Omega)\Psi^{-1}$. Using (5.49) we thus obtain $\widehat{W_{11}^\#}+\widehat{W_{12}^\#} = \Psi^{-1}$. Because the identity $W_{21}+W_{22} = \Phi^{-1}$ is also valid, we can see that $[\Phi, \Psi]$ is the left connected pair of $[H^2(\mathbb{D})]^{q\times q}$-functions associated with W. Lemma 2.5 and (5.46) show that V is the inner function which realizes this left connection. Obviously, $W_{22} - W_{21} = \Phi^{-1}\Omega$ and therefore $(W_{22}+W_{21})^{-1}(W_{22}-W_{21}) = \Omega$, i. e., Ω is the $q \times q$ Carathéodory function which is left generated by W. Finally, we observe that (5.42), (5.43), (5.44), (5.45), (5.48) and (5.50) provide immediately the representation (5.36) of the Potapov-Ginzburg transform S of W. □

Analogously to Theorem 5.12, the following result can be verified.

Theorem 5.13. *Let* $[\Phi, \Psi]$ *be a right connected pair of* $[H^2(\mathbb{D})]^{q\times q}$*-functions such that the functions* $\det \Phi$ *does not identically vanish. Further, let* $\Omega \in \mathcal{C}_{q,\langle \underline{\Phi}^* \underline{\Phi}\rangle}$. *Then* Ω *admits a pseudocontinuation* $\Omega^{\#}$ *and*

$$W := \frac{1}{2} \cdot \begin{bmatrix} I + \widehat{\Omega^{\#}} & I - \Omega \\ I - \widehat{\Omega^{\#}} & I + \Omega \end{bmatrix} \cdot \operatorname{diag}\left[(\widehat{\Psi^{\#}})^{-1}, \Phi^{-1}\right] \tag{5.51}$$

is a j_{qq}*-inner function. Moreover,* $[\Phi, \Psi]$ *and* Ω *are the right connected pair of* $[H^2(\mathbb{D})]^{q\times q}$*-functions and the right* $q \times q$ *Carathéodory function, respectively, generated by* W. *If* V *is the (unique) inner* $q \times q$ *Schur function which realizes the right connection of* $[\Phi, \Psi]$, *then the Potapov-Ginzburg transform* S *of* W *admits the representation*

$$S = \begin{bmatrix} 2(I+\Omega)^{-1}\Psi & (I+\Omega)^{-1}(I-\Omega) \\ -V + 2\Phi(I+\Omega)^{-1}\Psi & 2\Phi(I+\Omega)^{-1} \end{bmatrix}. \tag{5.52}$$

The following modification of Theorems 5.12 and 5.13 gives now a complete answer to the inverse question associated with ADD-parametrization.

Theorem 5.14. *Let* $[\Phi, \Psi]$ *be a left connected pair of* $[H^2(\mathbb{D})]^{q\times q}$*-functions such that the function* $\det \Phi$ *does not identically vanish, and let* Ω_s *be a singular* $q \times q$ *Carathéodory function. Then there is a unique* j_{qq}*-inner function* W *such that* $[\Phi, \Psi, \Omega_s]$ *is the left ADD-parametrization of* W.

Proof. Lemma 3.3 shows that $\Omega := \Omega_{\langle \underline{\Phi}\, \underline{\Phi}^*\rangle} + \Omega_s$ belongs to $\mathcal{C}_{q,\langle \underline{\Phi}\, \underline{\Phi}^*\rangle}$. Theorem 5.12 yields then that W given by (5.31) is a j_{qq}-inner function, that $[\Phi, \Psi]$ and Ω are the left connected pair of $[H^2(\mathbb{D})]^{q\times q}$-functions and the left $q \times q$ Carathéodory function, respectively, generated by W. Thus $[\Phi, \Psi, \Omega_s]$ is the left ADD-parametrization of W. On the other hand, Corollary 5.8 provides that there is at most one j_{qq}-inner function the left ADD-parametrization of which is exactly $[\Phi, \Psi, \Omega_s]$. □

Theorem 5.15. *Let* $[\Phi, \Psi]$ *be a right connected pair of* $[H^2(\mathbb{D})]^{q\times q}$*-functions such that the function* $\det \Phi$ *does not identically vanish, and let* Ω_s *be a singular* $q \times q$ *Carathéodory function. Then there is a unique* j_{qq}*-inner function* W *such that* $[\Phi, \Psi, \Omega_s]$ *is the right ADD-parametrization of* W.

Proof. Using Lemma 3.3, Theorem 5.13 and Corollary 5.10, one can easily prove Theorem 5.15 analogously to the proof of Theorem 5.14. □

6. Construction of j_{qq}-inner functions by given left and right Carathéodory functions

Let $\Delta : \mathbb{T} \longrightarrow \mathbb{C}^{q\times q}$ be a function which is integrable with respect to the linear Legesgue-Borel measure $\underline{\lambda}$ on $\mathbb{T}$ and which satisfies $\Delta(z) \in \mathbb{C}_{\geq}^{q\times q}$ for $\underline{\lambda}$-almost all $z \in \mathbb{T}$. Then a function $\Theta \in [H^2(\mathbb{D})]^{q\times q}$ is called a *left* (respectively, *right*) *spectral factor of* $\langle\Delta\rangle$ if $\underline{\Theta}\,\underline{\Theta}^* = \Delta$ $\underline{\lambda}$-almost everywhere on $\mathbb{T}$ (respectively, $\underline{\Theta}^*\underline{\Theta} = \Delta$ $\underline{\lambda}$-almost everywhere on $\mathbb{T}$). It is said to be *normalized* if $\Theta(0) \in \mathbb{C}_{\geq}^{q\times q}$. If

$$\frac{1}{2\pi}\int_{\mathbb{T}} \log(\det \Delta) d\underline{\lambda} > -\infty , \tag{6.1}$$

then Masani [25] proved that there are a unique normalized outer left spectral factor Φ_0 of Δ and a unique normalized outer right spectral factor Ψ_0 of Δ.

Remark 6.1. Let ρ be a function which belongs to $[\mathcal{NM}(\mathbb{D})]^{q\times q}$. Then $\det\rho$ does not identically vanish if and only if

$$\frac{1}{2\pi}\int_{\mathbb{T}} \log(|\det \underline{\rho}|) d\underline{\lambda} > -\infty$$

(see, e. g., [14, Theorems 2.1 and 2.2]).

We know from [20, Lemma 4] that, for each $\Omega \in \mathcal{C}_q(\mathbb{D})$, the function $\mathrm{Re}\,\underline{\Omega}$ belongs to $[\mathcal{L}^1(\mathbb{T})]^{q\times q}$. Thus a $q \times q$ Carathéodory function is called a $q\times q$ *Carathéodory function of finite entropy* if

$$\frac{1}{2\pi}\int_{\mathbb{T}} \log[\det(\mathrm{Re}\,\underline{\Omega})] d\underline{\lambda} > -\infty . \tag{6.2}$$

If $\Omega \in \mathcal{C}_q(\mathbb{D})$ admits a pseudocontinuation $\Omega^{\#}$, then we see from Remarks 1.1 and 6.1 that Ω is a $q\times q$ Carathéodory function of finite entropy if and only if $\det(\Omega + \widehat{\Omega^{\#}})$ does not identically vanish.

In some sense, the following lemma can be considered as converse statement to Lemmas 3.6 and 3.7.

Lemma 6.2. *Let Ω be a $q\times q$ Carathéodory function of finite entropy. Suppose that Ω admits a pseudocontinuation $\Omega^{\#}$. Then:*

(a) *Every outer left spectral factor Φ of $\langle\mathrm{Re}\,\underline{\Omega}\rangle$ admits a pseudocontinuation $\Phi^{\#}$, namely $\Phi^{\#} = \frac{1}{2}(\Omega^{\#} + \widehat{\Omega})\widehat{\Phi^{-1}}$.*

(b) *Every outer right spectral factor Ψ of $\langle\mathrm{Re}\,\underline{\Omega}\rangle$ admits a pseudocontinuation $\Psi^{\#}$, namely $\Psi^{\#} = \frac{1}{2}\widehat{\Psi^{-1}}(\Omega^{\#} + \widehat{\Omega})$.*

Proof. (a) Let Φ be an outer left spectral factor of $\langle\mathrm{Re}\,\underline{\Omega}\rangle$. Then Φ is a function which belongs to $[H^2(\mathbb{D})]^{q\times q}$ and which fulfills $\underline{\Omega} + \underline{\Omega}^* = 2\underline{\Phi}\,\underline{\Phi}^*$ $\underline{\lambda}$-almost everywhere on $\mathbb{T}$. Moreover, the function $\det\Phi$ does not identically vanish. Remark 1.1

shows that the function $g := \frac{1}{2}(\Omega^{\#} + \widehat{\Omega})\widehat{\Phi^{-1}}$ belongs to $[\mathcal{NM}(\mathbb{E})]^{q\times q}$ and satisfies $\underline{g} = \frac{1}{2}(\underline{\Omega}+\underline{\Omega}^*)\underline{\Phi}^{-*} = \underline{\Phi}$ $\underline{\lambda}$-almost everywhere on $\mathbb{T}$. Hence g is a pseudocontinuation of Φ.

(b) This part can be analogously proved. We omit the details. □

Theorem 6.3. *Let Ω be a $q \times q$ Carathéodory function of finite entropy. Suppose that Ω admits a pseudocontinuation $\Omega^{\#}$. Let Φ_l be an outer left spectral factor of $\langle \mathrm{Re}\ \underline{\Omega}\rangle$, and let V_l be a left denominator of $\widehat{\Phi_l^{\#}}$. Then*

$$W := \frac{1}{2}\mathrm{diag}\left[V_l\Phi_l^{-1}, \Phi_l^{-1}\right] \cdot \begin{bmatrix} I + \widehat{\Omega^{\#}} & I - \widehat{\Omega^{\#}} \\ I - \Omega & I + \Omega \end{bmatrix} \tag{6.3}$$

is a j_{qq}-inner function the left $q \times q$ Carathéodory function generated of which is Ω. Moreover, $[\Phi_l, \Psi_l]$ where $\Psi_l := V_l\widehat{\Phi_l^{\#}}$ is the left connected pair of $[H^2(\mathbb{D})]^{q\times q}$-functions generated by W.

Proof. Since Φ_l is a left spectral factor of $\langle \mathrm{Re}\ \underline{\Omega}\rangle$ we have $\Omega \in \mathcal{C}_{\langle \underline{\Phi_l}\, \underline{\Phi_l}^*\rangle}$. Because of the fact that Φ_l is an outer function in $[H^2(\mathbb{D})]^{q\times q}$, the function $\det \Phi_l$ does not identically vanish. From Proposition 2.6 we see that $[\Phi_l, \Psi_l]$ is a left connected pair of $[H^2(\mathbb{D})]^{q\times q}$-functions where V_l is an inner function which realizes this left connection. According to Proposition 2.3, then we infer that Ψ_l admits a pseudocontinuation which satisfies $(\widehat{\Psi_l^{\#}})^{-1} = V_l\Phi_l^{-1}$. The application of Theorem 5.12 completes the proof. □

Using Theorem 5.13 the following result can be analogously proved.

Theorem 6.4. *Let Ω be a $q \times q$ Carathéodory function of finite entropy. Suppose that Ω admits a pseudocontinuation $\Omega^{\#}$. Let Φ_r be an outer right spectral factor of $\langle \mathrm{Re}\ \underline{\Omega}\rangle$, and let V_r be a right denominator of $\widehat{\Phi_r^{\#}}$. Then*

$$W := \frac{1}{2}\begin{bmatrix} I + \widehat{\Omega^{\#}} & I - \Omega \\ I - \widehat{\Omega^{\#}} & I + \Omega \end{bmatrix} \cdot \mathrm{diag}\left[\Phi_r^{-1}V_r, \Phi_r^{-1}\right] \tag{6.4}$$

is a j_{qq}-inner function the right $q \times q$ Carathéodory function generated of which is Ω. Moreover, $[\Phi_r, \Psi_r]$ where $\Psi_r := \widehat{\Phi_r^{\#}}V_r$ is the right connected pair of $[H^2(\mathbb{D})]^{q\times q}$-functions generated by W.

Obviously, for each $f \in \mathcal{S}_{p\times q}(\mathbb{D})$, the functions $I - \underline{f}\,\underline{f}^*$ and $I - \underline{f}^*\underline{f}$ are bounded $\underline{\lambda}$-almost everywhere on $\mathbb{T}$. Hence both functions belong to $[\mathcal{L}^\infty(\mathbb{T})]^{p\times p}$ and $[\mathcal{L}^\infty(\mathbb{T})]^{q\times q}$, respectively. In view of this fact, a function $f \in \mathcal{S}_{p\times q}(\mathbb{D})$ is called a $p \times q$ *Schur function of finite entropy* if

$$\frac{1}{2\pi}\int_{\mathbb{T}} \log[\det(I - \underline{f}\,\underline{f}^*)]\mathrm{d}\underline{\lambda} > -\infty\,. \tag{6.5}$$

Because

$$\det(I - KK^*) = \det\begin{pmatrix} I_p & K \\ K^* & I_q \end{pmatrix} = \det(I - K^*K)$$

holds true for every contractive $p \times q$ complex matrix K, a function $f \in \mathcal{S}_{p\times q}(\mathbb{D})$ is a $p \times q$ Schur function of finite entropy if and only if

$$\frac{1}{2\pi}\int_{\mathbb{T}} \log[\det(I - \underline{f}^*\underline{f})]\mathrm{d}\underline{\lambda} \ > \ -\infty\ .$$

Remark 6.5. If f is a $p \times q$ Schur function which admits a pseudocontinuation $f^\#$, then the following statements are equivalent:

(i) $\det(I - f\widehat{f^\#})$ does not identically vanish.

(ii) $\det(I - \widehat{f^\#}f)$ does not identically vanish.

(iii) f is a $p \times q$ Schur function of finite entropy.

Lemma 6.6. *Let $\Omega \in \mathcal{C}_q(\mathbb{D})$. Then $\det(I + \Omega)$ nowhere vanishes in $\mathbb{D}$ and the function $f := (I - \Omega)(I + \Omega)^{-1}$ belongs to $\mathcal{S}_{q\times q}(\mathbb{D})$. The function Ω admits a pseudocontinuation if and only if f admits a pseudocontinuation. Moreover, Ω has finite entropy if and only if f has finite entropy.*

A proof of Lemma 6.6 is given in [17, Lemmas 1 – 3]. Further, we need the following construction of inner $(p+q) \times (p+q)$–Schur functions with given right upper $p \times q$ block (see [21, Proposition 4.6] or [3, Theorem 2]).

Proposition 6.7. *Let $f \in \mathcal{S}_{p\times q}(\mathbb{D})$. Suppose that f admits a pseudocontinuation $f^\#$. Let $\rho := I - f\widehat{f^\#}$ and $\sigma := I - \widehat{f^\#}f$. Assume that $\det\rho$ does not identically vanish. Let ϕ be the unique normalized left spectral factor of $\langle\rho\rangle$, and let ψ be the unique normalized right spectral factor of $\langle\sigma\rangle$. Further, let $b \in \mathcal{S}_{p\times p}(\mathbb{D})$ and $c \in \mathcal{S}_{q\times q}(\mathbb{D})$ be inner functions such that*

$$c\psi\widehat{f^\#}\rho^{-1}\phi b \in [\mathcal{N}_+(\mathbb{D})]^{q\times p}\ . \tag{6.6}$$

Then

$$U := \operatorname{diag}[I_p, c] \cdot \begin{bmatrix} \phi & f \\ -\psi\widehat{f^\#}\rho^{-1}\phi & \psi \end{bmatrix} \cdot \operatorname{diag}[b, I_q]$$

is an inner $(p+q) \times (p+q)$ Schur function.

Now we characterize the case that a given pair of $q \times q$ Carathéodory functions of finite entropy is the pair of $q \times q$ Carathéodory functions generated by some j_{qq}–inner functions in the sense of Definition 5.3.

Theorem 6.8. *Let Ω_1 and Ω_2 be $q \times q$ Carathéodory functions of finite entropy. Suppose that both functions Ω_1 and Ω_2 admit pseudocontinuations. Then $f_1 :=$*

$(I-\Omega_1)(I+\Omega_1)^{-1}$ and $f_2 := (I-\Omega_2)(I+\Omega_2)^{-1}$ are pseudocontinuable $q \times q$ Schur functions of finite entropy. In particular, $\rho_1 := I - f_1\widehat{f_1^\#}$, $\rho_2 := I - f_2\widehat{f_2^\#}$, $\sigma_1 := I - \widehat{f_1^\#} f_1$ and $\sigma_2 := I - \widehat{f_2^\#} f_2$ are functions whose determinants do not identically vanish. Let ϕ_1 and ϕ_2 be the unique normalized left spectral factors of $\langle \underline{\rho_1} \rangle$ and $\langle \underline{\rho_2} \rangle$, respectively, and let ψ_1 and ψ_2 be the unique normalized right spectral factors of $\langle \underline{\sigma_1} \rangle$ and $\langle \underline{\sigma_2} \rangle$, respectively. Then the following statements are equivalent:

(i) *There exists a j_{qq}-inner function W such that Ω_1 and Ω_2 are the left $q \times q$ Carathéodory function generated by W and the right $q \times q$ Carathéodory function generated by W, respectively.*
(ii) *There are inner $q \times q$ Schur functions b_2 and c_2 such that $f_1 = c_2\psi_2\widehat{f_2^\#}\rho_2^{-1}\phi_2 b_2$.*
(iii) *There are inner $q \times q$ Schur functions c_1 and b_1 such that $f_2 = c_1\psi_1\widehat{f_1^\#}\rho_1^{-1}\phi_1 b_1$.*

Proof. From Lemma 6.6 we see that f_1 and f_2 are pseudocontinuable $q \times q$ Schur functions of finite entropy. Thus we obtain from Remark 6.5 that the functions $\det \rho_1$, $\det \sigma_1$, $\det \rho_2$ and $\det \sigma_2$ do not identically vanish.

(i)⇒(ii): Let (i) be satisfied. From Propositions 4.2, 5.1 and 5.2 we obtain

$$\Omega_1 = (I - S_{21})(I + S_{21})^{-1} \quad \text{and} \quad \Omega_2 = (I - S_{12})(I + S_{12})^{-1} \tag{6.7}$$

where $S_{21} := -W_{22}^{-1}W_{21}$ and $S_{12} := W_{12}W_{22}^{-1}$ are strictly contractive $q \times q$ Schur functions. Using a property of the Cayley transform (see, e. g., [13, Lemma 1.3.12]) we thus infer

$$S_{21} = -(I-\Omega_1)(I+\Omega_1)^{-1} = -f_1 \quad \text{and} \quad S_{12} = (I-\Omega_2)(I+\Omega_2)^{-1} = f_2 \,. \tag{6.8}$$

Parts (b) and (e) of Proposition 4.2 show that the Potapov-Ginzburg transfrom S of W is an inner $2q \times 2q$ Schur function. Setting $S_{11} := W_{11} - W_{12}W_{22}^{-1}W_{21}$ and $S_{22} := W_{22}^{-1}$ then we see in particular that $\det S_{11}$ does not identically vanish and that

$$I - \underline{S_{12}}\,\underline{S_{12}^*} = \underline{S_{11}}\,\underline{S_{11}^*} \quad , \quad \underline{S_{21}} = -\underline{S_{22}}\,\underline{S_{12}^*}\,\underline{S_{11}^{-*}} \tag{6.9}$$

and

$$I - \underline{S_{12}^*}\,\underline{S_{12}} = \underline{S_{22}^*}\,\underline{S_{22}} \tag{6.10}$$

hold $\underline{\lambda}$-almost everywhere on $\mathbb{T}$. In view of Remark 1.1 and (6.8) thus it follows

$$\begin{aligned} \underline{S_{22}}\,\underline{f_2^*}\,\underline{\rho_2}^{-1}\underline{S_{11}} &= \underline{S_{22}}\,\underline{S_{12}^*}(I - \underline{S_{12}}\,\underline{S_{12}^*})^{-1}\underline{S_{11}} \\ &= \underline{S_{22}}\,\underline{S_{12}^*}\,\underline{S_{11}^{-*}} = -\underline{S_{21}} = \underline{f_1} \quad \underline{\lambda}\text{-a. e. on } \mathbb{T}. \end{aligned} \tag{6.11}$$

Hence we can conclude from Remark 1.1 that

$$S_{22}\widehat{f_2^\#}\rho_2^{-1}S_{11} = f_1 \,. \tag{6.12}$$

From part (b) of Proposition 4.2 we know that both functions S_{11} and S_{22} belong to $[H^2(\mathbb{D})]^{q\times q}$. According to a factorization theorem for the class $[H^2(\mathbb{D})]^{q\times q}$ (see, e.g., [25]), there are unique inner $q \times q$ Schur functions b_2 and c_2, and there are unique normalized outer functions $\phi_2^\square$ and $\psi_2^\square$ which belong to $[H^2(\mathbb{D})]^{q\times q}$ such that $S_{11} = \phi_2^\square b_2$ and $S_{22} = c_2\psi_2^\square$. Therefore the equation (6.12) can be written as

$$f_1 = c_2\psi_2^\square \widehat{f_2^\# \rho_2^{-1}} \phi_2^\square b_2 \,. \tag{6.13}$$

Using (6.9), (6.10) and Remark 1.1 we infer that

$$\underline{\phi_2^\square}\,(\underline{\phi_2^\square})^* = \underline{S_{11}}\ \underline{S_{11}^*} = I - \underline{S_{12}}\ \underline{S_{12}^*} = \underline{\rho_2}$$

and

$$(\underline{\psi_2^\square})^*\underline{\psi_2^\square} = \underline{S_{22}^*}\ \underline{S_{22}} = I - \underline{S_{12}^*}\ \underline{S_{12}} = \underline{\sigma_2}$$

is valid $\underline{\lambda}$-almost everywhere on $\mathbb{T}$. Hence $\phi_2^\square = \phi_2$ and $\psi_2^\square = \psi_2$. Thus the identity (6.13) yields (ii).

(ii)$\Rightarrow$(i): Suppose (ii). Then Proposition 6.7 shows that

$$U := \begin{bmatrix} \phi_2 b_2 & f_2 \\ -f_1 & c_2\psi_2 \end{bmatrix} \tag{6.14}$$

is an inner $2q \times 2q$ Schur function, where $\det(c_2\psi_2)$ does not identically vanish. In view of Proposition 4.3 we see that

$$W := \begin{bmatrix} \phi_2 b_2 + f_2(c_2\psi_2)^{-1} f_1 & f_2(c_2\psi_2)^{-1} \\ (c_2\psi_2)^{-1} f_1 & (c_2\psi_2)^{-1} \end{bmatrix} \tag{6.15}$$

is a j_{qq}-inner function. Let Ω_l and Ω_r be the left $q \times q$ Carathéodory function and the right $q\times q$ Carathéodory function, respectively, generated by W. Since Ω_1 and Ω_2 admit the representations

$$\Omega_1 = (I + f_1)^{-1}(I - f_1) \quad \text{and} \quad \Omega_2 = (I - f_2)(I + f_2)^{-1}$$

(see, e. g., [13, Lemma 1.3.12]), we have

$$\begin{aligned} \Omega_l &= [(c_2\psi_2)^{-1} + (c_2\psi_2)^{-1} f_1]^{-1}[(c_2\psi_2)^{-1} - (c_2\psi_2)^{-1} f_1] \\ &= (I + f_1)^{-1}(I - f_1) \ = \ \Omega_1 \end{aligned} \tag{6.16}$$

and

$$\begin{aligned} \Omega_r &= [(c_2\psi_2)^{-1} - f_2(c_2\psi_2)^{-1}][(c_2\psi_2)^{-1} + f_2(c_2\psi_2)^{-1}]^{-1} \\ &= (I - f_2)(I + f_2)^{-1} \ = \ \Omega_2 \,, \end{aligned} \tag{6.17}$$

i. e., Ω_1 and Ω_2 are exactly the left $q \times q$ Carathéodory function and the right $q \times q$ Carathéodory function, respectively, generated by the j_{qq}-inner function W. Hence (i) holds.

(i)$\Leftrightarrow$(iii): This equivalence can be analogously verified as the fact that (ii) is necessary and sufficient for (i). □

Assuming that condition (ii) of Theorem 6.8 is satisfied we construct a j_{qq}–inner function W with prescribed pair of $q \times q$ Carathéodory functions.

Theorem 6.9. *Let Ω_1 and Ω_2 be $q \times q$ Carathéodory functions of finite entropy. Suppose that both functions Ω_1 and Ω_2 admit pseudocontinuations. Let $f_1 := (I - \Omega_1)(I + \Omega_1)^{-1}$, $f_2 := (I - \Omega_2)(I + \Omega_2)^{-1}$, $\rho_2 := I - f_2 \widehat{f_2^{\#}}$ and $\sigma_2 := I - \widehat{f_2^{\#}} f_2$. Let ϕ_2 be the unique normalized left spectral factors of $\langle \rho_2 \rangle$, and let ψ_2 be the unique normalized right spectral factors of $\langle \underline{\sigma_2} \rangle$. Suppose that there exist inner $q \times q$ Schur functions b_2 and c_2 such that*

$$f_1 = c_2 \psi_2 \widehat{f_2^{\#}} \rho_2^{-1} \phi_2 b_2 \ . \tag{6.18}$$

Then

$$W := \begin{bmatrix} I & f_2 \\ \widehat{f_2^{\#}} & I \end{bmatrix} \cdot \operatorname{diag}\left[\rho_2^{-1}\phi_2, \psi_2^{-1}\right] \cdot \operatorname{diag}\left[b_2, c_2^{-1}\right] \tag{6.19}$$

is a j_{qq}-inner function such that Ω_1 is the left $q \times q$ Carathéodory function and Ω_2 is the right $q \times q$ Carathéodory function generated by W, and

$$V := \operatorname{diag}\left[c_2, b_2^{-1}\right] \cdot \operatorname{diag}\left[\psi_2 \sigma_2^{-1}, \phi_2^{-1}\right] \cdot \begin{bmatrix} I & \widehat{f_2^{\#}} \\ f_2 & I \end{bmatrix} \tag{6.20}$$

is a j_{qq}-inner function such that Ω_1 is the right $q \times q$ Carathéodory function and Ω_2 is the left $q \times q$ Carathéodory function generated by W.

Proof. Lemma 6.6 shows that f_2 admits a pseudocontinuation $f_2^{\#}$ and that f_2 has finite entropy. In particular, both functions $\det \rho_2$ and $\det \sigma_2$ does not identically vanish. Let (4.2) be the $q \times q$ block partition of W. Then we have

$$W_{22}^{-1} = c_2 \psi_2 \quad , \quad W_{12} W_{22}^{-1} = f_2 \ , \tag{6.21}$$

$$W_{11} - W_{12} W_{22}^{-1} W_{21} = \rho_2^{-1} \phi_2 b_2 - f_2 \widehat{f_2^{\#}} \rho_2^{-1} \phi_2 b_2 = \phi_2 b_2 \tag{6.22}$$

and, in view of (6.18),

$$-W_{22}^{-1} W_{21} = -c_2 \psi_2 \widehat{f_2^{\#}} \rho_2^{-1} \phi_2 b_2 = -f_1 \ . \tag{6.23}$$

Thus the Potapov-Ginzburg transform S of W has the shape

$$S = \operatorname{diag}\left[I_q, c_2\right] \cdot \begin{bmatrix} \phi_2 & f_2 \\ -c_2 \psi_2 \widehat{f_2^{\#}} \rho_2^{-1} \phi_2 b_2 & \psi_2 \end{bmatrix} \cdot \operatorname{diag}\left[b_2, I_q\right] \ . \tag{6.24}$$

Since $\mathcal{S}_{q \times q}(\mathbb{D})$ is a subset of $[\mathcal{N}_+(\mathbb{D})]^{q \times q}$ we obtain from (6.18), (6.24) and Proposition 6.7 that S is an inner $2q \times 2q$ Schur function. Using Proposition 4.3 then we can conclude that the Potapov-Ginzburg transform $W^{\square}$ of S is a j_{qq}-inner function. On the other hand, since S is the Potapov-Ginzburg transform of W, we have $W^{\square} = W$, i. e., W is a j_{qq}-inner function. Let Ω_l (respectively, Ω_r) denote the left (respectively, right) $q \times q$ Carathéodory function generated by W. Then we get from (6.21) and (6.23) that (6.16) and (6.17) hold true. Let

$$\Delta := \begin{bmatrix} 0 & I_q \\ I_q & 0 \end{bmatrix}.$$

Using the identity $\sigma_2^{-1}\widehat{f_2^{\#}} = \widehat{f_2^{\#}}\rho_2^{-1}$ and the fact that a $2q \times 2q$ Schur function T is inner if and only if $U := \Delta T \Delta$ is an inner $2q \times 2q$ Schur function, the other part of the assertion can be verified analogously. □

Acknowledgements

This paper was supported by the ROLLS project CHRX-CT93-0416. Furthermore, the third author was supported by the Deutsche Akademie der Naturforscher Leopoldina from funds of the BMBF.

References

[1] Arov, D. Z.: *Darlington realization of matrix-valued functions* (in Russian), Izv. Akad. Nauk SSSR, Ser. Mat. **37** (1973), 1299–1331; Math. USSR Izvestija **7** (1973), 1295–1326.

[2] Arov, D. Z.: *On functions of class* Π (in Russian), Zap. Nauc. Sem. LOMI **135** (1984), 5–30.

[3] Arov, D. Z.: *Stable dissipative linear stationary dynamical scattering systems* (in Russian), J. Operator Theory **2** (1979), 95–126.

[4] Arov, D. Z.: *γ−generating matrices, J−inner matrix-functions and related extrapolation problems* (in Russian), Teor. Funkcii, Funk. Anal. i Prilozen., Part I: **51** (1989), 61–67; Part II: **52** (1989), 103–109; Part III: **53** (1990), 57–64; J. Soviet Math. **52** (1990), 3487–3491; **52** (1990), 3421–3425; **58** (1992), 532–537.

[5] Arov, D. Z.; Fritzsche, B.; Kirstein, B.: *On some completion problems for various subclasses of j_{pq}−inner functions*, Zeitschr. Anal. Anw. **11** (1992), 489–508.

[6] Arov, D. Z.; Fritzsche, B.; Kirstein, B.: *Completion problems for j_{pq}−inner Functions*, Integral Equations and Operator Theory, I: **16** (1993), 155–185; II: **16** (1993), 453–495.

[7] Arov, D. Z.; Fritzsche, B.; Kirstein, B.: *On Block Completion Problems for Various Subclasses of j_{pq}−inner Functions*, in: Challenges of a Generalized System Theory (Eds.: P. Dewilde, M. A. Kaashoek, M. Verhagen), North-Holland, Amsterdam-Oxford-New York-Tokyo 1993, pp. 179–194.

[8] Arov, D. Z.; Fritzsche, B.; Kirstein, B.: *On some aspects of V. E. Katsnelson's investigations on interrelations between left and right Blaschke-Potapov products*, in: Operator Theory and Boundary Eigenvalue Problems (Eds.: I. Gohberg, H. Langer), Operator Theory: Advances and Applications, Vol. 80, Birkhäuser, Basel 1995, pp. 21–41.

[9] Ball, J. A.; Gohberg, I.; Rodman, L.: *Interpolation of Rational Matrix Functions*, Operator Theory Series, Vol. 45, Birkhäuser, Basel 1990.

[10] Dewilde, P.; Dym, H.: *Schur recursion error formulas and convergence of rational estimations for stationary stochastic processes*, IEEE Trans. Inf. Theory **27** (1981), 416–461.

[11] Dewilde, P.; Dym, H.: *Lossless chain scattering matrices and optimum linear prediction: the vector case*, Intern. J. Circuit Theory Appl. **9** (1981), 135–175.

[12] Dubovoj, V. K.: *Indefinite metric in the interpolation problem of Schur for analytic matrix functions* (in Russian), Teor. Funkcii, Funkcional. Anal. i Prilozen., Part I: **37** (1982), 14–26; Part II: **38** (1982), 32–39; Part III: **41** (1984), 55–64; Part IV: **42** (1984), 46–57; Part V: **45** (1986), 16–21; Part VI: **47** (1987), 112–119.

[13] Dubovoj, V. K.; Fritzsche, B.; Kirstein, B.: *Matricial Version of the Classical Schur Problem*, Teubner-Texte zur Mathematik, Bd. 129, B. G. Teubner, Stuttgart-Leipzig 1992.

[14] Duren, P. L.; *Theory of H^p Spaces*, Academic Press, New York 1970.

[15] Dym, H.: *$J-$contractive Matrix Functions, Reproducing Kernel Hilbert Spaces and Interpolation*, CBMS Regional Conf. Ser. Math. 71, Amer. Math. Soc., Providence, R. I. 1989.

[16] Fritzsche, B.; Fuchs, S.; Kirstein, B.: *Schur sequence parametrizations of Potapov-normalized full-rank $j_{pq}-$elementary factors*, Linear Algebra Appl. **191** (1993), 107–149.

[17] Fritzsche, B.; Fuchs, S.; Kirstein, B.: *An inverse entropy optimization problem for matrix-valued Carathéodory functions*, Optimization **28** (1994), 1–32.

[18] Fritzsche, B.; Fuchs, S.; Kirstein, B.: *On resolvent matrices for nondegenerate matricial Schur problems*, Linear Algebra Appl. **237** (1996), 191–220.

[19] Fritzsche, B.; Kirstein, B.: *A Schur type matrix extension problem*, Part I: Math. Nachr. **134** (1987), 257–271; Part II: Math. Nachr. **138** (1988), 195–216; Part III: Math. Nachr. **143** (1989), 227–247; Part IV: Math. Nachr. **147** (1990), 235–258.

[20] Fritzsche, B.; Kirstein, B.: *On the Largest Minorants Associated with a Matrix-valued Carathéodory Function and Spectral Factorization*, Zeitschr. Anal. Anw. **12** (1993), 471–490.

[21] Fritzsche, B.; Kirstein, B.; Müller, K.: *A block completion problem for matrix-valued inner functions*, J. Comp. Appl. Math. **77** (1997), 157–172.

[22] Galstjan, L. A.: *Analytic $J-$expansive matrix functions and the Schur problem* (in Russian), Izv. Akad. Nauk Armjan. SSR, Ser. Mat. **12** (1977), No. 3, 204–228.

[23] Ginzburg, J. P.: *On $J-$nonexpansive operators in Hilbert space* (in Russian), Naucn. Zap. Fiz.-Mat. Fak. Odessk. Gos. Ped. Inst. **22** (1958), 13–20.

[24] Kovalishina, I. V.: *Analytic theory of a class of interpolation problems* (in Russian), Izv. Akad. Nauk SSSR, Ser. Mat. **47** (1983), 455–497.

[25] Masani, P. R.: *Shift invariant spaces and prediction theory*, Acta Math. **107** (1962), 275–290.

[26] Nevanlinna, R.: *Eindeutige analytische Funktionen*, Springer, Berlin 1953.

Universität Leipzig
Mathematisches Institut
Augustusplatz 10
D-04109 Leipzig

Universität Leipzig
Mathematisches Institut
Augustusplatz 10
D-04109 Leipzig

Katholieke Universiteit Leuven
Department Computerwetenschappen
Celestijnenlaan 200A
B-3001 Leuven

1991 Mathematics Subject Classification. Primary 30E05; Secondary 47A57

Operator Theory:
Advances and Applications, Vol. 106

Selfadjoint extensions of the orthogonal sum of symmetric relations, II

SEPPO HASSI, MICHAEL KALTENBÄCK, AND HENK DE SNOO

Dedicated to Heinz Langer on the occasion of his 60th birthday

Let S_1 and S_2 be closed symmetric linear relations in Hilbert spaces $\mathcal{H}_1$ and $\mathcal{H}_2$ with finite and equal defect numbers. The selfadjoint extensions of the closed symmetric linear relation $S = S_1 \oplus S_2$ are studied and a description for these extensions in the Hilbert space $\mathcal{H}_1$ is given. The results are applied to a class of differential operators.

1. Introduction

Let $\mathcal{H}_1$ and $\mathcal{H}_2$ be Hilbert spaces and let S_1 and S_2 be closed symmetric linear relations in $\mathcal{H}_1$ and $\mathcal{H}_2$, respectively. Let the Hilbert space $\mathcal{H}$ and the relation S be defined by the orthogonal sums $\mathcal{H} = \mathcal{H}_1 \oplus \mathcal{H}_2$ and $S = S_1 \oplus S_2$. Then S is a closed symmetric linear relation in $\mathcal{H}$. Every selfadjoint extension $\tilde{A}$ of S is also a selfadjoint extension of S_1. Thus, $\tilde{A}$ induces a generalized resolvent $\tilde{P}_1(\tilde{A}-\ell)^{-1}|_{\mathcal{H}_1}$ of S_1 in $\mathcal{H}_1$. Here $\tilde{P}_1$ denotes the orthogonal projection of $\mathcal{H}$ onto $\mathcal{H}_1$. We are interested in the description of $\tilde{P}_1(\tilde{A}-\ell)^{-1}|_{\mathcal{H}_1}$ in the Hilbert space $\mathcal{H}_1$. In fact, our aim is to describe all generalized resolvents of S_1 in the Hilbert space $\mathcal{H}_1$, induced by those selfadjoint extensions of S_1 which also extend S.

It is well known that the family of all generalized resolvents of S_1 in $\mathcal{H}_1$ can be parametrized by means of Kreĭn's formula, cf. [12], [13], [14]. Here the parameter functions belong to the extended Nevanlinna class. The problem is to determine which of these parameter functions correspond to selfadjoint extensions of S_1 which also extend S. Of course in the Hilbert space $\mathcal{H}$ the selfadjoint extensions of S can also be parametrized via parameter functions of the extended Nevanlinna class (now acting on a different space). We will establish a correspondence between these two descriptions and, in particular, between the parameter functions. The special case of generalized resolvents of S_1 corresponding to the canonical selfadjoint extensions of S in $\mathcal{H}$ was considered in [9]. There we based our approach on so-called Nevanlinna pairs with a special structure and studied certain minimality properties of these extensions. For some applications of that situation see also [8]. In the present paper we will use a geometric approach and study all possible selfadjoint extensions of S.

At the end of this paper we apply the abstract results to a system of regular Sturm-Liouville operators and we give a concrete example.

2. Preliminaries

Let $\mathcal{G}$ be a Hilbert space with inner product $(\cdot,\cdot)$ and let T be a closed linear relation in $\mathcal{G}$, i.e., a closed linear subspace of $\mathcal{G}^2 = \mathcal{G} \oplus \mathcal{G}$. Then T is called dissipative, if $\operatorname{Im}(g,f) \geq 0$ for all $\{f,g\} \in T$, and maximal dissipative, if there is no proper dissipative extension of T. The Cayley transform $C_\mu(T)$, $\mu \in \mathbb{C} \setminus \mathbb{R}$, of T is defined by the formula

$$\{\{g - \mu f, g - \bar{\mu} f\} : \{f,g\} \in T\}.$$

Recall that for $\mu \in \mathbb{C}^-$, T is dissipative if and only if $C_\mu(T)$ is a contractive operator in $\mathbf{L}(\mathcal{G})$. For further details on linear relations and properties of Cayley transforms we refer to [2] and [7]. We denote by $\mathbf{N}(\mathcal{G})$ the set of all functions $N(\ell)$ defined on $\mathbb{C} \setminus \mathbb{R}$ and with values in the set of linear relations on $\mathcal{G}$ such that

(i) $N(\ell)$ is maximal dissipative for all $\ell \in \mathbb{C}^+$,
(ii) $N(\bar{\ell}) = N(\ell)^*$ for all $\ell \in \mathbb{C} \setminus \mathbb{R}$,
(iii) for some $\mu \in \mathbb{C}^-$ the Cayley transform $C_\mu(N(\ell))$ depends holomorphically on $\ell \in \mathbb{C}^+$.

Notice that all constant selfadjoint relations in $\mathcal{G}$ belong to $\mathbf{N}(\mathcal{G})$. If $N \in \mathbf{N}(\mathcal{G})$ then it follows from (i) and (ii) that $-N(\ell)$ is maximal dissipative for all $\ell \in \mathbb{C}^-$. The singular part

$$N_\infty(\ell) = \{\, \{0,\varphi\} \in \mathcal{G}^2 : \{0,\varphi\} \in N(\ell) \,\} = \{0\} \oplus \operatorname{mul} N(\ell)$$

of $N(\ell)$ turns out to be independent of $\ell \in \mathbb{C} \setminus \mathbb{R}$, so that $N(\ell)$ can be decomposed as $N_s(\ell) \oplus N_\infty$, where $N_s(\ell)$ is the operator part of $N(\ell)$, a densely defined operator in $(\operatorname{mul} N)^\perp$ and where N_∞ is the singular part of $N(\ell)$. If P is the orthogonal projection of $\mathcal{G}$ onto $(\operatorname{mul} N)^\perp$ then the decomposition of $N(\ell)$ can also be written as

$$N(\ell) = \{\, \{Ph, (N_s(\ell)P + I - P)h\} : Ph \in \operatorname{dom} N_s(\ell) \,\}, \tag{2.1}$$

or, equivalently, as

$$N(\ell) = (N_s(\ell)P + I - P)P^{-1}, \tag{2.2}$$

where the operations in the righthand side are in the sense of relations. In the case of a finite-dimensional space $\mathcal{G}$, we can interpret these identities in terms of Nevanlinna pairs, cf. [4], [8]. In the sequel also a certain subclass of $\mathbf{N}(\mathcal{G})$ will be important. For bounded selfadjoint operators $A, B \in \mathbf{L}(\mathcal{G})$ we write $A \geq B$ if $(Ah,h) \geq (Bh,h)$ for all $h \in \mathcal{G}$. Moreover, with $B >> 0$ we mean that $B \geq \beta I$ for some $\beta > 0$. We denote by $\mathbf{M}(\mathcal{G})$ the set of all functions $Q(\ell)$ defined on $\mathbb{C} \setminus \mathbb{R}$, with values in $\mathbf{L}(\mathcal{G})$, for which

(i) $\operatorname{Im} Q(\ell) >> 0$ for all $\ell \in \mathbb{C}^+$,
(ii) $Q(\bar{\ell}) = Q(\ell)^*$ for all $\ell \in \mathbb{C} \setminus \mathbb{R}$,
(iii) $Q(\ell)$ depends holomorphically on $\ell \in \mathbb{C} \setminus \mathbb{R}$.

Clearly, $\mathbf{M}(\mathcal{G})$ is a subclass of $\mathbf{N}(\mathcal{G})$. If $\mathcal{G}$ is one-dimensional then $\mathbf{M}(\mathcal{G})$ can be identified with the class of all Nevanlinna functions, which are not equal to a constant real constant. For arbitrary $Q \in \mathbf{M}(\mathcal{G})$ and $N \in \mathbf{N}(\mathcal{G})$ the relation $(Q(\ell) + N(\ell))^{-1}$ belongs to $\mathbf{L}(\mathcal{G})$ for all $\ell \in \mathbb{C} \setminus \mathbb{R}$, cf. [4], [8], [14].

Now let $\mathcal{H}$ be a Hilbert space with inner product $(\cdot, \cdot)$ and let S be a closed symmetric linear relation on $\mathcal{H}$. The defect subspaces of S are given by

$$\ker(S^* - \ell) = \operatorname{ran}(S - \bar{\ell})^\perp, \quad \ell \in \mathbb{C} \setminus \mathbb{R}.$$

The Hilbert space dimension of $\ker(S^* - \ell)$ is constant for $\ell \in \mathbb{C}^+$ and $\ell \in \mathbb{C}^-$; this determines the defect numbers (n_+, n_-) of S. In the following we will assume that $n_+ = n_-$ or, equivalently, that S has selfadjoint extensions in the Hilbert space $\mathcal{H}$, cf. e.g. [1], [7]. Such extensions are also called canonical extensions of S.

Let A be a canonical selfadjoint extension of S. Then $I + (\ell - \zeta)(A - \ell)^{-1}$ maps $\ker(S^* - \zeta)$ bijectively onto $\ker(S^* - \ell)$, $\ell, \zeta \in \rho(A)$. In fact, its inverse operator is given by $I + (\zeta - \ell)(A - \zeta)^{-1}$. By means of this operator function we introduce a holomorphic mapping onto the defect subspaces of S: let $\mathcal{G}$ be a Hilbert with the same Hilbert space dimension as $\ker(S^* - \ell)$, $\ell \in \mathbb{C} \setminus \mathbb{R}$, let Γ_i be a bijective linear mapping from $\mathcal{G}$ onto $\ker(S^* - i)$, and define

$$\Gamma_\ell = (I + (\ell - i)(A - \ell)^{-1})\Gamma_i, \quad \ell \in \rho(A).$$

Then Γ_ℓ is a bicontinuous linear mapping from $\mathcal{G}$ onto $\ker(S^* - \ell)$ and satisfies

$$\Gamma_\ell = (I + (\ell - \zeta)(A - \ell)^{-1})\Gamma_\zeta, \quad \ell, \zeta \in \rho(A). \tag{2.3}$$

A function $Q(\ell)$ defined on $\mathbb{C} \setminus \mathbb{R}$ and with values in $\mathbf{L}(\mathcal{G})$ is called a Q-function of the triple (S, A, Γ_i), if the relation

$$\frac{Q(\ell) - Q(\zeta)^*}{\ell - \bar{\zeta}} = \Gamma_\zeta^* \Gamma_\ell, \quad \ell, \zeta \in \rho(A), \tag{2.4}$$

holds. The function $Q(\ell)$ is uniquely determined by (2.4), up to a bounded selfadjoint operator C in $\mathcal{G}$:

$$Q(\ell) = C - i\Gamma_i^* \Gamma_i + (\ell + i)\Gamma_i^* \Gamma_\ell. \tag{2.5}$$

Since the choice of the mapping Γ_i does not essentially influence the structure of $Q(\ell)$, $Q(\ell)$ is also called a Q-function of the pair (S, A). It is easily seen that $Q(\ell)$ belongs to the class $\mathbf{M}(\mathcal{G})$, cf. [14].

Now let $\tilde{A}$ be another selfadjoint relation on a Hilbert space $\tilde{\mathcal{H}} \supset \mathcal{H}$ such that $\tilde{A} \supset S$. Let $\tilde{P}$ be the orthogonal projection from $\tilde{\mathcal{H}}$ onto $\mathcal{H}$. The operator function $R(\ell) = \tilde{P}(\tilde{A} - \ell)^{-1}|_{\mathcal{H}}$, $\ell \in \rho(\tilde{A})$, is called a generalized resolvent of $\tilde{A}$ in $\mathcal{H}$. The Štraus relation associated with $\tilde{A}$ is defined by

$$T(\ell) = R(\ell)^{-1} + \ell, \quad \ell \in \mathbb{C} \setminus \mathbb{R}. \tag{2.6}$$

It turns out that $T(\ell) \supset S$ for all $\ell \in \mathbb{C} \setminus \mathbb{R}$ and that $-T(\ell) \in \mathbf{N}(\mathcal{H})$. Conversely, given $-T(\ell) \in \mathbf{N}(\mathcal{H})$ such that $T(\ell) \supset S$ for all $\ell \in \mathbb{C} \setminus \mathbb{R}$, there exists a Hilbert space $\tilde{\mathcal{H}} \supset \mathcal{H}$ and a selfadjoint relation $\tilde{A} \supset S$ in $\tilde{\mathcal{H}}$ such that the identity

$$(T(\ell) - \ell)^{-1} = \tilde{P}(\tilde{A} - \ell)^{-1}|_{\mathcal{H}}$$

holds, cf. [7]. According to Kreĭn's formula there is a one-to-one correspondence between the generalized resolvents $R(\ell)$ of S and parameter functions $N(\ell) \in \mathbf{N}(\mathcal{G})$ given by

$$\tilde{P}(\tilde{A} - \ell)^{-1}|_{\mathcal{H}} = (A - \ell)^{-1} - \Gamma_\ell (Q(\ell) + N(\ell))^{-1} \Gamma_{\bar{\ell}}^*, \tag{2.7}$$

see [12], [13]. The selfadjoint extension $\tilde{A}$ is uniquely determined by $N(\ell)$ up to unitary equivalence if

$$\tilde{\mathcal{H}} \ominus \mathcal{H} = \overline{\operatorname{span}}\, \{ (I - \tilde{P})(\tilde{A} - \ell)^{-1} h : \, h \in \mathcal{H}, \ell \in \mathbb{C} \setminus \mathbb{R} \},$$

see [14]. In (2.7) $\tilde{A}$ is a canonical selfadjoint extension of S if and only if $N(\ell)$ is a constant selfadjoint relation in $\mathcal{G}$. The following proposition will present a more explicit version of (2.7).

Proposition 2.1. *Let $N(\ell) = (N_s(\ell)P + I - P)P^{-1}$ be the decomposition of $N(\ell)$ as in* (2.2). *Then* (2.7) *can be written as*

$$\tilde{P}(\tilde{A} - \ell)^{-1}|_{\mathcal{H}} = (A - \ell)^{-1} - \Gamma_\ell P (Q(\ell)P + N_s(\ell)P + (I - P))^{-1} \Gamma_{\bar{\ell}}^*, \tag{2.8}$$

where

$$(Q(\ell)P + N_s(\ell)P + (I - P))^{-1} \in \mathbf{L}(\mathcal{G}).$$

Moreover,

$$\begin{aligned} (Q(\ell) + N(\ell))^{-1} &= (PQ(\ell)|_{(\operatorname{mul} N)^\perp} + N_s(\ell))^{-1} P \\ &= P(PQ(\ell)P + N_s(\ell)P + I - P)^{-1} P. \end{aligned} \tag{2.9}$$

Proof. Considering all terms as linear relations the following equalities hold:

$$\begin{aligned} &(Q(\ell) + N(\ell))^{-1} \\ &\quad = \{ \{(Q(\ell)P + N_s(\ell)P + I - P)u; Pu\} \in \mathcal{G}^2 : \, Pu \in \operatorname{dom} N_s(\ell) \} \\ &\quad = P(Q(\ell)P + N_s(\ell)P + (I - P))^{-1}. \end{aligned} \tag{2.10}$$

Since $(Q(\ell) + N(\ell))^{-1}$ is a bounded linear operator defined on $\mathcal{G}$, (2.10) implies that the relation $(Q(\ell)P + N_s(\ell)P + (I - P))^{-1}$ has as domain all of $\mathcal{G}$. Suppose that $h \in \ker (Q(\ell)P + N_s(\ell)P + (I - P))$. Then (2.10) implies that $Ph = 0$ since $\ker (Q(\ell) + N(\ell)) = \{0\}$. This gives $(I - P)h = 0$ and hence $h = 0$. Therefore the relation $(Q(\ell)P + N_s(\ell)P + (I - P))^{-1}$ is an operator defined on $\mathcal{G}$ and since the relation $Q(\ell)P + N_s(\ell)P + (I - P)$ is closed, we see that the operator $(Q(\ell)P + N_s(\ell)P + (I - P))^{-1}$ belongs to $\mathbf{L}(\mathcal{G})$.

Now consider $Q(\ell)P + N_s(\ell)P + (I - P)$ as a 2×2 operator block matrix with respect to $\mathcal{G} = (\operatorname{mul} N)^\perp \oplus \operatorname{mul} N$. This matrix has domain $\operatorname{dom} N(\ell) \oplus \operatorname{mul} N$ and it is lower triangular with the identity operator in the lower right corner. Since $Q(\ell)P + N_s(\ell)P + (I-P)$ is invertible, its left upper corner $PQ(\ell)|_{(\operatorname{mul} N)^\perp} + N_s(\ell)$ is also invertible and $(PQ(\ell)|_{(\operatorname{mul} N)^\perp} + N_s(\ell))^{-1}$ is equal to the left upper corner of the lower triangular operator block matrix

$$(Q(\ell)P + N_s(\ell)P + (I - P))^{-1}$$

with respect to the same decomposition $\mathcal{G} = (\operatorname{mul} N)^\perp \oplus \operatorname{mul} N$. Hence,

$$\begin{aligned}(Q(\ell) + N(\ell))^{-1} &= P(Q(\ell)P + N_s(\ell)P + (I - P))^{-1} \\ &= P(Q(\ell)P + N_s(\ell)P + (I - P))^{-1}P \\ &= (PQ(\ell)|_{(\operatorname{mul} N)^\perp} + N_s(\ell))^{-1}P \\ &= P(PQ(\ell)P + N_s(\ell)P + I - P)^{-1}P.\end{aligned}$$

This completes the proof. □

The previous facts have a natural interpretation in terms of the Weyl functions, as introduced by Derkach and Malamud [5], [6]. Let $\mathcal{H}$ be a Hilbert space, and let S be a closed, densely defined, symmetric operator in $\mathcal{H}$ with equal defect numbers. A triplet $(\mathcal{G}, \Pi^1, \Pi^2)$, where $\mathcal{G}$ is a Hilbert space and Π^1, Π^2 are bounded linear mappings from $\operatorname{dom} S^*$ to $\mathcal{G}$, is called a boundary value space for S^*, if

(1) $(S^*f, g)_\mathcal{H} - (f, S^*g)_\mathcal{H} = (\Pi^1 f, \Pi^2 g)_\mathcal{G} - (\Pi^2 f, \Pi^1 g)_\mathcal{G}$ for all $f, g \in \operatorname{dom} S^*$,
(2) the mapping $\Pi : f \mapsto \{\Pi^2 f, -\Pi^1 f\}$ from $\operatorname{dom} S^*$ to $\mathcal{G}^2$ is surjective.

Let $A^j = S^*|_{\ker(\Pi^j)}$, $j = 1, 2$. Then A^j is a selfadjoint extension of S, $j = 1, 2$, cf. [5]. Since $\operatorname{dom} S^* = \operatorname{dom} A^2 \dot{+} \ker(S^* - \ell)$, $\ell \in \mathbb{C} \setminus \mathbb{R}$, the mapping $\Pi^2|_{\ker(S^* - \ell)}$ from $\ker(S^* - \ell)$ to $\mathcal{G}$ is bijective. Therefore we can define $\Gamma_\ell = (\Pi^2|_{\ker(S^* - \ell)})^{-1}$. The mapping $M(\ell)$, defined by $M(\ell) = \Pi^1 \Gamma_\ell$, $\ell \in \mathbb{C} \setminus \mathbb{R}$, is called the Weyl function of $(\mathcal{G}, \Pi^1, \Pi^2)$. It is shown in [5, Lemma 1, Theorem 1] that Γ_ℓ satisfies (2.3) and that $M(\ell)$ is a Q-function of (S, A^2, Γ_i). The Kreĭn's formula (2.7) can be formulated by means of boundary value spaces. Assume that $\tilde{A}$ is a selfadjoint extension of S in a Hilbert space $\tilde{\mathcal{H}} \supset \mathcal{H}$ and let $T(\ell)$ be the Štraus relation defined by (2.6). Then

$$\tilde{P}(\tilde{A} - \ell)^{-1}|_\mathcal{H} = (A^2 - \ell)^{-1} - \Gamma_\ell (M(\ell) + \mathcal{T}(\ell))^{-1} \Gamma_{\bar{\ell}}^*, \tag{2.11}$$

where $\mathcal{T}(\ell) = \Pi(\operatorname{dom} T(\ell)) \in \mathbf{N}(\mathcal{G})$. Moreover, the Štraus relation corresponding to (2.11) is given by $T(\ell) = S^*|_{\Pi^{-1}(\mathcal{T}(\ell))}$. For $g \in \mathcal{H}$ this means that $f = R(\ell)g = (T(\ell) - \ell)^{-1}g$ is the solution of the equation

$$(S^* - \ell)f = g, \quad \Pi(f) \in \mathcal{T}(\ell).$$

In terms of boundary value spaces the canonical selfadjoint extensions of S are in one-to-one correspondence between the selfadjoint relations in $\mathcal{G}$ via $\Theta = \Pi(\operatorname{dom} \tilde{A})$ (or $\tilde{A} = S^*|_{\Pi^{-1}(\Theta)}$). For further details, see [5], [6].

3. Orthogonal sums of symmetric relations

Let $\mathcal{H}_1, \mathcal{H}_2$ be two Hilbert spaces and let S_1, S_2 be closed symmetric linear relations in $\mathcal{H}_1$ and $\mathcal{H}_2$, respectively. Assume that S_1 and S_2 have finite and equal defect numbers (n_1, n_1) and (n_2, n_2), respectively. Fix canonical selfadjoint extensions A_1 of S_1 and A_2 of S_2, let Γ_i^j be a bijective linear mapping from $\mathbb{C}^{n_j}$ onto $\ker(S_j^* - i)$, and define $\Gamma_\ell^j = (I + (\ell - i)(A_j - \ell)^{-1})\Gamma_i^j$, $\ell \in \mathbb{C} \setminus \mathbb{R}$, for $j = 1, 2$. Then Γ_ℓ^j is a bijective linear mapping from $\mathbb{C}^{n_j}$ onto $\ker(S_j^* - \ell)$ and satisfies (2.3). We denote by $Q_j(\ell)$ the Q-function of (S_j, A_j, Γ_i^j) which clearly can be seen as an $n_j \times n_j$ matrix valued function. Let $\mathcal{H} = \mathcal{H}_1 \oplus \mathcal{H}_2$ and denote by $\tilde{P}_1, \tilde{P}_2$ the orthogonal projection of $\mathcal{H}$ onto $\mathcal{H}_1$ and $\mathcal{H}_2$, respectively. Define $S = S_1 \oplus S_2$, i.e.

$$S = \{ \{\{f_1, f_2\}, \{g_1, g_2\}\} \in \mathcal{H}^2 : \{f_1, g_1\} \in S_1, \ \{f_2, g_2\} \in S_2 \}, \tag{3.1}$$

and let $A = A_1 \oplus A_2$. Then clearly S is a closed symmetric relation in $\mathcal{H}$ with finite and equal defect numbers (n, n), $n = n_1 + n_2$, and A is a canonical selfadjoint extension of S. Moreover, $\Gamma_\ell = \Gamma_\ell^1 \oplus \Gamma_\ell^2$ is a bijective linear mapping from $\mathbb{C}^n$ onto $\ker(S_1^* - \ell) \oplus \ker(S_2^* - \ell) = \ker(S^* - \ell)$ and it satisfies (2.3). Finally, $Q(\ell) = Q_1(\ell) \oplus Q_2(\ell)$ satisfies (2.4) and hence it is a Q-function of (S, A, Γ_i). In the following we will identify the spaces $\mathbb{C}^{n_1}$ and $\mathbb{C}^{n_2}$ with the subspaces $\mathbb{C}^{n_1} \times \{0\}$ and $\{0\} \times \mathbb{C}^{n_2}$ of $\mathbb{C}^n = \mathbb{C}^{n_1} \oplus \mathbb{C}^{n_2}$. Moreover, we denote by P_1, P_2 the orthogonal projections from $\mathbb{C}^n$ onto $\mathbb{C}^{n_1}, \mathbb{C}^{n_2}$, respectively. Then it is elementary to check that $\Gamma_\ell^j P_j = \tilde{P}_j \Gamma_\ell$. With respect to the decomposition $\mathbb{C}^n = \mathbb{C}^{n_1} \oplus \mathbb{C}^{n_2}$ the 2×2 block matrix representation of $Q(\ell)$ has the diagonal form $Q(\ell) = \operatorname{diag}(Q_1(\ell), Q_2(\ell))$. The aim of this section is to describe by means of Krein's formula all generalized resolvents of S_1 in $\mathcal{H}_1$ induced by selfadjoint relations $\tilde{A}$ in a Hilbert space $\tilde{\mathcal{H}}$, such that $\mathcal{H} \subset \tilde{\mathcal{H}}$ and $S \subset \tilde{A}$. More precisely, let $N(\ell) \in \mathbf{N}(\mathbb{C}^n)$ be the parameter in

$$\tilde{P}(\tilde{A} - \ell)^{-1}|_{\mathcal{H}} = (A - \ell)^{-1} - \Gamma_\ell (Q(\ell) + N(\ell))^{-1} \Gamma_{\bar{\ell}}^*, \tag{3.2}$$

where $\tilde{P}$ is the orthogonal projection of $\tilde{\mathcal{H}}$ onto $\mathcal{H}$. Then we will determine $\mathcal{T}(\ell) \in \mathbf{N}(\mathbb{C}^{n_1})$ such that

$$\tilde{P}_1(\tilde{A} - \ell)^{-1}|_{\mathcal{H}_1} = (A_1 - \ell)^{-1} - \Gamma_\ell^1 (Q_1(\ell) + \mathcal{T}(\ell))^{-1} \Gamma_{\bar{\ell}}^{1*}, \tag{3.3}$$

where $\tilde{P}_1$ is the orthogonal projection of $\tilde{\mathcal{H}}$ onto $\mathcal{H}_1$.

Before giving a solution to the above question we will make some remarks. The first remark is of a geometric nature. Let $\mathcal{L}$ be a closed linear subspace of a Hilbert space $\mathcal{K}$ and let P be an orthogonal projection in $\mathcal{K}$. Then we have

$$\dim \mathcal{L} = \dim P\mathcal{L} + \dim(\mathcal{L} \cap \ker P) \tag{3.4}$$

while $\mathcal{L}$ and P decompose the space $\mathcal{K}$ into four orthogonal parts in a natural way, which will be used below. The next remark is well known: let $D = (D_{ij})_{i,j=1}^2$ be a 2×2 block matrix, such that

$$D_{11} \in \mathbb{C}^{k_1 \times k_1}, \ D_{12} \in \mathbb{C}^{k_1 \times k_2}, \ D_{21} \in \mathbb{C}^{k_2 \times k_1}, \ D_{22} \in \mathbb{C}^{k_2 \times k_2},$$

where $k_1, k_2 \in \mathbb{N}$, and assume that D_{22} is invertible. Then D is invertible if and only if the Schur complement $R = D_{11} - D_{12}D_{22}^{-1}D_{21}$ is invertible, and in this case

$$D^{-1} = \begin{pmatrix} R^{-1} & -R^{-1}D_{12}D_{22}^{-1} \\ -D_{22}^{-1}D_{21}R^{-1} & D_{22}^{-1} + D_{22}^{-1}D_{21}R^{-1}D_{12}D_{22}^{-1} \end{pmatrix}. \tag{3.5}$$

Next notice that if $Q(\ell)$ is a matrix Nevanlinna function of the form

$$Q(\ell) = \begin{pmatrix} Q_{11}(\ell) & Q_{12}(\ell) \\ Q_{21}(\ell) & Q_{22}(\ell), \end{pmatrix}$$

such that $Q_{22}(\ell)$ is invertible, then the Schur complement

$$\hat{Q}(\ell) = Q_{11}(\ell) - Q_{12}(\ell)Q_{22}(\ell)^{-1}Q_{21}(\ell)$$

is also a matrix Nevanlinna function, cf. [9].

Let $\tilde{A}$ be a selfadjoint extension of S in $\tilde{\mathcal{H}} \supset \mathcal{H}$ and let $N(\ell) \in \mathbf{N}(\mathbb{C}^n)$ be the parameter in (2.7), written in the form (2.2):

$$N(\ell) = (N_s(\ell)P + I - P)P^{-1}.$$

Here $N_s(\ell) = N(\ell)_s$ is the operator part of $N(\ell)$, $\ell \in \mathbb{C}\setminus\mathbb{R}$, and P denotes the orthogonal projection from $\mathbb{C}^n$ onto $(\operatorname{mul} N(\ell))^\perp$. Let Q be the orthogonal projection from $\mathbb{C}^n$ onto the range of P_1P, and let Q' be the orthogonal projection from $\mathbb{C}^n$ onto $P(\mathbb{C}^n)\cap\mathbb{C}^{n_2}$. Write P, $N_s(\ell)P$, and $Q(\ell)$ as 4×4 block matrices with respect to the decomposition $Q(\mathbb{C}^{n_1}) \oplus (I-Q)(\mathbb{C}^{n_1}) \oplus (I-Q')(\mathbb{C}^{n_2}) \oplus Q'(\mathbb{C}^{n_2})$ of $\mathbb{C}^n$. Since the range of P_1P equals the range of P_1PP_1, the block P_{11} is invertible in $Q(\mathbb{C}^n)$. The identity $PQ' = Q'$ implies $Q'P(I-Q') = 0$ and since $(I-Q)P_1P = 0$, we conclude that P has the following form:

$$P = \begin{pmatrix} P_{11} & 0 & P_{13} & 0 \\ 0 & 0 & 0 & 0 \\ P_{13}^* & 0 & P_{33} & 0 \\ 0 & 0 & 0 & I \end{pmatrix}. \tag{3.6}$$

The matrix function $N_s(\ell)P$ commutes with P and hence $N_s(\ell)P$ has the form

$$N_s(\ell)P = \begin{pmatrix} N_{11}(\ell) & 0 & N_{13}(\ell) & N_{14}(\ell) \\ 0 & 0 & 0 & 0 \\ N_{31}(\ell) & 0 & N_{33}(\ell) & N_{34}(\ell) \\ N_{41}(\ell) & 0 & N_{43}(\ell) & N_{44}(\ell) \end{pmatrix}. \tag{3.7}$$

The following representation is clear:

$$Q(\ell) = \begin{pmatrix} Q_{11}(\ell) & Q_{12}(\ell) & 0 & 0 \\ Q_{21}(\ell) & Q_{22}(\ell) & 0 & 0 \\ 0 & 0 & Q_{33}(\ell) & Q_{34}(\ell) \\ 0 & 0 & Q_{43}(\ell) & Q_{44}(\ell) \end{pmatrix}. \tag{3.8}$$

With these matrix results we are now able to state and prove the next theorem.

Theorem 3.1. *Let S be given by (3.1) and let $\tilde{A}$ be a selfadjoint extension of S in $\tilde{\mathcal{H}} \supset \mathcal{H}$ associated with the parameter function $N(\ell) \in \mathbf{N}(\mathbb{C}^n)$ in (2.7). Then the parameter function $\mathcal{T}(\ell) \in \mathbf{N}(\mathbb{C}^{n_1})$ in the representation (3.3) of the generalized resolvent $\tilde{P}_1(\tilde{A}-\ell)^{-1}|_{\mathcal{H}_1}$ has the form*

$$\mathcal{T}(\ell) = \mathcal{R}(\ell) \oplus \{\{0,(I-Q)x\} \in (\mathbb{C}^{n_1})^2 : x \in \mathbb{C}^{n_1}\} = (\mathcal{R}(\ell)Q + (I-Q))Q^{-1},$$

where $\mathcal{R}(\ell) \in \mathbf{N}(Q(\mathbb{C}^n))$ is given by

$$(3.9)\quad \begin{aligned}\mathcal{R}(\ell) = P_{11}^{-1}(&P_{13}Q_{33}(\ell)P_{13}^* + N_{11}(\ell)\\ &-(P_{13}Q_{34}(\ell)+N_{14}(\ell))(Q_{44}(\ell)+N_{44}(\ell))^{-1}(Q_{43}(\ell)P_{13}^*+N_{41}(\ell)))P_{11}^{-1}.\end{aligned}$$

Proof. From (2.8) and (3.3) we obtain

$$\tilde{P}_1(\tilde{A}-\ell)^{-1}|_{\mathcal{H}_1} = (A_1-\ell)^{-1} - \Gamma_\ell^1 P_1 P\;(Q(\ell)P + N_s(\ell)P + (I-P))^{-1}|_{\mathbb{C}^{n_1}}\;\Gamma_{\bar{\ell}}^{1*}.$$

If $Q=0$, or equivalently $P_1P=0$, then $\tilde{P}_1(\tilde{A}-\ell)^{-1}|_{\mathcal{H}_1} = (A_1-\ell)^{-1}$ and hence $\tilde{P}_1(\tilde{A}-\ell)^{-1}|_{\mathcal{H}_1}$ is of the form (3.3) with $\mathcal{T}(\ell) = \{\{0;x\} \in (\mathbb{C}^{n_1})^2 : x \in \mathbb{C}^{n_1}\}$.

Now assume that $Q \neq 0$. Then (2.9) shows that

$$(3.10)\quad P(Q(\ell)P + N_s(\ell)P + (I-P))^{-1} = P(PQ(\ell)P + N_s(\ell)P + I - P)^{-1}P.$$

According to (3.4) we have $\operatorname{rank} P = \operatorname{rank} P_{11} + \dim Q'(\mathbb{C}^{n_2})$ and hence the spectral representation of P is of the form $P = VJV^*$ with

$$V = \begin{pmatrix} V_{11} & 0 & V_{13} & 0\\ 0 & I & 0 & 0\\ V_{31} & 0 & V_{33} & 0\\ 0 & 0 & 0 & I\end{pmatrix}, \qquad J = V^*PV = \begin{pmatrix} I & 0 & 0 & 0\\ 0 & 0 & 0 & 0\\ 0 & 0 & 0 & 0\\ 0 & 0 & 0 & I\end{pmatrix}.$$

(This is seen e.g. by permutating the second and third column and row of P so that P and, hence V, takes a simple block diagonal form.) Here V is unitary and hence multiplying these two matrices shows that PV is of the same form as V with zeros in the second and third column. This implies that V_{11} is invertible and it is easy to check that $V_{31}V_{11}^{-1} = P_{13}^*P_{11}^{-1}$. Now a straightforward calculation shows that $V^*N_s(\ell)PV$ and $V^*PQ(\ell)PV$ are given by

$$\begin{pmatrix} N'_{11}(\ell) & 0 & 0 & N'_{14}(\ell)\\ 0 & 0 & 0 & 0\\ 0 & 0 & 0 & 0\\ N'_{41}(\ell) & 0 & 0 & N_{44}(\ell)\end{pmatrix}, \qquad \begin{pmatrix} Q'_{11}(\ell) & 0 & 0 & V_{31}^*Q_{34}(\ell)\\ 0 & 0 & 0 & 0\\ 0 & 0 & 0 & 0\\ Q_{43}(\ell)V_{31} & 0 & 0 & Q_{44}(\ell)\end{pmatrix},$$

respectively, such that

$$\begin{aligned} N'_{11}(\ell) &= V_{11}^*P_{11}^{-1}N_{11}(\ell)P_{11}^{-1}V_{11},\\ N'_{14}(\ell) &= V_{11}^*P_{11}^{-1}N_{14}(\ell),\\ N'_{41}(\ell) &= N_{41}(\ell)P_{11}^{-1}V_{11},\\ Q'_{11}(\ell) &= V_{11}^*Q_{11}(\ell)V_{11} + V_{31}^*Q_{33}(\ell)V_{31}.\end{aligned}$$

Here the forms of $N'_{11}(\ell)$, $N'_{14}(\ell)$, and $N'_{41}(\ell)$ are obtained by means of the identity $N_s(\ell)P = PN_s(\ell)PP$ using (3.6), (3.7). Notice that $P_1PV = QPVQ$ and that $P_1PV|_{\operatorname{ran} Q} = V_{11}$. Hence, the relation (3.10) and the above considerations imply

$$\begin{aligned} &P_1P\ (Q(\ell)P + N_s(\ell)P + (I-P))^{-1}|_{\mathbb{C}^{n_1}} \\ &\quad = P_1PV\ (V^*PQ(\ell)PV + V^*N_s(\ell)PV + I - V^*PV)^{-1}\ (PV)^*|_{\mathbb{C}^{n_1}} \\ &\quad = V_{11}R(\ell)V_{11}^*Q|_{\mathbb{C}^{n_1}}, \end{aligned}$$

where the inverse in the second expression has the block form

$$\begin{pmatrix} V_{11}^*Q_{11}(\ell)V_{11} + V_{31}^*Q_{33}(\ell)V_{31} + N'_{11}(\ell) & 0 & 0 & V_{31}^*Q_{34}(\ell) + N'_{14}(\ell) \\ 0 & I & 0 & 0 \\ 0 & 0 & I & 0 \\ Q_{43}(\ell)V_{31} + N'_{41}(\ell) & 0 & 0 & Q_{44}(\ell) + N_{44}(\ell) \end{pmatrix}^{-1},$$

and $R(\ell)$ corresponds to the left upper entry of this inverse. Since $\operatorname{Im} Q_{44}(\ell) > 0$, the 3×3 submatrix in the lower right corner is invertible. Hence, (3.5) shows that

$$V_{11}R(\ell)V_{11}^* = (Q_{11}(\ell) + \mathcal{R}(\ell))^{-1}, \tag{3.11}$$

where the operator $\mathcal{R}(\ell)$ acts on $Q(\mathbb{C}^n)$ and is of the form (3.9).

Clearly, $\mathcal{R}(\ell)$ forms the operator part of $\mathcal{T}(\ell)$, while the multivalued part of $\mathcal{T}(\ell)$ is $(I-Q)(\mathbb{C}^{n_1})$. By the remarks preceding this theorem $\mathcal{R} \in \mathbf{N}(Q(\mathbb{C}^n))$ and consequently $\mathcal{T} \in \mathbf{N}(\mathbb{C}^{n_1})$. This completes the proof. □

We may write (3.9) in another form as follows. Choose $\alpha \in \mathbb{C}$ such that $\mathcal{R}(\ell)+\alpha I$ is invertible (which is in particular the case if $|\alpha|$ is sufficiently large or if α belongs to the same half plane as ℓ), then $\mathcal{R}(\ell)$ can also be written in the form

$$\begin{aligned} \mathcal{R}(\ell) = P_{11}^{-1}((Q((Q+Q')(PQ(\ell)P_2P + N_s(\ell)P + \alpha Q)(Q+Q') \\ +I - (Q+Q'))^{-1}|_{Q(\mathbb{C}^{n_1})})^{-1} - \alpha I)P_{11}^{-1}. \end{aligned} \tag{3.12}$$

This is seen by means of (3.5).

At the end of this section we consider some special cases of Theorem 3.1. If P commutes with P_1, which is in particular the case if $N(\ell)$ is an operator valued function, then we have

$$\mathcal{R}(\ell) = N_{11}(\ell) - N_{14}(\ell)(Q_{44}(\ell) + N_{44}(\ell))^{-1}N_{41}(\ell). \tag{3.13}$$

Indeed, if P commutes with P_1 then in the block matrix representation (3.6) of P the entry P_{13} is zero and $P_{11} = I$ so that (3.9) reduces to (3.13).

If $Q' = 0$, then the fourth component in the decomposition $Q(\mathbb{C}^{n_1}) \oplus (I-Q)(\mathbb{C}^{n_1}) \oplus (I-Q')(\mathbb{C}^{n_2}) \oplus Q'(\mathbb{C}^{n_2})$ of $\mathbb{C}^n$ has dimension zero and hence the relation (3.9) reads as

$$\mathcal{R}(\ell) = P_{11}^{-1}(P_{13}Q_{33}(\ell)P_{13}^* + N_{11}(\ell))P_{11}^{-1}. \tag{3.14}$$

Finally we mention that a result similar to Theorem 3.1 was obtained in [9] in a different way. There the compressed resolvents of the canonical selfadjoint extensions of S in the space $\mathcal{H} = \mathcal{H}_1 \oplus \mathcal{H}_2$ were studied in detail.

4. Applications for a class of differential operators

In this section we apply the results of the previous section to differential operators which are obtained by combining a finite number of differential operators defined on compact intervals.

For $j = 1, \ldots, r$, we let $[a_j, b_j]$ be a compact interval and we assume that $l_j = -Dp_jD + q_j$ is a regular symmetric differential expression on $[a_j, b_j]$, i.e. the functions q_j and $\frac{1}{p_j}$ are absolutely summable on $[a_j, b_j]$. The minimal realization of l_j in $L^2[a_j, b_j]$ is denoted by L_j, so that L_j is a densely defined, symmetric operator with defect numbers $(2, 2)$. The adjoint L_j^* of L_j is the maximal realization of l_j in $L^2[a_j, b_j]$. Note that $\operatorname{dom} L_j$ is the set of all $h \in \operatorname{dom} L_j^*$ which satisfy $h(a_j) = h(b_j) = h'(a_j) = h'(b_j) = 0$. For details we refer to [15]. Define the linear mappings from $\operatorname{dom} L_j^*$ to $\mathbb{C}^2$ by

$$(4.1) \qquad \Pi_j^1(f) = \begin{pmatrix} f^{[1]}(a_j) \\ -f^{[1]}(b_j) \end{pmatrix}, \qquad \Pi_j^2(f) = \begin{pmatrix} f(a_j) \\ f(b_j) \end{pmatrix},$$

where the notation $h^{[1]} = p_j h'$, $h \in \operatorname{dom} L_j^*$ is used. It follows for example from [15] that the triplet $(\mathbb{C}^2, \Pi_j^1, \Pi_j^2)$ with Π_j^1 and Π_j^2 defined by (4.1) is a boundary value space for L_j^*.

The selfadjoint operators $A_j^1 = L_j^*|_{\ker \Pi_j^1}$ and $A_j^2 = L_j^*|_{\ker \Pi_j^2}$ are selfadjoint extensions of L_j whose domains are determined by the boundary conditions $f'(a_j) = f'(b_j) = 0$ and $f(a_j) = f(b_j) = 0$, respectively. The Weyl function M_j of $(\mathbb{C}^2, \Pi_j^1, \Pi_j^2)$ is given by $M_j(\ell) = \Pi_j^1 \Delta_\ell^j$, where $\Delta_\ell^j = (\Pi_j^2|_{\operatorname{ran}(L_j - \bar{\ell})^\perp})^{-1}$.

Now let $\mathcal{H} = \bigoplus_{j=1}^r L^2[a_j, b_j]$ and define $L = \bigoplus_{j=1}^r L_j$. Then L is a closed, densely defined, symmetric operator in $\mathcal{H}$ with defect numbers $(2r, 2r)$. Moreover, the adjoint operator L^* of L in $\mathcal{H}$ is given by $\bigoplus_{j=1}^r L_j^*$. Hence the domains of L and L^* satisfy $\operatorname{dom} L = \bigoplus_{j=1}^r \operatorname{dom} L_j$ and $\operatorname{dom} L^* = \bigoplus_{j=1}^r \operatorname{dom} L_j^*$. Define the linear mappings $\Pi^1(f)$ and $\Pi^2(f)$ from $\operatorname{dom} L^*$ into $\mathbb{C}^{2r}$ by

$$\Pi^1(f) = (\Pi_1^1(f_1)^\top, \ldots, \Pi_r^1(f_r)^\top)^\top, \quad \Pi^2(f) = (\Pi_1^2(f_1)^\top, \ldots, \Pi_r^2(f_r)^\top)^\top,$$

with $f = (f_j)_{j=1}^r \in \operatorname{dom} L^*$, and let

$$\Pi(f) = \{\Pi^2(f), -\Pi^1(f)\} \in \mathbb{C}^{2r} \oplus \mathbb{C}^{2r} = \mathbb{C}^{4r}, \quad f \in \operatorname{dom} L^*.$$

From the above considerations it follows that the triplet $(\mathbb{C}^{2r}, \Pi^1, \Pi^2)$ is a boundary value space for L^*. Moreover, the mapping Δ_ℓ and the Weyl function $M(\ell) = \Pi^1 \Delta_\ell$ of $(\mathbb{C}^{2r}, \Pi^1, \Pi^2)$ are given by

$$\Delta_\ell = (\Pi^2|_{\operatorname{ran}(L-\bar{\ell})^\perp})^{-1} = \Delta_\ell^1 \oplus \ldots \oplus \Delta_\ell^r,$$

and

$$M(\ell) = M_1(\ell) \oplus \ldots \oplus M_r(\ell), \quad \ell \in \mathbb{C} \setminus \mathbb{R}.$$

The selfadjoint extension $A^2 = L^*|_{\ker(\Pi^2)}$ is now given by $A^2 = \bigoplus_{j=1}^r A_j^2$, so that an element f of $\operatorname{dom} L^*$ belongs to $\operatorname{dom} A^2$ if and only if

$$f_1(a_1) = f_1(b_1) = \ldots = f_r(a_r) = f_r(b_r) = 0.$$

The results in the previous section are now applied in this situation by taking $n = 2r$, $n_1 = 2$, $n_2 = 2r-2$, and $\mathcal{H}_1 = L^2[a_1, b_1]$, $\mathcal{H}_2 = \bigoplus_{j=2}^r L^2[a_j, b_j]$,

$$S_1 = L_1, \quad S_2 = \bigoplus_{j=2}^{r} L_j, \quad A_1 = A_1^2, \quad A_2 = \bigoplus_{j=2}^{r} A_j^2,$$

$$\Gamma_\ell^1 = \Delta_\ell^1, \quad \Gamma_\ell^2 = \Delta_\ell^2 \oplus \ldots \oplus \Delta_\ell^r,$$
$$Q_1(\ell) = M_1(\ell), \quad Q_2(\ell) = M_2(\ell) \oplus \ldots \oplus M_r(\ell).$$

Then, clearly, $\mathcal{H} = \mathcal{H}_1 \oplus \mathcal{H}_2$, $S = S_1 \oplus S_2 = L$, $A = A_1 \oplus A_2 = A^2$, $\Gamma_\ell = \Gamma_\ell^1 \oplus \Gamma_\ell^2 = \Delta_\ell$, $Q(\ell) = Q_1(\ell) \oplus Q_2(\ell) = M(\ell)$. The operators S_1 and S_2 have defect numbers $(2,2)$ and $(2r-2, 2r-2)$, respectively. Moreover, Γ_ℓ^1 maps $\mathbb{C}^2$ bijectively onto $\operatorname{ran}(S_1 - \bar{\ell})^\perp$ and Γ_ℓ^2 maps $\mathbb{C}^{2r-2}$ bijectively onto $\operatorname{ran}(S_2 - \bar{\ell})^\perp$, and both satisfy (2.3). The functions $Q_1(\ell)$ and $Q_2(\ell)$ are Q-functions of (S_1, A_1, Γ_i^1) and (S_2, A_2, Γ_i^2), respectively.

Assume that $\tilde{A}$ is a canonical selfadjoint extension of S in $\mathcal{H}$. It follows from (3.2) with $\Theta = \Pi(\operatorname{dom} \tilde{A})$ that

$$(\tilde{A} - \ell)^{-1} = (A - \ell)^{-1} - \Gamma_\ell (M(\ell) + \Theta)^{-1} \Gamma_{\bar{\ell}}^*.$$

Compression of this resolvent operator to $\mathcal{H}_1$ leads to

$$\tilde{P}_1 (\tilde{A} - \ell)^{-1}|_{\mathcal{H}_1} = (A_1 - \ell)^{-1} - \Gamma_\ell^1 (Q_1(\ell) + \mathcal{T}(\ell))^{-1} (\Gamma_{\bar{\ell}}^1)^*, \tag{4.2}$$

for some $\mathcal{T} \in N(\mathbb{C}^2)$, cf. Section 3. From (2.11) we know that $\tilde{P}_1(\tilde{A} - z)^{-1} g = f$ is the solution of the ℓ-dependent boundary value problem

$$(L_1^* - z) f = g, \quad \Pi^1(f) \in \mathcal{T}(\ell).$$

In the following example we illustrate how the results in the abstract case can be applied to study the connection between the boundary conditions determining the selfadjoint extension $\tilde{A}$ of S in $\mathcal{H}$ and the induced ℓ-dependent boundary conditions when $\tilde{A}$ is considered as an extension of S_1 in $\mathcal{H}_1$. Conversely, this indicates how the ℓ-dependent boundary conditions determined by $\mathcal{T}(\ell)$ can be linearized by means of an exit space, cf. [9]. Our main task here is to calculate the parameter function $\mathcal{T}(\ell)$ from the parameter Θ by applying Theorem 3.1. Let $r = 3$ and let

$$M_j(z) = \begin{pmatrix} m_{11}^j(z) & m_{12}^j(z) \\ m_{21}^j(z) & m_{22}^j(z) \end{pmatrix}$$

be the Weyl function of the triplet $(\mathbb{C}^2, \Pi_j^1, \Pi_j^2)$, $j = 1,2,3$. Let $\tilde{A}$ be the restriction of L^* defined by $h \in \operatorname{dom} \tilde{A}$ if and only if $h \in \operatorname{dom} L^*$ and

$$h(a_2) = 0,\ 16h(b_1) + 12ih(b_2) = 0,\ 2h(a_1) = -h^{[1]}(a_1),\ h(a_3) = h^{[1]}(a_3),$$
$$-12h(b_1) + 16ih(b_2) + 625h(b_3) = 625h^{[1]}(b_3),$$
$$57(3h(b_1) - 4ih(b_2)) - 4h(b_3) = 75h^{[1]}(b_1) - 100ih^{[1]}(b_2).$$

In this case we have $\Pi(\operatorname{dom} \tilde{A}) = \Theta = NP^{-1}$, cf. (2.2), where

$$P = \begin{pmatrix} 1 & 0 & 0 & 0 & 0 & 0 \\ 0 & \frac{9}{25} & 0 & \frac{-12i}{25} & 0 & 0 \\ 0 & 0 & 0 & 0 & 0 & 0 \\ 0 & \frac{12i}{25} & 0 & \frac{16}{25} & 0 & 0 \\ 0 & 0 & 0 & 0 & 1 & 0 \\ 0 & 0 & 0 & 0 & 0 & 1 \end{pmatrix}$$

and

$$NP = \frac{1}{625} \begin{pmatrix} 1250 & 0 & 0 & 0 & 0 & 0 \\ 0 & 513 & 0 & -684i & 0 & -12 \\ 0 & 0 & 0 & 0 & 0 & 0 \\ 0 & 684i & 0 & 912 & 0 & -16i \\ 0 & 0 & 0 & 0 & -625 & 0 \\ 0 & -12 & 0 & 16i & 0 & 625 \end{pmatrix}.$$

Now Θ is a selfadjoint relation and again $\tilde{A}$ is a selfadjoint extension of L. We have $Q = P_1$, $Q'P_2 = \operatorname{diag}(0,0,1,1)$, and applying Theorem 3.1 we get

$$\mathcal{T}(z) = \mathcal{R}(z) = \begin{pmatrix} 2 & 0 \\ 0 & \frac{19}{3} + \frac{16}{9} m_{22}^2(\ell) \end{pmatrix} - \begin{pmatrix} 0 & 0 \\ 0 & -\frac{4}{75} \end{pmatrix} \times$$
$$\begin{pmatrix} m_{11}^3(\ell) - 1 & m_{12}^3(\ell) \\ m_{21}^3(\ell) & m_{22}^3(\ell) + 1 \end{pmatrix}^{-1} \begin{pmatrix} 0 & 0 \\ 0 & -\frac{4}{75} \end{pmatrix}$$
$$= \begin{pmatrix} 2 & 0 \\ 0 & \frac{19}{3} + \frac{16 m_{22}^2(\ell)}{9} - \frac{16(m_{11}^3(\ell)-1)}{5625((m_{11}^3(\ell)-1)(m_{22}^3(\ell)+1) - m_{12}^3(\ell) m_{21}^3(\ell))} \end{pmatrix}.$$

Hence $\tilde{P}_1(\tilde{A} - \ell)^{-1} g = f$ is the solution of

$$L_1^* f - \ell f = g,\ 2f(a_1) = -f^{[1]}(a_1),$$

$$f(b_1)\left(\frac{19}{3} + \frac{16 m_{22}^2(\ell)}{9} - \frac{16(m_{11}^3(\ell) - 1)}{5625((m_{11}^3(\ell) - 1)(m_{22}^3(\ell) + 1) - m_{12}^3(\ell) m_{21}^3(\ell))}\right) = f^{[1]}(b_1).$$

Acknowledgements

The first author was supported by the EC-programme "Human Capital and Mobility" and the second author was supported by "Fonds zur Förderung der wissenschaftlichen Forschung" of Austria, Project P 09832-MAT.

References

[1] N.I. Achieser and I.M. Glasman, "Theorie der linearen Operatoren im Hilbertraum", 8th edition, Akademie Verlag, Berlin, 1981.

[2] R. Arens, "Operational calculus of linear relations", Pacific J. Math., **11** (1961) 9–23.

[3] J. Bognar, "Indefinite Inner Product Spaces", Ergebnisse der Mathematik und ihre Grenzgebiete, Band 78, Springer-Verlag, 1974.

[4] P. Bruinsma, Interpolation problems for Schur and Nevanlinna pairs, Dissertation, Rijksuniversiteit Groningen, 1991.

[5] V.A. Derkach, M.M. Malamud, "Generalized resolvents and the boundary value problems for hermitian operators with gaps", J. Functional Analysis, 95 (1991), 1–95.

[6] V.A. Derkach, M.M. Malamud, "The extension theory of hermitian operators and the moment problem", J. Math. Sciences, 73 (1995), 141–242.

[7] A. Dijksma and H.S.V. de Snoo, "Selfadjoint extensions of symmetric subspaces", Pacific J. Math., **54** (1974), 71–100.

[8] S. Hassi, M. Kaltenbäck, and H.S.V. de Snoo, "The sum of matrix Nevanlinna functions and selfadjoint extensions in exit spaces", Oper. Theory: Adv. Appl., to appear.

[9] S. Hassi, M. Kaltenbäck, and H.S.V. de Snoo, "Selfadjoint extensions of the orthogonal sum of symmetric relations, I", 16th OT Conference Proceedings, to appear.

[10] S. Hassi, H. Langer, and H.S.V. de Snoo, "Selfadjoint extensions for a class of symmetric operators with defect numbers $(1,1)$", 15th OT Conference Proceedings, (1995), 115–145.

[11] M. Kaltenbäck, *Some questions related to symmetric operators in Hilbert spaces*, Dissertation, Technische Universität Wien, 1996.

[12] M.G. Krein, "On hermitian operators with defect one", Dokl. Akad. Nauk SSSR, **43**, No. 8 (1944), 339–342.

[13] M.G. Kreĭn and H. Langer, "Defect subspaces and generalized resolvents of an hermitian operator in the space Π_κ", Funct. Anal. Appl., **5** (1971), 136–146, 217–228.

[14] H. Langer and B. Textorius, "On generalized resolvents and Q-functions of symmetric linear relations (subspaces) in Hilbert space", Pacific J. Math., **72** (1977), 135–165.

[15] M.A. Neumark, "Lineare Differentialoperatoren", Mathematische Monographien, Band XI, Akademie-Verlag, Berlin 1960.

Department of Statistics
University of Helsinki
PL 54
00014 Helsinki
Finland

Institut für Analysis, Technische Mathematik und Versicherungsmathematik
Technische Universität Wien
Wiedner Hauptstrasse 8-10/114
A-1040 Wien Österreich

Department of Mathematics
University of Groningen
Postbus 800
9700 AV Groningen
Nederland

1991 Mathematics Subject Classification. Primary 47A20, 47B25; Secondary 46E22, 47A57

Operator Theory:
Advances and Applications, Vol. 106

Some interpolation problems of Nevanlinna-Pick type. The Kreĭn-Langer method.

SEPPO HASSI, HENK DE SNOO, AND HARALD WORACEK

Dedicated to Heinz Langer on the occasion of his 60th birthday

The method of M.G. Kreĭn and H. Langer to solve interpolation problems of Nevanlinna-Pick type is explored. The classical Nevanlinna-Pick problem and a version involving derivatives are treated. The data give rise to an indefinite inner product space and a symmetric operator in it. In general, the inner product space is degenerate.

1. Introduction

In this paper we consider some interpolation problems of Nevanlinna-Pick type with data which are not necessarily positive definite. An approach to such problems was proposed by M.G. Kreĭn and H. Langer [15], who adapted a construction for the case of positive definite data by B. Sz.-Nagy and A. Koranyi [16, 17]. The method consists of constructing an indefinite inner product space and a symmetric linear operator or relation in it, so that the solutions of the particular Nevanlinna-Pick problem correspond to selfadjoint extensions of the symmetric operator. The construction of the indefinite inner product space can be given abstractly [13, 14, 15, 16, 17] or via reproducing kernel spaces [3]. Several papers have appeared, where this method was applied to similar situations under different conditions on the data [1, 2, 7, 8, 9, 18]. The aim of our paper is expository: we show in detail the basic ideas of the method in conjunction with some interpolation problems. A similar approach with the Nevanlinna class on the upper half plane replaced by the Schur class on the unit disk and with selfadjoint relations replaced by unitary operators was discussed by J.A. Ball [5].

Some preliminary material about selfadjoint relations in Pontryagin spaces can be found in Section 2, cf. [10, 11]. In Sections 3 and 4 the basic constructions associated with such selfadjoint relations are presented [13, 14]. In Sections 5 and 6 we consider the classical indefinite Nevanlinna-Pick problem and a version involving derivatives. For each problem we associate a model, i.e. an indefinite inner product space and a symmetric operator or relation to the prescribed data. There are no restrictions on the data, so that the model spaces may be degenerate. We show that the solutions of these interpolation problems are in one-to-one correspondence with the selfadjoint relations which extend the model operator. In a sequel to this paper we will give parametrizations of the solutions in terms of resolvent matrices. Also certain special types of solutions will be considered and a connection of our results to the theory of Q-functions will be established. If the model space is nondegenerate or even a Hilbert space solutions of the Nevanlinna-Pick interpolation problems exist. In general, this need not be true and the existence of solutions depends on the structure of the model.

A central role in this paper is played by functions belonging to the so-called generalized Nevanlinna class. In order to define this class, we will introduce some terminology and notations. Let $\mathfrak{H}$ be a Hilbert space with inner product $(\cdot,\cdot)$. Let $f:\mathbb{C}\setminus\mathbb{R}\to\mathbf{L}(\mathfrak{H})$ be a meromorphic function; its domain of holomorphy in $\mathbb{C}\setminus\mathbb{R}$ is denoted by $\rho(f)$. The function f is called real, if for each $z\in\rho(f)$ also $\bar{z}\in\rho(f)$ and $f(\bar{z})=f(z)^*$. Let $\nu,\pi\in\overline{\mathbb{N}}=\mathbb{N}\cup\{0\}\cup\{\infty\}$, not both equal to ∞. Then $\mathbf{N}_\nu^\pi(\mathfrak{H})$ is the set of all real meromorphic functions $f:\mathbb{C}\setminus\mathbb{R}\to\mathbf{L}(\mathfrak{H})$, such that the Nevanlinna kernel

$$\mathsf{N}_f(z,w)=\frac{f(z)-f(w)^*}{z-\bar{w}},\quad z,w\in\rho(f),\quad z\neq\bar{w},$$

$$\mathsf{N}_f(z,\bar{z})=f'(z),\quad z\in\rho(f),$$

has precisely ν negative and π positive squares. In other words, for each $n\in\mathbb{N}$ and each choice of $z_1,\dots,z_n\in\rho(f)$ and $x_1,\dots,x_n\in\mathfrak{H}$, the quadratic form

$$\sum_{i,j=1}^{n}(\mathsf{N}_f(z_i,z_j)x_i,x_j)\xi_i\bar{\xi}_j$$

has at most ν negative and π positive squares, and there is an $n\in\mathbb{N}$ and a choice of $z_1,\dots,z_n\in\rho(f)$ and $x_1,\dots,x_n\in\mathfrak{H}$, such that if $\nu<\infty$ then the quadratic form has precisely ν negative and if $\pi<\infty$ it has precisely π positive squares. In this definition we may restrict ourselves to Hilbert spaces. For if $\mathfrak{K}$ is a Kreĭn space with fundamental symmetry J and $\mathfrak{H}$ is the associated Hilbert space, then J gives a bijective correspondence $f\to fJ$ between $\mathbf{N}_\nu^\pi(\mathfrak{K})$ and $\mathbf{N}_\nu^\pi(\mathfrak{H})$.

2. Selfadjoint relations in Pontryagin spaces

The indices (ν,π) of a Kreĭn space $\mathfrak{P}$ are the maximal dimensions of a negative and of a positive subspace of $\mathfrak{P}$. We will always assume that one of the indices is finite, in which case we speak of a Pontryagin space. Let $\mathfrak{P}$ be a Pontryagin space and let A be a selfadjoint relation in $\mathfrak{P}$. In general, the resolvent set $\rho(A)$ may be empty due to the structure of the multivalued part of A; if A is an operator, it is nonempty. In the sequel we will consider only selfadjoint relations A whose resolvent set is nonempty. In that case $\mathbb{C}\setminus\mathbb{R}\subset\rho(A)$ with a possible exception of finitely many points which are normal eigenvalues of A and which lie symmetrically with respect to the real axis, cf. [10, 11]. Two selfadjoint relations A_1, A_2 in Pontryagin spaces

$\mathfrak{P}_1$, $\mathfrak{P}_2$ with nonempty resolvent sets are unitarily equivalent if there exists a unitary operator U from $\mathfrak{P}_1$ onto $\mathfrak{P}_2$, such that $(A_2 - z)^{-1}U = U(A_1 - z)^{-1}$, $z \in \rho(A_1) \cap \rho(A_2)$. In this case, for $z \in \rho(A_1) \cap \rho(A_2)$,

$$\begin{aligned}\{\, \{U(A_1 - z)^{-1}h, U(I + z(A_1 - z)^{-1})x\} : x \in \mathfrak{P}\,\} \\ = \{\, \{(A_2 - z)^{-1}Uh, (I + z(A_2 - z)^{-1})Ux\} : x \in \mathfrak{P}\,\},\end{aligned}$$

which leads to $\rho(A_1) = \rho(A_2)$. Now we will discuss the reduction of a selfadjoint relation and the construction of a selfadjoint relation via a symmetric operator or relation in an indefinite inner product space.

Let $\mathfrak{M}$ be a subspace of $\mathfrak{P}$, not necessarily closed. The selfadjoint relation A induces a closed linear subspace $\mathfrak{L}_{\mathfrak{M}}$ of $\mathfrak{P}$ defined by

$$\mathfrak{L}_{\mathfrak{M}} = \overline{\operatorname{span}}\,\{\, (I + (z - z_0)(A - z)^{-1})a : a \in \mathfrak{M},\ z \in \rho(A)\,\}, \quad z_0 \in \rho(A).$$

Clearly, $\mathfrak{M} \subset \mathfrak{L}_{\mathfrak{M}}$. It follows from the resolvent identity and the continuity of $(A - w)^{-1}$, that

$$(A - w)^{-1}\mathfrak{L}_{\mathfrak{M}} \subset \mathfrak{L}_{\mathfrak{M}}, \quad w \in \rho(A). \tag{2.1}$$

The invariant subspace $\mathfrak{L}_{\mathfrak{M}}$ may be a proper subspace of $\mathfrak{P}$; it can even be degenerate. However, after factorization the invariant subspace $\mathfrak{L}_{\mathfrak{M}}$ and the selfadjoint relation A give rise to a Pontryagin space $\mathfrak{P}_{\mathfrak{M}}$ and a "minimal" selfadjoint relation $A_{\mathfrak{M}}$ in $\mathfrak{P}_{\mathfrak{M}}$ in the following way. The invariance property (2.1) implies that

$$\{\, \{(A - z)^{-1}x, (I + z(A - z)^{-1})x\} : x \in \mathfrak{L}_{\mathfrak{M}}\,\} \subset A \cap \mathfrak{L}_{\mathfrak{M}}^2.$$

Conversely, each element in $A \cap \mathfrak{L}_{\mathfrak{M}}^2$ is contained in the left side. Hence for each $z \in \rho(A)$,

$$A \cap \mathfrak{L}_{\mathfrak{M}}^2 = \{\, \{(A - z)^{-1}x, (I + z(A - z)^{-1})x\} : x \in \mathfrak{L}_{\mathfrak{M}}\,\}. \tag{2.2}$$

Let $\mathfrak{L}_{\mathfrak{M}}^0 = \mathfrak{L}_{\mathfrak{M}} \cap \mathfrak{L}_{\mathfrak{M}}^{\perp}$ be the isotropic part of $\mathfrak{L}_{\mathfrak{M}}$. Then the factor space

$$\mathfrak{P}_{\mathfrak{M}} = \mathfrak{L}_{\mathfrak{M}} / \mathfrak{L}_{\mathfrak{M}}^0$$

is a Pontryagin space, cf. [4, p.69] and [6, Theorem 2.6]. In $\mathfrak{P}_{\mathfrak{M}}$ we define the relation $A_{\mathfrak{M}}$ by $A_{\mathfrak{M}} = (A \cap \mathfrak{L}_{\mathfrak{M}}^2)/(\mathfrak{L}_{\mathfrak{M}}^0)^2$ or, more explicitly, by

$$A_{\mathfrak{M}} = \{\, \{\hat{x}, \hat{y}\} : \{x, y\} \in A \cap \mathfrak{L}_{\mathfrak{M}}^2\,\}.$$

It follows from (2.1) that $\mathfrak{L}_{\mathfrak{M}}^{\perp}$, hence also $\mathfrak{L}_{\mathfrak{M}}^0$, is invariant under $(A - w)^{-1}$, $w \in \rho(A)$. Therefore, the resolvent $(A - z)^{-1}$ induces a bounded linear mapping in $\mathfrak{P}_{\mathfrak{M}}$, which we denote by R_z. The identity $(A - z)^{-*} = (A - \bar{z})^{-1}$ implies that $R_z^* = R_{\bar{z}}$. It follows from the definition and (2.2) that

$$A_{\mathfrak{M}} = \{\, \{R_z x, (I + zR_z)x\} : x \in \mathfrak{P}_{\mathfrak{M}}\,\}, \quad z \in \rho(A).$$

These observations give the following result.

Theorem 2.1. *Let $\mathfrak{M}$ be a subspace of the Pontryagin space $\mathfrak{P}$ and let A be a selfadjoint relation in $\mathfrak{P}$ with a nonempty resolvent set. Then the relation $A_{\mathfrak{M}}$ is selfadjoint in $\mathfrak{P}_{\mathfrak{M}}$ and $\rho(A) \subset \rho(A_{\mathfrak{M}})$, so that the resolvent set of $A_{\mathfrak{M}}$ is nonempty. Moreover, the resolvent operator $(A_{\mathfrak{M}} - z)^{-1}$ coincides with the mapping induced by $(A - z)^{-1}$ in $\mathfrak{P}_{\mathfrak{M}}$.*

The selfadjoint relation A is called minimal with respect to $\mathfrak{M}$ if $\mathfrak{L}_{\mathfrak{M}} = \mathfrak{P}$, in which case $A_{\mathfrak{M}} = A$. Clearly, the relation $A_{\mathfrak{M}}$ is minimal with respect to the image of $\mathfrak{M}$ in $\mathfrak{P}_{\mathfrak{M}}$. We will call $A_{\mathfrak{M}}$ the minimal part of A.

Let $\mathfrak{L}$ be a linear space with inner product $[\cdot,\cdot]$. The indices (ν, π) of $\mathfrak{L}$ are the maximal dimensions of a negative and of a positive subspace of $\mathfrak{L}$. Assume that either ν or π is finite. A sequence $\{(a_n)\}_1^\infty$ of elements in $\mathfrak{L}$ is said to converge to an element $a \in \mathfrak{L}$ if

(i) $[a_n, b] \to [a, b]$ for all $b \in \mathfrak{L}$;

(ii) $[a_n, a_n] \to [a, a]$.

A linear subspace $\mathfrak{A}$ of $\mathfrak{L}$ is dense if every element of $\mathfrak{L}$ can be approximated in this sense by a sequence of elements of $\mathfrak{A}$. Since $\mathfrak{L}$ may be degenerate, i.e. the isotropic part $\mathfrak{L}^0 = \mathfrak{L} \cap \mathfrak{L}^\perp$ may be nontrivial, limits of sequences are not uniquely determined. If $a_n \to a$, then also $a_n \to a + h$ for any $h \in \mathfrak{L}^0$. Conversely, if $a_n \to a$ and $a_n \to a'$, then clearly $a - a' \in \mathfrak{L}^0$. The completion of the factor space $\mathfrak{L}/\mathfrak{L}^0$ is a Pontryagin space $\mathfrak{P}$ with indices (ν, π) in which the above notion of convergence is preserved [12, 13].

Theorem 2.2. *Let $\mathfrak{L}$ be an inner product space with indices (ν, π), where $\nu, \pi \in \overline{\mathbb{N}}$, not both equal to ∞. Let S be a symmetric relation in $\mathfrak{L}$, such that* ran $(S - z)$ *is dense in $\mathfrak{L}$ for some $z \in \mathbb{C}^+$ and some $z \in \mathbb{C}^-$. Then*

$$A = \overline{\operatorname{span}}\, \{ \{\hat{x}, \hat{y}\} \in \mathfrak{P}^2 : \{x, y\} \in S \},$$

is a selfadjoint relation in the Pontryagin space $\mathfrak{P}$ with a nonempty resolvent set.

Proof. It is easy to see that

$$A_1 = \operatorname{span}\, \{ \{\hat{x}, \hat{y}\} \in \mathfrak{P}^2 : \{x, y\} \in S \},$$

is a symmetric linear relation in $\mathfrak{P}$. Moreover, since ran $(S - z)$ is dense in $\mathfrak{L}$, also ran $(A_1 - z)$ is dense in $\mathfrak{P}$. As the closure of (the graph of) A_1, A is symmetric and ran $(A - z) = \mathfrak{P}$. The symmetry of A implies that ker $(A - \bar{z}) = \{0\}$. We conclude that A is selfadjoint and has a nonempty resolvent set [10, 11]. □

In various forms of the Nevanlinna-Pick interpolation problem we will encounter an indefinite inner product space $\mathfrak{G}$, a Pontryagin space $\mathfrak{P}$, and an isometry Φ from $\mathfrak{G}$ to $\mathfrak{P}$. Then a relation T in $\mathfrak{G}$ can be lifted to the relation $\Phi \circ T$ in $\mathfrak{P}$ by

$$\Phi \circ T = \{ \{\Phi x, \Phi y\} : \{x, y\} \in T \}.$$

If $\mathfrak{G}$ is degenerate, then the isometry Φ may have a nontrivial kernel $\ker \Phi \subset \mathfrak{G}^0$, in which case $\Phi \circ T$ may be a proper linear relation. Clearly, if T is a symmetric relation in $\mathfrak{G}$, then $\Phi \circ T$ is a symmetric relation in $\mathfrak{P}$.

3. Induced functions

Let $\mathfrak{P}$ be a Pontryagin space with inner product $[\cdot, \cdot]$ and let A be a selfadjoint relation in $\mathfrak{P}$ with a nonempty resolvent set. Let $\mathfrak{H}$ be a Hilbert space and assume that $\Gamma \in \mathbf{L}(\mathfrak{H}, \mathfrak{P})$. Fix $z_0 \in \rho(A)$ and let $B \in \mathbf{L}(\mathfrak{H})$ such that $\operatorname{Im} B = (\operatorname{Im} z_0)\, \Gamma^* \Gamma$. Define the function f_A by

$$f_A(z) = B^* + (z - \bar{z}_0)\Gamma^*(I + (z - z_0)(A - z)^{-1})\Gamma, \quad z \in \rho(A), \tag{3.1}$$

where the adjoints are taken in the corresponding spaces. Note that $f(z) \in \mathbf{L}(\mathfrak{H})$ and $f_A(z_0) = B$. We will consider f_A as a function defined on its domain of holomorphy $\rho(f_A)$ within $\mathbb{C} \setminus \mathbb{R}$. Clearly, $\rho(f_A) \supseteq \rho(A) \cap (\mathbb{C} \setminus \mathbb{R})$, so that $\rho(f_A)$ contains $\mathbb{C} \setminus \mathbb{R}$ with the possible exception of finitely many points. Using

$$B^* + (z - \bar{z}_0)\Gamma^*\Gamma = \operatorname{Re} B + (z - \operatorname{Re} z_0)\Gamma^*\Gamma,$$

we see that the function f_A also has the representation

$$f_A(z) = \operatorname{Re} B + (z - \operatorname{Re} z_0)\Gamma^*\Gamma + (z - z_0)(z - \bar{z}_0)\Gamma^*(A - z)^{-1}\Gamma. \tag{3.2}$$

Hence f_A is real. Moreover, by using the resolvent identity we see that

$$\mathsf{N}_f(z, w) = [(I + (z - z_0)(A - z)^{-1})\Gamma x, (I + (w - z_0)(A - w)^{-1})\Gamma y]. \tag{3.3}$$

This relation implies the following result.

Theorem 3.1. *Let $\mathfrak{P}$ be a Pontryagin space with indices (ν, π). Let A be a selfadjoint relation in $\mathfrak{P}$ with a nonempty resolvent set and let $\Gamma \in \mathbf{L}(\mathfrak{H}, \mathfrak{P})$. Then $f_A \in \mathbf{N}_{\nu'}^{\pi'}(\mathfrak{H})$ where $\nu' \leq \nu$ and $\pi' \leq \pi$. Moreover, if A is Γ-minimal, then $\nu' = \nu$ and $\pi' = \pi$.*

Let A_1 and A_2 be selfadjoint relations in Pontryagin spaces $\mathfrak{P}_1$, $\mathfrak{P}_2$, whose resolvent sets are nonempty, and let $\Gamma_1 \in \mathbf{L}(\mathfrak{H}, \mathfrak{P}_1)$ and $\Gamma_2 \in \mathbf{L}(\mathfrak{H}, \mathfrak{P}_2)$. The relations A_1 and A_2 are called Γ-unitarily equivalent, if A_1 and A_2 are unitarily equivalent and the associated unitary operator $U : \mathfrak{P}_1 \to \mathfrak{P}_2$ satisfies $U\Gamma_1 = \Gamma_2$.

Theorem 3.2. *If A_1 and A_2 are Γ-unitarily equivalent then $f_{A_1} = f_{A_2}$. Conversely, if A_1 is Γ_1-minimal, A_2 is Γ_2-minimal and $f_{A_1} = f_{A_2}$, then A_1 and A_2 are Γ-unitarily equivalent.*

Proof. Let A_1 and A_2 be Γ-unitarily equivalent and let U be the associated unitary operator. Due to $U\Gamma_1 = \Gamma_2$ we have $\Gamma_1^* U^* = \Gamma_2^*$. These identities together with $U^*U = I$ and $(A_2 - z)^{-1}U = U(A_1 - z)^{-1}$ immediately show that $f_{A_1}(z) = f_{A_2}(z)$ for all $z \in \rho(A_1) \cap \rho(A_2)$. (Note that we use the normalization $f_{A_1}(z_0) = f_{A_2}(z_0) = B$.) Since $(\mathbb{C} \setminus \mathbb{R}) \setminus (\rho(A_1) \cap \rho(A_2))$ is finite we conclude that $\rho(f_{A_1}) = \rho(f_{A_2})$, i.e. $f_{A_1} = f_{A_2}$.

To prove the converse part, let A_1 and A_2 be Γ-minimal selfadjoint relations and suppose that $f_{A_1} = f_{A_2}$. It follows from (3.3) that the relation

$$V(I + (z - z_0)(A_1 - z)^{-1})\Gamma_1 y = (I + (z - z_0)(A_2 - z)^{-1})\Gamma_2 y, \quad y \in \mathfrak{H},$$

defines a linear mapping from the dense subspace $\mathfrak{L}_{\operatorname{ran}\Gamma_1} \subset \mathfrak{P}_1$ onto the dense subspace $\mathfrak{L}_{\operatorname{ran}\Gamma_2} \subset \mathfrak{P}_2$, which is isometric. Moreover, $V\Gamma_1 = \Gamma_2$. It is also clear that $V(A_1 - z)^{-1}\Gamma_1 = (A_2 - z)^{-1}\Gamma_2$. We extend this identity from the range of Γ_1 by means of the resolvent identity as follows

$$\begin{aligned} V(A_1 - w)^{-1}&(I + (z - z_0)(A_1 - z)^{-1})\Gamma_1 \\ &= (A_2 - w)^{-1}(I + (z - z_0)(A_2 - z)^{-1})\Gamma_2 \\ &= (A_2 - w)^{-1}V(I + (z - z_0)(A_1 - z)^{-1})\Gamma_1. \end{aligned}$$

Then, by linearity the relation

$$V(A_1 - w)^{-1}y = (A_2 - w)^{-1}Vy, \quad w \in \rho(A_1) \cap \rho(A_2),$$

holds for all $y \in \mathfrak{L}_{\operatorname{ran}\Gamma_1}$. Sinve V maps a dense set from the Pontryagin space $\mathfrak{P}_1$ onto a dense subset of the Pontryagin space $\mathfrak{P}_2$, it has a unitary continuation U from $\mathfrak{P}_1$ onto $\mathfrak{P}_2$. Then clearly A_1 and A_2 are Γ-unitarily equivalent under U. □

In order to study the derivatives of the function f_A in (3.1) it is convenient to introduce the function ϕ_A by

$$\phi_A(z) = (I + (z - z_0)(A - z)^{-1})\Gamma, \quad z \in \rho(A). \tag{3.4}$$

Then the right side of (3.3) can be written as $[\phi_A(z)x, \phi_A(w)y]$. Hence when $z, w \in \rho(A)$, $x, y \in \mathfrak{H}$, and $k, l \geq 0$, this gives

$$[\phi_A^{(k)}(z)x, \phi_A^{(l)}(w)y] = \left(\frac{\partial^k}{\partial z^k} \frac{\partial^l}{\partial \bar{w}^l} \frac{f_A(z) - f_A(\bar{w})}{z - \bar{w}} x, y \right), \tag{3.5}$$

and differentiation of the function in the right side leads to

$$\begin{aligned} \sum_{h=0}^{k} \binom{k}{h} &\frac{(-1)^{k-h}(k+l-h)!}{(z - \bar{w})^{k+l+1-h}} (f_A^{(h)}(z)x, y) \\ &+ \sum_{h=0}^{l} \binom{l}{h} \frac{(-1)^{l-h}(k+l-h)!}{(\bar{w} - z)^{k+l+1-h}} (f_A^{(h)}(\bar{w})x, y). \end{aligned}$$

Lemma 3.3. *Let $z_1, \dots, z_n \in \rho(A)$ and $x_1, \dots, x_n \in \mathfrak{H}$, then*

$$\{\textstyle\sum_{i=1}^n \phi_A(z_i)x_i, \sum_{i=1}^n z_i\phi_A(z_i)x_i\} \in A \text{ if } \sum_{i=1}^n x_i = 0. \tag{3.6}$$

Let $z \in \rho(A)$, $x \in \mathfrak{H}$, and $k \geq 1$, then

$$\left\{\phi_A^{(k)}(z)x, z\phi_A^{(k)}(z)x + k\phi_A^{(k-1)}(z)x\right\} \in A. \tag{3.7}$$

Let $z, w \in \rho(A)$, $x, y \in \mathfrak{H}$, and $k, l \geq 1$, then

$$(f_A^{(k)}(z)x, y) = (z - \bar{w})[\phi_A^{(k)}(z)x, \phi_A(w)y] + k[\phi_A^{(k-1)}(z)x, \phi_A(w)y], \tag{3.8}$$

and

$$\begin{aligned}(z - \bar{w})&[\phi_A^{(k)}(z)x, \phi_A^{(l)}(w)y] \\ &= l[\phi_A^{(k)}(z)x, \phi_A^{(l-1)}(w)y] - k[\phi_A^{(k-1)}(z)x, \phi_A^{(l)}(w)y].\end{aligned} \tag{3.9}$$

Proof. The resolvent identity implies that

$$(A - z_0)^{-1}\phi(z) = (A - z)^{-1}\Gamma,$$

which leads to the inclusion (3.6). By differentiation of (3.4) we obtain

$$\phi_A^{(k)}(z) = (z - z_0)\frac{d^k}{dz^k}(A - z)^{-1}\Gamma + k\frac{d^{k-1}}{dz^{k-1}}(A - z)^{-1}\Gamma.$$

With

$$\frac{d^k}{dz^k}(A - z)^{-1} = k!(A - z)^{-(k+1)},$$

this gives

$$\phi_A^{(k)}(z) = k!(z - z_0)(A - z)^{-(k+1)}\Gamma + k!(A - z)^{-k}\Gamma = k!(A - z)^{-k}\phi_A(z). \tag{3.10}$$

The inclusion (3.7) can be seen from (3.10). Since

$$f_A(z) = f_A(z_0)^* + \Gamma^*(z - \bar{z}_0)\phi_A(z),$$

it follows for $k \geq 1$ that,

$$(f_A^{(k)}(z)x, y) = \left((z - \bar{z}_0)\phi_A^{(k)}(z)x + k\phi_A^{(k-1)}(z)x, \Gamma y\right), \tag{3.11}$$

so that (3.8) holds for $w = z_0$. The general case follows from the fact that the element in the left side of (3.7) belongs to A and that, according to (3.6), also the element

$$\{\phi_A(w)y - \phi_A(z_0)y, w\phi_A(w)y - z_0\phi_A(z_0)y\}$$

belongs to A. Since A is selfadjoint, these elements are adjoint which in conjunction with (3.11) implies (3.8). Similarly, (3.9) follows from (3.7) and

$$\left\{\phi_A^{(l)}(w)y, w\phi_A^{(l)}(w)y + l\phi_A^{(l-1)}(w)y\right\} \in A.$$

□

4. Induced models

Let $f \in \mathbf{N}_\nu^\pi(\mathfrak{H})$ where $\nu, \pi \in \overline{\mathbb{N}}$, not both equal to ∞. We will show that there exist a Pontryagin space $\mathfrak{P}$, a selfadjoint relation A in $\mathfrak{P}$ with a nonempty resolvent set, and a mapping $\Gamma \in \mathbf{L}(\mathfrak{H}, \mathfrak{P})$ such that $f = f_A$ as in (3.1). Define a linear space $\mathfrak{L}_f$ of formal finite sums by

$$\mathfrak{L}_f = \{ \textstyle\sum_{z \in \rho(f)} x_z e_z : x_z \in \mathfrak{H},\ x_z = 0 \text{ for almost all } z \},$$

and provide $\mathfrak{L}_f$ with an inner product given by

$$[xe_z, ye_w] = (\mathsf{N}_f(z,w)x, y), \quad z, w \in \rho(f), \quad x, y \in \mathbb{C}.$$

Then $\mathfrak{L}_f$ is an inner product space with ν negative and π positive squares. Convergence in $\mathfrak{L}_f$ is defined as in Section 2. Define the (graph of the) operator S_f in $\mathfrak{L}_f$ by

$$S_f = \{ \{ \textstyle\sum_{z \in \rho(f)} x_z e_z, \sum_{z \in \rho(f)} z x_z e_z \} : \sum_{z \in \rho(f)} x_z = 0 \},$$

so that S_f stands for pointwise multiplication.

Lemma 4.1. *The operator S_f in $\mathfrak{L}_f$ is symmetric without eigenvalues in $\rho(f)$. For each $z \in \rho(f)$, the range* ran $(S_f - z)$ *is dense in $\mathfrak{L}_f$. Moreover, for all $z, z_0 \in \rho(f)$ and all $x, y \in \mathfrak{H}$:*

(4.1) $(f(z)x, y) = (f(z_0)^* x, y) + (z - \bar{z}_0)[(I + (z - z_0)(S_f - z)^{-1}) x e_{z_0}, y e_{z_0}]$.

Proof. The operator S_f is symmetric, since it follows from $\sum_{z \in \rho(f)} x_z = 0$ and $\sum_{w \in \rho(f)} y_w = 0$, that

$$\begin{aligned}
&[\textstyle\sum_{z \in \rho(f)} z x_z e_z, \sum_{w \in \rho(f)} y_w e_w] - [\sum_{z \in \rho(f)} x_z e_z, \sum_{w \in \rho(f)} w y_w e_w] \\
&\quad = \textstyle\sum_{z, w \in \rho(f)} (z - \bar{w})[x_z e_z, y_w e_w] \\
&\quad = \textstyle\sum_{z, w \in \rho(f)} ((f(z)x_z, y_w) - (x_z, f(w)y_w)) \\
&\quad = (\textstyle\sum_{z \in \rho(f)} f(z)x_z, \sum_{w \in \rho(f)} y_w) - (\sum_{z \in \rho(f)} x_z, \sum_{w \in \rho(f)} f(w)y_w) \\
&\quad = 0.
\end{aligned}$$

The operator $S_f - z$, $z \in \rho(f)$, is injective, as no single component e_z is included in dom S_f. It follows directly from the definition that

$$(S_f - z)^{-1} x e_w = \frac{xe_z - xe_w}{z - w}, \quad z \neq w. \tag{4.2}$$

The relation (4.2) implies that for $z, z_0 \in \rho(f)$, $z \neq \bar{z}_0$,

$$[(I + (z - z_0)(S_f - z)^{-1}) x e_{z_0}, y e_{z_0}] = [xe_z, ye_{z_0}] = \left(\frac{f(z) - f(z_0)^*}{z - \bar{z}_0} x, y \right).$$

This gives (4.1). Furthermore, we obtain for each $z_0 \in \rho(f)$,

$$\mathfrak{L}_f = \operatorname{span}\{\, x e_{z_0}, x e_{z_0} + (z - z_0)(S_f - z)^{-1} x e_{z_0} \,:\, z \in \rho(f) \setminus \{z_0\}, x \in \mathfrak{H} \,\}. \tag{4.3}$$

We now show that $\operatorname{ran}(S_f - z)$ is dense in $\mathfrak{L}_f$. It follows from (4.2) that

$$\operatorname{ran}(S_f - z) = \operatorname{span}\{\, x e_w \,:\, w \in \rho(f) \setminus \{z\}, x \in \mathfrak{H} \,\}. \tag{4.4}$$

Let $z_n, z \in \rho(f)$, $z_n \to z$ in $\mathbb{C}$ and let $x_n, x \in \mathfrak{H}$, $x_n \to x$ in the norm of $\mathfrak{H}$. We claim that

$$x_n e_{z_n} \to x e_z \text{ in } \mathfrak{L}_f. \tag{4.5}$$

Then (4.5) and (4.4) imply that $\operatorname{ran}(S_f - z)$ is dense in $\mathfrak{L}_f$. To see (4.5), first note that $z_n \to z$ implies that

$$\frac{f(z_n) - f(z_0)^*}{z_n - \bar{z}_0} \to \frac{f(z) - f(z_0)^*}{z - \bar{z}_0} \text{ or } f'(\bar{z}_0),$$

depending on $z \neq \bar{z}_0$ or $z = \bar{z}_0$, respectively. Moreover, this sequence is uniformly bounded in the operator norm. Therefore, we conclude

$$\begin{aligned} [x_n e_{z_n}, y e_{z_0}] &= \left(\frac{f(z_n) - f(z_0)^*}{z_n - \bar{z}_0} x, y \right) + \left(\frac{f(z_n) - f(z_0)^*}{z_n - \bar{z}_0} (x_n - x), y \right) \\ &\to [x e_z, y e_{z_0}], \end{aligned}$$

and

$$\begin{aligned} [x_n e_{z_n}, x_n e_{z_n}] &= \left(\frac{f(z_n) - f(z_n)^*}{z_n - \bar{z}_n} x, x \right) \\ &\quad + \left(\frac{f(z_n) - f(z_n)^*}{z_n - \bar{z}_n} (x_n - x), x \right) \\ &\quad + \left(\frac{f(z_n) - f(z_n)^*}{z_n - \bar{z}_n} x, x_n - x \right) \\ &\quad + \left(\frac{f(z_n) - f(z_n)^*}{z_n - \bar{z}_n} (x_n - x), x_n - x \right) \\ &\to \left(\frac{f(z) - f(z)^*}{z - \bar{z}} x, x \right) \\ &= [x e_z, x e_z]. \end{aligned}$$

According to Section 2 these limiting relations imply (4.5). □

Theorem 4.2. *Let $f \in \mathbf{N}_\nu^\pi(\mathfrak{H})$ where $\nu, \pi \in \overline{\mathbb{N}}$, not both equal to ∞. Then there exist a Pontryagin space $\mathfrak{P}_f$ with indices (ν, π), a mapping $\Gamma \in \mathbf{L}(\mathfrak{H}, \mathfrak{P}_f)$, and a Γ-minimal selfadjoint relation A_f in $\mathfrak{P}_f$ with $\rho(A_f) \cap (\mathbb{C} \setminus \mathbb{R}) = \rho(f)$, such that*

$$f(z) = f(z_0)^* + (z - \bar{z}_0)\Gamma^*(I + (z - z_0)(A_f - z)^{-1})\Gamma, \quad z, z_0 \in \rho(f). \tag{4.6}$$

Proof. Denote the completion of $\mathfrak{L}_f$ by $\mathfrak{P}_f$. Since $f \in \mathbf{N}_\nu^\pi(\mathfrak{H})$, the indices of $\mathfrak{P}_f$ are given by (ν, π). By Lemma 4.1, S_f is a symmetric operator in $\mathfrak{L}_f$ and its range $\operatorname{ran}(S_f - z)$ is dense in $\mathfrak{L}_f$ for each $z \in \rho(f)$. Hence by Theorem 2.2, S_f in $\mathfrak{L}_f$ induces a selfadjoint relation A_f in $\mathfrak{P}_f$. The relation A_f has a nonempty resolvent set; in fact, $\rho(f) \subseteq \rho(A_f)$. Define the mapping $\Gamma : \mathfrak{H} \to \mathfrak{L}_f$ by

$$\Gamma x = x e_{z_0}, \quad x \in \mathfrak{H}. \tag{4.7}$$

For $x, y \in \mathfrak{H}$ it follows that

$$[\Gamma x, \Gamma x] = [x e_{z_0}, x e_{z_0}] = (\mathsf{N}_f(z_0, z_0)x, x), \qquad [\Gamma x, y e_w] = (\mathsf{N}_f(z_0, w)x, y).$$

Hence the operator Γ is bounded and by

$$\Gamma x = \widehat{x e}_{z_0}, \quad x \in \mathfrak{H},$$

it can be viewed as $\Gamma \in \mathbf{L}(\mathfrak{H}, \mathfrak{P})$. Due to (4.3), A_f is Γ-minimal. We have

$$[a, \widehat{y e}_{z_0}] = [a, \Gamma y] = (\Gamma^* a, y), \quad y \in \mathfrak{H}, \quad a \in \mathfrak{P}_f,$$

and, in particular,

$$\frac{\operatorname{Im} f(z_0)}{\operatorname{Im} z_0} = \frac{f(z_0) - f(z_0)^*}{z - \bar{z}_0} = \Gamma^* \Gamma.$$

Now the identity (4.6) follows from the corresponding identity (4.1). In other words, $f = f_{A_f}$ and we have $\rho(A_f) \cap (\mathbb{C} \setminus \mathbb{R}) = \rho(f)$. □

Note that if a selfadjoint relation A is Γ-minimal, then $\rho(f_A) = \rho(A) \cap (\mathbb{C} \setminus \mathbb{R})$.

5. A classical Nevanlinna-Pick interpolation problem

The classical Nevanlinna-Pick interpolation problem which we consider here is to find all solutions $f \in \mathbf{N}_\nu^\pi(\mathfrak{H})$ of

$$z_i \in \rho(f), \quad f(z_i) = W_i, \quad i = 1, \dots, n, \tag{5.1}$$

when for some $n \in \mathbb{N}$ the data $z_1, \dots, z_n \in \mathbb{C}^+$ and $W_1, \dots, W_n \in \mathbf{L}(\mathfrak{H})$ are given. In order to describe the solutions of this problem we will follow the approach used by Kreĭn and Langer [15]. Define the linear space $\mathfrak{G}$ of formal finite sums by

$$\mathfrak{G} = \{ \textstyle\sum_{i=1}^n x_i e_i : x_i \in \mathfrak{H} \},$$

and provide $\mathfrak{G}$ with an inner product given by

$$[x e_i, y e_j] = \left(\frac{W_i - W_j^*}{z_i - \bar{z}_j} x, y \right), \quad x, y \in \mathfrak{H}, \quad i, j = 1, \dots, n.$$

Define the (graph of the) operator T in $\mathfrak{G}$ by

$$T = \{ \{\sum_{i=1}^n x_i e_i, \sum_{i=1}^n z_i x_i e_i\} : \sum_{i=1}^n x_i = 0 \}.$$

Clearly, T is a symmetric operator in $\mathfrak{G}$, since it follows from $\sum_{i=1}^n x_i = 0$ and $\sum_{j=1}^n y_j = 0$, that

$$\begin{aligned}
&[\sum_{i=1}^n z_i x_i e_i, \sum_{j=1}^n y_j e_j] - [\sum_{i=1}^n x_i e_i, \sum_{j=1}^n z_j y_j e_j] \\
&= \sum_{i,j=1}^n (z_i - \bar{z}_j)[x_i e_i, y_j e_j] \\
&= \sum_{i,j=1}^n ((W_i - W_j^*)x_i, y_j) \\
&= (\sum_{i=1}^n W_i x_i, \sum_{j=1}^n y_j) - (\sum_{i=1}^n x_i, \sum_{j=1}^n W_j y_j) \\
&= 0.
\end{aligned}$$

Theorem 5.1. *Let $\nu, \pi \in \overline{\mathbb{N}}$, not both equal to ∞. The function $f \in \mathbf{N}_\nu^\pi(\mathfrak{H})$ satisfies (5.1) if and only if there exist a Pontryagin space $\mathfrak{P}$, a selfadjoint relation A in $\mathfrak{P}$, and an isometry $\Phi : \mathfrak{G} \to \mathfrak{P}$ with the following properties:*

(i) *the indices of $\mathfrak{P}$ are (ν, π);*

(ii) *$z_1, \ldots, z_n \in \rho(A)$;*

(iii) *A is Γ-minimal, when $\Gamma \in \mathbf{L}(\mathfrak{H}, \mathfrak{P})$ and $\Gamma x = \Phi(xe_1)$, $x \in \mathfrak{H}$;*

(iv) *A extends $\Phi \circ T$,*

such that

$$f(z) = W_1^* + (z - \bar{z}_1)\Gamma^*(I + (z - z_1)(A - z)^{-1})\Gamma. \tag{5.2}$$

Proof. Let $\mathfrak{P}$, A, and Φ be given, such that (i)–(iv) hold. Hence the function f defined by (5.2) is of the form (3.1) with $z_0 = z_1$: $f(z) = f_A(z)$. It follows from (i), (ii), (iii), and Theorem 3.1 that $f \in \mathbf{N}_\nu^\pi(\mathfrak{H})$ with $z_1, \ldots, z_n \in \rho(f)$. The equality (5.1) is equivalent to

$$\frac{W_i - W_1^*}{z_i - \bar{z}_1} = \Gamma^*(I + (z_i - z_1)(A - z_i)^{-1})\Gamma, \quad i = 1, \ldots, n. \tag{5.3}$$

For $i = 1$, the identity (5.3) follows from

$$\left(\frac{W_1 - W_1^*}{z_1 - \bar{z}_1} x, y\right) = [xe_1, ye_1] = [\Phi(xe_1), \Phi(ye_1)] = [\Gamma x, \Gamma y] = (\Gamma^* \Gamma x, y),$$

when $x, y \in \mathfrak{H}$. For $i > 1$, the definition of T implies that

$$\left\{xe_1, \frac{xe_i - xe_1}{z_i - z_1}\right\} \in (T - z_i)^{-1}.$$

It follows from (iv), that

$$\left\{\Gamma x, \Phi\left(\frac{xe_i - xe_1}{z_i - z_1}\right)\right\} = \left\{\Phi(xe_1), \Phi\left(\frac{xe_i - xe_1}{z_i - z_1}\right)\right\} \in (A - z_i)^{-1},$$

and, therefore, that

$$\Gamma^*(I + (z_i - z_1)(A - z_i)^{-1})\Gamma x = \Gamma^*\Phi(xe_i). \tag{5.4}$$

For all $y \in \mathfrak{H}$,

$$(\Gamma^*\Phi(xe_i), y) = [\Phi(xe_i), \Gamma y] = [\Phi(xe_i), \Phi(ye_1)] = [xe_i, ye_1] = \left(\frac{W_i - W_1^*}{z_i - \bar{z}_1} x, y\right),$$

so that (5.4) implies (5.3). Hence $f = f_A$ satisfies (5.1).

Conversely, let $f \in \mathbf{N}_\nu^\pi(\mathfrak{H})$ be a function with $z_1, \dots, z_n \in \rho(f)$. Let $\mathfrak{P}$, A, and Γ be as in Theorem 4.2, so that $f = f_A$ with $z_0 = z_1$ in (3.1), cf. (3.2). This establishes (i), (ii), and the first statement of (iii), that A is Γ-minimal. Define $\Phi : \mathfrak{G} \to \mathfrak{P}$ by

$$\Phi(xe_i) = \widehat{xe}_{z_i},$$

so that

$$[\Phi(xe_i), \Phi(ye_j)] = [\widehat{xe}_{z_i}, \widehat{ye}_{z_j}] = [xe_{z_i}, ye_{z_j}] = \left(\frac{f(z_i) - f(z_j)^*}{z_i - \bar{z}_j} x, y\right).$$

Hence, if in addition f satisfies (5.1), the right side equals $[xe_i, ye_j]$ and Φ is an isometry from $\mathfrak{G}$ to $\mathfrak{P}$. Moreover, from (4.7) with $z_0 = z_1$ it follows that $\Gamma x = \Phi(xe_1)$, which completes the proof of (iii). The definition of T and A imply (iv). □

6. A multiple point Nevanlinna-Pick interpolation problem

The multiple point Nevanlinna-Pick interpolation problem is to find all solutions $f \in \mathbf{N}_\nu^\pi(\mathfrak{H})$ of

$$z_i \in \rho(f), \quad f^{(k)}(z_i) = W_{ik}, \quad i = 1, \dots, n, \quad k = 0, \dots, k_i, \tag{6.1}$$

when for some $n \in \mathbb{N}$ the data $z_1, \dots, z_n \in \mathbb{C}^+$, $k_1, \dots, k_n \in \mathbb{N} \cup \{0\}$, and $W_{ik} \in \mathbf{L}(\mathfrak{H})$, $i = 1, \dots, n$, $k = 0, \dots, k_i$, are given. Closely connected to (6.1) is the following classical Nevanlinna-Pick problem: to find all solutions $f \in \mathbf{N}_\nu^\pi(\mathfrak{H})$ of

$$z_i \in \rho(f), \quad f(z_i) = W_{i0}, \quad i = 1, \dots, n. \tag{6.2}$$

In order to solve (6.2) we define the linear space $\mathfrak{G}_0$ as the set of all finite sums

$$\mathfrak{G}_0 = \left\{\textstyle\sum_{i=1}^n x_{i0}e_{i0} : x_{i0} \in \mathfrak{H}\right\},$$

provided with an inner product by

$$[x_{i0}e_{i0}, y_{j0}e_{j0}] = \left(\frac{W_{i0} - W_{j0}^*}{z_i - \bar{z}_j} x_{i0}, y_{j0} \right).$$

In $\mathfrak{G}_0$ the operator T_0 defined by

$$T_0 = \{ \{ \textstyle\sum_{i=1}^n x_{i0}e_{i0}, \sum_{i=1}^n z_i x_{i0} e_{i0} \} : \sum_{i=1}^n x_{i0} = 0 \}$$

is symmetric. In order to solve (6.1) we extend the space $\mathfrak{G}_0$ to the linear space $\mathfrak{G}$ of formal finite sums by

$$\mathfrak{G} = \left\{ \textstyle\sum_{i=1}^n \sum_{k=0}^{k_i} x_{ik} e_{ik} : x_{ik} \in \mathfrak{H} \right\},$$

and provide $\mathfrak{G}$ with the inner product inductively defined by

$$[xe_{ik}, ye_{jl}] = \frac{l[xe_{ik}, ye_{j,l-1}] - k[xe_{i,k-1}, ye_{jl}]}{z_i - \bar{z}_j}, \quad k, l \geq 1,$$

$$[xe_{ik}, ye_{j0}] = \frac{(W_{ik}x, y) - k[xe_{i,k-1}, ye_{j0}]}{z_i - \bar{z}_j}, \quad k \geq 1,$$

and

$$[xe_{i0}, ye_{jl}] = \frac{l[xe_{i0}, ye_{j,l-1}] - (x, W_{jl}y)}{z_i - \bar{z}_j}, \quad l \geq 1.$$

In $\mathfrak{G}$ we extend T_0 to the operator T by

$$\left\{ \left\{ \textstyle\sum_{i=1}^n \sum_{k=0}^{k_i} x_{ik} e_{ik}, \sum_{i=1}^n \sum_{k=0}^{k_i} (z_i x_{ik} e_{ik} + k x_{i,k-1} e_{i,k-1}) \right\} : \sum_{i=1}^n x_{i0} = 0 \right\},$$

where we formally put $x_{i,-1} = 0$, so that

$$T(\textstyle\sum_{i=1}^n x_{i0} e_{i0}) = \sum_{i=1}^n z_i x_{i0} e_{i0} \text{ if } \sum_{i=1}^n x_{i0} = 0,$$

and

$$T(x_{ik}e_{ik}) = z_i x_{ik} e_{ik} + k x_{ik} e_{i,k-1}, \quad k \geq 1.$$

The operator T is symmetric, since for $k, l \geq 1$,

$$\begin{aligned} &[T(x_{ik}e_{ik}), y_{jl}e_{jl}] - [x_{ik}e_{ik}, T(y_{jl}e_{jl})] \\ &= (z_i - \bar{z}_j)[x_{ik}e_{ik}, y_{jl}e_{jl}] + k[xe_{i,k-1}, y_{jl}e_{jl}] - l[x_{ik}e_{ik}, y_{jl}e_{j,l-1}] \\ &= 0, \end{aligned}$$

while, for instance for $l \geq 1$, $\sum_{i=1}^n x_{i0} = 0$ implies that

$$\begin{aligned} &[T(\textstyle\sum_{i=1}^n x_{i0}e_{i0}), y_{jl}e_{jl}] - [\sum_{i=1}^n x_{i0}e_{i0}, T(y_{jl}e_{jl})] \\ &= \textstyle\sum_{i=1}^n ((z_i - \bar{z}_j)[x_{i0}e_{i0}, y_{jl}e_{jl}] - [x_{i0}e_{i0}, l y_{jl} e_{j,l-1}]) \\ &= -(\textstyle\sum_{i=1}^n x_{i0}, W_{jl} y_{jl}) \\ &= 0, \end{aligned}$$

due to the definition of the inner product. Finally, note that the explicit form of the inner product $[\cdot,\cdot]$ is given by

$$(6.3)\quad [xe_{ik}, ye_{jl}] = \sum_{h=0}^{k} \binom{k}{h} \frac{(-1)^{k-h}(k+l-h)!}{(z_i-\bar{z}_j)^{k+l+1-h}} (W_{ih}x, y) + \sum_{h=0}^{l} \binom{l}{h} \frac{(-1)^{l-h}(k+l-h)!}{(\bar{z}_j-z_i)^{k+l+1-h}} (W_{jh}^{*}x, y), \quad x, y \in \mathfrak{H}.$$

Theorem 6.1. *Let $\nu, \pi \in \overline{\mathbb{N}}$, not both equal to ∞. The function $f \in \mathbf{N}_\nu^\pi(\mathfrak{H})$ satisfies (6.1) if and only if there exist a Pontryagin space $\mathfrak{P}$, a selfadjoint relation A in $\mathfrak{P}$, and an isometry $\Phi : \mathfrak{G} \to \mathfrak{P}$ with the following properties:*

(i) *the indices of $\mathfrak{P}$ are (ν, π);*

(ii) *$z_1, \dots, z_n \in \rho(A)$;*

(iii) *A is Γ-minimal, when $\Gamma \in \mathbf{L}(\mathfrak{H}, \mathfrak{P})$ and $\Gamma x = \Phi(xe_{10})$, $x \in \mathfrak{H}$;*

(iv) *A extends $\Phi \circ T$,*

such that

$$(6.4)\quad f(z) = W_{10}^{*} + (z-\bar{z}_1)\Gamma^{*}(I + (z-z_1)(A-z)^{-1})\Gamma.$$

Proof. Let A, $\mathfrak{P}$, and Φ be given such that (i)–(iv) hold. Hence the function f defined by (6.4) is of the form (3.1) with $z_0 = z_1$: $f(z) = f_A(z)$. It follows from (i), (ii), (iii), and Theorem 3.1 that $f \in \mathbf{N}_\nu^\pi(\mathfrak{H})$ with $z_1, \dots, z_n \in \rho(f)$. The inclusion $T_0 \subset T$ and Theorem 5.1 imply that (6.2) holds, i.e. (6.1) holds for $k = 0$. Now assume that $k \geq 1$. By the definition of T,

$$\{xe_{ik}, z_i xe_{ik} + kxe_{i,k-1}\} \in T,$$

and it follows from (iv) that

$$\{\Phi(kxe_{i,k-1}), \Phi(xe_{ik})\} \in (A - z_i)^{-1}.$$

Hence for all $k \geq 1$,

$$k(A - z_i)^{-1}\Phi(xe_{i,k-1}) = \Phi(xe_{ik}),$$

which implies that

$$(6.5)\quad k!(A - z_i)^{-k}\Phi(xe_{i,0}) = \Phi(xe_{ik}).$$

Since $T_0 \subset T$, it follows as in (5.4) that $(I + (z_i - z_1)(A - z_i)^{-1})\Gamma x = \Phi(xe_{i,0})$. Substituting this into (6.5) and using (3.10) we observe,

$$\phi_A^{(k)}(z_i)x = \Phi(xe_{ik}).$$

Hence according to (3.11) with $z = z_i$ and $z_0 = z_1$,

$$\begin{aligned}(f_A^{(k)}(z_i)x, y) &= (z_i - \bar{z_1})[\phi_A^{(k)}(z_i)x, \Gamma y] + k[\phi_A^{(k-1)}(z_i)x, \Gamma y] \\ &= (z_i - \bar{z_1})[\Phi(xe_{ik}), \Phi(ye_{10})] + k[\Phi(xe_{i,k-1}), \Phi(ye_{10})] \\ &= (z_i - \bar{z_1})[xe_{ik}, ye_{10}] + k[xe_{i,k-1}, ye_{10}].\end{aligned}$$

Therefore, using the definition of the inner product in $\mathfrak{G}$, we conclude that

$$(f_A^{(k)}(z_i)x, y) = (W_{ik}x, y).$$

Hence $f = f_A$ satisfies (6.1).

Conversely, assume that a function $f \in \mathbf{N}_\nu^\pi(\mathfrak{H})$ satisfies (6.1) with $z_1, \ldots, z_n \in \rho(f)$. Let $\mathfrak{P}$, A, and Γ be as in Theorem 4.2, so that $f = f_A$ with $z_0 = z_1$ in (3.1), cf. (3.2). This establishes (i), (ii), and the first statement of (iii), that A is Γ-minimal. Define the corresponding function $\phi_A(z)$ as in (3.4) and define $\Phi : \mathfrak{G} \to \mathfrak{P}$ by

$$\Phi(xe_{ik}) = \phi_A^{(k)}(z_i)x.$$

Since f is a solution of (6.1), it follows from Lemma 3.3 with $z = z_i$ and $w = z_j$ and the definition of the inner product in $\mathfrak{G}$, that

$$[\phi_A^{(k)}(z_i)x, \phi_A^{(l)}(z_j)y] = [xe_{ik}, ye_{jl}],$$

and hence Φ is isometric. Clearly, $\Gamma x = \Phi(xe_{10})$, which completes the proof of (iii). Since, in particular, f satisfies (6.2), Theorem 5.1 shows that $\Phi \circ T_0 \subset A$. Now let $k \geq 1$ and let $\{xe_{ik}, z_i xe_{ik} + kxe_{i,k-1}\} \in T$. Then

$$\{\Phi(xe_{ik}), \Phi(z_i xe_{ik} + kxe_{i,k-1})\} = \{\phi_A^{(k)}x, z_i\phi_A^{(k)}(z_i)x + k\phi_A^{(k-1)}(z_i)x\} \in A,$$

by (3.7) of Lemma 3.3. By the definition of T we conclude that $\Phi \circ T \subset A$, showing (iv). □

Acknowledgements

The first author was supported by the EC-programme "Human Capital and Mobility".

References

[1] D. Alpay, P. Bruinsma, A. Dijksma, and H.S.V. de Snoo, "Interpolation problems, extensions of symmetric operators and reproducing kernel spaces I", Oper. Theory: Adv. Appl., 50 (1991), 35–82.

[2] D. Alpay, P. Bruinsma, A. Dijksma, and H.S.V. de Snoo, "Interpolation problems, extensions of symmetric operators and reproducing kernel spaces II", Integral Equations Operator Theory, 14 (1991), 465–500; 15 (1992), 378–388.

[3] D. Alpay, A. Dijksma, J. Rovnyak, and H.S.V. de Snoo, *Schur functions, operator colligations, and reproducing kernel Pontryagin spaces*, Oper. Theory: Adv. Appl., Birkhäuser, Basel-Boston (to appear).

[4] T. Azizov and I.S. Iohvidov, *Linear operators in spaces with an indefinite metric*, John Wiley and Sons, New York 1989.

[5] J.A. Ball, "Interpolation problems of Pick-Nevanlinna and Loewner types for meromorphic matrix functions", Integral Equations Operator Theory, 6 (1983), 804–840.

[6] J. Bognar, *Indefinite inner product spaces*, Springer Verlag, Berlin 1974.

[7] P. Bruinsma, *Interpolation problems for Schur and Nevanlinna pairs*, Doctoral dissertation, Groningen, 1991.

[8] P. Bruinsma, "Degenerate interpolation problems for Nevanlinna pairs", Indag. Math., N.S., 2 (2) (1991), 179–200.

[9] A. Dijksma and H. Langer, "Notes on a Nevanlinna-Pick interpolation problem for generalized Nevanlinna functions", Proceedings of the Workshop on Schur analysis, Leipzig 1994 (to appear).

[10] A. Dijksma and H.S.V. de Snoo, "Symmetric and selfadjoint relations in Kreĭn spaces I", Operator Theory: Adv. Appl., 24 (1987), 145–166.

[11] A. Dijksma and H.S.V. de Snoo, "Symmetric and selfadjoint relations in Kreĭn spaces II", Ann. Acad. Sci. Fenn., Ser. A I Math., 12 (1987), 199–216.

[12] I.S. Iohvidov, M.G. Kreĭn, and H. Langer, *Introduction to the spectral theory of operators in spaces with an indefinite metric*, Akademie Verlag, Berlin 1982.

[13] M.G. Kreĭn and H. Langer, "On defect subspaces and generalized resolvents of a Hermitian operator in the space Π_κ", Funktsional. Anal. i Prilozhen., 5 (2) (1971), 59–71; 5 (3) (1971), 54-69 (Russian); English transl.: Functional Anal. Appl., 5 (1971/1972), 139–146, 217–228.

[14] M.G. Kreĭn and H. Langer, "Über die Q-Funktion eines π-hermiteschen Operators im Raume Π_κ", Acta Sci. Math. (Szeged), 34 (1973), 191–230.

[15] M.G. Kreĭn and H. Langer, "Über einige Fortsetzungsprobleme, die eng mit der Theorie hermitescher Operatoren im Raume Π_κ zusammenhängen. I. Einige Funktionenklassen und ihre Darstellungen", Math. Nachr., 77 (1977), 187–236

[16] B. Sz.-Nagy and A. Koranyi, "Relations d'un problème de Nevanlinna et Pick avec la théorie des opérateurs de l'espace hilbertien", Acta Math. Acad. Sci. Hungaricae, 7 (1957), 295–302.

[17] B. Sz.-Nagy and A. Koranyi, "Operatortheoretische Behandlung und Verallgemeinerung eines Problemkreises in der komplexen Funktionentheorie", Acta Math., 100 (1958), 171–202.

[18] H. Woracek, "Multiple point interpolation in Nevanlinna classes", Integral Equations Operator Theory, 28 (1997), 97–109.

Department of Statistics
University of Helsinki
PL 54
00014 Helsinki
Finland

Department of Mathematics
University of Groningen
Postbus 800
9700 AV Groningen
Nederland

Institut für Analysis, Technische Mathematik und Versicherungsmathematik
Technische Universität Wien
Wiedner Hauptstrasse 8-10/114
A-1040 Wien Österreich

1991 Mathematics Subject Classification. Primary 30E05, 47A57; Secondary 47B25, 47B50

Operator Theory:
Advances and Applications, Vol. 106

On the spectral representation for singular selfadjoint boundary eigenvalue problems

DON HINTON AND ALBERT SCHNEIDER

Dedicated to Professor Heinz Langer on the occasion of his 60th birthday

In this note we prove the spectral representation theorem for selfadjoint realizations A defined by formally selfadjoint (S-hermitian) differential systems and separated selfadjoint boundary conditions. We show, that A is isometrically isomorphic to the operator of multiplication with the independent variable in the Hilbert space $\mathcal{L}^2_\rho(\mathbb{R})$, where the matrix-valued mapping ρ can be calculated via the Stieltjes inversion formula from a Nevanlinna matrix, which is immediately determined by the Titchmarsh-Weyl matrices, defined by the selfadjoint boundary conditions. It is interesting to observe, that analoguously to the Sturm-Liouville operator in case of one regular endpoint (half-line operator) $\rho(t)$ is a $(m \times m)$ matrix, where m is half the order of the system, and in case of two singular endpoints (whole-line operator) $\rho(t)$ is a $(2m \times 2m)$ matrix. In the last section we illustrate our results by applying them to an ordinary 4th order differential operator.

0. Introduction

In the study of boundary value problems associated with Hamiltonian systems, certain difficulties arise when we attempt to associate an operator with the equation. For the moment consider the regular inhomogeneous problem,

$$J_m y' = [\lambda B(x) + C(x)]\, y + B(x) f(x), \quad a \le x \le b, \tag{0.1}$$

$$[\alpha_1, \alpha_2]\, y(a) = 0 = [\beta_1, \beta_2]\, y(b)\,, \tag{0.2}$$

where $B(x) = B^*(x) \ge 0$ and $C(x) = C^*(x)$ are $(2m \times 2m)$ matrices of continuous functions, $J_m = \begin{pmatrix} 0 & -E_m \\ E_m & 0 \end{pmatrix}$, E_m is the $(m \times m)$ identity matrix, $f(x)$ is a $2m$-vector of continuous functions, and α_1, α_2, β_1, β_2 are $(m \times m)$ matrices satisfying the selfadjoint conditions

$$\operatorname{rank}[\alpha_1, \alpha_2] = m = \operatorname{rank}[\beta_1, \beta_2], \quad \alpha_1\alpha_2^* = \alpha_2\alpha_1^*, \quad \beta_1\beta_2^* = \beta_2\beta_1^*\,.$$

If we make the positive definiteness assumption

$$\int_a^b Y^*(x,\lambda) B(x) Y(x,\lambda)\, dx > 0 \tag{0.3}$$

for a fundamental matrix $Y(\cdot,\lambda)$ of the associated homogeneous equation, then a rather complete theory of (0.1)-(0.2) can be given without the introduction of function spaces or operators; for example this development is carried out in the texts by Atkinson [1] and Reid [15]. The assumption (0.3) is needed for a number of properties, e.g., completeness in eigenfunction expansions.

If (0.1) is considered on an infinite interval, difficulties arise with the "equation point of view," and it is desirable to study the equation by associating with it a selfadjoint operator acting in a Hilbert space. Returning to (0.1) a natural space to consider is the space $\mathbb{E}$ consisting of all equivalence classes of Lebesgue measurable functions f satisfying

$$\|f\| := \left(\int_a^b f(x)^* B(x) f(x) dx\right)^{\frac{1}{2}} < \infty .$$

Here f equivalent to g means $\|f-g\| = 0$; denote by $[f]$ the equivalence class containing f. Suppose (0.1) holds for pairs y, f_1 and y, f_2 where y is absolutely continuous on $[a,b]$. Then $B(x)[f_1(x) - f_2(x)] = 0$ a.e. and hence $\|f_1 - f_2\| = 0$. Thus one might hope to define an operator A by $[y] \to [f]$ where y, f is a pair satisfying (0.1)-(0.2). However, such a mapping is in general a relation and not a function. We illustrate this with a simple example. Take $m = 2$ so E_2 is the (2×2) identity matrix, and let $\lambda = 0$,

$$B = \begin{pmatrix} E_2 & 0 \\ 0 & 0 \end{pmatrix}, \quad C = \begin{pmatrix} 0 & c_{12} \\ c_{12}^\top & c_{22} \end{pmatrix}, \quad c_{12} = \begin{pmatrix} -1 & 0 \\ 0 & -2 \end{pmatrix}, \quad c_{22} = \begin{pmatrix} 1 & -1 \\ -1 & 1 \end{pmatrix}.$$

If $y = (y_1, y_2)^\top$, with y_1, y_2 2-vectors, is a solution of (0.1) with $f = 0$, $\lambda = 0$, and $\|y\| = 0$, then $y_1 \equiv 0$, and for y_2 we have $-y_2' = c_{12}y_2$, $0 = c_{22}y_2$. A calculation shows also $y_2 \equiv 0$. Hence $\|y\| = 0$ implies $y(x) = 0$ a.e. so the positive definiteness condition (0.3) holds. Define $f(x) = (1,2,0,0)^\top$ and $y(x) = (0,0,1,1)^\top$. Then y satisfies (0.1) with $\lambda = 0$. Moreover y satisfies (0.2) with $\alpha_1 = \beta_1 = E_2$ and $\alpha_2 = \beta_2 = 0$. Since $\|y\| = 0$ and $\|f\| \neq 0$, we see that in general we cannot define an operator A for (0.1)-(0.2) by $[y] \to [f]$.

In [10] the problem of defining A was solved by first defining a resolvent operator, and then taking the orthogonal complement of the null space of this resolvent. This avoids the problem of the above example. The boundary value problem studied in [10] was singular. In [10] it was shown how to define selfadjoint boundary conditions with one or two singular endpoints. In the case of one singular endpoint b, the resolvent operator was explicitly constructed and the operator A defined and shown to be equivalent to the boundary value problem. A unique feature of this case was defining the Titchmarsh-Weyl matrix $M_b(\lambda)$ directly in terms of the boundary conditions and not through a limiting process applied to regular problems. Indeed, a limit of regular selfadjoint problems may not be selfadjoint as proved in [11] and also in [14]. In [10] $M_b(\lambda)$ was defined for arbitrary, but equal, deficiency indices at the singular endpoint. To complete the theory of [10], it is desirable to prove that the generalized Fourier transform associated with the boundary value problem is a unitary map that makes A unitarily equivalent to the multiplication operator in $\mathcal{L}^2_\rho(\mathbb{R})$ where ρ is the spectral matrix. This would give a theory for singular systems analogous to the theory of the singular Sturm-Liouville operator as developed in chapter 9 of [2].

The work of [10] was carried out for a positive definite S-hermitian system of which (0.1) is a special case. This is a system of the form $Fy = \lambda Gy$ where

$$\begin{aligned}(Fy)(x) &:= C_1(x)y'(x) + D_1(x)y(x)\,,\\ (Gy)(x) &:= D_2(x)y(x)\,,\\ (Sy)(x) &:= D_3(x)y(x)\end{aligned}$$

with $(2m \times 2m)$ matrix-valued coefficients and F is symmetric with respect to S, i.e.,

$$\int_\alpha^\beta (Sz)^*(x)(Fy)(x)dx - \int_\alpha^\beta (Fz)^*(x)(Sy)(x)dx = [y,z](\beta) - [y,z](\alpha)$$

with Lagrange form $[y,z](x) := z^*(x)H(x)y(x)$. Here $H(x)$ can be defined by $H(x) = -C_1^*(x)D_3(x)$ and S-hermitian requires $D_2^*(x)D_3(x) = D_3^*(x)D_2(x)$, $H^*(x) = -H(x)$, and $H'(x) = D_3^*(x)D_1(x) - D_1^*(x)D_3(x)$ and positive definite means $D_3^*(x)D_2(x) \geq 0$. We recall some of the notations of [10]. For an arbitrary interval I with endpoints a and b let $E(I)$ denote the set of all Lebesgue measurable functions from I into $\mathbb{C}^{2m}$ with finite norm where the inner product is

$$(y,z) := \int_I z^*(x)D_3^*(x)D_2(x)y(x)dx\,.$$

Selfadjoint boundary conditions for a function u are defined at the endpoints a, b of I by requiring that certain Lagrange bilinear forms vanish; notationally indicated by $u \in R_\phi^a$, $u \in R_\phi^b$. The resolvent operator R_λ was constructed for the case $I = [a,b)$, a regular, for the problem

$$(F - \lambda G)u = Gv, \quad u \in R_\phi^a \cap R_\phi^b, \quad \mathrm{Im}\lambda \neq 0\,. \tag{0.4}$$

The solution $u_\lambda(\cdot, v) = R_\lambda(v)$ took the form

$$u_\lambda(x,v) = \int_a^b \mathcal{A}(t,x,\lambda)^* D_3^*(t)D_2(t)v(t)dt$$

where

$$\mathcal{A}(t,x,\lambda) = \begin{cases} W_{\overline{\lambda}}(t)\begin{pmatrix} 0 & 0 \\ E_m & M_b(\overline{\lambda})\end{pmatrix} W_\lambda^*(x) \quad , & t \leq x \\ W_{\overline{\lambda}}(t)\begin{pmatrix} 0 & E_m \\ 0 & M_b(\overline{\lambda})\end{pmatrix} W_\lambda^*(x) \quad , & t > x \end{cases},$$

and $W_\lambda(x) = (\Theta(x,\lambda), \Phi(x,\lambda))$ is a fundamental matrix for $Fu = \lambda Gu$ with $W_\lambda(a)$ chosen so that $\Phi(a,\lambda)$ satisfies the boundary conditions at a. By defining $\mathbb{E} := E(I)/N$, $N := \{u \in E(I) | \|u\| = 0\}$ and denoting by π the canonical homomorphism from $E(I)$ onto $\mathbb{E}$, it followed that a resolvent Γ_λ on $\mathbb{E}$ was defined by $\Gamma_\lambda(\pi(u)) := \pi(R_\lambda(u))$. Setting $\mathbb{U}$ to be the kernel of Γ_λ and $W = R^a_\phi \cap R^b_\phi$, it was proved that Γ_λ defines an injective endomorphism from $\mathbb{U}^\perp = \pi(\overline{W})$, and consequently $A := \Gamma_\lambda^{-1} + \lambda \mathrm{id}_{\pi(\overline{W})}$ is a selfadjoint operator in the Hilbert space $\mathbb{H} := \pi(\overline{W})$.

The Titchmarsh-Weyl matrix $M_b(\lambda)$ is a Nevanlinna matrix and has an integral representation of the form

$$M_b(\lambda) = A_b + \lambda B_b + \int_{\mathbb{R}} \left(\frac{1}{t-\lambda} - \frac{t}{1+t^2} \right) d\rho(t)$$

with a matrix-valued weight function ρ which is unique up to normalization. It is this spectral matrix that we will use in proving that A is unitarily equivalent to the operator of multiplication with the independent variable in the Hilbert space $\mathcal{L}^2_\rho(\mathbb{R})$.

It is interesting to observe, that in case of one regular endpoint (half-line operator) analoguously to the Sturm-Liouville operator the number of rows and columns of $\rho(t)$ is identical with half the order of the system and in case of two singular endpoints (whole-line operator) it is identical with the order of the system.

In section 1, we prove theorem 1.20 which contains the representation for the spectral projections E_λ of the operator A and from this theorem we conclude that the generalized Fourier transform defined by formula (1.27) yields a linear mapping into the space $\mathcal{L}^2_\rho(\mathbb{R})$. In section 2, we use this transform to generate a mapping $\hat{T}$, and in theorem 2.13 we show that $\hat{T}$ defines a surjective isometry from the basic Hilbert space $\mathbb{H}$ onto the space $\mathcal{L}^2_\rho(\mathbb{R})$ and $\hat{T}$ transforms the operator A into the operator $\mathcal{M}_{id}$ of multiplication with the independent variable in the space $\mathcal{L}^2_\rho(\mathbb{R})$. Section 2 finishes with some remarks on the inversion formula (2.26) following known arguments of [13].

In section 3, we consider the case of two singular endpoints and we start in proving theorem 3.8, where we derive the representation of the unique solution for the inhomogeneous boundary value problem for nonreal λ using the Titchmarsh-Weyl matrices with respect to both singular endpoints. This representation leads to the definition of a $2m$-dimensional Nevanlinna matrix $F(\lambda)$ given by formula (3.18) and the integral representation in Corollary 3.22 defines a $(2m \times 2m)$ matrix-valued weight function ρ. In section 4, we consider the associated operator A and the main result is theorem 4.16, proving that the generalized Fourier transform T defined by (4.11) generates a surjective isometry $\hat{T}$ from the basic Hilbert space $\mathbb{H}$ onto the Lebesgue-Stieltjes space $\mathcal{L}^2_\rho(\mathbb{R})$. Again $\hat{T}$ transforms A into the operator of multiplication with the independent variable in the space $\mathcal{L}^2_\rho(\mathbb{R})$, that is:

$$A = \hat{T}^{-1} \mathcal{M}_{id} \hat{T}\,.$$

Finally we give a representation for the inverse $\hat{T}^{-1}$ in theorem 4.22.

In section 5 we work out the details in applying the results of sections 3 and 4 to a 4th order differential operator defined on $\mathbb{R}$.

General theories exist for the construction of selfadjoint extensions of a symmetric linear relation in a Hilbert space as can be found in [4] or [5]. Also the treatment of very general boundary conditions, depending analytically on the eigenparameter have been treated in [6,7,8]. These general theories as applied to Hamiltonian systems usually require the limit point hypothesis at a singular point particularly when a Titchmarsh-Weyl matrix is discussed. The surjectivity of the mapping $\hat{T}$ above is similar to that of [3] developed for ordinary differential operators.

In this paper we use all the notations and definitions introduced in [10] and we will not repeat them here. Usually they are clear from the context, or we refer to the corresponding formulas of [10].

1. Representation of the spectral projections E_λ

Let A be the selfadjoint operator defined in [10, section 6] by the selfadjoint resolvent of a singular boundary eigenvalue problem in the Hilbert space $\mathbb{H} = \pi(\overline{W})$. Further let P be the orthogonal projection from $\mathbb{E} = \pi(E(I_b))$, $I_b := [c, b)$, onto the subspace $\mathbb{H}$. For $u, v \in E(I_b)$ we have the relation - for the definition of Γ_λ see [10, eq. (6.1)] - $(\Gamma_\lambda P(\pi(u)), P(\pi(v)) = (R_\lambda u, v)$, and using Stone's formula (see e.g. [17, p.324]) we get for the right continuous spectral projections E_λ of A at points α and β of continuity the representation

$$((E_\beta - E_\alpha)P(\pi(u)), P(\pi(u))) = \lim_{\varepsilon \searrow 0} \left(\frac{1}{\pi} \int_\alpha^\beta \operatorname{Im}(R_{\lambda+i\varepsilon}u, u) d\lambda \right).$$

Using the integral representation of R_λ with the kernel $\mathcal{A}(t, x, \lambda)$ of [10, eqns. (5.5), (5.6)], we get for functions $u \in E(I_b)$ with compact support in $I_b = [c, b)$ the relation

$$\begin{aligned}
(1.1)\quad & ((E_\beta - E_\alpha)P(\pi(u)), P(\pi(u))) \\
& = \lim_{\varepsilon \searrow 0} \left(\frac{1}{2\pi i} \int_\alpha^\beta \int_{I_b} \int_{I_b} u^*(x) D_3^*(x) D_2(x) \times \right. \\
& \qquad \left. \times \left(\mathcal{A}^*(t, x, \lambda + i\varepsilon) - \mathcal{A}^*(t, x, \lambda - i\varepsilon)\right) D_3^*(t) D_2(t) u(t) dt dx d\lambda \right)
\end{aligned}$$

where the order of integration is arbitrary due to the Fubini-Tonelli theorem.

In order to determine the limit in (1.1) we consider for $\varepsilon > 0$ the kernel

$$\begin{aligned}
(1.2)\quad & K_\varepsilon(t, x, \lambda) \\
& := \mathcal{A}^*(t, x, \lambda + i\varepsilon) - \mathcal{A}^*(t, x, \lambda - i\varepsilon) - W_\lambda(x) \begin{pmatrix} 0 & 0 \\ 0 & 2i\operatorname{Im} M_b(\lambda + i\varepsilon) \end{pmatrix} W_\lambda^*(t)
\end{aligned}$$

where $W_\lambda(x)$ is the fundamental matrix of $Fy = \lambda Gy$ introduced in [10, section 4]. We prove

Lemma 1.3. *For all $u \in E(I_b)$ with compact support in $I_b = [c, b)$, we have*

$$(1.4)\quad \lim_{\varepsilon \searrow 0} \left(\int_\alpha^\beta \int_{I_b} \int_{I_b} u^*(x) D_3^*(x) D_2(x) K_\varepsilon(t, x, \lambda) D_3^*(t) D_2(t) u(t) dt dx d\lambda \right) = 0 .$$

Proof. For $u \in E(I_b)$, $\bar{b} \in \mathbb{R}$ with $\operatorname{supp}(u) \subset [c, \bar{b}] \subset I_b$ and $\varepsilon_0 > 0$ we prove for $\varepsilon \in (0, \varepsilon_0]$:

(1.5) K_ε is measurable on $Q := [c, \bar{b}] \times [c, \bar{b}] \times [\alpha, \beta]$,

(1.6) There exists a constant $M > 0$ with $|K_\varepsilon(t, x, \lambda)| \le M$ on $(0, \varepsilon_0] \times Q$,

(1.7) $\lim_{\varepsilon \searrow 0} (K_\varepsilon(t, x, \lambda)) = 0$ a.e. on Q.

To this end we first consider points $(t, x, \lambda) \in Q$ with $t \le x$. Then with respect to (4.19), (5.5) of [10], we have

$$\begin{aligned}(1.8)\quad K_\varepsilon(t, x, \lambda) \;=\; & \varepsilon^{-1}(W_{\lambda+i\varepsilon}(x) - W_\lambda(x)) \begin{pmatrix} 0, & \varepsilon E_m \\ 0, & \varepsilon M_b(\lambda + i\varepsilon) \end{pmatrix} W^*_{\lambda - i\varepsilon}(t) \\ & + W_\lambda(x) \begin{pmatrix} 0, & \varepsilon E_m \\ 0, & \varepsilon M_b(\lambda + i\varepsilon) \end{pmatrix} \varepsilon^{-1}(W^*_{\lambda - i\varepsilon}(t) - W^*_\lambda(t)) \\ & - \varepsilon^{-1}(W_{\lambda - i\varepsilon}(x) - W_\lambda(x)) \begin{pmatrix} 0, & \varepsilon E_m \\ 0, & \varepsilon M_b(\lambda - i\varepsilon) \end{pmatrix} W^*_{\lambda + i\varepsilon}(t) \\ & - W_\lambda(x) \begin{pmatrix} 0, & \varepsilon E_m \\ 0, & \varepsilon M_b(\lambda - i\varepsilon) \end{pmatrix} \varepsilon^{-1}(W^*_{\lambda + i\varepsilon}(t) - W^*_\lambda(t)) .\end{aligned}$$

Obviously, the righthand side is continuous on Q for $\varepsilon \in (0, \varepsilon_0]$, and therefore K_ε is measurable on $Q_1 = \{(t, x, \lambda) \in Q | t \le x\}$. From the equation $FW_{\lambda \pm i\varepsilon} = \lambda G W_{\lambda \pm i\varepsilon} \pm i\varepsilon G W_{\lambda \pm i\varepsilon}$ we get by the method of variation of constants using (1.2), (4.6) of [10] the relation

$$W_{\lambda \pm i\varepsilon}(x) = W_\lambda(x) \left\{ E_{2m} \mp i\varepsilon \int_c^x J_m W^*_\lambda(s) D_3^*(s) D_2(s) W_{\lambda \pm i\varepsilon}(s) ds \right\},$$

and thus the equation

$$(1.9)\quad \varepsilon^{-1}(W_{\lambda \pm i\varepsilon}(x) - W_\lambda(x)) = \mp i W_\lambda(x) \int_c^x J_m W^*_\lambda(s) D_3^*(s) D_2(s) W_{\lambda \pm i\varepsilon}(s) ds .$$

Since $D_3^* D_2$ is locally integrable and $W_\lambda(x)$ is continuous with respect to (x, λ), $\varepsilon^{-1}(W_{\lambda \pm i\varepsilon}(x) - W_\lambda(x))$ is bounded on $(0, \varepsilon_o] \times [c, \bar{b}] \times [\alpha, \beta]$. Further by known theorems on the boundary behaviour of Nevanlinna functions we have for $\lambda \in \mathbb{R}$,

$$(1.10)\quad \lim_{\varepsilon \searrow 0} (\varepsilon \cdot M_b(\lambda + i\varepsilon)) = i(\rho(\lambda + 0) - \rho(\lambda - 0)).$$

Here ρ is the hermitian nondecreasing matrix function in the representation of $M_b(\lambda)$ (see e.g. [10, p. 324]). Further $\varepsilon \cdot M_b(\lambda + i\varepsilon)$ is bounded for $(\varepsilon, \lambda) \in (0, \varepsilon_0] \times [\alpha, \beta]$ and with $\varepsilon \cdot M_b(\lambda - i\varepsilon) = (\varepsilon \cdot M_b^*(\lambda + i\varepsilon))^*$ the same is true for $\varepsilon \cdot M_b(\lambda - i\varepsilon)$. Finally the monotonicity of ρ also proves

$$\lim_{\varepsilon \searrow 0} (\varepsilon \cdot M_b(\lambda + i\varepsilon)) = 0 \text{ a.e. on } [\alpha, \beta]. \tag{1.11}$$

Therefore (1.5) and (1.7) are valid on Q_1 and (1.6) on $(0, \varepsilon_0] \times Q_1$. By (5.10) of [10] we get

$$K_\varepsilon(t, x, \lambda) = (-K_\varepsilon(x, t, \lambda))^*,$$

hence (1.5), (1.6), (1.7) are also true for $t \geq x$.

Now the mapping $D_3^* D_2$ is locally integrable and thus for $u \in E(I_b)$ with compact support in I_b the mapping $D_3^* D_2 u$ is locally integrable and using Fubini-Tonelli's theorem we conclude that $u^*(x) D_3^*(x) D_2(x) K_\varepsilon(t, x, \lambda) D_3^*(t) D_2(t) u(t)$ is integrable on Q. Using (1.6) and (1.7) we can apply Lebesgue's theorem on dominated convergence and hence the assertion (1.4) follows. □

With the partition $W_\lambda(x) = (\Theta(x, \lambda), \Phi(x, \lambda))$ lemma 1.3 together with the polar formula yields

Corollary 1.12. *For $u, v \in E(I_b)$ with compact support in I_b and at points of continuity α and β of E_λ we get from* (1.1) *that*

$$\begin{aligned}
&((E_\beta - E_\alpha) P(\pi(u)), P(\pi(v))) \\
&= \lim_{\varepsilon \searrow 0} \Big(\frac{1}{\pi} \int_\alpha^\beta \int_{I_b} \int_{I_b} v^*(x) D_3^*(x) D_2(x) \Phi(x, \lambda) \mathrm{Im}(M_b(\lambda + i\varepsilon)) \times \\
&\qquad\qquad \times \Phi^*(t, \lambda) D_3^*(t) D_2(t) u(t) dt dx d\lambda \Big) \\
&= \lim_{\varepsilon \searrow 0} \int_\alpha^\beta \left(\int_{I_b} \Phi^*(x, \lambda) D_3^*(x) D_2(x) v(x) dx \right)^* \frac{1}{\pi} \mathrm{Im}(M_b(\lambda + i\varepsilon)) \times \\
&\qquad\qquad \times \left(\int_{I_b} \Phi^*(t, \lambda) D_3^*(t) D_2(t) u(t) dt \right) d\lambda.
\end{aligned}$$

Next we need

Lemma 1.13. *Let*

$$\rho_\varepsilon(\lambda) := \frac{1}{\pi} \int_0^\lambda \mathrm{Im}(M_b(\mu + i\varepsilon)) d\mu \quad \text{and} \quad \tau(\lambda) := \frac{1}{2}(\rho(\lambda + 0) + \rho(\lambda - 0)).$$

Then we have

(1.14) ρ_ε *is continuously differentiable,*

(1.15) $\rho_\varepsilon(\lambda) \to \tau(\lambda)$ *for* $\varepsilon \searrow 0$,

(1.16) $\rho_\varepsilon(\lambda)$ *is hermitian and monotone nondecreasing.*

Proof. (1.14) is obvious by the continuity of $M_b(\mu + i\varepsilon)$ for $\varepsilon > 0$. (1.15) is just the Stieltjes inversion formula and as $\mathrm{Im}(M_b(\mu + i\varepsilon))$ is hermitian and positive semidefinite, (1.16) follows too. □

The elements $(\rho_\varepsilon)_{ij}$ of the matrix ρ_ε are locally of bounded variation and the elements $(\rho_\varepsilon)_{ii}$ of the diagonal are monotone nondecreasing. Hence by (1.15) there exists a constant C, such that for the total variations we have

$$\int_\alpha^\beta |d(\rho_\varepsilon)_{ij}| \leq C \quad (i,j = 1,2,\ldots,m,\ \varepsilon \in (0,\varepsilon_0]). \tag{1.17}$$

Here the assertion is obvious for $i = j$ but this implies the case of arbitrary i and j.

Because of (1.17) we can use a theorem concerning limits with Riemann-Stieltjes integrals. Exactly we have

Lemma 1.18. *If the mappings $h, k : \mathbb{R} \to \mathbb{C}^m$ are continuous and α and β are points of continuity of ρ, then we have*

$$\lim_{\varepsilon \searrow 0} \left(\frac{1}{\pi} \int_\alpha^\beta k^*(\lambda) \mathrm{Im}(M_b(\lambda + i\varepsilon)) h(\lambda) d\lambda \right) = \int_\alpha^\beta k^*(\lambda) d\rho(\lambda) h(\lambda). \tag{1.19}$$

Proof.

$$\begin{aligned}
\int_\alpha^\beta k^*(\lambda) \left(\frac{1}{\pi} \mathrm{Im}(M_b(\lambda + i\varepsilon)) \right) h(\lambda) d\lambda &= \int_\alpha^\beta k^*(\lambda) d(\rho_\varepsilon(\lambda)) h(\lambda) \\
\underset{\varepsilon \searrow 0}{\longrightarrow} \int_\alpha^\beta k^*(\lambda) d\tau(\lambda) h(\lambda) &= \int_\alpha^\beta k^*(\lambda) d\rho(\lambda) h(\lambda).
\end{aligned}$$

For the last equality we have used, that ρ and τ are equal on a dense subset of $[\alpha, \beta]$ including the endpoints α and β. □

For $u \in E(I_b)$ with compact support in I_b the mapping $\tilde{u}$, defined by

$$\tilde{u}(\lambda) := \int_{I_b} \Phi^*(t,\lambda) D_3^*(t) D_2(t) u(t) dt$$

is continuous on $\mathbb{R}$ and thus we finally get

Theorem 1.20. *If $u, v \in E(I_b)$ have compact support in I_b, then for $\alpha, \beta \in \mathbb{R}$ which are points of continuity of E_λ and $\rho(\lambda)$, we have the representation:*

$$((E_\beta - E_\alpha) P(\pi(u)), P(\pi(v))) = \int_\alpha^\beta \tilde{v}^*(\lambda) d\rho(\lambda) \tilde{u}(\lambda). \tag{1.21}$$

Proof. As $\tilde{u}$ and $\tilde{v}$ are continuous on $\mathbb{R}$ we get from 1.12 using (1.19) that

$$\begin{aligned}((E_\beta - E_\alpha)P(\pi(u)), P(\pi(v))) &= \lim_{\varepsilon \searrow 0} \left(\int_\alpha^\beta \tilde{v}^*(\lambda) \left(\frac{1}{\pi} \mathrm{Im} M_b(\lambda + i\varepsilon) \right) \tilde{u}(\lambda) d\lambda \right) \\ &= \int_\alpha^\beta \tilde{v}^*(\lambda) d\rho(\lambda) \tilde{u}(\lambda).\end{aligned}$$

□

Formula (1.21) coincides with the corresponding result (4.13) of [13, p. 139], but the proof there is based on the very general expansion theorem proven by H.D. Niessen in [12, section 6].

Theorem 1.20 is the basis for the following conclusions. If $u \in E(I_b)$ and J is a measurable subset with compact closure in I_b, then $\chi_J \cdot u \in E(I_b)$ with compact support in I_b. Here χ_J is the characteristic function of J. Thus we can apply Theorem 1.20 and we have

Corollary 1.22. *If we define for $u \in E(I_b)$ and a measurable subset J of I_b with compact closure in I_b,*

$$u_J(\lambda) := \int_J \Phi^*(t, \lambda) D_3^*(t) D_2(t) u(t) dt,$$

then for all $u, v \in E(I_b)$ and all measurable subsets J_1, J_2 of I_b with compact closure in I_b and at points α and β of continuity of E_λ and $\rho(\lambda)$ we have

$$((E_\beta - E_\alpha)P(\pi(\chi_{J_1} \cdot u)), P(\pi(\chi_{J_2} \cdot v))) = \int_\alpha^\beta v_{J_2}^*(\lambda) d\rho(\lambda) u_{J_1}(\lambda). \tag{1.23}$$

From (1.23) we get in the same way as in [13, section 4] the corresponding results about the general Fourier transform (integral transform). The space $\mathcal{L}_{\mathcal{P}}^2(\mathbb{R})$ (see [13, section 2]) has to be replaced by the space $\mathcal{L}_\rho^2(\mathbb{R})$. Denoting the scalar product by $(\cdot, \cdot)_\rho$ and the norm by $\| \cdot \|_\rho$ we get from (1.23)

Corollary 1.24. *Let $u \in E(I_b)$ and J be a measurable subset of I_b with compact closure in I_b. Then $u_J \in \mathcal{L}_\rho^2(\mathbb{R})$ and*

$$\|u_J\|_\rho = \|P(\pi(\chi_J \cdot u))\|. \tag{1.25}$$

The proof immediately follows from (1.23) taking the limits $\alpha \to -\infty$ and $\beta \to +\infty$ since it is sufficient to consider the dense set of points of continuity of E_λ and $\rho(\lambda)$.

Analoguously as in [13] we deduce from (1.25)

Theorem 1.26. *Let $u \in E(I_b)$. Then the integral*

$$\int_c^b \Phi^*(t,\lambda)D_3^*(t)D_2(t)u(t)dt$$

is convergent in $\mathcal{L}_\rho^2(\mathbb{R})$. If we define for $u \in E(I_b)$

$$(Tu)(\lambda) := \int_c^b \Phi^*(t,\lambda)D_3^*(t)D_2(t)u(t)dt, \tag{1.27}$$

then (1.27) *defines a linear mapping T from $E(I_b)$ into $\mathcal{L}_\rho^2(\mathbb{R})$ with*

$$\|Tu\|_\rho = \|P(\pi(u))\| \tag{1.28}$$

and

$$\|Tu\|_\rho = \|u\|; \quad u \in \overline{W}. \tag{1.29}$$

Proof. Let $J_1 := [c, b_1]$, $J_2 := [c, b_2]$ be intervals with $b_1 < b_2 < b$. Then u_{J_1} and u_{J_2} belong to $\mathcal{L}_\rho^2(\mathbb{R})$. Since $u_{J_2} - u_{J_1} = u_{J_2 \setminus J_1}$, we get from (1.25)

$$\|u_{J_2} - u_{J_1}\|_\rho = \|P(\pi(\chi_{J_2\setminus J_1} \cdot u))\| \to 0$$

for $J_1 \to I_b$. Hence $u_{[c,\overline{b}]} = \int_c^{\overline{b}} \Phi^*(t,\cdot)D_3^*(t)D_2(t)u(t)dt$ is a Cauchy net in $\mathcal{L}_\rho^2(\mathbb{R})$ and by (1.25) the assertions follow. □

Remark. Using the polar formula we get from (1.28), (1.29) the relations

$$(Tu, Tv)_\rho = (P(\pi(u)), P(\pi(v))); \quad u, v \in E(I_b) \tag{1.30}$$

and

$$(Tu, Tv)_\rho = (u, v); \quad u, v \in \overline{W}. \tag{1.31}$$

Using the integral transformation T, equation (1.21) can be written in the following form:

$$((E_\beta - E_\alpha)P(\pi(u)), P(\pi(v))) = \int_\alpha^\beta (Tv)^*(\lambda)d\rho(\lambda)(Tu)(\lambda), \tag{1.32}$$

valid for $u, v \in E(I_b)$ with compact support in I_b and points α and β of continuity of E_λ and $\rho(\lambda)$. But this representation is also valid for arbitrary $u, v \in E(I_b)$. This can be seen easily by the following argument. Define for $f \in \mathcal{L}_\rho^2(\mathbb{R})$ the seminorm $\phi(f)$ by

$$\phi(f) := \left(\int_\alpha^\beta f^*(\lambda)d\rho(\lambda)f(\lambda)\right)^{\frac{1}{2}}.$$

Obviously

$$|\phi(f)| \le \left(\int_{\mathbb{R}} f^*(\lambda)d\rho(\lambda)f(\lambda)\right)^{\frac{1}{2}} = \|f\|_\rho,$$

proving the continuity of ϕ. Therefore if $u \in E(I_b)$, we choose a sequence $u_n \in E(I_b)$ with compact support in I_b and $\|u_n - u\| \to 0$ for $n \to \infty$. Since T is continuous, we have $\|Tu_n - Tu\|_\rho \to 0$ and thus $\phi(Tu_n) \to \phi(Tu)$, that is,

$$\begin{aligned}\int_\alpha^\beta (Tu)^*(\lambda)d\rho(\lambda)(Tu)(\lambda) &= \lim_{n\to\infty}\left(\int_\alpha^\beta (Tu_n)^*(\lambda)d\rho(\lambda)(Tu_n)(\lambda)\right) \\ &= \lim_{n\to\infty}\left((E_\beta - E_\alpha)P(\pi(u_n)), P(\pi(u_n))\right) \\ &= \left((E_\beta - E_\alpha)P(\pi(u)), P(\pi(u))\right).\end{aligned}$$

Thus by the polar formula (1.32) is then valid for arbitrary $u, v \in E(I_b)$.

Remark. The assumption that α and β have to be points of continuity of E_λ and $\rho(\lambda)$ was used when proving (1.21), since our investigations are based on the Riemann-Stieltjes integral. But (1.32) is valid for arbitrary half-open intervals of the form $(\alpha, \beta]$, if we take the Lebesgue-Stieltjes integral. We will make some remarks on this point.

The mapping ρ from $\mathbb{R}$ into the set of all hermitian $(m\times m)$ matrices is monotone nondecreasing and right continuous. Hence $\tau := \operatorname{trace}(\rho)$ defines a real-valued function on $\mathbb{R}$, which is monotone nondecreasing and right continuous. Thus τ generates a nonnegative σ-additive measure μ_τ on the σ-algebra $\mathbb{B}$ of all Borel sets in $\mathbb{R}$. For the elements ρ_{ij} of the matrix ρ the monotonicity of ρ yields for points $\lambda_1, \lambda_2 \in \mathbb{R}$ with $\lambda_1 < \lambda_2$ the inequality

$$|\rho_{ij}(\lambda_2) - \rho_{ij}(\lambda_1)| \le \tau(\lambda_2) - \tau(\lambda_1). \tag{1.33}$$

Hence the elements of the matrix ρ are complex-valued, right continuous functions locally of bounded variation. Therefore if $J \subset \mathbb{R}$ is a compact interval and $\mathbb{B}(J)$ is the σ-algebra of all Borel sets contained in J, then ρ_{ij} generates a complex measure $\mu_{\rho_{ij}}$ on $\mathbb{B}(J)$ which is absolutely continuous with respect to the nonnegative measure $\mu_{\tau|_{\mathbb{B}(J)}}$. Now the Radon-Nikodym theorem can be applied and since J is arbitrary we get a matrix-valued function ρ' on $\mathbb{R}$, which is locally integrable with respect to μ_τ such that for each bounded Borel set $A \in \mathbb{B}$ we have

$$\mu_\rho(A) = (\mu_{\rho_{ij}}(A)) = \int_A \rho'(\lambda)d\mu_\tau(\lambda).$$

ρ' is uniquely determined up to a μ_τ-nullset, hermitian and positive semidefinite. Especially for $\alpha, \beta \in \mathbb{R}$ with $\alpha < \beta$ we have

$$\mu_\rho((\alpha, \beta]) = \rho(\beta) - \rho(\alpha) = \int_{(\alpha,\beta]} \rho'(\lambda)d\mu_\tau(\lambda). \tag{1.34}$$

Using (1.34) we easily get a representation for the completion $\mathcal{L}^2_\rho(\mathbb{R})$ of the space $C^2_\rho(\mathbb{R})$ as introduced in [13, p. 135]. We consider the space $\mathcal{L}^2_{\rho'}(\mathbb{R}, \mu_\tau)$ of all mappings $f : \mathbb{R} \to \mathbb{C}^m$, which are μ_τ-measurable and $f^*\rho' f \in \mathcal{L}^1(\mathbb{R}, \mu_\tau)$. We define the scalar product

$$(f,g)_\rho := \int_{\mathbb{R}} g^*(\lambda)\rho'(\lambda)f(\lambda)d\mu_\tau(\lambda)$$

and the seminorm

$$\|f\|_\rho := (f,f)_\rho^{\frac{1}{2}}. \tag{1.35}$$

Observe that with the Euclidian norm $\|\cdot\|_2$ in $\mathbb{C}^m$ the relation

$$f^*(\lambda)\rho'(\lambda)f(\lambda) = \|(\rho'(\lambda))^{\frac{1}{2}}f(\lambda)\|_2^2$$

is valid. With respect to the seminorm (1.35) the space $\mathcal{L}^2_{\rho'}(\mathbb{R},\mu_\tau)$ is complete (see e.g. [9] XIII 5.10).

Now if $f:\mathbb{R}\to\mathbb{C}^m$ is continuous and J is a bounded interval, then $\chi_J{\cdot}f$ belongs to $\mathcal{L}^2_{\rho'}(\mathbb{R},\mu_\tau)$ and

$$\begin{aligned}\|\chi_J{\cdot}f\|_\rho^2 &= \int_{\mathbb{R}} (\chi_J{\cdot}f)^*(\lambda)\rho'(\lambda)(\chi_J{\cdot}f)(\lambda)d\mu_\tau(\lambda)\\ &= \int_J f^*(\lambda)\rho'(\lambda)f(\lambda)d\mu_\tau(\lambda).\end{aligned}$$

But now with respect to (1.34) we can show, that the Riemann-Stieltjes integral

$$\int_J f^*(\lambda)d\rho(\lambda)f(\lambda)$$

coincides with the foregoing Lebesgue integral and thus the space $C^2_\rho(\mathbb{R})$ as defined in [13, p. 135] is contained in $\mathcal{L}^2_{\rho'}(\mathbb{R},\mu_\tau)$. Hence $\mathcal{L}^2_\rho(\mathbb{R})$ can be represented by the closure of $C^2_\rho(\mathbb{R})$ in $\mathcal{L}^2_{\rho'}(\mathbb{R},\mu_\tau)$. Further using density arguments it follows, that this closure is given by the mappings of $\mathcal{L}^2_{\rho'}(\mathbb{R},\mu_\tau)$, which are Borel measurable. For details we refer to [9, chap. 13] particularly 5.10.

Now let $u\in E(I_b)$ and define $e_u(\lambda) := (E_\lambda P(\pi(u)), P(\pi(u)))$. e_u is a real-valued function on $\mathbb{R}$, which is monotone nondecreasing and right continuous. Hence e_u generates a nonnegative (finite) measure μ_{e_u} on the σ-algebra $\mathbb{B}$ of all Borel sets in $\mathbb{R}$. The mapping $\nu:\mathbb{B}\to\mathbb{R}$ defined by

$$\begin{aligned}\nu(A) &:= \int_A (Tu)^*(\lambda)d\rho(\lambda)(Tu)(\lambda)\\ &= \int_A (Tu)^*(\lambda)\rho'(\lambda)(Tu)(\lambda)d\mu_\tau(\lambda)\end{aligned}$$

also defines a (finite) nonnegative measure on $\mathbb{B}$ and for all intervals $[\alpha,\beta]\in\mathbb{B}$ with α and β from the dense set of points of continuity of E_λ and $\rho(\lambda)$, (1.32) yields that

$$\mu_{e_u}([\alpha,\beta]) = \nu([\alpha,\beta]). \tag{1.36}$$

But then μ_{e_u} and ν coincide on all of $\mathbb{B}$ and for $J = (\alpha, \beta]$ we have

$$((E_\beta - E_\alpha)P(\pi(u)), P(\pi(u))) = \int_J (Tu)^*(\lambda)d\rho(\lambda)(Tu)(\lambda).$$

Now again with the polar formula we get for $u, v \in E(I_b)$ the equation

$$\begin{aligned} ((E_\beta - E_\alpha)P(\pi(u)), \pi(v)) &= ((E_\beta - E_\alpha)P(\pi(u)), P(\pi(v))) \\ &= \int_{(\alpha,\beta]} (Tv)^*(\lambda)d\rho(\lambda)(Tu)(\lambda) \\ &= \left(\chi_{(\alpha,\beta]}\cdot Tu, Tv\right)_\rho \\ &= \left(\chi_{(\alpha,\beta]}\cdot Tu, \chi_{(\alpha,\beta]}\cdot Tv\right)_\rho . \end{aligned} \tag{1.37}$$

Here $u = v$ yields

$$\|(E_\beta - E_\alpha)P(\pi(u))\|^2 = \|\chi_{(\alpha,\beta]}\cdot Tu\|_\rho^2 . \tag{1.38}$$

2. The spectral representation of the operator A

For $u_1, u_2 \in E(I_b)$ with $\pi(u_1) = \pi(u_2)$ we have from (1.28) that $Tu_1 = Tu_2$. Hence T induces a linear mapping $\hat{T}$ from $\mathbb{E} = \pi(E(I_b))$ into the space $\mathcal{L}^2_\rho(\mathbb{R})$ with

$$\hat{T}\pi(u) = Tu; \quad u \in E(I_b) \tag{2.1}$$

and

$$(\hat{T}x, \hat{T}y)_\rho = (Px, Py); \quad x, y \in \mathbb{E}. \tag{2.2}$$

We prove

Lemma 2.3. *For $w \in \mathbb{E}$ and $J := (\alpha, \beta]$ the equation*

$$\chi_J\cdot\hat{T}w = \hat{T}(E_\beta - E_\alpha)Pw \tag{2.4}$$

is valid.

Proof. If $u, v \in E(I_b)$ we get using (1.37)

$$\begin{aligned} \left(\hat{T}(E_\beta - E_\alpha)P(\pi(u)), \hat{T}\pi(v)\right)_\rho &= ((E_\beta - E_\alpha)P(\pi(u)), P(\pi(v))) \\ &= ((E_\beta - E_\alpha)P(\pi(u)), \pi(v)) \\ &= \left(\chi_J\cdot Tu, Tv\right)_\rho . \end{aligned}$$

Therefore

$$\chi_J\cdot Tu = \hat{T}(E_\beta - E_\alpha)P(\pi(u)) + \xi$$

with $\xi \in T(E(I_b))^{\perp}$ and respecting (1.38) we get

$$\begin{aligned} \|\chi_J \cdot Tu\|_\rho^2 &= \|\hat{T}(E_\beta - E_\alpha)P(\pi(u))\|_\rho^2 + \|\xi\|_\rho^2 \\ &= \|(E_\beta - E_\alpha)P(\pi(u))\|^2 + \|\xi\|_\rho^2 \\ &= \|\chi_J \cdot Tu\|_\rho^2 + \|\xi\|_\rho^2. \end{aligned}$$

Hence $\xi = 0$ and Lemma 2.3 is proved. □

In the following we denote by

$$\mathcal{M}_{id}$$

the operator of multiplication with the independent variable λ, that is: $(\mathcal{M}_{id}f)(\lambda) := \lambda f(\lambda)$ for $f \in \mathcal{L}_\rho^2(\mathbb{R})$. The domain will be denoted by $D_{\mathcal{M}_{id}}$. Then:

Theorem 2.5. *For $w \in D_A$ we have $\hat{T}w \in D_{\mathcal{M}_{id}}$ and*

$$\|\mathcal{M}_{id}\hat{T}w\|_\rho = \|\hat{T}(Aw)\|_\rho.$$

Proof. Choose $w_1 \in W$ with $w = \pi(w_1)$. Then

$$\|\hat{T}(Aw)\|_\rho^2 = \|Aw\|^2 = \int_{\mathbb{R}} \lambda^2 d(E_\lambda \pi(w_1), \pi(w_1)).$$

For every Borel set $B \in \mathbb{B}$ we have

$$\begin{aligned} \mu_{e_{w_1}}(B) &= \int_B (Tw_1)^*(\lambda) d\rho(\lambda)(Tw_1)(\lambda) \\ &= \int_B \|(\rho'(\lambda))^{\frac{1}{2}}(Tw_1)(\lambda)\|_2^2 d\mu_\tau(\lambda). \end{aligned}$$

Hence

$$\begin{aligned} \int_{\mathbb{R}} \lambda^2 d(E_\lambda \pi(w_1), \pi(w_1)) &= \int_{\mathbb{R}} \lambda^2 d\mu_{e_{w_1}}(\lambda) \\ &= \int_{\mathbb{R}} \lambda^2 \|\rho'(\lambda)^{\frac{1}{2}}(Tw_1)(\lambda)\|_2^2 d\mu_\tau(\lambda) \\ &= \int_{\mathbb{R}} \|(\rho'(\lambda))^{\frac{1}{2}}(\lambda(Tw_1)(\lambda))\|_2^2 d\mu_\tau(\lambda). \end{aligned}$$

Thus $\mathcal{M}_{id}\hat{T}w \in \mathcal{L}_\rho^2(\mathbb{R})$ and

$$\|\mathcal{M}_{id}\hat{T}w\|_\rho^2 = \|Aw\|^2 = \|\hat{T}(Aw)\|_\rho^2,$$

and the proof of 2.5 is complete. □

Remark. For $v = \pi(v_1) \in D_A$ we have

$$\begin{aligned}(Av, v) &= \int_{\mathbb{R}} \lambda d(E_\lambda v, v)\\ &= \int_{\mathbb{R}} \lambda d\mu_{e_{v_1}}(\lambda)\\ &= \int_{\mathbb{R}} \lambda (\hat{T}v)^*(\lambda) d\rho(\lambda)(\hat{T}v)(\lambda)\\ &= \left(\mathcal{M}_{id}\hat{T}v, \hat{T}v\right)_\rho .\end{aligned}$$

Then the polar formula shows for $w, v \in D_A$ the equation

$$(2.6) \qquad (Av, w) = \left(\mathcal{M}_{id}\hat{T}v, \hat{T}w\right)_\rho .$$

Since P is the orthogonal projection from $\mathbb{E}$ onto $\mathbb{H}$, we get for $v \in D_A$, $w \in \mathbb{E}$ that

$$(Av, w) = (Av, Pw),$$

and with respect to (2.2),

$$\hat{T}w = \hat{T}(Pw).$$

Hence equation (2.6) is valid for $v \in D_A$ and $w \in \mathbb{E}$.

Now with the same arguments as used in Lemma 2.3 we prove

Theorem 2.7. *For $w \in D_A$ we have*

$$(2.8) \qquad \hat{T}(Aw) = \mathcal{M}_{id}\hat{T}w .$$

Proof. Let $v \in \mathbb{E}$. Then

$$\begin{aligned}\left(\hat{T}(Aw), \hat{T}v\right)_\rho &= (Aw, Pv) = (Aw, v)\\ &= \left(\mathcal{M}_{id}\hat{T}w, \hat{T}v\right)_\rho .\end{aligned}$$

Thus

$$\mathcal{M}_{id}\hat{T}w = \hat{T}(Aw) + \xi$$

with $\xi \in (\hat{T}(\mathbb{E}))^\perp = (T(E(I_b)))^\perp$ and from Theorem 2.5 we get

$$\begin{aligned}\|\hat{T}(Aw)\|_\rho^2 &= \|\mathcal{M}_{id}\hat{T}w\|_\rho^2\\ &= \|\hat{T}(Aw)\|_\rho^2 + \|\xi\|_\rho^2.\end{aligned}$$

Hence $\xi = 0$ and the assertion follows. □

By means of Lemma 2.3 we now obtain

Corollary 2.9. *For $J = (\alpha, \beta]$ and $u \in \mathbb{E}$ we have the relations*

$$\chi_J \cdot \hat{T}u \in \hat{T}(\mathbb{E}) \tag{2.10}$$

and

$$\mathcal{M}_{id}\chi_J \cdot (\hat{T}u) = \mathcal{M}_{id}\hat{T}(E_\beta - E_\alpha)Pu \in \hat{T}(\mathbb{E}). \tag{2.11}$$

Now we can prove, that the mapping $\hat{T}$ is surjective and we want to point out, that the definiteness assumption III of [10] with respect to the interval I_b is essential for the proof. Since the range of $\hat{T}$ is closed in $\mathcal{L}^2_\rho(\mathbb{R})$ it is sufficient to show

Theorem 2.12. *If $\xi \in \mathcal{L}^2_\rho(\mathbb{R})$ and $(\hat{T}w, \xi)_\rho = 0$ for all $w \in \mathbb{E}$, then $\xi = 0$.*

Proof. We choose a point $\bar{b} \in \operatorname{int}(I_b)$ such that

$$\int_c^{\bar{b}} W_0^*(x) D_3^*(x) D_2(x) W_0(x) dx > 0,$$

which is possible by the definiteness assumption III of [10]. Then for $y \in \mathbb{C}^m$ let $k \in \mathbb{C}^{2m}$ be such that

$$\begin{pmatrix} y \\ 0 \end{pmatrix} + \left(\int_c^{\bar{b}} J_m^{-1} W_0^*(x) D_3^*(x) D_2(x) W_0(x) dx \right) k = 0.$$

The mapping $h : I_b \to \mathbb{C}^{2m}$, defined by

$$h(x) := \begin{cases} W_0(x) \cdot k, & x \in [c, \bar{b}) \\ 0, & x \geq \bar{b} \end{cases}$$

is contained in $E(I_b)$. Define f by

$$f(x) := W_0(x) \left\{ \begin{pmatrix} y \\ 0 \end{pmatrix} + \int_c^x J_m^{-1} W_0^*(s) D_3^*(s) D_2(s) h(s) ds \right\}.$$

Then f is locally absolutely continuous in I_b and fulfills the equation $Ff = Gh$. Further we have $f(c) = \Theta(c, 0)y = \Theta(c, \lambda)y$ and $f(x) \equiv 0$ for $x \geq \bar{b}$. For this f and h we get

$$\begin{aligned} \lambda \cdot (Tf)(\lambda) &= \int_c^{\bar{b}} \lambda \Phi^*(t, \lambda) D_3^*(t) D_2(t) f(t) dt \\ &= \int_c^{\bar{b}} (\lambda D_2(t) \Phi(t, \lambda))^* (D_3(t) f(t)) dt \end{aligned}$$

$$\begin{aligned}
&= \int_c^{\bar{b}} ((F\Phi)(t,\lambda))^*(Sf)(t)dt \\
&= \int_c^{\bar{b}} ((S\Phi)(t,\lambda))^*(Ff)(t)dt + [f,\Phi(\cdot,\lambda)](c) \\
&= \int_c^{\bar{b}} \Phi^*(t,\lambda)D_3^*(t)D_2(t)h(t)dt + \Phi^*(c,\lambda)H(c)\Theta(c,\lambda)y \\
&= (Th)(\lambda) + y
\end{aligned}$$

from equation (4.5) of [10]. Now let $J := (\alpha,\beta]$ be an arbitrary half-open interval in $\mathbb{R}$. Then we get

$$\mathcal{M}_{id}(\chi_J\cdot\hat{T}(\pi(f))) = \chi_J\cdot\hat{T}(\pi(h)) + \chi_J\cdot y\,,$$

and therefore with respect to (2.10), (2.11) we have

$$(\chi_J\cdot y,\xi)_\rho = \left(\mathcal{M}_{id}(\chi_J\cdot\hat{T}(\pi(f))),\xi\right)_\rho - \left(\chi_J\cdot\hat{T}(\pi(h)),\xi\right)_\rho = 0\,.$$

Thus the continuous linear functional $(\cdot,\xi)_\rho$ vanishes on the dense subspace of stepfunctions in $\mathcal{L}^2_\rho(\mathbb{R})$, hence $\xi = 0$ and the proof is accomplished. □

As a consequence of the foregoing arguments we have

Theorem 2.13. *The mapping $\hat{T}$ from $\mathbb{E}$ into $\mathcal{L}^2_\rho(\mathbb{R})$ is a linear and surjective mapping onto $\mathcal{L}^2_\rho(\mathbb{R})$ with*

$$\|\hat{T}u\|_\rho = \|Pu\|;\quad u\in\mathbb{E} \tag{2.14}$$

$$\hat{T}(Aw) = \mathcal{M}_{id}\hat{T}w;\quad w\in D_A \tag{2.15}$$

$$\hat{T}(E_\beta - E_\alpha)Pw = \chi_{(\alpha,\beta]}\cdot\hat{T}w;\quad w\in\mathbb{E}. \tag{2.16}$$

Especially for the subspace $\mathbb{H} = P(\mathbb{E})$ the restriction of $\hat{T}$ to $\mathbb{H}$ is a surjective isometry from $\mathbb{H}$ onto $\mathcal{L}^2_\rho(\mathbb{R})$ with

$$E_\beta - E_\alpha = \hat{T}^{-1}\chi_{(\alpha,\beta]}\cdot\hat{T}\,, \tag{2.17}$$

$$A = \hat{T}^{-1}\mathcal{M}_{id}\hat{T}\,. \tag{2.18}$$

We shall conclude this section with some remarks on the representation of $\hat{T}^{-1}$, which defines the inversion formula for the integral transform $\hat{T}$. The arguments are already given in [13], and thus we can confine ourself to the essentials.

The starting point is the relation (1.37). For $u, v\in E(I_b)$, compact subintervals $J, \tilde{J}$ of I_b and an interval $J_1 = (\alpha,\beta]\subset\mathbb{R}$ we have the equation

$$((E_\beta - E_\alpha)P(\pi(\chi_J\cdot u)),\pi(\chi_{\tilde{J}}\cdot v)) = \int_{J_1} (T(\chi_{\tilde{J}}\cdot v))^*(\lambda)d\rho(\lambda)(T(\chi_J\cdot u))(\lambda) \tag{2.19}$$

$$= \int_{J_1}\left(\int_{\tilde{J}} v^*(s)D_3^*(s)D_2(s)\Phi(s,\lambda)ds\right)d\rho(\lambda)\left(\int_J \Phi^*(t,\lambda)D_3^*(t)D_2(t)u(t)dt\right)$$

$$= \int_{\tilde J} v^*(s) D_3^*(s) D_2(s) \left(\int_{J_1} \Phi(s,\lambda) d\rho(\lambda) \int_J \Phi^*(t,\lambda) D_3^*(t) D_2(t) u(t) dt \right) ds$$
$$= \int_{\tilde J} v^*(s) D_3^*(s) D_2(s) h_{J_1,J}(s) ds$$

where

$$\text{(2.20)} \qquad h_{J_1,J}(s) := \int_{J_1} \Phi(s,\lambda) d\rho(\lambda) \int_J \Phi^*(t,\lambda) D_3^*(t) D_2(t) u(t) dt = \left(T(\chi_J \cdot u), \chi_{J_1} \cdot \Phi^*(s,\cdot) \right)_\rho .$$

Obviously $h_{J_1,J}$ is continuous and therefore $w := \chi_{\tilde J} \cdot h_{J_1,J} \in E(I_b)$. Then from (2.19) we get

$$\int_{\tilde J} h_{J_1,J}^*(s) D_3^*(s) D_2(s) h_{J_1,J}(s) ds = \|w\|^2 = ((E_\beta - E_\alpha) P(\pi(\chi_J \cdot u)), \pi(w)) \le \|u\| \cdot \|w\|$$

and therefore

$$\|w\| = \|\chi_{\tilde J} \cdot h_{J_1,J}\| \le \|u\| .$$

Since $\tilde J$ is arbitrary, $h_{J_1,J} \in E(I_b)$ and $\|h_{J_1,J}\| \le \|u\|$. Further from (2.19) we have

$$\Big((E_\beta - E_\alpha) P(\pi(\chi_J \cdot u)), \pi(\chi_{\tilde J} \cdot v) \Big) = (h_{J_1,J}, \chi_{\tilde J} \cdot v) = (\pi(h_{J_1,J}), \pi(\chi_{\tilde J} \cdot v))$$

and taking the limit $\tilde J \to I_b$, we have for all $v \in E(I_b)$

$$((E_\beta - E_\alpha) P(\pi(\chi_J \cdot u)) - \pi(h_{J_1,J}), \pi(v)) = 0 ,$$

hence

$$\text{(2.21)} \qquad (E_\beta - E_\alpha) P(\pi(\chi_J \cdot u)) = \pi(h_{J_1,J}) .$$

Now with $J \to I_b$ we have $\|\chi_J \cdot u - u\| \to 0$ and by continuity of T this implies $\|T(\chi_J \cdot u) - Tu\|_\rho \to 0$. Thus from (2.20) we get

$$h_{J_1,J}(s) \to h_{J_1}(s) := \left(Tu, \chi_{J_1} \cdot \Phi^*(s,\cdot) \right)_\rho = \int_{J_1} \Phi(s,\lambda) d\rho(\lambda) (Tu)(\lambda) ,$$

and the convergence is locally uniform. Therefore h_{J_1} is continuous and for each compact subinterval $\tilde J \subset I_b$ we have $\|\chi_{\tilde J} \cdot h_{J_1,J} - \chi_{\tilde J} \cdot h_{J_1}\| \to 0$ with $J \to I_b$. Observe that

$$\|\chi_{\tilde J} \cdot h_{J_1,J}\| \le \|h_{J_1,J}\| \le \|u\|$$

and therefore

$$\|\chi_{\tilde{J}} h_{J_1}\| = \lim_{J\to I_b} \|\chi_{\tilde{J}} h_{J_1,J}\| \leq \|u\|$$

proving that $h_{J_1} \in E(I_b)$. Formula (2.21) implies

$$\|\pi(h_{J_1,J}) - (E_\beta - E_\alpha)P(\pi(u))\| \to 0 \quad \text{for } J \to I_b\,. \tag{2.22}$$

Now take $z \in W$ with $(E_\beta - E_\alpha)P(\pi(u)) = \pi(z)$. Then for each compact subinterval $\tilde{J} \subset I_b$ we have the estimation

$$\begin{aligned}\|\pi(\chi_{\tilde{J}} h_{J_1}) - \pi(\chi_{\tilde{J}} z)\| &\leq \|\pi(\chi_{\tilde{J}} h_{J_1}) - \pi(\chi_{\tilde{J}} h_{J_1,J})\| + \|\pi(\chi_{\tilde{J}} h_{J_1,J}) - \pi(\chi_{\tilde{J}} z)\| \\ &\leq \|\chi_{\tilde{J}} h_{J_1} - \chi_{\tilde{J}} h_{J_1,J}\| + \|\pi(h_{J_1,J}) - \pi(z)\| \\ &\leq \|\chi_{\tilde{J}} h_{J_1} - \chi_{\tilde{J}} h_{J_1,J}\| + \|\pi(h_{J_1,J}) - (E_\beta - E_\alpha)P(\pi(u))\|.\end{aligned}$$

Now the righthand side of this inequality converges to zero for $J \to I_b$ and therefore

$$\|\pi(\chi_{\tilde{J}} h_{J_1}) - \pi(\chi_{\tilde{J}} z)\| = 0\,,$$

and then taking the limits for $\tilde{J} \to I_b$ we have the representation

$$\pi(h_{J_1}) = \pi(z) = (E_\beta - E_\alpha)P(\pi(u))\,. \tag{2.23}$$

Here we take $J_1 \to \mathbb{R}$ and we get

$$\|\pi(h_{J_1}) - P(\pi(u))\| \to 0\,. \tag{2.24}$$

For $u \in \overline{W}$ we have $P(\pi(u)) = \pi(u)$, and then (2.24) yields that

$$\int_{\mathbb{R}} \Phi(s,\lambda) d\rho(\lambda)(Tu)(\lambda) \tag{2.25}$$

converges in $E(I_b)$ to $u(s)$, which may be written as

$$(\hat{T}^{-1}v)(s) = \int_{\mathbb{R}} \Phi(s,\lambda) d\rho(\lambda) v(\lambda)\,; \quad v \in \mathcal{L}^2_\rho(\mathbb{R})\,. \tag{2.26}$$

3. The case of two singular endpoints

In this section we consider the formally selfadjoint (S-hermitian) system $Fy = \lambda Gy$ on an interval $I = (a,b)$ with two singular endpoints a and b, and we will start our investigations with some remarks on the case of a half-open interval of the form $I_a := (a,c]$ with singular endpoint a and regular endpoint c. Corresponding to the general assumptions of [10, section 1] we assume that

$$\tau_a(i) = \tau_a(-i) =: \tau_a\,, \quad s = t = m\,,$$

and that the definiteness condition III for the interval I_a is given. Let

$$\varphi_1^a, \ldots, \varphi_{\tau_a}^a\,; \quad \varphi_1^c, \ldots, \varphi_m^c$$

be separated selfadjoint boundary conditions in a and c respectively and let $W_\lambda(x)$ be the fundamental matrix of $Fy = \lambda Gy$ with $W_\lambda(c) = A$, as defined in [10, section 4]. Then with the partition $W_\lambda(x) = (\Theta(x,\lambda), \Phi(x,\lambda))$ we get analoguously to Theorem 4.7 of [10].

Theorem 3.1. *There exists a unique $(m \times m)$ matrix $M_a(\lambda)$ defined on $\mathbb{C}\backslash\mathbb{R}$ with the following properties:*

(3.2) $\Theta(\cdot,\lambda) + \Phi(\cdot,\lambda)M_a(\lambda) \in (E(I_a))^m$ *for all* $\lambda \in \mathbb{C}\backslash\mathbb{R}$

(3.3) $[\Theta(\cdot,\lambda) + \Phi(\cdot,\lambda)M_a(\lambda), \varphi_\nu^a](a) = 0 \quad (\nu = 1,2,\ldots,\tau_a)$

(3.4) $[\Theta(\cdot,\lambda) + \Phi(\cdot,\lambda)M_a(\lambda), \Theta(\cdot,\mu) + \Phi(\cdot,\mu)M_a(\mu)](a) = 0; \quad \lambda,\mu \in \mathbb{C}\backslash\mathbb{R}.$

The matrix $M_a(\lambda)$ is holomorphic on $\mathbb{C}\backslash\mathbb{R}$, and it holds that $M_a^*(\lambda) = M_a(\overline{\lambda})$ and $-(\mathrm{Im}\lambda)^{-1}\mathrm{Im}M_a(\lambda) > 0$; hence $M_a(\lambda)$ is invertible.

The proof of Theorem 3.1 follows step by step that of Theorem 4.7 changing the corresponding denotations with the only difference, that here from Green's formula the relation

$$-(\mathrm{Im}\lambda)^{-1}\mathrm{Im}M_a(\lambda) = \|\Theta(\cdot,\lambda) + \Phi(\cdot,\lambda)M_a(\lambda)\|_{I_a}^2 \tag{3.5}$$

follows.

Now we consider the interval $I := (a,b)$ and we assume as in 3.1 of [10] that

$$\tau_a(i) = \tau_a(-i) =: \tau_a; \quad \tau_b(i) = \tau_b(-i) =: \tau_b \tag{3.6}$$

and further

$$s = t = m\,. \tag{3.7}$$

From (3.7) we have $n = 2m$ and thus the differential system is of even order. This condition always holds for real systems.

Remark. From (3.6) we cannot conclude that (3.7) is fulfilled. This can be seen by the simple one-dimensional system

$$-iy' = \lambda y \text{ on } I = (-\infty, +\infty)\,.$$

This equation is S-hermitian with $Sy = y$ and $H(x) \equiv -i$. Since $i \cdot H(x) = 1$ we have $s = 0$ and $t = 1$. Further $y(x,\lambda) = \exp(i\lambda x)$ is the fundamental matrix with $y(0,\lambda) = 1$. Hence the matrix $A(x,\lambda)$ defined by (2.1) of [10] is

$$A(x,i) = -\frac{1}{2}\exp(-2x); \quad A(x,-i) = \frac{1}{2}\exp(2x)\,.$$

Therefore

$$\tau_{-\infty}(i) = \tau_{-\infty}(-i) = 0; \quad \tau_\infty(i) = \tau_\infty(-i) = 0\,.$$

Hence (3.6) is true but (3.7) does not hold.

Now let $c \in \mathbb{R}$ be a point with $a < c < b$ and we assume, that the definiteness condition is given for both the intervals $I_a = (a, c]$ and $I_b = [c, b)$. We choose selfadjoint boundary conditions

$$\varphi_1^a, \ldots, \varphi_{\tau_a}^a; \quad \varphi_1^b, \ldots, \varphi_{\tau_b}^b; \quad \varphi_1^c, \ldots, \varphi_m^c$$

in a, b, and c respectively and we take the fundamental matrix $W_\lambda(x) = (\Theta(x,\lambda), \Phi(x,\lambda))$ with initial condition defined by the boundary conditions in c. Then the Titchmarsh-Weyl matrices $M_a(\lambda)$ and $M_b(\lambda)$ are uniquely determined by these data.

In order to define the selfadjoint operator generated by the system and the boundary conditions in a and b, we prove the following.

Theorem 3.8. *Let $\lambda \in \mathbb{C}\backslash\mathbb{R}$. Then for each $v \in E(I)$ there exists a uniquely determined solution $u_\lambda(\cdot, v)$ of the inhomogeneous equation*

$$(F - \lambda G)u = Gv \tag{3.9}$$

with

$$u_\lambda(\cdot, v) \in R_\varphi^a \cap R_\varphi^b . \tag{3.10}$$

This solution is given by

$$\begin{aligned} u_\lambda(x,v) &= W_\lambda(x) \int_a^x P(\lambda) W_{\overline{\lambda}}^*(t) D_3^*(t) D_2(t) v(t) dt \\ &\quad + W_\lambda(x) \int_x^b Q(\lambda) W_{\overline{\lambda}}^*(t) D_3^*(t) D_2(t) v(t) dt \end{aligned} \tag{3.11}$$

with

$$P(\lambda) = \begin{pmatrix} (M_a(\lambda) - M_b(\lambda))^{-1} & ; & (M_a(\lambda) - M_b(\lambda))^{-1} M_a(\lambda) \\ M_b(\lambda)(M_a(\lambda) - M_b(\lambda))^{-1} & ; & M_b(\lambda)(M_a(\lambda) - M_b(\lambda))^{-1} M_a(\lambda) \end{pmatrix}$$

$$Q(\lambda) = \begin{pmatrix} (M_a(\lambda) - M_b(\lambda))^{-1} & ; & (M_a(\lambda) - M_b(\lambda))^{-1} M_b(\lambda) \\ M_a(\lambda)(M_a(\lambda) - M_b(\lambda))^{-1} & ; & M_a(\lambda)(M_a(\lambda) - M_b(\lambda))^{-1} M_b(\lambda) \end{pmatrix}.$$

The proof is in all details nearly identical with that of Theorem 5.1 of [10]. Therefore we will confine ourselves to the essential points.

First observe, that the matrix $M_a(\lambda) - M_b(\lambda)$ is invertible since

$$(\mathrm{Im}\lambda)^{-1} \cdot \mathrm{Im}(M_a(\lambda) - M_b(\lambda)) = (\mathrm{Im}\lambda)^{-1}\mathrm{Im}M_a(\lambda) - (\mathrm{Im}\lambda)^{-1}\mathrm{Im}M_b(\lambda) < 0 .$$

Further the integrals in (3.11) exist due to the behaviour of $\Theta(\cdot,\lambda) + \Phi(\cdot,\lambda)M_a(\lambda)$ and $\Theta(\cdot,\lambda) + \Phi(\cdot,\lambda)M_b(\lambda)$ near a and b respectively. By a simple calculation we get the relation

$$P(\lambda) - Q(\lambda) = -J_m \tag{3.12}$$

from which the equation

$$(3.13)\quad u'_\lambda(x,v) = W'_\lambda(x)W_\lambda(x)^{-1}u_\lambda(x,v) + H(x)^{-1}D_3^*(x)D_2(x)v(x) \quad \text{(a.e.)}$$

follows. Hence $(F-\lambda G)u_\lambda = Gv$ as in [10, p. 337].

To prove that $u_\lambda(\cdot,v) \in R_\varphi^a \cap R_\varphi^b$, we start with an element $v \in E(I)$ with compact support contained in $\text{int}(I)$. Then by definition of $M_a(\lambda)$ and $M_b(\lambda)$ (3.10) is clear. Further we get the inequality

$$(3.14)\qquad \|u_\lambda(\cdot,v)\|_I \leq |\text{Im}\lambda|^{-1}\|v\|_I$$

and with the aid of this inequality (3.10) can be extended to all $v \in E(I)$. For the details we may refer to [10, section 5].

Remark 3.15. If we define the Green's matrix $\mathcal{G}(t,x,\lambda)$ by

$$\mathcal{G}(t,x,\lambda) := \begin{cases} W_{\overline{\lambda}}(t)P^*(\lambda)W_{\overline{\lambda}}^*(x) = W_{\overline{\lambda}}(t)Q(\overline{\lambda})W_{\overline{\lambda}}^*(x); & t \leq x \\ W_{\overline{\lambda}}(t)Q^*(\lambda)W_{\overline{\lambda}}^*(x) = W_{\overline{\lambda}}(t)P(\overline{\lambda})W_{\overline{\lambda}}^*(x); & t > x \end{cases}$$

then the solution $u_\lambda(\cdot,v)$ can be written as

$$(3.16)\qquad u_\lambda(x,v) = (v, \mathcal{G}(\cdot,x,\lambda))_I\,.$$

Further we get

$$(3.17)\qquad \mathcal{G}^*(t,x,\lambda) = \mathcal{G}(x,t,\overline{\lambda})\,.$$

Now consider the matrix $F(\lambda)$ defined on $\mathbb{C}\backslash\mathbb{R}$ by

$$(3.18)\qquad F(\lambda) := P(\lambda) + \frac{1}{2}J_m = Q(\lambda) - \frac{1}{2}J_m\,,$$

where the last equality comes from (3.12). Then we have

Theorem 3.19. *The matrix $F(\lambda)$ is holomorphic in $\mathbb{C}\backslash\mathbb{R}$ and fulfills the relation*

$$(\text{Im}\lambda)^{-1}\text{Im}(F(\lambda)) > 0\,.$$

Proof. Since $P(\lambda)$ is holomorphic in $\mathbb{C}\backslash\mathbb{R}$ so is $F(\lambda)$. By definition of $M_b(\lambda)$ we get from

$$Z_\lambda(x) := W_\lambda(x)P(\lambda) = (\Theta(x,\lambda) + \Phi(x,\lambda)M_b(\lambda))N(\lambda)$$

that $Z_\lambda \in (R_\varphi^b(I_b))^m$, and applying Green's formula we have

$$\begin{aligned} \lambda(Z_\lambda,Z_\lambda)_{I_b} - \overline{\lambda}(Z_\lambda,Z_\lambda)_{I_b} &= [FZ_\lambda, SZ_\lambda]_{I_b} - [SZ_\lambda, FZ_\lambda]_{I_b} \\ &= [Z_\lambda,Z_\lambda](b) - [Z_\lambda,Z_\lambda](c) \\ &= -[Z_\lambda,Z_\lambda](c) \\ &= -P^*(\lambda)J_mP(\lambda)\,, \end{aligned}$$

hence

$$2i\mathrm{Im}\lambda(Z_\lambda, Z_\lambda)_{I_b} = -P^*(\lambda)J_m P(\lambda)\,. \tag{3.20}$$

Considering

$$V_\lambda(x) := W_\lambda(x)Q(\lambda)$$

we get in the same way the equation

$$2i\mathrm{Im}\lambda(V_\lambda, V_\lambda)_{I_a} = Q^*(\lambda)J_m Q(\lambda)\,. \tag{3.21}$$

Now adding (3.20) and (3.21) we get

$$\begin{aligned} 2i\mathrm{Im}\lambda\{(Z_\lambda, Z_\lambda)_{I_b} + (V_\lambda, V_\lambda)_{I_a}\} &= Q^*(\lambda)J_m Q(\lambda) - P^*(\lambda)J_m P(\lambda) \\ &= (F(\lambda) + \frac{1}{2}J_m)^* J_m (F(\lambda) + \frac{1}{2}J_m) - (F(\lambda) - \frac{1}{2}J_m)^* J_m (F(\lambda) - \frac{1}{2}J_m) \\ &= F(\lambda) - F^*(\lambda)\,, \end{aligned}$$

that is

$$(\mathrm{Im}\lambda)^{-1}\mathrm{Im}F(\lambda) = (V_\lambda, V_\lambda)_{I_a} + (Z_\lambda, Z_\lambda)_{I_b} > 0\,.$$

□

By virtue of Theorem 3.19 we can use Nevanlinna's representation theorem (see eg. [16], Appendix B) and we get

Corollary 3.22. *There exists a mapping ρ from $\mathbb{R}$ into the set of the hermitian $(2m \times 2m)$ matrices, monotone nondecreasing such that*

$$F(\lambda) = A + \lambda B + \int_{\mathbb{R}} \left(\frac{1}{t-\lambda} - \frac{t}{1+t^2} \right) d\rho(t) \tag{3.23}$$

with hermitian matrices A and B.

Remark. If we make ρ right continuous and take $\rho(0) = 0$, then ρ is uniquely determined by the Stieltjes inversion formula.

4. The associated selfadjoint operator

In order to generate the selfadjoint operator A, associated with the singular boundary eigenvalue problem we start with the mapping

$$R_\lambda : E(I) \to R(I), \quad \lambda \in \mathbb{C}\backslash\mathbb{R}$$

defined by

$$(R_\lambda v)(x) := u_\lambda(x, v)\,,$$

where $u_\lambda(\cdot, v)$ is the uniquely determined solution $u_\lambda(\cdot, v)$ in Theorem 3.8. Then we have

Theorem 4.1. *The linear mapping $R_\lambda : E(I) \to R(I)$ has the following properties:*

$$R_\lambda(E(I)) = R^a_\varphi \cap R^b_\varphi \tag{4.2}$$

$$(R_\lambda u, v)_I = (u, R_{\overline{\lambda}} v)_I \tag{4.3}$$

and for $\lambda, \mu \in \mathbb{C}\backslash\mathbb{R}$ the Hilbert relation

$$R_\lambda = R_\mu + (\lambda - \mu) R_\lambda R_\mu \tag{4.4}$$

is valid.

The proof is identical with that of Theorem 5.11 of [10] and can be omitted.

Next we consider in $E(I)$ again the closed subspace $N := \{u \in E(I) | \|u\|_I = 0\}$ as in section 6 of [10], define the Hilbert space $\mathbb{E} := E(I)/N$ and the mapping Γ_λ using the selfadjoint resolvent R_λ of Theorem 4.1 and get finally the selfadjoint operator

$$A := \Gamma_\lambda^{-1} + \lambda id_{\pi(\overline{W})}$$

with $W = R^a_\varphi \cap R^b_\varphi$ in the Hilbert space $\mathbb{H} = \pi(\overline{W})$. The only difference concerning the case of the half-open interval $I_b = [c, b)$ is, that we have to replace I_b by $I = (a, b)$. P shall be again the orthogonal projector from $\mathbb{E}$ onto $\mathbb{H}$.

To determine the representation of the spectral projection E_λ for the operator A, we start again from Stone's formula and using the representation (3.16) for R_λ we have

Theorem 4.5. *For $u \in E(I)$ with compact support in I and points α and β of continuity of E_λ we have*

$$\begin{aligned}(4.6)\quad &((E_\beta - E_\alpha)P(\pi(u)), P(\pi(v))) \\ &= \lim_{\varepsilon \searrow 0} \Big(\frac{1}{2\pi i} \int_\alpha^\beta \int_I \int_I u^*(x) D_3^*(x) D_2(x) \times \\ &\qquad \times (\mathcal{G}^*(t, x, \lambda + i\varepsilon) - \mathcal{G}^*(t, x, \lambda - i\varepsilon)) D_3^*(t) D_2(t) u(t) dt dx d\lambda \Big)\end{aligned}$$

the existence of the limit included.

In order to determine the limit in (4.6), we have similar to Lemma 1.3 now the following

Lemma 4.7. *Let*

$$L_\varepsilon(t, x, \lambda) := \mathcal{G}^*(t, x, \lambda + i\varepsilon) - \mathcal{G}^*(t, x, \lambda - i\varepsilon) - W_\lambda(x)(2i \mathrm{Im} F(\lambda + i\varepsilon)) W_\lambda^*(t) . \tag{4.8}$$

Then we have for $u \in E(I)$ with compact support in I

$$\lim_{\varepsilon \searrow 0} \left(\int_\alpha^\beta \int_I \int_I u^*(x) D_3^*(x) D_2(x) L_\varepsilon(t,x,\lambda) D_3^*(t) D_2(t) u(t) dt dx d\lambda \right) = 0 .$$

The proof is completely analoguous to that of Lemma 1.3, observing that the properties (1.5), (1.6) and (1.7) hold for $L_\varepsilon(t,x,\lambda)$ instead of $K_\varepsilon(t,x,\lambda)$. The only difference is, that for $t \leq x$ we get the relation

$$\begin{aligned} L_\varepsilon(t,x,\lambda) &= \varepsilon^{-1}(W_{\lambda+i\varepsilon}(x) - W_\lambda(x))(\varepsilon F(\lambda+i\varepsilon))W^*_{\lambda-i\varepsilon}(t) \\ &\quad + W_\lambda(x)(\varepsilon F(\lambda+i\varepsilon))\varepsilon^{-1}(W^*_{\lambda-i\varepsilon}(t) - W^*_\lambda(t)) \\ &\quad - \varepsilon^{-1}(W_{\lambda-i\varepsilon}(x) - W_\lambda(x))(\varepsilon F(\lambda-i\varepsilon))W^*_{\lambda+i\varepsilon}(t) \\ &\quad - W_\lambda(x)(\varepsilon F(\lambda-i\varepsilon))\varepsilon^{-1}(W^*_{\lambda+i\varepsilon}(t) - W^*_\lambda(t)) \\ &\quad - \frac{1}{2} W_{\lambda+i\varepsilon}(x) J_m W^*_{\lambda-i\varepsilon}(t) + \frac{1}{2} W_{\lambda-i\varepsilon}(x) J_m W^*_{\lambda+i\varepsilon}(t) . \end{aligned}$$

Observing that $W_\mu(x)$ is continuous in (μ, x), the sum of the last two terms converges locally uniformly to zero with $\varepsilon \searrow 0$. Together with $L_\varepsilon(t,x,\lambda) = -L_\varepsilon^*(x,t,\lambda)$ the proof follows now as in Lemma 1.3.

Theorem 4.5 is the basis for proving the spectral representation of the operator A. Following the arguments in section 1, we get as first result

Theorem 4.9. *Let $u \in E(I)$. Then the integral*

$$\int_I W^*_\lambda(t) D_3^*(t) D_2(t) u(t) dt \tag{4.10}$$

is convergent in $\mathcal{L}^2_\rho(\mathbb{R})$. Defining $(Tu)(\lambda)$ by

$$(Tu)(\lambda) = \int_I W^*_\lambda(t) D_3^*(t) D_2(t) u(t) dt , \tag{4.11}$$

we get

$$\|Tu\|_\rho = \|P(\pi(u))\|, \tag{4.12}$$

$$\|Tu\|_\rho = \|u\| \quad \textit{for } u \in \overline{W} . \tag{4.13}$$

Remarks.

1) The space $\mathcal{L}^2_\rho(\mathbb{R})$ in 4.9 is built up with the $(2m \times 2m)$ weight matrix ρ in (3.23) for $F(\lambda)$.
2) With the polar formula we have

$$(Tu, Tv)_\rho = (P(\pi(u)), P(\pi(v))); \quad u, v \in E(I) \tag{4.14}$$

$$(Tu, Tv)_\rho = (u, v); \quad u, v \in \overline{W} . \tag{4.15}$$

By virtue of property (4.12) the transformation T induces a mapping $\hat{T} : \mathbb{E} \to \mathcal{L}^2_\rho(\mathbb{R})$ by $\hat{T}(\pi(u)) := Tu$. Then all conclusions of section 2 work, and thus we get the following

Theorem 4.16. (Spectral representation) *The linear mapping $\hat{T} : \mathbb{E} \to \mathcal{L}^2_\rho(\mathbb{R})$ is surjective and fulfills the following relations*

$$\|\hat{T}u\|_\rho = \|Pu\|; \quad u \in \mathbb{E} \tag{4.17}$$

$$\hat{T}(Aw) = \mathcal{M}_{id}\hat{T}w; \quad w \in D_A \tag{4.18}$$

$$\hat{T}(E_\beta - E_\alpha)Pw = \chi_{(\alpha,\beta]}\cdot\hat{T}w; \quad w \in \mathbb{E}\,. \tag{4.19}$$

Especially $\hat{T}$ restricted to the subspace $\mathbb{H} = P(\mathbb{E})$ is a surjective isometry from $\mathbb{H}$ onto $\mathcal{L}^2_\rho(\mathbb{R})$ with

$$E_\beta - E_\alpha = \hat{T}^{-1}\chi_{(\alpha,\beta]}\cdot\hat{T}, \tag{4.20}$$

$$A = \hat{T}^{-1}\mathcal{M}_{id}\hat{T}\,. \tag{4.21}$$

To determine a representation for $\hat{T}^{-1}$ we can use all the arguments used in the corresponding conclusions of section 2 and we summarize the result in

Theorem 4.22. *For $u \in \overline{W}$ the integral*

$$\int_{\mathbb{R}} W_\lambda(s)d\rho(\lambda)(Tu)(\lambda) \tag{4.23}$$

converges to $u(s)$ in $E(I)$. Hence

$$(\hat{T}^{-1}v)(s) = \int_{\mathbb{R}} W_\lambda(S)d\rho(\lambda)v(\lambda); \quad v \in \mathcal{L}^2_\rho(\mathbb{R})\,.$$

5. Final remarks

In this last section we will discuss our results by applying them to a simple example showing simultaneously how ordinary selfadjoint differential operators are included.

But first we make a remark on the spectral representation 4.16. The selfadjoint operator A defined in section 4 is uniquely determined by the chosen boundary conditions in the singular endpoints a and b and so is the spectrum of A. This spectrum is described by Theorem 4.16 via the spectral distribution matrix ρ. But the construction of this matrix ρ depends on the choice of the selfadjoint boundary conditions in the interior point c of the interval (a, b), given by a $(2m \times m)$ matrix $\varphi(c)$. Thus $\rho = \rho_\varphi$. But choosing other selfadjoint boundary conditions $\tilde{\varphi}(c)$ we get a spectral distribution matrix $\tilde{\rho} = \rho_{\tilde{\varphi}}$, and we have to show that ρ and $\tilde{\rho}$ are equivalent for describing the properties of the spectrum.

In order to prove this we denote by $W_\lambda(x)$ the fundamental matrix defined in section 4 of [10] to the initial condition

$$W_\lambda(c) = A := (-B^{-1}J_m B\varphi(c), \varphi(c))$$

and by $\tilde{W}_\lambda(x)$ we denote the fundamental matrix with the initial condition

$$\tilde{W}_\lambda(c) = \tilde{A} := (-B^{-1}J_m B\tilde{\varphi}(c), \tilde{\varphi}(c)) .$$

Here the $(2m \times 2m)$ matrix B is defined by the relation (4.3) in [10]. Obviously

$$\tilde{W}_\lambda(x) = W_\lambda(x)A^{-1}\tilde{A}$$

and hence the associated generalized Fourier transformations defined by formula (4.11) satisfy the equation

$$(\tilde{T}u)(\lambda) = \tilde{A}^* A^{*-1}(Tu)(\lambda) . \tag{5.1}$$

Now by Theorem 4.16 we have for all u and v in $E(I)$ the relation

$$\int_{\mathbb{R}} (\tilde{T}v)^*(\lambda)d\tilde{\rho}(\lambda)(\tilde{T}u)(\lambda) = \int_{\mathbb{R}} (Tv)^*(\lambda)d\rho(\lambda)(Tu)(\lambda)$$

and with respect to (5.1) we get

$$\int_{\mathbb{R}} (Tv)^*(\lambda)(A^{-1}\tilde{A})d\tilde{\rho}(\lambda)(\tilde{A}^*A^{*-1})(Tu)(\lambda) = \int_{\mathbb{R}} (Tv)^*(\lambda)d\rho(\lambda)(Tu)(\lambda) .$$

The mapping T is surjective, hence for $\alpha < \beta$ we get

$$A^{-1}\tilde{A}(\tilde{\rho}(\beta) - \tilde{\rho}(\alpha))\tilde{A}^*A^{*-1} = \rho(\beta) - \rho(\alpha)$$

or

$$\tilde{A}(\tilde{\rho}(\beta) - \tilde{\rho}(\alpha))\tilde{A}^* = A(\rho(\beta) - \rho(\alpha))A^* .$$

A and $\tilde{A}$ are invertible and thus the description of the spectrum of the operator A is independent of the choice of the boundary conditions at c.

Remark. In the special case of Hamiltonian systems which contain formally selfadjoint differential equations of higher order, the matrix $H(x) \equiv J_m$ and thus B can be chosen as E_{2m}. In this case we easily verify that A and $\tilde{A}$ are unitary and therefore the matrix-valued measures μ_ρ and $\mu_{\tilde{\rho}}$ generated by ρ and $\tilde{\rho}$ respectively are unitarily equivalent. This result is well known for Sturm-Liouville operators.

As an application of our considerations we consider the differential operator L in $\mathcal{L}^2(\mathbb{R})$ defined on the manifold

$$D_L := \{\eta \in \mathcal{L}^2(\mathbb{R}) | \eta, \eta', \eta'', \eta''' \in AC_{loc}(\mathbb{R}) \text{ and } \eta^{(4)} \in \mathcal{L}^2(\mathbb{R})\} \text{ by } L\eta := \eta^{(4)} .$$

It is well known that L is selfadjoint. In order to study the spectrum of L we consider on $I := \mathbb{R}$ the Hamiltonian system $Fy = \lambda Gy$ defined by

$$(5.2)\quad \begin{pmatrix} 0 & 0 & -1 & 0 \\ 0 & 0 & 0 & -1 \\ 1 & 0 & 0 & 0 \\ 0 & 1 & 0 & 0 \end{pmatrix} y' + \begin{pmatrix} 0 & 0 & 0 & 0 \\ 0 & 0 & -1 & 0 \\ 0 & -1 & 0 & 0 \\ 0 & 0 & 0 & -1 \end{pmatrix} y = \lambda \begin{pmatrix} 1 & 0 & 0 & 0 \\ 0 & 0 & 0 & 0 \\ 0 & 0 & 0 & 0 \\ 0 & 0 & 0 & 0 \end{pmatrix} y\,.$$

(5.2) fulfills the basic assumptions I, II, III of Section 1 in [10]. For this example, we determine all spaces and mappings introduced in Section 6 of [10]. Writing the points in $\mathbb{C}^4$ as $y = (y_1, y_2, y_3, y_4)^\top$ we get for the space $E(I)$ and the scalar product $(\cdot,\cdot)_I$

$$\begin{aligned} E(I) &= \{y : \mathbb{R} \to \mathbb{C}^4 \text{ mb.} \mid \int_{\mathbb{R}} |y_1(x)|^2 dx < \infty\} \\ (u,v)_I &= \int_{\mathbb{R}} u_1(x)\overline{v_1(x)}dx\,. \end{aligned}$$

The system (5.2) has real coefficients, and it is in the limit point case in both endpoints. Hence

$$\tau_{-\infty} = \tau_{\infty} = 0\,,$$

and from Theorem 2.7 [10] we get for the space $R(I) = F^{-1}(G(E(I))) \cap E(I)$:

$$W = R_\varphi^{-\infty} \cap R_\varphi^{+\infty} = R(I)\,.$$

Now we introduce the mapping $\tau : D_L \to E(I)$ by

$$\tau(\eta) := (\eta, \eta', -\eta''', \eta'')^\top\,.$$

Then τ defines an isometric isomorphism from D_L onto $R(I)$ with

$$F(\tau(\eta)) = (L\eta, 0, 0, 0)^\top\,.$$

Since D_L is dense in $\mathcal{L}^2(\mathbb{R})$, we have

$$E(I) = \overline{R(I)} = \overline{W}\,,$$

and thus

$$\mathbb{H} = \pi(\overline{W}) = \pi(E(I)) = \mathbb{E}\,.$$

Further if $y \in R(I)$ and $\pi(y) = 0$, then

$$\|y\|_I^2 = \int_{\mathbb{R}} |y_1(x)|^2 dx = 0\,.$$

This implies $y_1 = 0$ and thus $y = \tau(y_1) = 0$. Hence π is injective on $R(I)$. Then $j := \pi \circ \tau$ also defines an isometric isomorphism from D_L onto D_A.

Now let $x \in D_A$ and $y := (A-\lambda I)x$. Choose $u \in E(I)$ and $w \in W$ with $y = \pi(u)$ and $x = \pi(w)$. Then the equation

$$\begin{aligned}(A-iI)x &= (A-\lambda I)x + (\lambda - i)x \\ &= y + (\lambda - i)x = \pi(u + (\lambda - i)w)\end{aligned}$$

yields

$$\begin{aligned}\pi(w) = x &= \Gamma_i(\pi(u + (\lambda - i)w)) \\ &= \pi(R_i(u + (\lambda - i)w)).\end{aligned}$$

π is injective on W and this implies that

$$w = R_i(u + (\lambda - i)w).$$

Therefore

$$(F - iG)w = G(u + (\lambda - i)w)$$

and

$$(F - \lambda G)w = Gu = (u_1, 0, 0, 0)^\top.$$

Finally defining $w_1 := \tau^{-1}(w)$ we get

$$(u_1, 0, 0, 0)^\top = (F - \lambda G)(\tau(w_1)) = ((L - \lambda)w_1, 0, 0, 0)^\top$$

proving that

$$\begin{aligned}\|(L-\lambda)w_1\|^2_{\mathcal{L}^2(\mathbb{R})} &= \|u_1\|^2_{\mathcal{L}^2(\mathbb{R})} = \|u\|_I^2 \\ &= \|\pi(u)\|^2 = \|y\|^2 \\ &= \|(A - \lambda I)x\|^2 \\ &= \|(A - \lambda I)j(w_1)\|^2 .\end{aligned}$$

Hence the spectrum of L is identical with that of the operator A. In order to determine a spectral distribution matrix $\rho(\lambda)$ we choose the (4×2) matrix

$$\varphi(0) := \begin{pmatrix} 0 & 0 \\ 0 & -1 \\ 1 & 0 \\ 0 & 0 \end{pmatrix}. \tag{5.3}$$

Since $\operatorname{rank}(\varphi(0)) = 2$ and $\varphi^*(0)J_2\varphi(0) = 0$, (5.3) defines selfadjoint boundary conditions at 0. Next we determine the fundamental matrix $W_\lambda(x)$ with the initial condition

$$W_\lambda(0) = (-J_2\varphi(0), \varphi(0)) = \begin{pmatrix} 1 & 0 & 0 & 0 \\ 0 & 0 & 0 & -1 \\ 0 & 0 & 1 & 0 \\ 0 & 1 & 0 & 0 \end{pmatrix} =: A.$$

For $\lambda = |\lambda| e^{i\psi}$ with $|\psi| < \pi$ we define the fourth root of λ by

$$\alpha := \sqrt[4]{\lambda} := |\lambda|^{\frac{1}{4}} e^{i\psi/4}$$

and we consider the four functions

$$\begin{aligned}
w_1(x,\lambda) &:= \frac{1}{4}\left\{e^{\alpha x} + e^{i\alpha x} + e^{-\alpha x} + e^{-i\alpha x}\right\}\\
w_2(x,\lambda) &:= \frac{1}{4\alpha^2}\left\{e^{\alpha x} - e^{i\alpha x} + e^{-\alpha x} - e^{-i\alpha x}\right\}\\
w_3(x,\lambda) &:= -\frac{1}{4\alpha^3}\left\{e^{\alpha x} + ie^{i\alpha x} - e^{-\alpha x} - ie^{-i\alpha x}\right\}\\
w_4(x,\lambda) &:= -\frac{1}{4\alpha}\left\{e^{\alpha x} - ie^{i\alpha x} - e^{-\alpha x} + ie^{-i\alpha x}\right\}.
\end{aligned}$$

Let $Y(x,\lambda)$ be the matrix, defined by

$$Y(x,\lambda) := (\tau(w_1)(x,\lambda), \tau(w_2)(x,\lambda), \tau(w_3)(x,\lambda), \tau(w_4)(x,\lambda)).$$

Since the functions $w_i(\cdot,\lambda)$ are solutions of the differential equation $y^{(4)} = \lambda y$, the columns of the matrix $Y(\cdot,\lambda)$ are solutions of (5.2). Further $Y(0,\lambda) = A$ and therefore $Y(x,\lambda)$ is the matrix $W_\lambda(x)$.

For the system (5.2) we have $\tau_{+\infty} = 0$. Hence the Titchmarsh-Weyl matrix

$$M_\infty(\lambda) =: \begin{pmatrix} m_{11}(\lambda) & m_{12}(\lambda) \\ m_{21}(\lambda) & m_{22}(\lambda) \end{pmatrix}$$

is uniquely determined by the condition that both columns of

$$W_\lambda(\cdot)\begin{pmatrix} E_2 \\ M_\infty(\lambda) \end{pmatrix} \tag{5.4}$$

belong to $E([0,\infty))$ and by definition of the space $E([0,\infty))$ in our example this is equivalent to the condition that the first component of both columns in (5.4) is contained in $\mathcal{L}^2([0,\infty))$, that is

$$w_1(\cdot,\lambda) + m_{11}(\lambda)w_3(\cdot,\lambda) + m_{21}(\lambda)w_4(\cdot,\lambda) \in \mathcal{L}^2([0,\infty)) \tag{5.5}$$

and

$$w_2(\cdot,\lambda) + m_{12}(\lambda)w_3(\cdot,\lambda) + m_{22}(\lambda)w_4(\cdot,\lambda) \in \mathcal{L}^2([0,\infty)). \tag{5.6}$$

In the sequel we first consider the case $\mathrm{Im}\lambda > 0$. Then the condition (5.5) is just the condition, that

$$\begin{aligned}
&\left(1 - \frac{1}{\alpha^3}m_{11}(\lambda) - \frac{1}{\alpha}m_{21}(\lambda)\right)e^{\alpha x} + \left(1 - \frac{i}{\alpha^3}m_{11}(\lambda) + i\frac{1}{\alpha}m_{21}(\lambda)\right)e^{i\alpha x}\\
&+ \left(1 + \frac{1}{\alpha^3}m_{11}(\lambda) + \frac{1}{\alpha}m_{21}(\lambda)\right)e^{-\alpha x} + \left(1 + \frac{i}{\alpha^3}m_{11}(\lambda) - i\frac{1}{\alpha}m_{21}(\lambda)\right)e^{-i\alpha x}\\
&\in \mathcal{L}^2([0,\infty)),
\end{aligned}$$

and by definition of $\lambda^{\frac{1}{4}}$ this is true if and only if

$$\begin{aligned} (5.7) \qquad 1 - \frac{1}{\alpha^3} m_{11}(\lambda) - \frac{1}{\alpha} m_{21}(\lambda) &= 0 \\ (5.8) \qquad 1 + \frac{i}{\alpha^3} m_{11}(\lambda) - \frac{i}{\alpha} m_{21}(\lambda) &= 0\,. \end{aligned}$$

The unique solution of this linear system is

$$m_{11}(\lambda) = \frac{1}{2}(1+i)\alpha^3, \quad m_{21}(\lambda) = \frac{1}{2}(1-i)\alpha\,.$$

Condition (5.6) is equivalent to

$$\left(\frac{1}{\alpha^2} - \frac{1}{\alpha^3} m_{12}(\lambda) - \frac{1}{\alpha} m_{22}(\lambda)\right) e^{\alpha x} + \left(-\frac{1}{\alpha^2} - i\frac{1}{\alpha^3} m_{12}(\lambda) + i\frac{1}{\alpha} m_{22}(\lambda)\right) e^{i\alpha x}$$
$$+ \left(\frac{1}{\alpha^2} + \frac{1}{\alpha^3} m_{12}(\lambda) + \frac{1}{\alpha} m_{22}(\lambda)\right) e^{-\alpha x} + \left(-\frac{1}{\alpha^2} + i\frac{1}{\alpha^3} m_{12}(\lambda) - i\frac{1}{\alpha} m_{22}(\lambda)\right) e^{-i\alpha x}$$
$$\in \mathcal{L}^2([0,\infty))\,,$$

and this is true if and only if the equations

$$\begin{aligned} (5.9) \qquad \frac{1}{\alpha^2} - \frac{1}{\alpha^3} m_{12}(\lambda) - \frac{1}{\alpha} m_{22}(\lambda) &= 0 \\ (5.10) \qquad -\frac{1}{\alpha^2} + i\frac{1}{\alpha^3} m_{12}(\lambda) - \frac{i}{\alpha} m_{22}(\lambda) &= 0 \end{aligned}$$

are fulfilled. The uniquely determined solution of these equations is

$$m_{12}(\lambda) = \frac{1}{2}(1-i)\alpha; \quad m_{22}(\lambda) = \frac{1}{2}(1+i)\frac{1}{\alpha}\,.$$

Hence for $\mathrm{Im}\lambda > 0$ we get

$$(5.11) \qquad M_\infty(\lambda) = \frac{1}{2}\begin{pmatrix} \alpha^3 & \alpha \\ \alpha & \frac{1}{\alpha} \end{pmatrix} + \frac{1}{2} i \begin{pmatrix} \alpha^3 & -\alpha \\ -\alpha & \frac{1}{\alpha} \end{pmatrix}.$$

Since $\tau_{-\infty} = 0$ the Titchmarsh-Weyl matrix $M_{-\infty}(\lambda)$ is uniquely determined by the condition that both columns of the matrix

$$(5.12) \qquad W_\lambda(\cdot)\begin{pmatrix} E_2 \\ M_{-\infty}(\lambda) \end{pmatrix} = \Theta(\cdot,\lambda) + \Phi(\cdot,\lambda) M_{-\infty}(\lambda)$$

are in $E((-\infty,0])$. The calculation of $M_{-\infty}(\lambda)$ can be achieved in principle in the same way as with $M_\infty(\lambda)$, but here we use the fact that the two solutions $\Theta(\cdot,\lambda)$

and $\Phi(\cdot,\lambda)$ satisfy the relation $\Theta(-x,\lambda) = \Theta(x,\lambda)$ and $\Phi(-x,\lambda) = -\Phi(x,\lambda)$. Hence the matrix $M_{-\infty}(\lambda)$ is determined by the condition, that the columns of

$$\Theta(-x,\lambda) + \Phi(-x,\lambda)M_{-\infty}(\lambda) = \Theta(x,\lambda) + \Phi(x,\lambda)(-M_{-\infty}(\lambda))$$

are in $E([0,\infty))$. Thus

$$M_{-\infty}(\lambda) = -M_{\infty}(\lambda)\,. \tag{5.13}$$

Remark. Since the Titchmarsh-Weyl matrices fulfill the relation $M(\lambda)^* = M(\overline{\lambda})$, $M_\infty(\lambda)$ and $M_{-\infty}(\lambda)$ are determined for $\mathrm{Im}\lambda \neq 0$ and (5.13) is valid for all $\lambda \in \mathbb{C}\backslash\mathbb{R}$.

From (5.13) we get immediately

$$(M_{-\infty}(\lambda) - M_\infty(\lambda))^{-1} = -\frac{1}{2}(M_\infty(\lambda))^{-1} \tag{5.14}$$

$$(M_{-\infty}(\lambda) - M_\infty(\lambda))^{-1}M_{-\infty}(\lambda) = \frac{1}{2}E_2 \tag{5.15}$$

$$M_\infty(\lambda)(M_{-\infty}(\lambda) - M_\infty(\lambda))^{-1} = -\frac{1}{2}E_2 \tag{5.16}$$

$$M_\infty(\lambda)(M_{-\infty}(\lambda) - M_\infty(\lambda))^{-1}M_{-\infty}(\lambda) = \frac{1}{2}M_\infty(\lambda)\,. \tag{5.17}$$

Therefore with respect to (3.11) and (3.18) we get

$$F(\lambda) = \frac{1}{2}\begin{pmatrix} -(M_\infty(\lambda))^{-1} & 0 \\ 0 & M_\infty(\lambda) \end{pmatrix}.$$

In order to calculate the matrix $\rho(t)$ we use the Titchmarsh-Kodaira formula and for points α and β of continuity of ρ with $\alpha < \beta$ we have

$$\begin{aligned} \text{(5.18)}\quad & \rho(\beta) - \rho(\alpha) \\ &= \frac{1}{\pi}\lim_{\varepsilon\searrow 0}\int_\alpha^\beta \mathrm{Im}(F(t+i\varepsilon))dt \\ &= \frac{1}{2\pi}\begin{pmatrix} -\lim\limits_{\varepsilon\searrow 0}\left(\mathrm{Im}\int_\alpha^\beta (M_\infty(t+i\varepsilon))^{-1}dt\right) & 0 \\ 0 & \lim\limits_{\varepsilon\searrow 0}\left(\mathrm{Im}\int_\alpha^\beta M_\infty(t+i\varepsilon)dt\right) \end{pmatrix} \\ &=: \begin{pmatrix} \rho_1(\beta) - \rho_1(\alpha) & 0 \\ 0 & \rho_2(\beta) - \rho_2(\alpha) \end{pmatrix}. \end{aligned}$$

By definition of $\lambda^{\frac{1}{4}}$ for $\mathrm{Im}\lambda > 0$ we see that $M_\infty(\lambda)$ has a uniquely determined continuous extension to $\{\lambda \in \mathbb{C}|\mathrm{Im}\lambda \geq 0\}\backslash\{0\}$.

First we consider the case $0 < \alpha < \beta$. Then

$$\begin{aligned}
\rho_1(\beta) - \rho_1(\alpha) &= -\frac{1}{2\pi} \lim_{\varepsilon \searrow 0} \left(\operatorname{Im} \int_\alpha^\beta M_\infty(t + i\varepsilon)^{-1} dt \right) \\
&= -\frac{1}{2\pi} \int_\alpha^\beta \operatorname{Im} \left(M_\infty(t)^{-1} \right) dt \\
&= \frac{1}{4\pi} \int_\alpha^\beta \begin{pmatrix} t^{-\frac{3}{4}} & -t^{-\frac{1}{4}} \\ -t^{-\frac{1}{4}} & t^{\frac{1}{4}} \end{pmatrix} dt \\
&= \frac{1}{\pi} \begin{pmatrix} \beta^{\frac{1}{4}} - \alpha^{\frac{1}{4}} & -\frac{1}{3}(\beta^{\frac{3}{4}} - \alpha^{\frac{3}{4}}) \\ -\frac{1}{3}(\beta^{\frac{3}{4}} - \alpha^{\frac{3}{4}}) & \frac{1}{5}(\beta^{\frac{5}{4}} - \alpha^{\frac{5}{4}}) \end{pmatrix}
\end{aligned}$$

and

$$\begin{aligned}
\rho_2(\beta) - \rho_2(\alpha) &= \frac{1}{2\pi} \lim_{\varepsilon \searrow 0} \left(\operatorname{Im} \left(\int_\alpha^\beta M_\infty(t + i\varepsilon) dt \right) \right) \\
&= \frac{1}{2\pi} \int_\alpha^\beta \operatorname{Im}(M_\infty(t)) dt \\
&= \frac{1}{4\pi} \int_\alpha^\beta \begin{pmatrix} t^{\frac{3}{4}} & -t^{\frac{1}{4}} \\ -t^{\frac{1}{4}} & t^{-\frac{1}{4}} \end{pmatrix} dt \\
&= \frac{1}{\pi} \begin{pmatrix} \frac{1}{7}(\beta^{\frac{7}{4}} - \alpha^{\frac{7}{4}}) & -\frac{1}{5}(\beta^{\frac{5}{4}} - \alpha^{\frac{5}{4}}) \\ -\frac{1}{5}(\beta^{\frac{5}{4}} - \alpha^{\frac{5}{4}}) & \frac{1}{3}(\beta^{\frac{3}{4}} - \alpha^{\frac{3}{4}}) \end{pmatrix}.
\end{aligned}$$

Since

$$\det(\rho_1(\beta) - \rho_1(\alpha)) = \frac{1}{45}(\beta^{\frac{1}{4}} - \alpha^{\frac{1}{4}})^4 (4\beta^{\frac{1}{2}} + 7\alpha^{\frac{1}{4}}\beta^{\frac{1}{4}} + 4\alpha^{\frac{1}{2}})$$

and

$$\begin{aligned}
\det(\rho_2(\beta) - \rho_2(\alpha)) = \frac{1}{525}(\beta^{\frac{1}{4}} - \alpha^{\frac{1}{4}})^4 \times \\
\times (4\beta^{\frac{3}{2}} + 16\alpha^{\frac{1}{4}}\beta^{\frac{5}{4}} + 40\alpha^{\frac{1}{2}}\beta + 55\alpha^{\frac{3}{4}}\beta^{\frac{3}{4}} + 40\alpha\beta^{\frac{1}{2}} + 16\alpha^{\frac{5}{4}}\beta^{\frac{1}{4}} + 4\alpha^{\frac{3}{2}}),
\end{aligned}$$

both determinants are positive for $\beta > \alpha > 0$, and hence $\rho(t)$ is strongly monotone increasing for $t > 0$. Thus $(0, \infty)$ is contained in the absolutely continuous spectrum of L (ρ is even a C^∞-matrix in $t > 0$).

Now let $\alpha < \beta < 0$. Then the continuous extension of $M_\infty(\lambda)$ for $t < 0$ is

$$M_\infty(t) = \frac{1}{\sqrt{2}} \begin{pmatrix} -|t|^{\frac{3}{4}} & |t|^{\frac{1}{4}} \\ |t|^{\frac{1}{4}} & |t|^{-\frac{1}{4}} \end{pmatrix}$$

and therefore $\operatorname{Im}(M_\infty(t)) \equiv 0$. Analoguously we get

$$(M_{-\infty}(t))^{-1} = \frac{1}{\sqrt{2}} \begin{pmatrix} -|t|^{-\frac{3}{4}} & |t|^{-\frac{1}{4}} \\ |t|^{-\frac{1}{4}} & |t|^{\frac{1}{4}} \end{pmatrix}$$

implying $\mathrm{Im}(M_\infty(t)^{-1}) \equiv 0$ in $t < 0$. Therefore $\rho(t) \equiv$ const in $t < 0$. Hence we get in accordance with the well known result, that $(-\infty, 0)$ belongs to the resolvent set of the operator L and $(0, \infty)$ to the absolutely continuous part of the spectrum of L (ρ is a C^∞-matrix in $\mathbb{R}\backslash\{0\}$).

Finally we verify immediately that $\mathrm{Im}(F(t+i\varepsilon))$ converges in (α, β) with $\alpha < 0 < \beta$ for $\varepsilon \searrow 0$ to an integrable limit matrix $H(t)$ and using Lebesgue's theorem on dominated convergence we get

$$\rho(\beta) - \rho(\alpha) = \frac{1}{\pi} \int_\alpha^\beta H(t)dt\,.$$

Thus ρ is continuous at 0 which coincides with the fact, that 0 is not an eigenvalue of L.

Acknowledgements

The first named author would like to thank the Deutsche Forschungsgemeinschaft (DFG) for support of this research.

References

[1] F.V. ATKINSON, Discrete and Continuous Boundary Value Problems, Academic Press, New York, 1964.

[2] E.A. CODDINGTON AND N. LEVINSON, Theory of Ordinary Differential Equations, McGraw-Hill, New York, 1955.

[3] E.A. CODDINGTON AND A. DIJKSMA, Self-adjoint subspaces and eigenfunction expansions for ordinary differential subspaces, J. Diff. Eqs. **20** (1976), 473–526.

[4] E.A. CODDINGTON AND A. DIJKSMA, Adjoint subspaces in Banach spaces with applications to ordinary differential subspaces, Annali di Matematica pura ed applicata **118** (1978), 1–118.

[5] A. DIJKSMA, H. LANGER, AND H.S.V. DE SNOO, Unitary colligations in π_k-Spaces, characteristic functions and Štraus extensions, Pacific J. Math. **125** (1986), 347–362.

[6] A. DIJKSMA, H. LANGER, AND H.S.V. DE SNOO, Symmetric Sturm-Liouville Operators with eigenvalue depending boundary conditions, Can. Math. Soc. Conference Proc. **8** (1987), 87–116.

[7] A. DIJKSMA, H. LANGER, AND H.S.V. DE SNOO, Hamiltonian systems with eigenvalue depending boundary conditions, Operator Theory: Adv. Appl. **35** (1988), 37–83.

[8] A. DIJKSMA, H. LANGER, AND H.S.V. DE SNOO, Eigenvalues and pole functions of Hamiltonian systems with eigenvalue depending boundary conditions, Math. Nachr. **161** (1993), 107–154.

[9] N. DUNFORD AND J. SCHWARTZ, Linear Operators, part II, Interscience, New York, 1963.

[10] D.B. Hinton and A. Schneider, On the Titchmarsh-Weyl coefficients for singular S-Hermitian systems I, Math. Nachr. **163** (1993), 323–342.

[11] D.B. Hinton and A. Schneider, On the Titchmarsh-Weyl coefficients for singular S-Hermitian systems II, Math. Nachr. **185** (1997), 67–84.

[12] H.D. Niessen, Singuläre S-hermitesche Rand-Eigenwertprobleme, manuscripta math. **3** (1970), 35–68.

[13] H.D. Niessen and A. Schneider, Integraltransformationen zu singulären S-hermiteschen Rand-Eigenwertproblemen, manuscripta math. **5** (1971), 133–145.

[14] C. Remling, Geometric characterization of singular self-adjoint boundary conditions for Hamiltonian systems, (to appear).

[15] W.T. Reid, Ordinary Differential Equations, John Wiley & Sons, New York, 1971.

[16] J. Weidmann, Linear Operators in Hilbert Space, Springer-Verlag, Berlin, 1980.

[17] K. Yosida, Functional Analysis, Springer-Verlag, Berlin, 1974.

University of Tennessee
Knoxville, TN 37996-1300
USA

Universität Dortmund
44221 Dortmund
Germany

1991 Mathematics Subject Classification. Primary 34B20; Secondary 47E05, 34B05, 34L05

Operator Theory:
Advances and Applications, Vol. 106

Some characteristics of a linear manifold in a Kreĭn space and their applications

E.I. IOKHVIDOV

Dedicated to Heinz Langer on the occasion of his 60th birthday

We introduce two characteristics $\varepsilon_{\pm}(\mathcal{L})$ of a linear manifold $\mathcal{L}$ in a Kreĭn space and give two applications, one in the geometry of Kreĭn spaces and the other one in the theory of operators in these spaces.

1. Introduction

Let $\mathcal{H}$ be a Kreĭn space with indefinite inner product $[\cdot,\cdot]$; see [AI]. We fix a fundamental decomposition $\mathcal{H} = \mathcal{H}_+ \oplus \mathcal{H}_-$ and corresponding symmetry $J = P_+ - P_-$, where $P_\pm$ are the orthogonal projections on $\mathcal{H}$ onto $\mathcal{H}_\pm$. Denote by $(x,y) = [Jx,y]$, $x,y \in \mathcal{H}$, the inner product which makes $(\mathcal{H},(\cdot,\cdot))$ a Hilbert space and set $\|x\| = \sqrt{(x,x)}$, $x \in \mathcal{H}$. For a linear manifold $\mathcal{L} \subset \mathcal{H}$ we define the numbers

$$\varepsilon_-(\mathcal{L}) = \inf_{x\in\mathcal{L},x\neq 0} \frac{[x,x]}{\|x\|^2} \qquad \text{and} \qquad \varepsilon_+(\mathcal{L}) = \sup_{x\in\mathcal{L},x\neq 0} \frac{[x,x]}{\|x\|^2}.$$

They satisfy the inequalities

$$-1 \leq \varepsilon_-(\mathcal{L}) \leq \varepsilon_+(\mathcal{L}) \leq 1$$

and have the following properties

1° $\varepsilon_-(\mathcal{L}) = \varepsilon_+(\mathcal{L}) = -1 \Leftrightarrow \mathcal{L} \subset \mathcal{H}_-$.
2° $\varepsilon_-(\mathcal{L}) = \varepsilon_+(\mathcal{L}) = 1 \Leftrightarrow \mathcal{L} \subset \mathcal{H}_+$.
3° $\varepsilon_+(\mathcal{L}) < 0 \Leftrightarrow \mathcal{L}$ is uniformly negative.
4° $\varepsilon_-(\mathcal{L}) > 0 \Leftrightarrow \mathcal{L}$ is uniformly positive.
5° $\varepsilon_-(\mathcal{L}) > -1 \Rightarrow \mathcal{L} \cap \mathcal{H}_- = \{0\}$.
6° $\varepsilon_+(\mathcal{L}) < 1 \Rightarrow \mathcal{L} \cap \mathcal{H}_+ = \{0\}$.

The converse implications in 5° and 6° are not true in general. In fact, if $\mathcal{L}$ is a linear manifold in a Pontryagin space $\mathcal{H}$ with $\dim \mathcal{H}_+ < \infty$, then the conditions

$$\mathcal{L} \cap \mathcal{H}_- = \{0\} \quad \text{and} \quad \varepsilon_-(\mathcal{L}) = -1$$

are equivalent. A similar result holds related to 6°.

The characteristic numbers $\varepsilon_{\pm}(\mathcal{L})$ provide a criterion for the boundedness of the angle operator of a linear manifold $\mathcal{L}$ in $\mathcal{H}$; see Section 2. In Section 3 we show for some classes of linear operators V on $\mathcal{H}$, the relation between $\varepsilon_+(\mathcal{L})$ and $\varepsilon_+(V\mathcal{L})$.

2. A boundedness criterion

Let $\mathcal{L}$ be an arbitrary linear manifold in a Kreĭn space $\mathcal{H}$. It is well-known (see [AI, p. 79]) that the decomposition

$$x = P_+x + KP_+x, \qquad \forall x \in \mathcal{L},$$

is possible if and only if the condition $\mathcal{L} \cap \mathcal{H}_- = \{0\}$ holds. In this case the operator $P_+|\mathcal{L}$ is invertible and

$$K = P_-(P_+|\mathcal{L})^{-1}$$

is called the angle operator of a linear manifold $\mathcal{L}$. We have the following new criterion for the boundedness of K.

Theorem 1. *The angle operator K of a linear manifold $\mathcal{L}$ exists and is bounded if and only if the condition*

$$\varepsilon_-(\mathcal{L}) > -1$$

holds. Moreover, under this condition the norm of the operator K satisfies the relation

$$\|K\| = \sqrt{\frac{1-\varepsilon_-(\mathcal{L})}{1+\varepsilon_-(\mathcal{L})}}.$$

Proof. Suppose that $\mathcal{L}$ admits the representation

$$x = P_+x + KP_+x, \qquad \forall x \in \mathcal{L}, \tag{1}$$

with bounded angle operator K. This means that

$$\|P_-x\| = \|KP_+x\| \le \|K\|\,\|P_+x\|, \qquad \forall x \in \mathcal{L}.$$

The inequality is equivalent to the inequality

$$\frac{\|P_+x\|^2 - \|P_-x\|^2}{\|P_+x\|^2 + \|P_-x\|^2} \ge \frac{1-\|K\|^2}{1+\|K\|^2}, \qquad \forall x \in \mathcal{L},\ x \ne 0.$$

Hence we get

$$\frac{[x,x]}{\|x\|^2} \ge \frac{1-\|K\|^2}{1+\|K\|^2}, \qquad \forall x \in \mathcal{L},\ x \ne 0,$$

and, by the definition of $\varepsilon_-(\mathcal{L})$,

$$\varepsilon_-(\mathcal{L}) \geq \frac{1 - \|K\|^2}{1 + \|K\|^2}. \tag{2}$$

On the other hand, the function $f(t) = (1 - t^2)(1 + t^2)^{-1}$ satisfies the condition $f(t) > -1 \quad \forall t \in \mathbf{R}$. This and (2) imply the desired inequality $\varepsilon_-(\mathcal{L}) > -1$.

Suppose, conversely, that $\mathcal{L}$ satisfies the condition $\varepsilon_-(\mathcal{L}) > -1$. Then by property 5° we get $\mathcal{L} \cap \mathcal{H}_- = \{0\}$, hence $\mathcal{L}$ admits the representation (1) with the angle operator K. Let us show the boundedness of K. From the definition of $\varepsilon_-(\mathcal{L})$ we deduce:

$$\varepsilon_-(\mathcal{L}) \leq \frac{[x, x]}{\|x\|^2} = \frac{\|P_+x\|^2 - \|P_-x\|^2}{\|P_+x\|^2 + \|P_-x\|^2}, \qquad \forall x \in \mathcal{L},\ x \neq 0,$$

or, equivalently,

$$\|P_-x\|^2 \leq \frac{1 - \varepsilon_-(\mathcal{L})}{1 + \varepsilon_-(\mathcal{L})} \|P_+x\|^2, \qquad \forall x \in \mathcal{L}.$$

Since $P_-x = KP_+x$, we get

$$\|K\|^2 \leq \frac{1 - \varepsilon_-(\mathcal{L})}{1 + \varepsilon_-(\mathcal{L})}. \tag{3}$$

On the other hand, (2) may be written in the equivalent form as

$$\|K\|^2 \geq \frac{1 - \varepsilon_-(\mathcal{L})}{1 + \varepsilon_-(\mathcal{L})}. \tag{4}$$

(3) and (4) imply the equality

$$\|K\| = \sqrt{\frac{1 - \varepsilon_-(\mathcal{L})}{1 + \varepsilon_-(\mathcal{L})}}.$$

□

3. Characteristic numbers and J-contractions

In this section we establish a connection beween the numbers $\varepsilon_+(\mathcal{L})$ and $\varepsilon_+(V\mathcal{L})$ for some classes of linear operators V in a Kreĭn space $\mathcal{H}$. We recall some notations and definitions (see [AI]):

$$\mathcal{P}_- = \{x \in \mathcal{H} \mid [x, x] \leq 0\}, \qquad \mathcal{P}_+ = \{x \in \mathcal{H} \mid [x, x] \geq 0\},$$

$$\mathcal{P}_{--} = \{x \in \mathcal{H} \mid [x, x] < 0\};$$

a linear operator V is called J^-–contractive, if ($\mathcal{D}_V$ denotes domain of V)

$$[Vx, Vx] \leq [x, x], \qquad \forall x \in \mathcal{D}_V \cap \mathcal{P}_-;$$

a linear operator V is called J-contractive, if

$$[Vx, Vx] \leq [x, x], \qquad \forall x \in \mathcal{D}_V.$$

Theorem 1. *Let V be a J-contractive operator, and let $\mathcal{L}$ be a linear manifold in a Kreĭn space $\mathcal{H}$, satisfying the conditions:*

1) *The operator V^{-1} exists and is bounded.*
2) $\mathcal{L} \subset \mathcal{D}_V$.
3) $\mathcal{L} \cap \mathcal{P}_+ \neq \{0\}$.

Then the inequality

$$\varepsilon_+(V\mathcal{L}) \leq \|V^{-1}\|^2 \, \varepsilon_+(\mathcal{L}) \tag{1}$$

holds.

Proof. By the definition of $\varepsilon_+(V\mathcal{L})$ and using the existence of V^{-1}, we have

$$\varepsilon_+(V\mathcal{L}) = \sup_{x \in \mathcal{L},\ x \neq 0} \frac{[Vx, Vx]}{\|Vx\|^2}. \tag{2}$$

But V is a J-contractive operator, hence

$$\sup_{x \in \mathcal{L},\ x \neq 0} \frac{[Vx, Vx]}{\|Vx\|^2} \leq \sup_{x \in \mathcal{L},\ x \neq 0} \frac{[x, x]}{\|Vx\|^2}. \tag{3}$$

The condition 3) of the theorem implies the relation

$$\sup_{x \in \mathcal{L},\ x \neq 0} \frac{[x, x]}{\|Vx\|^2} = \sup_{x \in \mathcal{L} \cap \mathcal{P}_+,\ x \neq 0} \frac{[x, x]}{\|Vx\|^2}. \tag{4}$$

The boundedness of V^{-1} now implies

$$\sup_{x \in \mathcal{L} \cap \mathcal{P}_+,\ x \neq 0} \frac{[x, x]}{\|Vx\|^2} \leq \sup_{x \in \mathcal{L} \cap \mathcal{P}_+,\ x \neq 0} \frac{\|V^{-1}\|^2 [x, x]}{\|x\|^2}. \tag{5}$$

Again by 3)

$$\sup_{x \in \mathcal{L} \cap \mathcal{P}_+,\ x \neq 0} \frac{\|V^{-1}\|^2 [x, x]}{\|x\|^2} = \sup_{x \in \mathcal{L} \cap \mathcal{P}_+,\ x \neq 0} \frac{\|V^{-1}\|^2 [x, x]}{\|x\|^2} = \|V^{-1}\|^2 \, \varepsilon_+(\mathcal{L}). \tag{6}$$

Hence (5) follows from (6)–(10). □

Corollary 2. *Every J-semiunitary operator V satisfies the condition*

$$\varepsilon_+(V\mathcal{H}) \leq \|V^{-1}\|^2.$$

In particular, every J-unitary operator V satisfies the conditions

$$\|V\| \geq 1 \qquad \text{and} \qquad \|V^{-1}\| \geq 1.$$

The next theorem may be proved similarly to the previous one.

Theorem 3. *Let V be a bounded J^--contractive operator, and let $\mathcal{L}$ be a linear manifold in $\mathcal{H}$ satisfying the conditions:*

a) $\mathcal{L} \subset \mathcal{D}_V$.

b) $\mathcal{L} \subset \mathcal{P}_{--} \cup \{0\}$.

Then the inequality

$$\varepsilon_+(V\mathcal{L})\,\|V\|^2 \leq \varepsilon_+(\mathcal{L})$$

holds.

Corollary 4.[Io, § 3, Proposition 1°]. *Let V be a bounded J^--contractive operator, and let $\mathcal{L}$ be a uniformly negative linear manifold with $\mathcal{L} \subset \mathcal{D}_V$. Then the linear manifold $V\mathcal{L}$ is uniformly negative as well.*

References

[AI] T.YA. AZIZOV, I.S. IOKHVIDOV, Foundation of the theory of linear operators in spaces with an indefinite metric, "Nauka", Moscow, 1986 (Russian); English transl.: Linear operators in spaces with an indefinite metric, Wiley, New York, 1989.

[Io] I.S. IOKHVIDOV, Banach spaces with a J-metric and certain classes of linear operators in these spaces, Bul. Akad. Stiinnce RSS Moldoven, **1**(1968), 60–80 (Russian).

Perevjortkina str. 28
Apt. 53
394063 Voronezh
Russia

1991 Mathematics Subject Classification. Primary 46C20, 47B50

Operator Theory:
Advances and Applications, Vol. 106

Riggings and relatively form bounded perturbations of nonnegative operators in Kreĭn spaces

PETER JONAS

Dedicated to Heinz Langer on the occasion of his 60th birthday

For nonnegative operators in Kreĭn spaces we give conditions for the preservation of the nonemptiness of the resolvent set and the preservation of the regularity of critical points under relatively form bounded perturbations.

0. Introduction

In this note we study symmetric relatively form bounded perturbations of a nonnegative operator in a Kreĭn space with nonempty resolvent set such that the perturbed operator is still nonnegative.

We recall that a nonnegative operator A in a Kreĭn space $(\mathcal{H}, [\cdot,\cdot])$ with $\rho(A) \neq \emptyset$ possesses a spectral function $E(\cdot)$ (cf. [11]) defined for all bounded intervals whose endpoints are different from zero and for their complements with respect to $\overline{\mathbf{R}}$. If there is no $\varepsilon > 0$ such that $[\cdot,\cdot]$ restricted to $E((-\varepsilon,\varepsilon))\mathcal{H}$ $(E(\overline{\mathbf{R}} \setminus (-\varepsilon,\varepsilon))\mathcal{H})$ is positive or negative definite, the point 0 (∞, resp.) is called a *critical point* of A. The set of critical points of A, which is a subset of $\{0,\infty\}$, is denoted by $c(A)$. A critical point α of A is called *regular* if for some open neighbourhood Δ of α in $\overline{\mathbf{R}}$ we have

$$\sup\{\|E(\delta)\| : \delta \text{ closed bounded interval, } \delta \subset \Delta \setminus \{\alpha\}\} < \infty.$$

The set of regular critical points is denoted by $c_r(A)$. The elements of $c_s(A) := c(A) \setminus c_r(A)$ are called *singular* critical points. The operator A is of scalar type if and only if $c_s(A) = \emptyset$ and the root space of A corresponding to the point zero coincides with the kernel of A.

Relatively form bounded perturbations of a nonnegative operator A in a Kreĭn space $\mathcal{H}$ were first studied by K. Veselić. He proved in [17] that if the unpertubed operator A is of scalar type and uniformly positive and if the form bound of the perturbation is smaller than one then the perturbed operator B is also of scalar type. In [18] the uniform positivity of A was replaced by a more general assumption. In [5] it is assumed that A is an arbitrary positive operator in $\mathcal{H}$ of scalar type. In that paper for a subclass of relatively form bounded perturbations (characterized by $\alpha' = \alpha'' = 0$ in relation (2.1) below) such that the negative part has a form

bound < 1 it was proved that ∞ is no singular critical point of B. A condition for $0 \notin c_s(B)$ was also given. [5] deals also with a special class of relatively form compact perturbations of arbitrary size. In this case the perturbed operators are still definitizable. In [6] the unperturbed operator A fulfils only a local positivity assumption and general relatively form compact perturbations are admitted.
In the recent paper [4] B. Ćurgus and B. Najman consider an arbitrary positive operator A with nonempty resolvent set and the subclass of relatively form bounded perturbations mentioned above such that the negative part of the perturbation has a relative form bound $< \frac{1}{2}$. Under the assumption that B has nonempty resolvent set it is proved in [4] that ∞ (the point 0) is a singular critical point of A if and only if ∞ (0, resp.) is a singular critical point of B.
The main objective of the present note is to show that $\rho(B) \neq \emptyset$ follows already from relative boundedness assumptions. We establish conditions for the preservation of the regularity or singularity of the critical points which are a bit weaker than those in [4].
In the proofs we make use of some riggings of the Krein space $\mathcal{H}$ taken from a scale of Hilbert spaces which is associated with every nonnegative operator in $\mathcal{H}$ (cf. [8], [3]). For the case of an operator of Klein-Gordon type such a scale was introduced by B. Najman in [13]. In [13] and the subsequent papers [14], [15], [19] the operator is considered as acting in each of the scale spaces. In those papers certain properties of the operator such as similarity to a selfadjoint operator in Hilbert space and their dependence on the scale parameter are studied.
In Section 1 we collect definitions and simple facts on the riggings associated to a nonnegative operator in a Kreĭn space and on the associated operator A_r, which are needed for the study of form bounded perturbations in Section 2. The results of Section 1 are not new. We shall present this material here with more details than in [6] and not only as a reference for Section 2 but also to rely on it later on.

1. Scales of spaces associated with a nonnegative selfadjoint operator in a Kreĭn space

1.1. The scale of spaces associated with a nonnegative selfadjoint operator in a Hilbert space

Let $(\mathcal{H}, (\cdot,\cdot))$ be a Hilbert space and let H be a nonnegative operator in $\mathcal{H}$. We denote by $\mathcal{H}_s(H)$, $s \in \mathbf{R}$, the scale of Hilbert spaces corresponding to H: For $s \in [0,\infty)$ we set $\mathcal{H}_s(H) = \mathcal{D}(H^s)$, $s \in \mathbf{R}$, and equip this linear space with the Hilbert scalar product

$$(x,y)_s := ((1+H)^s x, (1+H)^s y), \qquad x,y \in \mathcal{H}_s(H).$$

$\mathcal{H}_{-s}(H)$, $s \in [0,\infty)$, is defined as the completion of $\mathcal{H}$ with respect to the scalar product

$$(x,y)_{-s} := ((1+H)^{-s} x, (1+H)^{-s} y), \qquad x,y \in \mathcal{H}.$$

We have

$$\|x\|_{-s} = \sup\{|(x,y)| : y \in \mathcal{H}_s(H), \|y\|_s \le 1\}, \quad s \in [0,\infty), \quad x \in \mathcal{H},$$

or, in other words, the triplets

$$(1.1) \qquad \mathcal{H}_s(H) \subset \mathcal{H} \subset \mathcal{H}_{-s}(H), \qquad s \in (0,\infty),$$

are riggings of $\mathcal{H}$. The extension by continuity of $(\cdot,\cdot)$ to $\mathcal{H}_s(H) \times \mathcal{H}_{-s}(H)$, $s \in \mathbf{R}$, is also denoted by $(\cdot,\cdot)$. The operator $(1+H)^s$, $s \in [0,\infty)$, is an isometric isomorphism of $\mathcal{H}_s(H)$ onto $\mathcal{H}_0(H)$, and $(1+H)^{2s}$ can be extended by continuity to an isometric isomorphism of $\mathcal{H}_s(H)$ onto $\mathcal{H}_{-s}(H)$ which will be denoted by $((1+H)^{2s})^{\sim}$. In the following the extension by continuity of an operator T to a larger linear space is often simply denoted by $\widetilde{T}$ if it is clear from the context which extension is meant. We have

$$(1.2) \qquad (x,y)_s = (x, ((1+H)^{2s})^{\sim} y), \qquad x,y \in \mathcal{H}_s(H).$$

Therefore $\mathcal{H}_{-s}(H)$ may be considered as the dual space of $\mathcal{H}_s(H)$. The natural embedding of $\mathcal{H}_s(H)$ in $\mathcal{H}_{-s}(H)$ will be denoted by E. If the basic space is not $\mathcal{H}$ but another space, say $\mathcal{K}$, the scale associated with a selfadjoint operator T in $\mathcal{K}$ will be denoted by $\mathcal{K}_s(T)$.
Let $\mathcal{H}$ be the direct sum of subspaces $\mathcal{H}'$ and $\mathcal{H}''$, $\mathcal{H} = \mathcal{H}' \dot{+} \mathcal{H}''$. Let on $\mathcal{H}'$ and $\mathcal{H}''$ scalar products be given which are equivalent to $(\cdot,\cdot)$. Assume that H is the direct sum of a nonnegative selfadjoint operator H' in $(\mathcal{H}', (\cdot,\cdot)')$ and a bounded operator H'' in $(\mathcal{H}'', (\cdot,\cdot)'')$. Then

$$(1.3) \qquad \mathcal{H}_s(H) = (\mathcal{H}')_s(H') \times \mathcal{H}''.$$

Here and in the following for some Hilbert spaces $\mathcal{K}_1$ and $\mathcal{K}_2$ we write $\mathcal{K}_1 = \mathcal{K}_2$ if $\mathcal{K}_1$ and $\mathcal{K}_2$ coincide as linear topological spaces.
If $(\cdot,\cdot)^\circ$ is a Hilbert scalar product on $\mathcal{H}$ equivalent to $(\cdot,\cdot)$, and $\overset{\circ}{H}$ is a selfadjoint operator in $(\mathcal{H}, (\cdot,\cdot)^\circ)$ with $\mathcal{D}(H) = \mathcal{D}(\overset{\circ}{H})$, then we have

$$(1.4) \qquad \mathcal{H}_s(H) = \mathcal{H}_s(\overset{\circ}{H}), \qquad s \in [-1,1].$$

This holds, in particular, if $(\cdot,\cdot) = (\cdot,\cdot)^\circ$ and $\overset{\circ}{H} = H + C$, where C is a bounded selfadjoint operator in $(\mathcal{H}, (\cdot,\cdot))$.
We put $\mathcal{L}^{(H)} := \mathcal{L}(\mathcal{H}_{\frac{1}{2}}(H), \mathcal{H}_{-\frac{1}{2}}(H))$. An operator $V \in \mathcal{L}^{(H)}$ is called *symmetric* (resp. *nonnegative*) if, for all $x \in \mathcal{H}_{\frac{1}{2}}(H)$, (Vx,x) is real (resp. nonnegative). In view of (1.2) with $s = \frac{1}{2}$, a nonnegative $V \in \mathcal{L}^{(H)}$ is an isomorphism if and only if it is *uniformly positive*, that is, there exists an $m > 0$ such that $m\|x\|^2_{\frac{1}{2}} \le (Vx,x)$, $x \in \mathcal{H}_{\frac{1}{2}}(H)$. For example, $\widetilde{H} \in \mathcal{L}^{(H)}$ is nonnegative, and $\widetilde{H}$ is an isomorphism if and only if $0 \in \rho(H)$.

1.2. The scale of spaces associated with a nonnegative selfadjoint operator in a Kreĭn space

Let now $(\mathcal{H}, [\cdot,\cdot])$ be a Kreĭn space. Let G be a positive selfadjoint bounded operator in $\mathcal{H}$ with $0 \in \rho(G)$. We will consider the following two Hilbert scalar products on $\mathcal{H}$:

$$(x,y)^H := [G^{-1}x, y], \quad (x,y)^K := [Gx, y], \qquad x, y \in \mathcal{H}.$$

G^{-1} maps $(\mathcal{H}, (\cdot,\cdot)^H)$ isometrically onto $(\mathcal{H}, (\cdot,\cdot)^K)$. Let A be a nonnegative selfadjoint operator in $\mathcal{H}$ with $\rho(A) \neq \emptyset$. Define

$$H := GA, \quad K := AG.$$

The operator H is selfadjoint and nonnegative in $(\mathcal{H}, (\cdot,\cdot)^H)$, and K is selfadjoint and nonnegative in $(\mathcal{H}, (\cdot,\cdot)^K)$. We have

$$H = GKG^{-1}. \tag{1.5}$$

Now we consider the scales of Hilbert spaces associated with H and K, $\mathcal{H}_s(H)$ and $\mathcal{H}_s(K)$, where $\mathcal{H}_0(H) = (\mathcal{H}, (\cdot,\cdot)^H)$ and $\mathcal{H}_0(K) = (\mathcal{H}, (\cdot,\cdot)^K)$. It follows immediately from (1.5) that G^{-1} maps $\mathcal{H}_s(H)$ isometrically onto $\mathcal{H}_s(K)$, $s \in [0,\infty)$. The adjoint of $G \in \mathcal{L}(\mathcal{H}_s(K), \mathcal{H}_s(H))$ is the extension by continuity of G^{-1}. This extension $\widetilde{G^{-1}}$ is an isometric isomorphism of $\mathcal{H}_{-s}(H)$ onto $\mathcal{H}_{-s}(K)$, $s \in [0,\infty)$. For $x \in \mathcal{H}_s(H)$, $y \in \mathcal{H}_{-s}(K)$, $s \in [0,\infty)$, we define

$$[x,y] = (G^{-1}x, y)^K = (x, \widetilde{G}y)^H.$$

This duality is the extension by continuity of the Kreĭn space inner product. By (1.2) we have

$$(x,y)^H_s = [x, \widetilde{G^{-1}}((1+H)^{2s})^{\sim} y], \quad x, y \in \mathcal{H}_s(H), \quad s \in [0,\infty).$$

Since $\widetilde{G^{-1}}((1+H)^{2s})^{\sim}$ is an isometric isomorphism of $\mathcal{H}_s(H)$ onto $\mathcal{H}_{-s}(K)$ we may regard $\mathcal{H}_{-s}(K)$ as the dual of $\mathcal{H}_s(H)$ with respect to the $[\cdot,\cdot]$-duality. We define

$$\mathcal{H}_s(A) := \mathcal{H}_s(H), \quad \mathcal{H}_{-s}(A) := \mathcal{H}_{-s}(K), \quad s \in (0,\infty).$$

(For a more general setting and equivalent definitions see [8] and [3].) For every positive s the scalar products of the spaces $\mathcal{H}_s(A) \subset \mathcal{H} \subset \mathcal{H}_{-s}(A)$ are connected with one another in the same way as in Section 1.1 but with the Hilbert scalar product of the middle space replaced by the Kreĭn inner product $[\cdot,\cdot]$. Such a triplet is called a *rigging* of the Kreĭn space $(\mathcal{H}, [\cdot,\cdot])$. The natural embedding of $\mathcal{H}_s(A)$ in $\mathcal{H}_{-s}(A)$ will again be denoted by E. Observe that by (1.4) the spaces $\mathcal{H}_s(A))$, $s \in [-1,1]$, regarded as (Hilbertable) topological linear spaces depend only on A and not on the special choice of G.

Let $\mathcal{L}^{(A)}$ denote the set $\mathcal{L}(\mathcal{H}_{\frac{1}{2}}(A), \mathcal{H}_{-\frac{1}{2}}(A))$ of bounded linear operators. An operator $W \in \mathcal{L}^{(A)}$ is called *symmetric* (resp. *nonnegative*) if for all $x \in \mathcal{H}_{\frac{1}{2}}(A)$, $[Wx, x]$ is real (resp. nonnegative). Similarly to the Hilbert space case, a nonnegative $W \in \mathcal{L}^{(A)}$ is an isomorphism if and only if it is *uniformly positive*, that is, there exists an $m > 0$ such that $m\|x\|^2_{\frac{1}{2}} \leq [Wx, x],\ x \in \mathcal{H}_{\frac{1}{2}}(A)$. The operator $\widetilde{A} := \widetilde{G^{-1}}\widetilde{H} \in \mathcal{L}^{(A)}$ can be obtained by extension by continuity of A. $\widetilde{A}$ is nonnegative. It is easy to see, by interpolation, that the operator $\widetilde{A} - zE,\ z \in \mathbf{C}$, which is the extension by continuity of $A - z$, is an isomorphism of $\mathcal{H}_{\frac{1}{2}}(A)$ onto $\mathcal{H}_{-\frac{1}{2}}(A)$ if and only if $z \in \rho(A)$.

1.3. The special case of a positive operator

Assume, in addition, that A is positive, i.e. $[Ax, x] > 0$ for $x \in \mathcal{D}(A),\ x \neq 0$. We denote the completion of $\mathcal{D}(A) = \mathcal{H}_1(H)$ ($\mathcal{D}(A^{-1}) = \mathcal{H}_1(K^{-1})$) with respect to the scalar product $(\cdot,\cdot)'_{\frac{1}{2}} := [A\cdot,\cdot] = (H\cdot,\cdot)^H$ ($(\cdot,\cdot)'_{-\frac{1}{2}} := [A^{-1}\cdot,\cdot] = (K^{-1}\cdot,\cdot)^K$, resp.) by $\mathcal{H}'_{\frac{1}{2}}(A)$ ($\mathcal{H}'_{\frac{1}{2}}(A^{-1})$, resp.). The extension by continuity of A to an operator $\widehat{A}$ of $\mathcal{H}'_{\frac{1}{2}}(A)$ into $\mathcal{H}'_{\frac{1}{2}}(A^{-1})$ is an isometric isomorphism. Since for $x \in \mathcal{D}(A),\ y \in \mathcal{D}(A^{-1})$ we have

$$|[x, y]| = |[Ax, A^{-1}y]| \leq [Ax, x]^{\frac{1}{2}}[A^{-1}y, y]^{\frac{1}{2}}$$

the form $[\cdot,\cdot]$ can be extended to $\mathcal{H}'_{\frac{1}{2}}(A) \times \mathcal{H}'_{\frac{1}{2}}(A^{-1})$, and for all $x, y \in \mathcal{H}'_{\frac{1}{2}}(A)$ we have

$$[\widehat{A}x, y] = (x, y)'_{\frac{1}{2}}.$$

We emphasize that, generally, there is no natural embedding of $\mathcal{H}'_{\frac{1}{2}}(A)$ or $\mathcal{H}'_{\frac{1}{2}}(A^{-1})$ into $\mathcal{H}$. Since on $\mathcal{H}_1(A)$ the norm of $\mathcal{H}_{\frac{1}{2}}(A)$ is greater than the norm of $\mathcal{H}'_{\frac{1}{2}}(A)$ and these norms are coordinated, there is a natural embedding i' of $\mathcal{H}_{\frac{1}{2}}(A)$ in $\mathcal{H}'_{\frac{1}{2}}(A)$ of norm ≤ 1. On $\mathcal{D}(A^{-1})$ the norm of $\mathcal{H}'_{\frac{1}{2}}(A^{-1})$ is greater than the norm of $\mathcal{H}_{-\frac{1}{2}}(A)$ and again these norms are coordinated. Hence the natural embedding i'' of $\mathcal{H}'_{\frac{1}{2}}(A^{-1})$ in $\mathcal{H}_{-\frac{1}{2}}(A)$ has a norm ≤ 1. Evidently, we have $\widetilde{A} = i''\widehat{A}i'$.
We remark that an element $y \in \mathcal{H}_{-\frac{1}{2}}(A)$ belongs to $\mathcal{H}'_{\frac{1}{2}}(A^{-1})$ if and only if $[\cdot, y]$ is continuous on $\mathcal{H}'_{\frac{1}{2}}(A)$.

1.4. The operator A_r

Assume first, in addition, that $\sigma(A) \subset \mathbf{R} \setminus (-1, 1)$. Then there exists a $\delta > 0$ such that

(1.6) $$[Ax, x] \geq \delta\|x\|^2$$

for all $x \in \mathcal{D}(A)$. Hence on $\mathcal{D}(A)$ the scalar products $(\cdot,\cdot)'_{\frac{1}{2}} = [A\cdot,\cdot]$ and $(\cdot,\cdot)_{\frac{1}{2}} = [A\cdot,\cdot]+(\cdot,\cdot)^H$ are equivalent and we have $\mathcal{H}'_{\frac{1}{2}}(A) = \mathcal{H}_{\frac{1}{2}}(A)$. Let $E(\cdot)$ be the spectral function of A. Since for any compact interval Δ,

$$[AE(\Delta)x, E(\Delta)x] \leq [AE(\Delta)x, E(\Delta)x]^{\frac{1}{2}}[Ax,x]^{\frac{1}{2}}, \qquad x \in \mathcal{D}(A),$$

and, hence,

$$(E(\Delta)x, E(\Delta)x)'_{\frac{1}{2}} \leq (x,x)'_{\frac{1}{2}}, \qquad x \in \mathcal{H}_{\frac{1}{2}}(A),$$

the projections $E(\Delta)$ restricted to $\mathcal{H}_{\frac{1}{2}}(A)$ are uniformly bounded. They are selfadjoint with respect to $(\cdot,\cdot)'_{\frac{1}{2}}$. We have $E(\Delta)\mathcal{H}_{\frac{1}{2}}(A) = E(\Delta)\mathcal{H}$ and on this range $(\cdot,\cdot)$ and $(\cdot,\cdot)'_{\frac{1}{2}}$ are equivalent. Since the sequences $(E([1,n)))$ and $(E((-n,-1]))$ are monotone with respect to $(\cdot,\cdot)'_{\frac{1}{2}}$ the strong limits

$$s-\lim_{n\to\infty} E([1,n)) =: E_+, \qquad s-\lim_{n\to\infty} E((-n,-1]) =: E_-$$

exist in $\mathcal{H}_{\frac{1}{2}}(A)$. Hence, by (1.6), for $x \in \mathcal{H}_{\frac{1}{2}}(A)$, the sequences $(E([1,n))x)$ and $(E((-n,-1])x)$ converge also in $\mathcal{H}$. Evidently,

$$E_+^2 = E_+, \quad E_-^2 = E_-, \quad E_+E_- = E_-E_+ = 0. \tag{1.7}$$

As the linear space

$$\mathcal{L}'_{(\infty)} := \bigcup_{n=2}^{\infty} E((-n,n))\mathcal{H}$$

is dense in $\mathcal{D}(A)$ with respect to the graph norm and, hence, in $\mathcal{H}_{\frac{1}{2}}(A)$, and $\lim_{n\to\infty} E([1,n))y + \lim_{n\to\infty} E((-n,-1])y = y$ for $y \in \mathcal{L}'_{(\infty)}$, we obtain

$$E_+ + E_- = I. \tag{1.8}$$

Now it follows that $E(\cdot)$ restricted to $(\mathcal{H}'_{\frac{1}{2}}(A), (\cdot,\cdot)'_{\frac{1}{2}})$ is the spectral function of a selfadjoint operator. Let J_A be the operator in $\mathcal{H}_{\frac{1}{2}}(A)$ defined by

$$J_A := E_+ - E_-.$$

Then by (1.7) and (1.8) we have $J_A^2 = 1$. The inner product $(\cdot,\cdot)_r$ on $\mathcal{H}_{\frac{1}{2}}(A)$ defined by

$$(x,y)_r = [J_A x, y], \qquad x, y \in \mathcal{H}_{\frac{1}{2}}(A),$$

is positive definite, and J_A is symmetric with respect to $(\cdot,\cdot)_r$.
Let $\mathcal{H}_r(A)$ or, shortly, $\mathcal{H}_r$ denote the completion of $\mathcal{H}_{\frac{1}{2}}(A)$ with respect to $(\cdot,\cdot)_r$. If the basic space is not $\mathcal{H}$ but another Kreĭn space, say $\mathcal{K}$, we shall write $\mathcal{K}_r$ instead of $\mathcal{H}_r$. The space $\mathcal{H}_r$ was introduced in [2] in a way similar to Lemma 1.1 below (see also [6]). For other equivalent descriptions of $\mathcal{H}_r$ see [3]. We extend J_A by

continuity to a selfadjoint and unitary operator in $\mathcal{H}_r$, which also will be denoted by J_A. Then the relation

$$[x, y] = (J_A x, y)_r, \qquad x, y \in \mathcal{H}_r,$$

extends the indefinite form from $\mathcal{H}_{\frac{1}{2}}(A)$ to $\mathcal{H}_r$, and $(\mathcal{H}_r, [\cdot,\cdot])$ is a Kreĭn space. Let $[X, Y]_{\frac{1}{2}}$ denote the interpolated space between X and Y with parameter $\frac{1}{2}$ where X and Y are Hilbert spaces satisfying the usual assumptions (see [12]).

Lemma 1.1. *The space $\mathcal{H}_r$ regarded as a linear topological space coincides with $[\mathcal{H}_{\frac{1}{2}}(A), \mathcal{H}_{-\frac{1}{2}}(A)]_{\frac{1}{2}}$.*

Proof. Since $E(\cdot)$ is a spectral function on $(\mathcal{H}_{\frac{1}{2}}(A), (\cdot,\cdot)'_{\frac{1}{2}})$, the left hand sides of the relations

$$\begin{aligned} &\int d(E(t)x, x)'_{\frac{1}{2}} = (x, x)'_{\frac{1}{2}}, \qquad \int |t|^{-1} d(E(t)x, x)'_{\frac{1}{2}} = (x, x)_r, \\ &\int |t|^{-2} d(E(t)x, x)'_{\frac{1}{2}} = [A^{-1}x, x] \end{aligned} \tag{1.9}$$

are continuous quadratic forms on $\mathcal{H}_{\frac{1}{2}}(A)$. As the left hand sides of the relations (1.9) coincide with the corresponding right hand sides on $\mathcal{L}'_{(\infty)}$, these relations are true for all $x \in \mathcal{H}_{\frac{1}{2}}(A)$. Then the lemma follows from the equivalence of the scalar products $[A^{-1}\cdot,\cdot]$ and $(\cdot,\cdot)_{-\frac{1}{2}}$ on $\mathcal{H}_{\frac{1}{2}}(A)$. □

Lemma 1.2.([2]) *The scalar products $(\cdot,\cdot)^H$ and $(\cdot,\cdot)_r$ on $\mathcal{H}_{\frac{1}{2}}(A)$ are topologically equivalent (or, equivalently, $\mathcal{H} = \mathcal{H}_r$) if and only if $\infty \notin c_s(A)$.*

Proof. If $\infty \notin c_s(A)$ then $J'_A := s - \lim_{n\to\infty} (E([1, n)) - E((-n, -1]))$ exists in $\mathcal{H}$ and is a fundamental symmetry of $\mathcal{H}$. Hence, on $\mathcal{H}_{\frac{1}{2}}(A)$, $[J'_A\cdot,\cdot] = (\cdot,\cdot)_r$ and $(\cdot,\cdot)^H$ are equivalent. Let, on the other hand, $(\cdot,\cdot)_r$ and $(\cdot,\cdot)^H$ be equivalent on $\mathcal{H}_{\frac{1}{2}}(A)$. By the Kreĭn Lemma from the definition of $(\cdot,\cdot)_r$ it follows that the spectral function of A restricted to $\mathcal{H}_{\frac{1}{2}}(A)$ is uniformly bounded with respect to $(\cdot,\cdot)_r$. Hence it is uniformly bounded with respect to $(\cdot,\cdot)^H$ and $\infty \notin c_s(A)$. □

Now we define an operator in $\mathcal{H}_r$, which is closely connected with the operator A. The operator A maps $\mathcal{L}'_{(\infty)}$ bijectively onto itself and $\mathcal{L}'_{(\infty)}$ is dense in $\mathcal{H}_r$. The operator $A|\mathcal{L}'_{(\infty)}$ is symmetric in $(\mathcal{H}_r, (\cdot,\cdot)_r)$ and nonnegative in $(\mathcal{H}_r, [\cdot,\cdot])$. Let A_r denote the closure of $A|\mathcal{L}'_{(\infty)}$ in $\mathcal{H}_r$.

Lemma 1.3. $A_r = \widetilde{A}\,|\{x \in \mathcal{H}_{\frac{1}{2}}(A) : \widetilde{A}x \in \mathcal{H}_r\}$.

Proof. By (1.6) $\widetilde{A}$ is an isomorphism of $\mathcal{H}_{\frac{1}{2}}(A)$ onto $\mathcal{H}_{-\frac{1}{2}}(A)$. Since $\mathcal{H}_r$ is continuously embedded in $\mathcal{H}_{-\frac{1}{2}}(A)$, the right side of the above relation is a closed operator with bounded everywhere defined inverse. Then, in view of the density of $\mathcal{L}'_{(\infty)}$ in $\mathcal{H}_r$, the lemma follows. □

Lemma 1.4.([6]) *The boundedly invertible operator A_r is nonnegative and selfadjoint in $(\mathcal{H}_r, [\cdot,\cdot])$ and selfadjoint in $(\mathcal{H}_r, (\cdot,\cdot)_r)$, hence $\infty \notin c_s(A_r)$. Moreover, the following holds.*

(i) $\sigma(A) = \sigma(A_r)$.

(ii) *If E_r denotes the spectral function of A_r, then for every $s > 0$ we have*

$$E((-s,s))\mathcal{H} = E_r((-s,s))\mathcal{H}_r \subset \mathcal{H}_{\frac{1}{2}}(A)$$

and A and A_r coincide on $E((-s,s))\mathcal{H}$.

Proof. That A_r is boundedly invertible was shown in the proof of Lemma 1.3. The first assertions follow from the properties of $A^{-1}|\mathcal{L}'_{(\infty)}$. Then statement (i) is a consequence of (ii).
By the definition of A_r the restrictions of $(A-z)^{-1}$ and $(A_r - z)^{-1}$, $z \neq \overline{z}$, to $\mathcal{H}_{\frac{1}{2}}(A)$ coincide. It is easy to see that this restriction is the resolvent of a selfadjoint operator in $(\mathcal{H}_{\frac{1}{2}}(A), (\cdot,\cdot)'_{\frac{1}{2}})$. Since the spectral projections of A and A_r can be written as strong limits of contour integrals over the resolvent (see [11]), it follows that for every $x \in \mathcal{H}_{\frac{1}{2}}(A)$ we have

$$E((-s,s))x = E_r((-s,s))x \in \mathcal{H}_{\frac{1}{2}}(A). \tag{1.10}$$

Now let $y \in E((-s,s))\mathcal{H} = E((-s,s))(E((-s,s))\mathcal{H})$ then there exists an $x \in \mathcal{H}_{\frac{1}{2}}(A)$ such that $y \in E_r((-s,s))x$ and, hence, by (1.10), $y \in E_r((-s,s))\mathcal{H}_r$. The converse inclusion follows by a similar reasoning. By the definition of A_r, A and A_r coincide on $E((-s,s))\mathcal{H}$. □

We consider the scale of Hilbert spaces connected with A_r. For the operator G from Section 1.2 we choose J_A. Then $H_r := J_A A_r$ and $K_r := A_r J_A$ coincide and we have $(\cdot,\cdot)^{H_r} = (\cdot,\cdot)^{K_r} = (\cdot,\cdot)_r$. Hence J_A is an isometric isomorphism of $(\mathcal{H}_r)_s(A_r)$, $s \geq 0$, and can be extended by continuity to an isometric isomorphism of

$$(\mathcal{H}_r)_{-s}(A_r), \qquad s \geq 0.$$

Lemma 1.5.([6])

$$(\mathcal{H}_r)_{\frac{1}{2}}(A_r) = \mathcal{H}_{\frac{1}{2}}(A), \quad (\mathcal{H}_r)_{-\frac{1}{2}}(A_r) = \mathcal{H}_{-\frac{1}{2}}(A).$$

Proof. $(\mathcal{H}_r)_{\frac{1}{2}}(A_r)$ is the completion of $\mathcal{D}(A_r)$ with respect to the scalar product $[A_r\cdot,\cdot]$. Since $\mathcal{L}'_{(\infty)}$ is dense in $\mathcal{H}_r$ and we have $0 \in \rho(A_r)$, $\mathcal{L}'_{(\infty)}$ is dense in $\mathcal{D}(A_r)$.

But for $x \in \mathcal{L}'_{(\infty)}$ we have $[A_r x, x] = [Ax, x]$. This implies the first relation. The second relation follows from the fact that $(\mathcal{H}_r)_{-\frac{1}{2}}(A_r)$ $(\mathcal{H}_{-\frac{1}{2}}(A))$ is the completion of $(\mathcal{H}_r)_{\frac{1}{2}}(A_r)$ $(\mathcal{H}_{\frac{1}{2}}(A), \text{resp.})$ with respect to the dual norm. □

In this section up to now we have been working under the assumption that $\sigma(A) \subset \mathbf{R} \setminus (-1,1)$. For the general case, set $E^0 := E((-1,1))$, $E^\infty := (1 - E((-1,1)))$, $A^0 := A|E^0\mathcal{H}$, $A^\infty := A|E^\infty\mathcal{H}$. The operator A^∞ fulfils the above assumptions. We define the Kreĭn space $\mathcal{H}_r(A)$ to be the product of the Kreĭn spaces $E^0\mathcal{H}$ and $(E^\infty\mathcal{H})_r(A^\infty)$,

$$\mathcal{H}_r(A) = E^0\mathcal{H} \times (E^\infty\mathcal{H})_r(A^\infty),$$

and the definite scalar product $(\cdot,\cdot)_r$ by

$$(x^0, y^0)^H + (x^\infty, y^\infty)_r, \qquad \begin{pmatrix} x^0 \\ x^\infty \end{pmatrix}, \begin{pmatrix} y^0 \\ y^\infty \end{pmatrix} \in \mathcal{H}_r(A).$$

In this general case Lemmas 1.1 and 1.2 remain valid.

We define A_r to be the orthogonal sum of A^0 and $(A^\infty)_r$. A_r is selfadjoint in $\mathcal{H}_r(A)$ with respect to the Kreĭn space inner product but, in general, not necessarily with respect to $(\cdot,\cdot)_r$. Evidently, Lemmas 1.3 and 1.4 remain true with the exception of the statement on bounded invertibility. By (1.3) the scale corresponding to A_r is

$$(\mathcal{H}_r)_s(A_r) = E^0\mathcal{H} \times ((E^\infty\mathcal{H})_r)_s((A^\infty)_r), \qquad s \in \mathbf{R}.$$

By relation (1.3) Lemma 1.5 remains valid in the general case. We emphasize that by the first relation of Lemma 1.5 for the general case there exist positive numbers m and M such that

$$\begin{aligned} m(\|x\|^{H2} + [\widetilde{A}x, x]) &\le (x,x)_r + [A_r x, x] \\ &\le M(\|x\|^{H2} + [\widetilde{A}x, x]), \quad x \in \mathcal{D}(A_r). \end{aligned} \tag{1.11}$$

2. Relatively form bounded perturbations of nonnegative operators in Kreĭn spaces

2.1. Definition of relatively form bounded perturbations

Let $(\mathcal{H}, [\cdot,\cdot])$ be the a Kreĭn space and let A, G, H, K be as in Section 1.2. For simplicity we write here $(x,y) := (x,y)^H = [G^{-1}x, y]$, $x, y \in \mathcal{H}$. The scalar products $(H^{\frac{1}{2}}\cdot, H^{\frac{1}{2}}\cdot) = (\widetilde{H}\cdot,\cdot) = [\widetilde{A}\cdot,\cdot] = (\cdot,\cdot)'_{\frac{1}{2}}$ coincide on $\mathcal{H}_{\frac{1}{2}}(A) = \mathcal{H}_{\frac{1}{2}}(H)$. The corresponding closed quadratic form will be denoted by h:

$$h(x) = (\widetilde{H}x, x), \qquad x \in \mathcal{H}_{\frac{1}{2}}(H).$$

We denote by v a quadratic form which is relatively bounded with respect to h, that is, v is defined on the domain $\mathcal{H}_{\frac{1}{2}}(H)$ of h and there exists positive numbers α', α'', β', $\beta'' \geq 0$ such that

$$(2.1) \qquad -\alpha' \|x\|^2 - \beta' h(x) \leq v(x) \leq \alpha'' \|x\|^2 + \beta'' h(x), \qquad x \in \mathcal{H}_{\frac{1}{2}}(H).$$

This holds if and only if v is continuous on $\mathcal{H}_{\frac{1}{2}}(H)$ or, equivalently, v has the form

$$v(x) = (Vx, x) \qquad \text{or} \qquad v(x) = [Wx, x], \qquad x \in \mathcal{H}_{\frac{1}{2}}(H),$$

with some $V \in \mathcal{L}^{(H)}$ or some $W \in \mathcal{L}^{(A)}$. In the following we assume, in addition, that $h + v$ is positive semidefinite and (2.1) holds with some $\beta' < 1$. Writing (2.1) in the form

$$(2.2) \qquad \begin{aligned} \|x\|^2 + (1-\beta')h(x) &\leq h(x) + v(x) + (\alpha' + 1)\|x\|^2 \\ &\leq (\alpha' + \alpha'' + 1)\|x\|^2 + (\beta'' + 1)h(x) \end{aligned}$$

we see that our additional assumptions are fulfilled if and only if the following conditions (i) and (ii) hold:

(i) $\widetilde{H} + V$ is nonnegative with respect to $(\cdot, \cdot)$.

(ii) There exist γ, $\delta > 0$ such that

$$(2.3) \qquad \gamma \|x\|^2_{\frac{1}{2}} \leq ((\widetilde{H} + V + \delta E)x, x), \qquad x \in \mathcal{H}_{\frac{1}{2}}(H).$$

Conditions (i) and (ii) are equivalent to the following conditions (i′) and (ii′), respectively.

(i′) $\widetilde{A} + W$ is nonnegative with respect to $[\cdot, \cdot]$.

(ii′) There exist γ, $\delta > 0$ such that

$$\gamma \|x\|^2_{\frac{1}{2}} \leq [(\widetilde{A} + W + \delta \widetilde{G^{-1}} E)x, x], \qquad x \in \mathcal{H}_{\frac{1}{2}}(A).$$

From (ii) it follows that $\widetilde{H} + V + \delta E$ is an isomorphism of $\mathcal{H}_{\frac{1}{2}}(H)$ onto $\mathcal{H}_{-\frac{1}{2}}(H)$. Then we may define a closed operator $H \dot{+} V$ in $\mathcal{H}$ by

$$\begin{aligned} \mathcal{D}(H \dot{+} V) &:= \{x \in \mathcal{H}_{\frac{1}{2}}(H) : (\widetilde{H} + V)x \in \mathcal{H}\}, \\ (H \dot{+} V)x &:= (\widetilde{H} + V)x, \qquad x \in \mathcal{D}(H \dot{+} V). \end{aligned}$$

The operator $H \dot{+} V$ is selfadjoint and nonnegative in the Hilbert space $(\mathcal{H}, (\cdot, \cdot))$ and we have $-\delta \in \rho(H \dot{+} V)$. $H \dot{+} V$ is the operator which is connected with the perturbed closed (by (2.3)) quadratic form $h+v$ (see e.g. [9], [16]). On account of (2.3) we have

(2.4) $$\mathcal{H}_{\frac{1}{2}}(H\dot{+}V) = \mathcal{H}_{\frac{1}{2}}(H), \quad \mathcal{H}_{-\frac{1}{2}}(H\dot{+}V) = \mathcal{H}_{-\frac{1}{2}}(H).$$

The operator

$$B := G^{-1}(H\dot{+}V),$$

which is a nonnegative and selfadjoint operator in the Kreĭn space $\mathcal{H}$, will be regarded as the operator obtained from $A = G^{-1}H$ by the form perturbation v. As Example 2.2 below shows B may have an empty resolvent set. Since $\widetilde{A} = \widetilde{G^{-1}}\widetilde{H}$, $W = \widetilde{G^{-1}}V$ and G is an isomorphism,

$$\begin{aligned} \mathcal{D}(B) &= \{x \in \mathcal{H}_{\frac{1}{2}}(A) : (\widetilde{A} + W)x \in \mathcal{H}\}, \\ Bx &= (\widetilde{A} + W)x, \qquad x \in \mathcal{D}(B), \end{aligned}$$

i.e., B coincides with the operator $A\dot{+}W$ introduced and studied in [6] (see also [8], [7]).

2.2. The critical point ∞

The following theorem gives conditions for B to have a nonempty resolvent set and for the preservation of regularity or singularity of the critical point ∞. For $\beta' < \frac{1}{2}$ and $\alpha' = \alpha'' = 0$ the assertions on the critical point ∞ were proved in [4].

Theorem 2.1.

(a) *If $\infty \notin c_s(A))$ holds, then the resolvent set of B is nonempty and we have $\infty \notin c_s(B)$.*

(b) *Assume, in addition, that $\alpha' = 0$ holds (see (2.1)). Then the resolvent set of B is nonempty and we have $\infty \in c_s(A)$ if and only if $\infty \in c_s(B)$ holds.*

Proof. (a) By relation (ii′) there exists a $\delta > 0$ such that $\widetilde{A} + W + \delta\widetilde{G^{-1}}E$ is an isomorphism of $\mathcal{H}_{\frac{1}{2}}(A)$ onto $\mathcal{H}_{-\frac{1}{2}}(A)$. Hence $B + \delta G^{-1}$ is a uniformly positive selfadjoint operator in $(\mathcal{H}, [\cdot,\cdot])$, $0 \in \rho(B + \delta G^{-1})$. In view of (2.4) and (1.4) we have

(2.5) $$\mathcal{H}_{\frac{1}{2}}(A) = \mathcal{H}_{\frac{1}{2}}(B) = \mathcal{H}_{\frac{1}{2}}(B + \delta G^{-1}).$$

Then by the criterion of B. Ćurgus [2] it follows from $\infty \notin c_s(A)$ that $\infty \notin c_s(B + \delta G^{-1})$. Therefore, we have $\lim_{\eta\to\infty} \|(B + \delta G^{-1} \mp i\eta)^{-1}\| = 0$, and, since δG^{-1} is bounded, the resolvent set of B is not empty. Using again (2.5) and the Ćurgus result we find $\infty \notin c_s(B)$.

(b) Let now $\alpha' = 0$. Since the norms $x \mapsto (\|x\|^2 + h(x))^{\frac{1}{2}}$ and $x \mapsto (\|x\|_r^2 + h(x))^{\frac{1}{2}}$ are equivalent on $\mathcal{H}_{\frac{1}{2}}(A) = (\mathcal{H}_r)_{\frac{1}{2}}(A_r)$ (see (1.11)) relation (2.1) with $\alpha' = 0$ implies

$$-\beta'\|x\|_r^2 - \beta' h(x) \le v(x) \le \alpha'''\|x\|_r^2 + \beta''' h(x), \qquad x \in \mathcal{H}_{\frac{1}{2}}(A),$$

with some positive α''' and β'''. Since $\infty \notin c_s(A_r)$, by the first part of the proof the nonnegative selfadjoint operator $B^{(r)}$ in $(\mathcal{H}_r, [\cdot,\cdot])$ defined by

$$\begin{aligned} \mathcal{D}(B^{(r)}) &= \{x \in \mathcal{H}_{\frac{1}{2}}(A) : (\widetilde{A} + W)x \in \mathcal{H}_r\}, \\ B^{(r)}x &= (\widetilde{A} + W)x, \quad x \in \mathcal{D}(B^{(r)}), \end{aligned}$$

has a nonempty resolvent set, in particular $i, -i \in \rho(B^{(r)})$. By (2.4) and Lemma 1.5 (for the general case, see the end of Section 1)

$$(\mathcal{H}_r)_{\pm\frac{1}{2}}(B^{(r)}) = (\mathcal{H}_r)_{\pm\frac{1}{2}}(A_r) = \mathcal{H}_{\pm\frac{1}{2}}(A).$$

Hence the operators $\widetilde{A} + W \pm iE$ are isomorphisms. It follows that $i, -i \in \rho(B)$. Then the first equality in (2.5) implies that we have $\infty \in c_s(A)$ if and only if $\infty \in c_s(B)$. □

The following example shows that without the assumption $\alpha' = 0$ Theorem 2.1, (b), is in general not true.

Example 2.2. Let $(\mathcal{H}', (\cdot,\cdot)')$ be a Hilbert space and H' an unbounded positive selfadjoint operator in $\mathcal{H}'$. Assume that $\min \sigma(H') =: m > 0$. Let

$$\mathcal{H} := \mathcal{H}' \times \mathcal{H}', \; J := \begin{pmatrix} 0 & I \\ I & 0 \end{pmatrix}, \left[\begin{pmatrix} x_1 \\ x_2 \end{pmatrix}, \begin{pmatrix} y_1 \\ y_2 \end{pmatrix}\right] := \left(J\begin{pmatrix} x_1 \\ x_2 \end{pmatrix}, \begin{pmatrix} y_1 \\ y_2 \end{pmatrix}\right)_{\mathcal{H}}.$$

The operator $A = \left(\begin{smallmatrix} 0 & m \\ H' & 0 \end{smallmatrix}\right)$, $m > 0$, is a positive selfadjoint operator in the Kreĭn space $(\mathcal{H}, [\cdot,\cdot])$ and we have $0 \in \rho(A)$. Now let $v(x) = -m\|x\|_{\mathcal{H}}^2$, $x \in \mathcal{H}_{\frac{1}{2}}(A)$. Then $B = A - mJ = \left(\begin{smallmatrix} 0 & 0 \\ H' - m & 0 \end{smallmatrix}\right)$, and we have $\rho(B) = \emptyset$ (cf. [10, Section 1.2]).

By well-known relations between relative boundedness and relative form boundedness (see [9, Theorem VI.1.38, Theorem V.4.12]) Theorem 2.1 implies the following corollary. We recall the definition of the Friedrichs extension of a nonnegative operator T in $(\mathcal{H}, [\cdot,\cdot])$. Let T^+ be the Krein space adjoint of T and let $\mathcal{D}[T]$ be the completion of $\mathcal{D}(T)$ with respect to the scalar product $(\cdot,\cdot)^H + [T\cdot,\cdot]$. Then $T_F := T^+|\mathcal{D}[T] \cap \mathcal{D}(T^+)$ is a nonnegative selfadjoint operator in $\mathcal{H}$. It is called the *Friedrichs extension* of T. We have $T_F = G^{-1}(GT)_F$ where $(GT)_F$ is the Friedrichs extension of the nonnegative operator GT in the Hilbert space $(\mathcal{H}, (\cdot,\cdot))$.

Corollary 2.3. *Let W_0 be a symmetric A-bounded operator in the Kreĭn space $(\mathcal{H}, [\cdot,\cdot])$. Assume that there exist nonnegative H-bounded operators $V_{0,+}$ and $V_{0,-}$ in $(\mathcal{H}, (\cdot,\cdot))$ such that*

$$GW_0 = V_{0,+} - V_{0,-},$$

the H-bound of $V_{0,-}$ is less than one and $H - V_{0,-}$ is nonnegative in $(\mathcal{H}, (\cdot,\cdot))$. Then for the nonnegative selfadjoint operator $(A+W_0)_F$ the following holds.
(a) *If $\infty \notin c_s(A)$ then the resolvent set of $(A+W_0)_F$ is nonempty and we have $\infty \notin c_s((A+W_0)_F)$.*
(b) *Assume, in addition, that, for some $b_- < 1$,*

$$\|V_{0,-}x\| \leq b_- \|Hx\|, \quad x \in \mathcal{D}(H).$$

Then the resolvent set of $(A+W_0)_F$ is nonempty and we have $\infty \in c_s(A)$ if and only if $\infty \in c_s((A+W_0)_F)$.

2.3. The critical point 0

Now we return to the assumptions of Section 2.1. The following theorem was proved in [4] under the assumptions $\rho(B) \neq \emptyset$ and $\beta' < \frac{1}{2}$.

Theorem 2.4. *Assume, in addition, that $[Ax, x] > 0$ for all $x \in \mathcal{D}(A)$, $x \neq 0$, and that the constants α' and α'' in (2.1) are zero. Then $0 \in c_s(A)$ if and only if $0 \in c_s(B)$.*

Proof. By Theorem 2.1 B is a nonnegative selfadjoint operator in $\mathcal{H}$ with nonempty resolvent set. By (2.1) for $x \in \mathcal{H}_{\frac{1}{2}}(A)$ we have

$$-\beta'[\widetilde{A}x, x] \leq [Wx, x] \leq \beta''[\widetilde{A}x, x]$$

and, hence,

$$(1-\beta')[\widetilde{A}x, x] \leq [(\widetilde{A}+W)x, x] \leq (1+\beta'')[\widetilde{A}x, x], \tag{2.6}$$

which implies that B is injective. By (2.6) there exists an $M > 0$ such that

$$|[(\widetilde{A}+W)x, y]| \leq M[\widetilde{A}x, x]^{\frac{1}{2}}[\widetilde{A}y, y]^{\frac{1}{2}}, \quad x, y \in \mathcal{H}_{\frac{1}{2}}(A). \tag{2.7}$$

Thus for every $x \in \mathcal{H}_{\frac{1}{2}}(A)$ the functional $[\cdot, (\widetilde{A}+W)x]$ is continuous with respect to $(\cdot,\cdot)'_{\frac{1}{2}}$, hence $(\widetilde{A}+W)x$ belongs to $\mathcal{H}'_{\frac{1}{2}}(A^{-1})$. By (2.7) $\widetilde{A}+W$ can be extended by continuity to a bounded linear operator $(\widetilde{A}+W)\widehat{}$ from $\mathcal{H}'_{\frac{1}{2}}(A)$ in $\mathcal{H}'_{\frac{1}{2}}(A^{-1})$. Relation (2.6) shows that $(\widetilde{A}+W)\widehat{}$ is an isomorphism of $\mathcal{H}'_{\frac{1}{2}}(A)$ onto $\mathcal{H}'_{\frac{1}{2}}(A^{-1})$. Since $\mathcal{D}(B)$ is dense in $\mathcal{H}_{\frac{1}{2}}(A)$ and, hence, in $\mathcal{H}'_{\frac{1}{2}}(A)$, $(\widetilde{A}+W)\mathcal{D}(B) = \mathcal{R}(B)$ is dense in $\mathcal{H}'_{\frac{1}{2}}(A^{-1})$. Let $y \in \mathcal{R}(B)$ and $x = B^{-1}y$. We have $x \in \mathcal{H}_{\frac{1}{2}}(A)$ and $x = (\widetilde{A}+W)\widehat{}^{-1}y$. Then by (2.6)

$$\begin{aligned}(1-\beta')[\widehat{A}(\widetilde{A}+W)\widehat{}^{-1}y, (\widetilde{A}+W)\widehat{}^{-1}y] &\leq [y, B^{-1}y] \\ &\leq (1+\beta'')[\widehat{A}(\widetilde{A}+W)\widehat{}^{-1}y, (\widetilde{A}+W)\widehat{}^{-1}y].\end{aligned}$$

Therefore, the scalar products $[B^{-1}\cdot, \cdot]$ and $(\cdot,\cdot)'_{-\frac{1}{2}}$ are equivalent on $\mathcal{R}(B)$. Hence $\mathcal{H}_{\frac{1}{2}}(B^{-1}) = \mathcal{H}_{\frac{1}{2}}(A^{-1})$ and by the Ćurgus criterion Theorem 2.4 is proved. □

Remark 2.5. An example given by R.V.Akopyan [1] shows that there exists a bounded positive operator A with a regular critical point 0 such that the regularity of 0 can be destroyed by a bounded nonnegative or nonpositive rank one perturbation of arbitrarily small norm. Hence neither the assumption $\alpha' = 0$ nor $\alpha'' = 0$ can be dropped.

As a consequence of Theorem 2.4 with the help of [9, Theorem V.4.12] we obtain the following corollary.

Corollary 2.6. *Assume that* $[Ax, x] > 0$ *for all* $x \in \mathcal{D}(A)$, $x \neq 0$. *Let* W_0 *be a symmetric* A*-bounded operator in* $(\mathcal{H}, [\cdot, \cdot])$ *fulfilling the assumptions of* Corollary 2.3, (b). *Assume, further, that for some positive* b_+

$$\|V_{0,+}x\| \leq b_+ \|Hx\|, \quad x \in \mathcal{D}(A).$$

Then $0 \in c_s(A)$ *if and only if* $0 \in c_s((A + W_0)_F)$.

References

[1] R.V. Akopyan, On the theory of the spectral function of a J-nonnegative operator, Izv. Akad. Nauk Armyanskoi SSR, **13**(1978), 114–121.

[2] B. Ćurgus, On the regularity of the critical point infinity of definitizable operators, Integral Equations and Operator Theory, **8**(1985), 462–488.

[3] B. Ćurgus and B. Najman, A Kreĭn space approach to elliptic eigenvalue problems with indefinite weights, Differential and Integral Equations, **7**(1994), 1241-1252.

[4] B. Ćurgus and B. Najman, Perturbations of range, Proc. Amer. Math. Soc. (to appear)

[5] P. Jonas, Compact perturbations of definitizable operators. II, J. Operator Theory, **8**(1982), 3–18.

[6] P. Jonas, On a problem of the perturbation theory of selfadjoint operators in Kreĭn space, J. Operator Theory, **25**(1991), 183–211.

[7] P. Jonas, On the spectral theory of operators associated with perturbed Klein–Gordon and wave type equations, J. Operator Theory, **29**(1993), 207–224.

[8] P. Jonas and H. Langer, Some questions in the perturbations theory of J-nonnegative operators in Kreĭn space, Math. Nachr. **114**(1983), 205–226.

[9] T. Kato, Perturbation theory for linear operators, Springer Verlag, New York, 1966.

[10] H. Langer, Verallgemeinerte Resolventen eines J-nichtnegativen Operators mit endlichem Defekt, J. Functional Analysis, **8**(1971), 287–320.

[11] H. Langer, Spectral functions of definitizable operators in Kreĭn spaces, Functional Analysis, Proceedings of a conference held at Dubrovnik, Lecture Notes in Mathematics, **948**, Springer Verlag, Berlin–Heidelberg–New York, 1982, 1–46.

[12] J.-L. LIONS, E. MAGENES, Problèmes aux limites non homogènes et applications, Vol. I, Paris, 1968.

[13] B. NAJMAN, Solution of a differential equation in a scale of spaces, Glasnik Matematički, **14**(1979),119-127.

[14] B. NAJMAN, Trace class perturbations and scattering theory for the equations of Klein-Gordon type, Glasnik Matematički, **15**(1980), 79-86.

[15] B. NAJMAN, Spectral properties of the operators of Klein-Gordon type, Glasnik Matematički, **15**(1980), 97-112.

[16] M. REED, B. SIMON, Methods of Modern Mathematical Physics, II: Fourier Analysis, Selfadjointness, Academic Press, New York, San Francisco, London 1975.

[17] K. VESELIĆ, On spectral properties of a class of J-selfadjoint operators, I, Glasnik Matematički, **7**(1972), 229–247.

[18] K. VESELIĆ, On spectral properties of a class of J-selfadjoint operators, II, Glasnik Matematički, **7**(1972), 249–254.

[19] K. VESELIĆ, A spectral theory of the Klein-Gordon equation involving a homogeneous electric field, J. Operator Theory, **25** (1991), 319-330.

Neltestr. 12
D-12489 Berlin
Germany

1991 Mathematics Subject Classification. Primary 47B50, 47A55

Operator Theory:
Advances and Applications, Vol. 106

Norm bounds for Volterra integral operators and time–varying linear systems with finite horizon

M.A. KAASHOEK AND A.C.M. RAN

Dedicated to Heinz Langer on the occasion of his sixtieth birthday.

Norm bounds are given for Volterra integral operators that appear as input-output operators of finite-dimensional time-varying linear systems on a finite interval. As an application a known result on state feedback H^∞-control for such systems is derived.

0. Introduction

In this note we study norm bounds of input-output operators of time-varying causal linear systems. More precisely, let

$$\begin{cases} \dot{x}(t) & = A(t)x(t) + B(t)u(t) \qquad 0 \le t \le \tau \\ y(t) & = C(t)x(t) \\ x(0) & = 0 \end{cases} \tag{0.1}$$

be a causal time varying linear system. Here A is assumed to be an integrable $n \times n$ matrix function of t on the interval $0 \le t \le \tau < \infty$, B and C are assumed to be square integrable on $0 \le t \le \tau$. We shall take inputs $u \in L_2^p[0, \tau]$. Given such u the system produces as output a function $y \in L_2^m[0, \tau]$. The input-output operator corresponding to this system is the linear operator $G : L_2^p[0, \tau] \to L_2^m[0, \tau]$ defined by $(Gu)(t) = y(t)$. More precisely, G is given by

$$y(t) = (Gu)(t) = C(t)U(t) \int_0^t U(s)^{-1} B(s) u(s)\, ds, \qquad 0 \le t \le \tau,$$

where $U(t)$ is the solution of $\dot{U}(t) = A(t)U(t), 0 \le t \le \tau, U(0) = I$. In this paper we are interested in bounds on the induced operator norm $\|G\|$. The following theorem is our main result.

Theorem 0.1. *The following are equivalent*
(a) $\|G\| < \gamma$,
(b) *there is a solution* $P(t)$, $0 \le t \le \tau$, *of the Riccati differential equation*

$$\begin{aligned} - \quad & \dot{P}(t) = A(t)^* P(t) + P(t)A(t) + \gamma^{-2} P(t)B(t)B(t)^* P(t) + C(t)^* C(t), \\ & P(\tau) = 0, \end{aligned} \tag{0.2}$$

(c) *there is a function $P(t)$, $0 \leq t \leq \tau$, satisfying the Riccati differential inequality*

$$\begin{aligned}&\dot{P}(t) + A(t)^* P(t) + P(t)A(t) + \gamma^{-2} P(t)B(t)B(t)^* P(t) + C(t)^* C(t) < 0,\\ &P(\tau) = 0.\end{aligned} \tag{0.3}$$

Our approach to the proof of Theorem 0.1 is based on the following observation. We have $\|G\| < \gamma$ if and only if $I - \gamma^{-2} G^* G > 0$, which in turn is equivalent to a factorization

$$I - \gamma^{-2} G^* G = (I - V^*)(I - V), \tag{0.4}$$

with $I - V$ invertible. We employ the results of [GK1] to obtain this factorization explicitly, given that the Riccati differential equation (0.2) has a solution. In fact, denoting by $P(t)$ the solution of (0.2) we show that (0.4) holds with V given by

$$(Vf)(t) = \gamma^{-2} B(t)^* P(t) U(t) \int_0^t U(s)^{-1} B(s) f(s) ds, \qquad 0 \leq t \leq \tau.$$

Conversely, also using [GK1], we shall show that if $\|G\| < \gamma$, then (0.2) has a solution.

Theorem 0.1 is the analogue for time varying sytems on a finite horizon of the bounded real lemma. A related result is Lemma 2.2 in [LAH], which is stated and proved in terms of standard optimal control theory and concerns the implication (a) implies (b).

The paper consists of three sections. Theorem 0.1 is proved in Section 1. In Section 2 we illustrate the results of Theorem 0.1 on a number of examples. In particular, notice that for $\|G\| = \gamma$ the Riccati equation (0.2) does not have a solution. This fact will also be illustrated on the examples in Section 2. In the final section a new proof of the state feedback H^∞-control problem for time varying linear causal systems on a finite horizon is given.

1. Proof of the main Theorem

In this section we give the proof of Theorem 0.1. In the sequel the time varying system with boundary values

$$\begin{cases} \dot{x}(t) = A(t)x(t) + B(t)u(t) & 0 \leq t \leq \tau \\ z(t) = C(t)x(t) + D(t)u(t) \\ N_1 x(0) + N_2 x(\tau) = 0 \end{cases}$$

will be denoted by $\Theta = (A(t), B(t), C(t), D(t); N_1, N_2)$. Note that such a system may be neither causal nor anti-causal.

Proof. We shall first prove the equivalence of (a) a nd (b). Note that $\|G\| < \gamma$ if and only if $I - \gamma^{-2} G^* G > 0$. Now

$$(G^* u)(t) = B(t)^* U(t)^{-*} \int_t^\tau U(s)^* C(s)^* u(s) ds,$$

in other words, G is the input-output operator of the system

$$\Theta = (A(t), B(t), C(t), 0; I, 0)$$

and G^* is the input-output operator of the system

$$\Theta^* = (-A(t)^*, C(t)^*, -B(t)^*, 0; 0, I).$$

Note that

$$\frac{d}{dt}U(t)^{-*} = -A(t)^* U(t)^{-*}, \qquad 0 \le t \le \tau \qquad U(0)^{-*} = I.$$

By [GK], Theorem II 1.1 a system having $I - \gamma^{-2}G^*G$ as its input-output operator is given by

$$\hat{\Theta} = (\hat{A}(t), \hat{B}(t), \hat{C}(t), I; N_1, N_2) \tag{1.1}$$

where

$$\begin{aligned} \hat{A}(t) &= \begin{pmatrix} -A(t)^* & C(t)^*C(t) \\ 0 & A(t) \end{pmatrix}, \qquad \hat{B}(t) = \tfrac{1}{\gamma}\begin{pmatrix} 0 \\ B(t) \end{pmatrix} \\ \hat{C}(t) &= \tfrac{1}{\gamma}\begin{pmatrix} B(t)^* & 0 \end{pmatrix}, \qquad N_1 = \begin{pmatrix} 0 & 0 \\ 0 & I \end{pmatrix}, \qquad N_2 = \begin{pmatrix} I & 0 \\ 0 & 0 \end{pmatrix}. \end{aligned} \tag{1.2}$$

First we show that (b) implies (a). Let $P(t)$ be a solution of (0.2), and put $S(t) = \begin{pmatrix} I & P(t) \\ 0 & I \end{pmatrix}$. Applying $S(t)$ as a state space similarity transformation on $\hat{\Theta}$ we obtain that $I - \gamma^{-2}G^*G$ is the input-output operator of the system

$$\tilde{\Theta} = (\tilde{A}(t), \tilde{B}(t), \tilde{C}(t), I; \tilde{N}_1, \tilde{N}_2)$$

where

$$\begin{aligned} \tilde{A}(t) &= S(t)\hat{A}(t)S(t)^{-1} + \dot{S}(t)S(t)^{-1} = \begin{pmatrix} -A(t)^* & -\gamma^{-2}P(t)B(t)B(t)^*P(t) \\ 0 & A(t) \end{pmatrix}, \\ \tilde{B}(t) &= S(t)\hat{B}(t) = \tfrac{1}{\gamma}\begin{pmatrix} P(t)B(t) \\ B(t) \end{pmatrix}, \\ \tilde{C}(t) &= \hat{C}(t)S(t) = \tfrac{1}{\gamma}\begin{pmatrix} B(t)^* & -B(t)^*P(t) \end{pmatrix}, \\ \tilde{N}_1 &= N_1 S(0)^{-1} = N_1, \qquad \tilde{N}_2 = N_2 S(\tau)^{-1} = N_2. \end{aligned}$$

Computing $\tilde{A}^\times(t) = \tilde{A}(t) - \tilde{B}(t)\tilde{C}(t)$ yields

$$\tilde{A}^\times(t) = \begin{pmatrix} -A(t)^* - \gamma^{-2}P(t)B(t)B(t)^* & 0 \\ -\gamma^{-2}B(t)B(t)^* & A(t) + \gamma^{-2}B(t)B(t)^*P(t) \end{pmatrix}.$$

Now put $\Pi = \begin{pmatrix} 0 & 0 \\ 0 & I \end{pmatrix}$ Then $\tilde{A}(t)(\operatorname{Ker}\Pi) \subset \operatorname{Ker}\Pi$, $\tilde{A}^\times(t)(\operatorname{Im}\Pi) \subset \operatorname{Im}\Pi$, while $N_1 = \Pi$, $N_2 = I - \Pi$. Thus we may apply the anti-causal/causal version of [GK],

Theorem III 2.1 and Corollary III 2.2. These imply that $I-\gamma^{-2}G^*G$ may be factorised as

$$I-\gamma^{-2}G^*G=(I-V_+)(I-V_-), \tag{1.3}$$

where $I-V_+$ is the (anti-causal) input-output operator of the system

$$\Theta_+=(-A(t)^*, \frac{1}{\gamma}P(t)B(t), \frac{1}{\gamma}B(t)^*, I; 0, I)$$

and $I-V_-$ is the (causal) input-output operator of

$$\Theta_-=(A(t), \frac{1}{\gamma}B(t), -\frac{1}{\gamma}B(t)^*P(t), I; I, 0),$$

i.e.,

$$(V_-f)(t)=\gamma^{-2}B(t)^*P(t)U(t)\int_0^t U(s)^{-1}B(s)f(s)ds. \tag{1.4}$$

Note that $\Theta_-=\Theta_+^*$, so $V_+=V_-^*$. Thus $I-\gamma^{-2}G^*G=(I-V_-^*)(I-V_-)\leq 0$. So $\|G\|\leq\gamma$. However, as V_- is a Volterra integral operator with Hilbert-Schmidt kernel, $I-V_-$ is invertible. So actually $I-\gamma^{-2}G^*G>0$, and hence $\|G\|<\gamma$.

Next, we show (a) implies (b). Assume $\|G\|<\gamma$, then $I-\gamma^{-2}G^*G>0$. Consider the system $\hat{\Theta}$ of (1.1), (1.2). Let $\hat{U}(t)$ be the solution of

$$\frac{d}{dt}\hat{U}(t)=\hat{A}(t)\hat{U}(t), \qquad 0\leq t\leq\tau, \qquad \hat{U}(0)=I.$$

Then one easiliy checks that

$$\hat{U}(t)=\begin{pmatrix} U(t)^{-*} & X(t) \\ 0 & U(t)\end{pmatrix}, \tag{1.5}$$

where

$$X(t)=U(t)^{-*}\int_0^t U(s)^*C(s)^*C(s)U(s)ds \tag{1.6}$$

is the solution of

$$\dot{X}(t)=-A(t)^*X(t)+C(t)^*C(t)U(t), \quad 0\leq t\leq\tau, \qquad X(0)=0. \tag{1.7}$$

The canonical boundary operator P_0 of $\hat{\Theta}$, defined by $P_0=(N_1+N_2\hat{U}(\tau))^{-1}N_2\hat{U}(\tau)$ (see [GK]), equals

$$P_0=\begin{pmatrix} I & U(\tau)^*X(\tau) \\ 0 & 0\end{pmatrix}. \tag{1.8}$$

Clearly P_0 is a projection, so in terms of [GK1], Section III 3, $\hat{\Theta}$ is an SB-system. Note that $\operatorname{Ker} P_0=\operatorname{Im}\begin{pmatrix} -U(\tau)^*X(\tau) \\ I\end{pmatrix}$.

As $I - \gamma^{-2}G^*G > 0$, there is an anti-causal/causal factorization

$$I - \gamma^{-2}G^*G = (I - V_-^*)(I - V_-)$$

(see [GKr], Section IV.7, also [K]). Using [GK1], Theorem I.8.2, also Theorem II.9.2 (which may be applied as $\hat{\Theta}$ is an SB-system) this implies the existence of a solution $R(t) : \operatorname{Ker} P_0 \to \operatorname{Im} P_0$ on $[0, \tau]$ of the Riccati equation

$$\begin{aligned} &\dot{R}(t) = -(P_0 + R(t)(I - P_0))\hat{U}(t)^{-1}\hat{B}(t)\hat{C}(t)\hat{U}(t)(R(t)(I - P_0) - (I - P_0)) \\ &R(\tau) = 0. \end{aligned} \tag{1.9}$$

Put

$$P(t) = -\begin{pmatrix} U(t)^{-*} & X(t) \end{pmatrix}(I - P_0 - R(t)(I - P_0))\begin{pmatrix} -U(\tau)^*X(\tau) \\ I \end{pmatrix} U(t)^{-1}. \tag{1.10}$$

We shall show that $P(t)$ solves (0.2). (Compare [GK1], formula I.8.9 with (1.4) to understand why one would try (1.10) as a solution of (0.2).) As $R(\tau) = 0$ and $\operatorname{Ker} P_0 = \operatorname{Im} \begin{pmatrix} -U(\tau)^*X(\tau) \\ I \end{pmatrix}$ one sees $P(\tau) = 0$. Furthermore, compute

$$\frac{d}{dt}(P(t)U(t)) = \dot{P}(t)U(t) + P(t)\dot{U}(t) = (\dot{P}(t) + P(t)A(t))U(t).$$

On the other hand $\frac{d}{dt}(P(t)U(t))$ equals

$$\begin{aligned} &-\{\tfrac{d}{dt}\begin{pmatrix} U(t)^{-*} & X(t) \end{pmatrix}\}(I - P_0 - R(t)(I - P_0))\begin{pmatrix} -U(\tau)^*X(\tau) \\ I \end{pmatrix} - \\ &\begin{pmatrix} U(t)^{-*} & X(t) \end{pmatrix}(-\dot{R}(t)(I - P_0))\begin{pmatrix} -U(\tau)^*X(\tau) \\ I \end{pmatrix}. \end{aligned}$$

Using (1.7) and (1.9) one sees that this equals

$$\begin{aligned} &A(t)^*\begin{pmatrix} U(t)^{-*} & X(t) \end{pmatrix}(I - P_0 - R(t)(I - P_0))\begin{pmatrix} -U(\tau)^*X(\tau) \\ I \end{pmatrix} \\ &-\begin{pmatrix} 0 & C(t)^*C(t)U(t) \end{pmatrix}\begin{pmatrix} -U(\tau)^*X(\tau) \\ I \end{pmatrix} - \\ &-\begin{pmatrix} U(t)^{-*} & X(t) \end{pmatrix}(P_0 + R(t)(I - P_0))\hat{U}(t)^{-1}\hat{B}(t)\times \\ &\times\hat{C}(t)\hat{U}(t)(R(t)(I - P_0) - (I - P_0))\begin{pmatrix} -U(\tau)^*X(\tau) \\ I \end{pmatrix} = \\ &= -A(t)^*P(t)U(t) - C(t)^*C(t)U(t) - Z(t), \end{aligned}$$

where

$$\begin{aligned} &Z(t) = \begin{pmatrix} U(t)^{-*} & X(t) \end{pmatrix}(P_0 + R(t)(I - P_0))\hat{U}(t)^1\hat{B}(t)\hat{C}(t)\hat{U}(t)\times \\ &\times(R(t)(I - P_0) - (I - P_0))\begin{pmatrix} -U(\tau)^*X(\tau) \\ I \end{pmatrix}. \end{aligned}$$

It remains to show that $Z(t) = \gamma^{-2}P(t)B(t)B(t)^*P(t)U(t)$. Note that

$$\hat{B}(t)\hat{C}(t)\hat{U}(t)\{R(t)(I-P_0)-(I-P_0)\}\begin{pmatrix} -U(\tau)^*X(\tau) \\ I \end{pmatrix} =$$
$$\gamma^{-2}\begin{pmatrix} 0 & 0 \\ B(t)B(t)^* & 0 \end{pmatrix}\begin{pmatrix} P(t)U(t) \\ * \end{pmatrix} = \begin{pmatrix} 0 \\ \gamma^{-2}P(t)B(t)B(t)^*P(t)U(t) \end{pmatrix}.$$

So with $\Gamma(t) = \begin{pmatrix} 0 \\ \gamma^{-2}P(t)B(t)B(t)^*P(t)U(t) \end{pmatrix}$,

$$\begin{aligned} Z(t) &= (\,U(t)^{-*} \quad X(t)\,)\,(P_0 + R(t)(I-P_0))\hat{U}(t)^{-1}\Gamma(t) = \\ &= -\,(\,U(t)^{-*} \quad X(t)\,)\,(I - P_0 - R(t)(I-P_0))\hat{U}(t)^{-1}\Gamma(t) + \\ &\quad + (\,U(t)^{-*} \quad X(t)]\,)\,\hat{U}(t)^{-1}\Gamma(t). \end{aligned}$$

The last term is equal to zero because of (1.5), and thus

$$Z(t) = -\,(\,U(t)^{-*} \quad X(t)\,)\,(I - P_0 - R(t)(I-P_0))\hat{U}(t)^{-1}\Gamma(t).$$

Now by (1.8) and (1.5)

$$\begin{aligned} &(I-P_0)\hat{U}(t)^{-1} = \\ &= (I-P_0)\begin{pmatrix} U(t)^* & -U(t)^*X(t)U(t)^{-1} \\ 0 & U(t)^{-1} \end{pmatrix} \\ &= \begin{pmatrix} 0 & -U(\tau)^*X(\tau)U(t)^{-1} \\ 0 & U(t)^{-1} \end{pmatrix}. \end{aligned}$$

From (1.10) we have $Z(t) = \gamma^{-2}P(t)B(t)B(t)^*P(t)U(t)$ as desired.

In the remainder of the proof we show (a) implies (c), and (c) implies (b). First we prove the latter implication. Let P be a solution of the Riccati differential inequality (0.3), and denote the left hand side of the inequality in (0.3) by $-D(t)D(t)^*$. Put $A_P = A - \gamma^{-2}BB^*P$. Then P also satisfies

$$\begin{aligned} &\dot{P} = -A_P^*P - PA_P + \gamma^{-2}PBB^*P - (\,C^* \quad D^*\,)\begin{pmatrix} C \\ D \end{pmatrix} \\ &P(\tau) = 0. \end{aligned}$$

From standard optimal control theory (see, e.g., [S]) we have $P(t) \geq 0$ on $[0,\tau]$. Let $Q(t)$ be the solution backward in time of the initial value problem (0.2), which exists on some interval $(\tau-\epsilon,\tau]$. Then $Q(t) \leq 0$ on $(\tau-\epsilon,\tau]$ using a similar argument as above. Moreover, it is a straightforward computation to show that $0 \leq x^*Q(t)x \leq x^*P(t)x$ for all t in $[0,\tau]$ for which $Q(t)$ exists and for all vectors x. (Also this is known from optimal control theory.) Therefore Q cannot have an escape time in the interval $[0,\tau]$, i.e., the solution of (0.2) exists on this interval.

Finally, we show (a) implies (c). Take $\alpha > 0$ and define

$$(G_\alpha f)(t) = \alpha U(t)\int_0^t U(s)^{-1}B(s)f(s)ds.$$

Note that $G_\alpha = \alpha G_1$, and G_1 is bounded. Consider the operator $\begin{pmatrix} G \\ G_\alpha \end{pmatrix}$. Taking α small enough we have $\| \begin{pmatrix} G \\ G_\alpha \end{pmatrix} \| < \gamma$, as $\|G_\alpha\| \to 0$ as $\alpha \downarrow 0$, and $\|G\| < \gamma$. As (a) implies (b) we may apply (b) to the operator $\begin{pmatrix} G \\ G_\alpha \end{pmatrix}$ which is the input-output operator of the system $\Theta_\alpha = \left(A(t), B(t), \begin{pmatrix} C(t) \\ \alpha I \end{pmatrix}, 0; I, 0\right)$. Thus there is a solution $P(t)$ on the interval $0 \le t \le \tau$ of the equation

$$\begin{aligned} - \quad & \dot{P}(t) = A(t)^*P(t) + P(t)A(t) + \gamma^{-2}P(t)B(t)B(t)^*P(t) + C(t)^*C(t) + \alpha^2 I \\ & P(\tau) = 0 \end{aligned} \tag{1.11}$$

Clearly, this P satisfies (0.3). □

From the theorem we have the following corollary.

Corollary 1.1. *The following are equivalent*

(a) $\|G\| < \gamma$

(b) *For any $\alpha > 0$ small enough there is a function $P_\alpha(t)$ satisfying* (1.11).

Also we have the following proposition.

Proposition 1.2. *Let $P(t)$ be a function satisfying* (0.3)*, then there is a solution $Q(t)$ of* (0.2)*, and moreover, for $0 \le t \le \tau$ we have $0 \le Q(t) \le P(t)$.*

2. Examples

Example 2.1. Theorem 0.1 states that for $\|G\| = \gamma$ the Riccati equation (0.2) does not have a solution. Let us illustrate this fact on the following system, which is in a sense the simplest example one can think of

$$\begin{aligned} \dot{x}(t) &= u(t) \\ y(t) &= x(t), \quad x(0) = 0 \end{aligned} \qquad 0 \le t \le 1.$$

Its input-output operator is the operator of integration

$$y(t) = (Gu)(t) = \int_0^t u(s)ds.$$

It is known that $\|G\| = \frac{2}{\pi}$ (compare [GGK], page 99). The corresponding Riccati equation is

$$-\dot{P}(t) = \gamma^{-2}P(t)^2 + 1, \qquad P(1) = 0. \tag{2.1}$$

This may easily be solved by separation of variables, giving $P(t) = \gamma \tan(\frac{1}{\gamma}(1-t))$. Let us compute for which values of γ the solution exists over the interval $[0,1]$. Clearly, if $\gamma > \frac{2}{\pi}$, then $P(t)$ is well defined on $0 \leq t \leq 1$. However, if $\gamma = \frac{2}{\pi}$, then $P(0) = \infty$. So, (2.1) has a solution if and only if $\gamma > \|G\|$, which is precisely the equivalence of (a) and (b) in Theorem 0.1 for the case considered here.

Example 2.2. Next we consider another example, which may be of interest. Let us consider the following integral operator on $L_2[0,1]$

$$y(t) = (G_a u)(t) = \int_0^t e^{a(t-s)} u(s) ds.$$

The corresponding system is

$$\begin{aligned} &\dot{x}(t) = ax(t) + u(t) \\ &y(t) = x(t), \quad x(0) = 0 \qquad 0 \leq t \leq 1. \end{aligned}$$

To find the norm of G_a we have to consider the Riccati differential equation

$$-\dot{P}(t) = \gamma^{-2} P(t)^2 + 2aP(t) + 1 = \gamma^{-2}((P(t) + a\gamma^2)^2 + (\gamma^2 - a^2\gamma^4)), \qquad P(1) = 0.$$

Again we can use separation of variables to solve this. To do this we have to distinguish between the cases $\gamma < \frac{1}{|a|}$ and $\gamma > \frac{1}{|a|}$, and $\gamma = \frac{1}{|a|}$.

For $\gamma < \frac{1}{|a|}$ we obtain by elementary calculus that

$$P(t) = -a\gamma + \gamma\sqrt{1-(a\gamma)^2} \tan(\sqrt{1-(a\gamma)^2}(-\frac{1}{\gamma}t + c)),$$

where c is determined from $P(1) = 0$, by

$$c = \frac{1}{\sqrt{1-(a\gamma)^2}} \arctan \frac{a}{\sqrt{1-(a\gamma)^2}} + \frac{1}{\gamma}.$$

We see that $P(t)$ exists for $t \in [0,1]$ if and only if for all $t \in [0,1]$

$$\sqrt{1-(a\gamma)^2}(-\frac{1}{\gamma}t + c) \in (-\frac{\pi}{2}, \frac{\pi}{2}).$$

As this is a decreasing function of t, and for $t = 1$ this requirement is satisfied by the definition of c, we obtain that $P(t)$ exists for $t \in [0,1]$ if and only if $\sqrt{1-(a\gamma)^2}c < \frac{\pi}{2}$, i.e., if and only if

$$\arctan \frac{a}{\sqrt{1-(a\gamma)^2}} + \frac{\sqrt{1-(a\gamma)^2}}{\gamma} < \frac{\pi}{2}.$$

Thus, the norm of G_a is detemined as the solution γ of

$$\arctan \frac{a}{\sqrt{1-(a\gamma)^2}} + \frac{\sqrt{1-(a\gamma)^2}}{\gamma} = \frac{\pi}{2}. \tag{2.2}$$

Recall that all this holds only under the assumption $\gamma < \frac{1}{|a|}$.

In case $\gamma = \frac{1}{|a|}$ the solution becomes

$$P(t) = -\gamma + \frac{1}{\frac{1}{\gamma} + \frac{1}{\gamma^2}(t-1)}.$$

This escapes to infinity for $t = 1 - \gamma$. For $a < 1$ we have $\gamma = \frac{1}{|a|} > 1$, and so the function $P(t)$ is defined on $[0,1]$ for these values of a. It follows that for $a < 1$ the norm of G_a is less than $\frac{1}{|a|}$, i.e., the situation of the previous paragraph occurs for all such a. Also, for $a > 1$ the function $P(t)$ is not defined on the whole of $[0,1]$, and thus the norm of G_a is larger then $\frac{1}{|a|}$, and can be found as in the next paragraph. For $a = 1$ we have $\gamma = 1$, and we see that the function $P(t)$ is defined on $(0,1]$, but not on $[0,1]$. So, $\|G_1\| = 1$.

It remains to consider the case $\gamma > \frac{1}{|a|}$. By the results of the previous paragraph, we can assume that $a > 1$. Solving the Riccati differential equation now yields that $P(t)$ is implicitly given by

$$\frac{1}{2\sqrt{(a\gamma)^2-1}} \ln \left(\frac{|\frac{1}{\gamma}P(t) + a - \sqrt{(a\gamma)^2-1}|}{\frac{1}{\gamma}P(t) + a + \sqrt{(a\gamma)^2-1}} \right) = -\frac{1}{\gamma}t + c, \tag{2.3}$$

where c is determined from $P(1) = 0$, by

$$c = \frac{1}{\gamma} + \frac{1}{2\sqrt{(a\gamma)^2-1}} \ln \left| \frac{a - \sqrt{(a\gamma)^2-1}}{a + \sqrt{(a\gamma)^2-1}} \right|.$$

From (2.3) we see that $P(t)$ is defined on $[0,1]$ provided $-\frac{1}{\gamma}t + c < 0$ for all $t \in [0,1]$. As this is the case for $t = 1$, and $c - \frac{1}{\gamma}t$ is decreasing in t, we see that this holds for all t provided $c < 0$. It follows that $\|G_a\|$ is determined as the solution of the equation

$$\frac{1}{\gamma} + \frac{1}{2\sqrt{(a\gamma)^2-1}} \ln \left| \frac{a - \sqrt{(a\gamma)^2-1}}{a + \sqrt{(a\gamma)^2-1}} \right| = 0. \tag{2.4}$$

Recall again that this holds only provided $\gamma > \frac{1}{a}$. However, it is easily seen that for $a > 1$ we have $\|G_a 1\| > \frac{1}{a}$, from which it follows that $\gamma > \frac{1}{a}$ in this case. Indeed, $G_a 1 = \frac{1}{a}(e^{at} - 1)$, so $\|G_a 1\|^2 = \int_0^1 \frac{1}{a^2}(e^{at}-1)^2\, dt = \frac{1}{a^2}(\frac{1}{2a}(e^{2a}-2)^2 + (\frac{1}{2} - \frac{1}{2a}) + \frac{1}{2})$. One easily checkes that for $a > 1$ we have $e^{2a} - 2 > 2a$, and moreover, clearly also $\frac{1}{2} - \frac{1}{2a} > 0$, so $\|G_a 1\|^2 > \frac{1}{a^2}(2a + \frac{1}{2}) > \frac{1}{a^2}$.

Figure 1 below shows the solution of (2.2) as a function of a for $a \in [-1,1)$, and the solution of (2.4) for $a \in (1,2]$.

Example 2.3. Next we consider the following operator on $L_2^2[0,1]$:

$$(G\begin{pmatrix} u_1 \\ u_2 \end{pmatrix})(t) = \begin{pmatrix} \int_0^t u_1(s)\, ds \\ \int_0^t (t-s)u_1(s)\, ds + \int_0^t u_2(s)\, ds \end{pmatrix}.$$

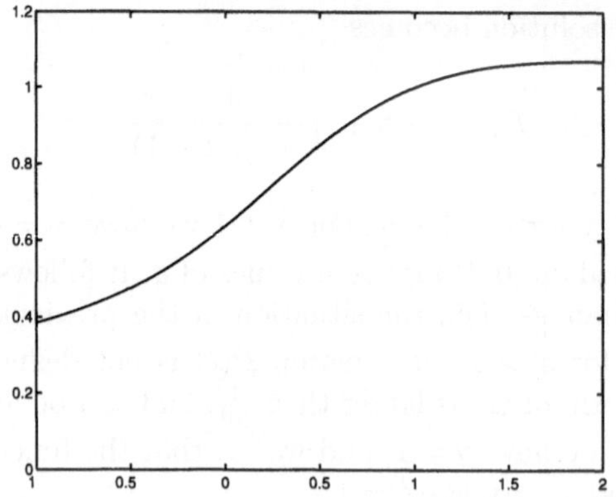

Figure 1: The norm of the integral operator $(G_a u)(t) = \int_0^t e^{a(t-s)}u(s)ds$.

The corresponding input-output system is given by

$$\begin{aligned} \dot{x}(t) &= \begin{pmatrix} 0 & 1 \\ 0 & 0 \end{pmatrix} x(t) + \begin{pmatrix} 1 & 0 \\ 0 & 1 \end{pmatrix} u(t), \\ y(t) &= \begin{pmatrix} 1 & 0 \\ 0 & 1 \end{pmatrix} x(t). \end{aligned}$$

Thus the corresponding Riccati equation is given by

$$\begin{aligned} - \quad \dot{P}(t) = \tfrac{1}{\gamma^2} P(t)^2 + \begin{pmatrix} 0 & 0 \\ 1 & 0 \end{pmatrix} P(t) + P(t) \begin{pmatrix} 0 & 1 \\ 0 & 0 \end{pmatrix} + \begin{pmatrix} 1 & 0 \\ 0 & 1 \end{pmatrix}, \\ P(1) = \begin{pmatrix} 0 & 0 \\ 0 & 0 \end{pmatrix}. \end{aligned}$$

Obviously, we have that $\|G\| \geq \frac{2}{\pi}$. Solving for γ from 1 downwards (using Matlab) we see that the norm is between 0.75 and 0.76 (see the left hand side of Figure 2). The right hand side of Figure 2 now shows for values of γ running from 0.756 to 0.761 $\max |P_{ij}(t)|$, where the maximum is taken over $i, j = 1, 2$. Note the difference in scale between the left hand side of Figure 2 and the right hand side. It can be shown that we have $0.75773 \leq \|G\| < 0.75774$.

Example 2.4. Let us next consider the operator G given by

$$(Gu)(t) = \int_0^t \cos(t+s)u(s)\,ds, \qquad u \in L_2[0,1].$$

Observe that this operator is in the class under consideration as $\cos(t+s) = \cos t \cos s - \sin t \sin s$. So we have that this operator is the input-output operator of a system with system matrices given by

$$A = \begin{pmatrix} 0 & 0 \\ 0 & 0 \end{pmatrix}, \quad B(t) = \begin{pmatrix} \cos t \\ \sin t \end{pmatrix}, \quad C(t) = (\, \cos t \quad -\sin t \,).$$

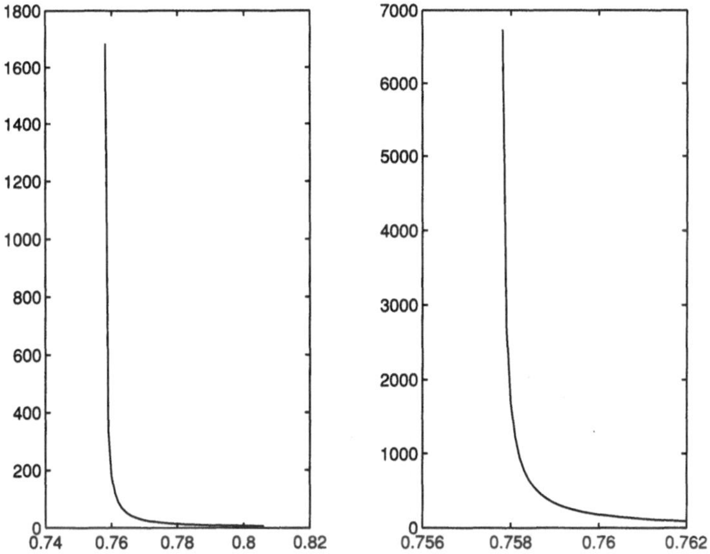

Figure 2: The maximal absolute value of an entry in $P(t)$ for different values of γ.

The corresponding Riccati equation is

$$\begin{aligned} -\quad \dot{P}(t) &= \tfrac{1}{\gamma^2}P(t)\begin{pmatrix} \cos^2 t & \frac{1}{2}\sin 2t \\ \frac{1}{2}\sin 2t & \sin^2 t \end{pmatrix}P(t) + \begin{pmatrix} \cos^2 t & -\frac{1}{2}\sin 2t \\ -\frac{1}{2}\sin 2t & \sin^2 t \end{pmatrix}. \\ P(1) &= 0. \end{aligned}$$

Writing this out in terms of the entries $p_{ij}(t)$ in $P(t)$ we obtain (using $p_{12}(t) = p_{21}(t)$ as the solution is known to be real symmetric)

$$\begin{aligned} -\quad \dot{p}_{11}(t) = {} & \tfrac{1}{\gamma^2}((\cos^2 t)(p_{11}^2(t)+1) + (\sin 2t)p_{11}(t)p_{12}(t) + (\sin^2 t)p_{12}^2(t)), \\ -\quad \dot{p}_{12}(t) = {} & \tfrac{1}{\gamma^2}((\cos^2 t)p_{11}(t)p_{12}(t) + (\sin 2t)(p_{11}(t)p_{22}(t) + p_{12}^2(t) - 1) + \\ & (\sin^2 t)p_{12}(t)p_{22}(t)), \\ -\quad \dot{p}_{22}(t) = {} & \tfrac{1}{\gamma^2}((\cos^2 t)p_{12}^2(t) + (\sin 2t)p_{12}(t)p_{22}(t) + (\sin^2 t)(p_{22}^2(t)+1)). \end{aligned}$$

This we solved numerically using Matlab for decreasing values of γ. As it is easily seen that $\|G\| < 1$, it is a good idea to start with $\gamma = 1$, and let γ go down from there. Doing this in steps of 0.01 and plotting the maximal value of an entry of $P(t)$, we obtain that the norm is between 0.38 and 0.39. This is plotted in the left hand side of Figure 3 below.

The right hand side of the figure was obtained by plotting the maximal absolute value of an entry in $P(t)$ for γ decreasing from 0.39 in steps of $\frac{1}{10^5}$. Again, note the difference in scale between the two graphs in Figure 3. We obtain that up to $\gamma = 0.38924$ the Riccati equation has a solution over the whole interval $[0, 1]$. Then

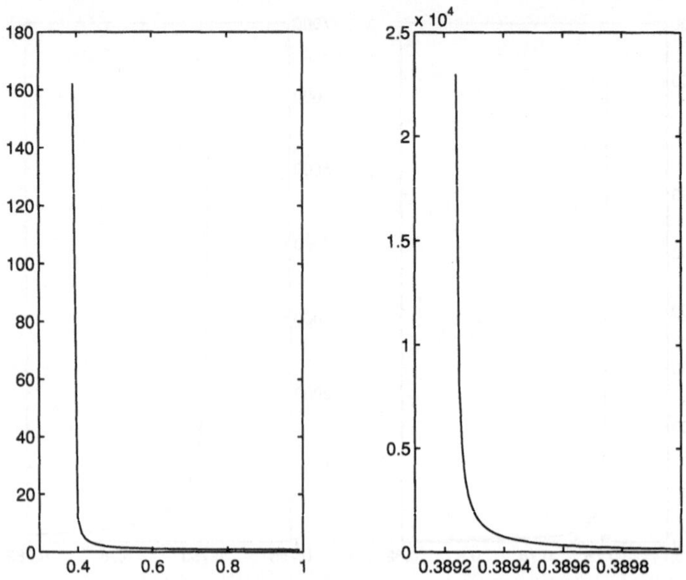

Figure 3: The maximal absolute value of an entry in $P(t)$ for different values of γ.

a singularity occurs for $\gamma = 0.38923$ at $t = 0.000006$. The maximal value of the solution of the Riccati equation for $\gamma = 0.38924$ is given as 22978. We arrive at the conclusion that $0.38923 \leq \|G\| < 0.38934$.

Example 2.5.

The final example we consider is the operator $G : L_2[0,1] \to L_2[0,1]$ given by

$$(Gu)(t) = \int_0^t \ln(ts)u(s)\,ds, \qquad 0 \leq t \leq 1.$$

As $\ln(ts) = \ln(t) + \ln(s) = \begin{pmatrix} \ln(t) & 1 \end{pmatrix} \begin{pmatrix} 1 \\ \ln(s) \end{pmatrix}$, we see that G falls in the class of operators considered here. The corresponding system matrices can be taken as $A(t) = \begin{pmatrix} 0 & 0 \\ 0 & 0 \end{pmatrix}$, $B(t) = \begin{pmatrix} 1 \\ \ln(t) \end{pmatrix}$ and $C(t) = \begin{pmatrix} \ln(t) & 1 \end{pmatrix}$. The corresponding Riccati equation is given by

$$\begin{aligned} -\dot{P}(t) &= \tfrac{1}{\gamma^2} P(t) \begin{pmatrix} 1 & \ln(t) \\ \ln(t) & \ln^2(t) \end{pmatrix} P(t) + \begin{pmatrix} \ln^2(t) & \ln(t) \\ \ln(t) & 1 \end{pmatrix}, \\ P(1) &= 0. \end{aligned}$$

Solving this numerically for γ decreasing from 2 in steps of 0.01 and plotting the maximal absolute value of an entry in $P(t)$, we obtain Figure 4 below. One arrives at the conclusion that $1.62 \leq \|G\| \leq 1.63$. Obviously, one could easily obtain more accurate approximations of $\|G\|$ in the manner explained in the previous examples.

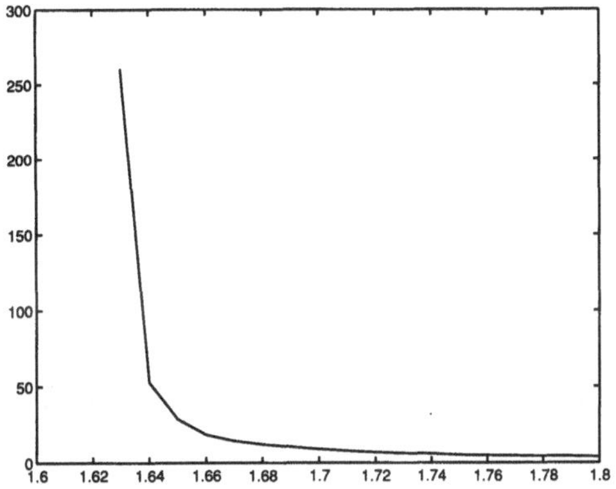

Figure 4: The maximal absolute value of an entry in $P(t)$.

3. State feedback H^∞-control for a finite horizon time varying system

In this section we consider the following problem: given is a causal time varying linear system

$$\begin{cases} \dot{x}(t) &= A(t)x(t) + B(t)u(t) + E(t)w(t) \qquad 0 \le t \le \tau \\ z(t) &= C(t)x(t) + D(t)u(t) \\ x(0) &= 0 \end{cases} \tag{3.1}$$

Here A is assumed to be an integrable $n \times n$ matrix function of t on the interval $0 \le t \le \tau$, B, E, and C are assumed to be square integrable on $0 \le t \le \tau$, and D is assumed to be measurable and essentially bounded. We shall take inputs $u \in L_2^p[0,\tau]$, and disturbances $w \in L_2^q[0,\tau]$. Given such u and w, the system produces as output a function $z \in L_2^m[0,\tau]$.

The problem is to find a state feedback time varying linear controller $u(t) = F(t)x(t)$ such that the input-output operator $G_F(w) = z$ of the closed loop system has operator norm $\|G_F\|$, where G_F is considered as an operator from $L_2^q[0,\tau]$ to $L_2^m[0,\tau]$, less than a given tolerance level $\gamma > 0$. We shall assume throughout that we are in the so-called regular case, i.e., $D(t)$ is injective for $0 \le t \le \tau$. Furthermore we consider only the case $\tau < \infty$ here.

For the time-invariant case an elegant approach to the H^∞-control problem is the one which employs the bounded real lemma combined with results on Riccati inequalities and equations. The continuous time case is treated in this way in [PAJ], the discrete time case in [SX]. For the time-varying case we believe that this approach has not been tried before. Here we shall employ Theorem 0.1 to treat the state feedback H^∞-control problem for time varying linear causal systems on a finite horizon. The following theorem gives the solution to this problem.

Theorem 3.1. *Given the system* (3.1) *there is a state feedback* $u(t) = F(t)x(t)$ *such that the closed loop input-output operator* G_F *satisfies* $\|G_F\| < \gamma$ *if and only if the Riccati differential equation*

$$\begin{aligned} -\quad & \dot{P}(t) = (A(t) - B(t)(D(t)^*D(t))^{-1}D(t)^*C(t))^*P(t)+ \\ & +P(t)(A(t) - B(t)(D(t)^*D(t))^{-1}D(t)^*C(t))+ \\ & +P(t)(\gamma^{-2}E(t)E(t)^* - B(t)(D(t)^*D(t))^{-1}B(t)^*)P(t)+ \\ & +C(t)^*(I - D(t)(D(t)^*D(t))^{-1}D(t)^*)C(t) \qquad 0 \le t \le \tau \\ & P(\tau) = 0 \end{aligned} \tag{3.2}$$

has a solution $P(t)$ *on the interval* $[0, \tau]$. *Moreover, if* $P(t)$ *is the solution of* (3.2), *then* $F(t)$ *given by*

$$F(t) = -(D(t)^*D(t))^{-1}(B(t)^*P(t) + D(t)^*C(t)) \tag{3.3}$$

is a state feedback such that $\|G_F\| < \gamma$.

Proof. First, assume $P(t)$ solves the Riccati differential equation (3.2) and put $F(t) = -(D(t)^*D(t))^{-1}(B(t)^*P(t) + D(t)^*C(t))$. Introduce

$$A_F(t) = A(t) + B(t)F(t). \tag{3.4}$$

Now it is a matter of straightforward calculation to see that

$$\begin{aligned} & A_F(t)^*P(t) + P(t)A_F(t) + \gamma^{-2}P(t)E(t)E(t)^*P(t)+ \\ & +(C(t) + D(t)F(t))^*(C(t) + D(t)F(t)) = \\ & = (A(t) - B(t)(D(t)^*D(t))^{-1}D(t)^*C(t))^*P(t)+ \\ & +P(t)(A(t) - B(t)(D(t)^*D(t))^{-1}D(t)^*C(t))+ \\ & +P(t)(\gamma^{-2}E(t)E(t)^* - B(t)(D(t)^*D(t))^{-1}B(t)^*)+ \\ & +C(t)^*(I - D(t)(D(t)^*D(t))^{-1}D(t)^*)C(t) = -\dot{P}(t). \end{aligned}$$

Furthermore, note that the closed loop input-output operator G_F mapping w to z is the input-output operator of the system $(A_F(t), E(t), C(t) + D(t)F(t), 0; I, 0)$, i.e.,

$$(G_F(w))(t) = (C(t) + D(t)F(t))U_F(t)\int_0^t U_F(s)^{-1}E(s)w(s)ds$$

for $0 \le t \le \tau$, where $U_F(t)$ is the solution of

$$\begin{aligned} & \dot{U}_F(t) = (A(t) + B(t)F(t))U_F(t) \qquad 0 \le t \le \tau \\ & U_F(0) = I. \end{aligned}$$

Thus, applying Theorem 0.1 to G_F yields that $\|G_F\| < \gamma$.

Conversely, assume $\|G_F\| < \gamma$. Let $\alpha > 0$ be small enough such that there is a solution $P(t)$ of

$$\begin{aligned} -\quad & \dot{P}(t) = A_F(t)^*P(t) + P(t)A_F(t) + \gamma^{-2}P(t)E(t)E(t)^*P(t) \\ & +(C(t) + D(t)F(t))^*(C(t) + D(t)F(t)) + \alpha^2 I \\ & P(\tau) = 0 \qquad 0 \le t \le \tau. \end{aligned} \tag{3.5}$$

Use (3.4) to rewrite this as

$$\begin{aligned}
&\dot{P}(t) + A(t)^*P(t) + P(t)A(t) + \gamma^{-2}P(t)E(t)E(t)^*P(t) + C(t)^*C(t) = \\
&= -F(t)^*B(t)^*P(t) - P(t)B(t)F(t) - \\
&-(C(t) + D(t)F(t))^*(C(t) + D(t)F(t)) + C(t)^*C(t) - \alpha^2 I = \\
&= -(F(t)^*D(t)^*D(t) + P(t)B(t) + C(t)^*D(t))(D(t)^*D(t))^{-1}\times \\
&\times(D(t)^*D(t)F(t) + B(t)^*P(t) + D(t)^*C(t)) + \\
&+(P(t)B(t) + C(t)^*D(t))(D(t)^*D(t))^{-1}(B(t)^*P(t) + D(t)^*C(t)) - \alpha^2 I.
\end{aligned} \tag{3.6}$$

Define

$$\begin{aligned}
R(P) = \quad &\dot{P} + (A - B(D^*D)^{-1}D^*C)^*P + P(A - B(D^*D)^{-1}D^*C) + \\
&P(\gamma^{-2}EE^* - B(D^*D)^{-1}B^*)P + C^*(I - D(D^*D)^{-1}D^*)C.
\end{aligned}$$

Then (3.6) implies $R(P(t)) < 0$ on $[0, \tau]$. Also by Proposition 1.2, $P(t) \geq 0$.

Let $Q(t)$ be the solution of the initial value problem (3.2), backwards in time. Define $A_Q = A - B(D^*D)^{-1}D^*C + \gamma^{-2}EE^*Q$, then

$$\begin{aligned}
&-\dot{Q} = A_Q^*Q + QA_Q - Q(\gamma^{-2}EE^* + B(D^*D)^{-1}B^*)Q + C^*(I - D(D^*D)^{-1}D^*)C, \\
&Q(\tau) = 0.
\end{aligned}$$

Again, by a standard result in optimal control theory $Q(t) \geq 0$ on the maximal interval where Q exists. As in the proof of the implication (b) implies (c) of Theorem 0.1 one sees that $0 \leq Q(t) \leq P(t)$, and so Q exists on the interval $[0, \tau]$. □

H^∞-control problems for time varying systems have been studied widely before, even on a finite horizon (see, for instance [T, RNK] and [BB], Section 4.2.2). The particular result presented here in Theorem 3.1 is contained in [LAKG], Theorem 1.1, and [ST], Corollary 2.4. However, the approach we take here is new, and, in our view, sufficiently interesting to deserve presentation.

References

[BB] T. Basar and P. Bernhard: *H^∞-Optimal Control and Related Minimax Design Problems.* Birkhäuser, Boston, 1991

[GGK] I. Gohberg, S. Goldberg and M.A. Kaashoek: *Classes of Linear Operators* Vol. I, Birkhäuser OT **49**, Basel

[GK] I. Gohberg and M.A. Kaashoek: Time varying linear systems with boundary conditions and integral operators, I. The transfer operator and its properties. *Integral Equations and Operator Theory* **7**, 325–391 (1984)

[GK1] I. Gohberg and M.A. Kaashoek: Minimal factorizations of integral operators and cascade decompositions of systems, in: *Constructive Methods of Wiener-Hopf factorization* (eds. I. Gohberg and M.A. Kaashoek) Birkhäuser OT **21**, Basel, 157–230 (1986)

[GKr] I.C. Gohberg and M.G. Krein: *Theory and Applications of Volterra Operators in Hilbert Space.* AMS Transl of Mathematical Monographs Vol **24**, Providence, Rhode Island, 1970

[K] T. Kailath: Fredholm resolvents, Wiener-Hopf equations and Riccati differential equations. *IEEE Trans. Information Theory*, Vol **IT 15 (6)**, 665–672 (1969)

[LAKG] D.J.N. Limebeer, B.D.O. Anderson, P.P. Khargonekar and M. Green: A game theoretic approach to H_∞-control for time varying systems. *SIAM J. Control and Optimization* **30**, 262–283 (1992)

[LAH] D.J.N. Limebeer, B.D.O. Anderson and B. Hendel: Nash games and mixed H_2/H_∞ control. *Lecture Notes in Control and Inform. Sci.* **183**, Springer Verlag, Berlin 1992

[PAJ] I.A. Petersen, B.D.O. Anderson and E.A. Jonckheere: A first principles solution to the non-singular H^∞ control problem. *Int. J. Robust and Nonlinear Control* **1**, 171–185 (1991)

[RNK] R. Ravi, K.N. Nagpal and P.P. Khargonekar: H_∞-control of linear time varying systems: A state space approach. *SIAM J. Control and Optimization* **29** (1991)

[S] E.D. Sontag: *Mathematical Control Theory, Deterministic Finite Dimensional Systems.* Springer Verlag, New York etc., 1990

[ST] A.A. Stoorvogel and H.L. Trentelman: The finite horizon singular time-varying H_∞ control problem with dynamic measurement feedback. *Linear Algebra and Aplications* **187**, 113–161 (1993)

[SX] C.E. de Souza and Lihua Xie: On the discrete-time bounded real lemma with application in the characterization of static state feedback H_∞ controllers. *Systems & Control Letters* **18**, 61–71 (1992)

[T] G. Tadmor: H_∞ in the time domain: the standard four blocks problem. *Mathematics of Control, Signals and Systems* **3**, 301–324 (1990)

Faculteit Wiskunde en Informatica,
Vrije Universiteit
De Boelelaan 1081a
1081 HV Amsterdam
The Netherlands

1991 Mathematics Subject Classification. Primary 47A30, 47A68, 47G10; Secondary 93B36

Operator Theory:
Advances and Applications, Vol. 106

The numerical range of selfadjoint matrix polynomials [1]

PETER LANCASTER, JOHN MAROULAS, AND PETER ZIZLER

Dedicated to Heinz Langer on the occasion of his 60th birthday

The numerical range of a selfadjoint matrix polynomial $L(\lambda) = \Sigma_{j=0}^{\ell} \lambda^j A_j$ is the set of points $\mu \in \mathbb{C}$ for which $x^* L(\mu)x = 0$ for some nonzero vector x. As for the classical eigenvalue problem (when $L(\lambda) = \lambda I - A$), the spectrum of $L(\lambda)$ is contained in its numerical range. Properties of the numerical range are investigated with special emphasis on the cases when $L(\lambda)$ has only real spectrum (and, possibly, the point at infinity) and when the coefficients of the matrix polynomial are real symmetric matrices.

1 Introduction

Let $A_0, A_1, \ldots, A_\ell$ be $n \times n$ complex hermitian matrices with $A_\ell \neq 0$. The matrix valued function

$$L(\lambda) = \sum_{j=0}^{\ell} \lambda^j A_j \tag{1.1}$$

is called a *selfadjoint matrix polynomial* of *degree* ℓ, and $L(\lambda)$ is said to be *regular* if there is at least one $\lambda \in \mathbb{C}$ for which $\det L(\lambda) \neq 0$.

The *numerical range* of such a polynomial is

$$w(L(\lambda)) = \{\lambda \in \mathbb{C} \mid x^* L(\lambda)x = 0 \text{ for some nonzero } x \in \mathbb{C}^n\},$$

and the *spectrum* $\sigma(L(\lambda))$, is the set of zeros of $\det L(\lambda)$; a discrete set which includes the point at infinity if $\det A_\ell = 0$. Thus, the finite points of $\sigma(L(\lambda))$ have the property that $L(\lambda)x = 0$ for some nonzero $x \in \mathbb{C}^n$. In this case λ is said to be an *eigenvalue* of $L(\lambda)$ and x is a corresponding *eigenvector*. When $\det A_\ell = 0$ eigenvectors corresponding to the infinite eigenvalue are the nonzero solutions of $A_\ell x = 0$. It is clear that $\sigma(L(\lambda)) \subseteq w(L(\lambda))$.

Li and Rodman [9] have made some fundamental contributions to the study of numerical ranges of matrix polynomials. Building on their results, a closer examination is made here of the selfadjoint case with some emphasis on the question: What can be said about $w(L(\lambda))$ when $\sigma(L(\lambda))$ is known to be real?

---

[1]Research supported in part by a grant from the Natural Sciences and Engineering Research Council of Canada.

In Section 2 some tools commonly used in the study of selfadjoint matrix polynomials are described and in Section 3 there is a discussion of hyperbolic and the more general quasihyperbolic polynomials (QHP). Section 4 contains results on the numerical range of QHP with a more detailed discussion of the class of gyroscopically stabilized (quadratic) polynomials. In discussions of the spectra of matrix polynomials a *linearization* often plays an important part. It is a polynomial of first degree but (in the case (1.1), for example) of size ℓn. For a selfadjoint matrix polynomial $L(\lambda)$ there is a selfadjoint linearization $\lambda\mathcal{A} - \mathcal{B}$. The relationship between $w(L(\lambda))$ and $w(\lambda\mathcal{A} - \mathcal{B})$ is the topic of Section 5. Section 6 contains some results on the numerical range of matrix polynomials whose coefficients are real symmetric matrices.

2 Some preliminaries

The following notions of "eigenvalue types" play an important part in the study of selfadjoint matrix polynomials. A real number $\lambda_0 \in \sigma(L(\lambda))$ is said to have *positive type* if $x^*L'(\lambda_0)x > 0$ for all nonzero $x \in \operatorname{Ker} L(\lambda_0)$, and a similar definition applies for eigenvalues of negative type. Eigenvalues of either positive or negative type are said to have *definite type*, and real eigenvalues which are not of definite type are said to be of *mixed type*.

When $\det A_\ell = 0$, $L(\lambda)$ has an eigenvalue at infinity corresponding to the zero eigenvalue of $M(\mu) := \mu^\ell L(\alpha + \mu^{-1})$ where $\alpha \in \mathbb{R}$ and $\det L(\alpha) \neq 0$. When one of these eigenvalues is definite so is the other, and they both have the same type (see §7 of [7], for example).

A theorem of Rellich (see Theorem S6.3 of [6], for example) provides a widely useful tool in the study of functions taking selfadjoint matrix values. For any fixed $\lambda \in \mathbb{R}$ let $\mu_1(\lambda), \mu_2(\lambda), \ldots, \mu_n(\lambda)$ be the (real) eigenvalues of $L(\lambda)$. Now consider $\mu_1(\lambda), \ldots, \mu_n(\lambda)$ as real valued functions on $\mathbb{R}$ which we call the *eigenfunctions* of $L(\lambda)$. Rellich's theorem says that eigenfunctions can be ordered in such a way that, for all $\lambda \in \mathbb{R}$,

$$L(\lambda) = U(\lambda)^* D(\lambda) U(\lambda) \tag{2.1}$$

where $D(\lambda) = \operatorname{diag}[\mu_1(\lambda), \ldots, \mu_n(\lambda)]$, $U(\lambda)$ is unitary, and $D(\lambda)$, $U(\lambda)$ are analytic matrix functions of λ on $\mathbb{R}$. Clearly, the eigenvalues of $L(\lambda)$ can then be identified with the zeros of the eigenfunctions and, for any $\lambda_0 \in \mathbb{R}$ the signature of $L(\lambda_0)$ is determined by the number of eigenfunctions taking positive, zero, and negative values at λ_0.

For our purposes it is important to observe that λ_0 is an eigenvalue of $L(\lambda)$ of positive (of negative) type with multiplicity m if and only if there are exactly m eigenfunctions $\mu_{j_1}, \ldots, \mu_{j_m}$ for which $\mu_{j_k}(\lambda_0) = 0$ and $\mu'_{j_k}(\lambda_0) > 0$ (< 0), $k = 1, 2, \ldots, m$ and the remaining $n - m$ eigenfunctions are nonzero at λ_0. (See Theorem 12.5 of [6].)

3 Quasihyperbolic polynomials

A regular selfadjoint matrix polynomial is said to be *quasihyperbolic* (is a QHP) if all of its eigenvalues are real and have definite type and, if the point at infinity is an eigenvalue, it also has definite type. The study of the numerical range of such polynomials is one of the main concerns of this paper.

Hyperbolic matrix polynomials form a relatively well-understood class of QHP. They are defined by the conditions that $A_\ell > 0$ and that, for any nonzero $x \in \mathbb{C}^n$, the scalar polynomial

$$p_x(\lambda) = x^* L(\lambda) x$$

has ℓ real and distinct zeros (see §31 of [10]). Consequently, if $L(\lambda)$ is hyperbolic, then $w(L(\lambda)) \subseteq \mathbb{R}$. Indeed, there are some real eigenvalues $a_1 \leq b_1 < a_2 \leq b_2 < \cdots < a_\ell \leq b_\ell$ such that

$$w(L(\lambda)) = \bigcup_{j=1}^{\ell} [a_j, b_j], \tag{3.1}$$

$[a_j, b_j]$ contains exactly n eigenvalues (counting multiplicities) all of the same type (either positive or negative), and these types alternate as j increases. The intervals $[a_j, b_j]$ are known as *root zones.*

A selfadjoint matrix polynomial with $A_\ell > 0$ is said to be *weakly hyperbolic* if the polynomials $p_x(\lambda)$ (for any nonzero $x \in \mathbb{C}^n$) have only real zeros but they are not necessarily distinct. (In this case the "root zones" defined above may not be disjoint. See §31 of [10].) Observe that, from the definition (and when $A_\ell > 0$), $L(\lambda)$ of (1.1) is weakly hyperbolic if and only if $w(L(\lambda)) \subseteq \mathbb{R}$. Thus, if $A_\ell > 0$ in (1.1) and $L(\lambda)$ is *not* weakly hyperbolic there must be non-real points in $w(L(\lambda))$.

Using the properties of eigenfunctions mentioned above, it is easy to see that $a_j = b_j$ is possible in (3.1) when $L(\lambda)$ is hyperbolic. In this case a_j is an eigenvalue of definite type with algebraic multiplicity n, and $L(\lambda)$ factorizes in the form

$$L(\lambda) = (\lambda - a_j)\hat{L}(\lambda),$$

where $\hat{L}(\lambda)$ is a hyperbolic polynomial of degree $\ell - 1$.

We develop analogous ideas for QHP. They all follow directly from the definition of a QHP:

If det $A_\ell \neq 0$ then $\infty \notin \sigma(L(\lambda))$ and eigenvalues $a_0, \ldots a_v$ and $b_0, \ldots, b_v$ of $L(\lambda)$ can be defined in such a way that:

a) $-\infty < a_0 \leq b_0 < a_1 \leq b_1 < \cdots < a_v \leq b_v < \infty$

b)

$$\sigma(L(\lambda)) \subset \bigcup_{j=0}^{v} [a_j, b_j] \tag{3.2}$$

c) For $j = 0, 1, \ldots, v$ all eigenvalues in $[a_j, b_j]$ have the same type (either positive or negative).

d) For $j = 0, 1, \ldots, v-1$, the eigenvalues b_j and a_{j+1} have different types.

If det $A_\ell = 0$ then $\infty \in \sigma(L(\lambda))$ and write $a_0 = -\infty$, $b_v = \infty$. Define eigenvalues $a_1, \ldots, a_v$ and $b_0, \ldots, b_{v-1}$ of $L(\lambda)$ in such a way that

a) $-\infty = a_0 \leq b_0 < a_1 \leq b_1 < \cdots < a_v \leq b_v = \infty$,

and conditions b), c), and d) above are satisfied.

With this construction, v is the number of changes in type of the eigenvalues of the QHP as λ traverses the whole real axis from $-\infty$ to ∞. The parameter v is called the *variation* of the QHP. In particular, $v = \ell - 1$ for a hyperbolic polynomial. Note also that, when $\det A_\ell = 0$, a_0 and b_v have the same type. (The parameter v can also be interpreted as the minimal degree for an associated definitizing polynomial, see [7].) The intervals $[a_j, b_j]$ are known as the *quasi-zones* of the QHP.

Proposition 3.1. *Let $L(\lambda)$ be a* QHP *of degree ℓ with variation v. Then $\ell - 1 \leq v \leq \ell n - 1$.*

Proof. The upper bound is trivial as there are exactly ℓn real eigenvalues when counted with multiplicities. For the lower bound observe that (for *any* selfadjoint matrix polynomial) as λ increases through a positive type eigenvalue of multiplicity m, $L(\lambda)$ "gains" m positive eigenvalues and "loses" m negative eigenvalues. Application of this principle shows that the number of eigenvalues in a quasi-zone $[a_j, b_j]$ cannot exceed n. Hence the number of quasi-zones cannot be less than ℓ and the lower bound on v follows. □

Note that both bounds of Proposition 3.1 are attained when $n = 1$ and, for any n, the lower bound is attained if and only if $L(\lambda)$ is hyperbolic.

An interesting and useful class of QHP has been studied elsewhere, known as gyroscopically stabilized (GS) systems. They give rise to quadratic polynomials,

$$L(\lambda) = \lambda^2 I + \lambda B + C \tag{3.3}$$

where $C > 0$, $B^* = B$ and is indefinite, and

$$|B| > kI + k^{-1}C \tag{3.4}$$

for some $k > 0$ (see [1] and [2]). Such systems are known to have $v = 3$. For the purposes of this paper a matrix is said to be *indefinite* if it has at least one positive and at least one negative eigenvalue. Then condition (3.4) ensures that B is both indefinite and invertible.

Suppose B has p positive eigenvalues (in (3.3)) and (3.4) holds. Then it is known (see [1] or [2]) that there are four quasi-zones as follows:

$$-\infty < a_1 \le b_1 < a_2 \le b_2 < 0 < a_3 \le b_3 < a_4 \le b_4 < \infty$$

where

$[a_1, b_1]$ contains p eigenvalues of negative type.
$[a_2, b_2]$ contains p eigenvalues of positive type.
$[a_3, b_3]$ contains $n-p$ eigenvalues of negative type.
$[a_4, b_4]$ contains $n-p$ eigenvalues of positive type.

The first example serves to show that notions of convexity and connectivity applying in the classical case $L(\lambda) = \lambda I - A$ must be re-examined in this context. It will also be useful in the sequel.

Example 3.2. Consider the system (3.3) defined by

$$B = \begin{bmatrix} 0 & 2.8i \\ -2.8i & 0 \end{bmatrix}, \quad C = \begin{bmatrix} 2 & 1 \\ 1 & 2 \end{bmatrix}.$$

This is a gyroscopic system which is stable (is a QHP) although the sufficient condition of (3.4) is not satisfied. The numerical range is indicated in Figure 1.

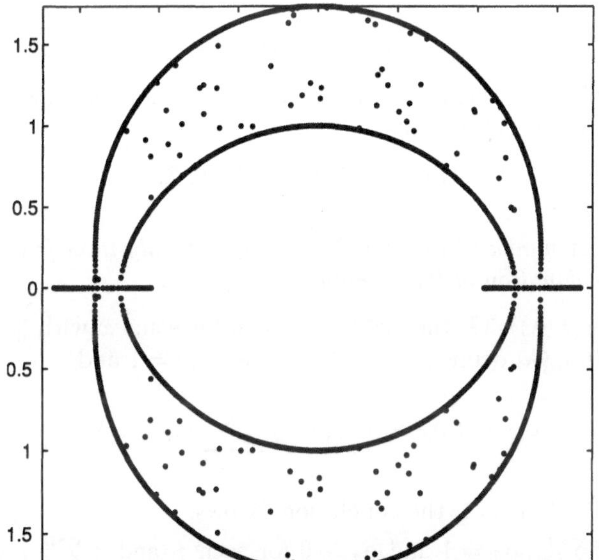

Figure 1: Numerical range for a QHP

4 The numerical range of selfadjoint matrix polynomials

As above, $L(\lambda)$ denotes a selfadjoint matrix polynomial and $w(L(\lambda))$ is its numerical range. Note first of all the basic properties of $w(L(\lambda))$ established by Li and Rodman [9]: $w(L(\lambda))$ is closed and $w(L(\lambda))$ is bounded if and only if $A_\ell > 0$ or $A_\ell < 0$, and when it is bounded the number of its connected components does not exceed ℓ. Also, it is easily seen that, when $L(\lambda)$ is selfadjoint, $w(L(\lambda))$ is symmetric with respect to the real line. For this reason, the intersection of $w(L(\lambda))$ with the real line is of special interest (see also Theorem 31.6 of [10]):

Proposition 4.1. *If $L(\lambda)$ is a selfadjoint matrix polynomial, then*

$$w(L(\lambda)) \cap \mathbb{R} = \{\lambda \in \mathbb{R} \mid L(\lambda) \textit{ is indefinite or semidefinite}\}.$$

Proof. It is clear that if $L(\lambda_0) > 0$ or $L(\lambda_0) < 0$ then $\lambda_0 \notin w(L(\lambda))$. Otherwise $\lambda_0 \in w(L(\lambda))$. □

A simple illustration is given by the GS systems described above, when

$$w(L(\lambda)) \cap \mathbb{R} = [a_1, b_2] \cup [a_3, b_4]. \tag{4.1}$$

Example 4.2. If $L(\lambda) \geq 0$ for all $\lambda \in \mathbb{R}$ then Theorem 5.1 of [5] shows that $L(\lambda) = M(\lambda)^* M(\lambda)$ when $\lambda \in \mathbb{R}$ for some matrix polynomial $M(\lambda)$. In this case, for $\lambda \in \mathbb{R}$,

$$x^* L(\lambda) x = \|M(\lambda)x\|^2 \geq 0$$

for all $x \in \mathbb{C}^n$. Thus $\lambda \in w(L(\lambda))$ implies $M(\lambda)x_0 = 0$ for some $x_0 \neq 0$ and hence $L(\lambda)x_0 = 0$, i.e. $\lambda \in \sigma(L(\lambda))$. Thus,

$$w(L(\lambda)) \cap \mathbb{R} = \sigma(L(\lambda)) \cap \mathbb{R}.$$

Proposition 4.3. *A number $\lambda_0 \in w(L(\lambda)) \cap \mathbb{R}$ if and only if λ_0 is a real zero of a convex linear combination of the eigenfunctions $\mu_1(\lambda), \dots, \mu_n(\lambda)$.*

Proof. If $\lambda_0 \in w(L(\lambda)) \cap \mathbb{R}$ then $x^* L(\lambda_0) x = 0$ for some x with $\|x\| = 1$. Write $y = U(\lambda_0)x$ (with $U(\lambda)$ defined as in (2.1)), then $\|y\| = 1$ and

$$0 = x^* L(\lambda_0) x = y^* D(\lambda_0) y = \sum_{j=1}^{n} |y_j|^2 \mu_j(\lambda_0).$$

Since $\|y\|^2 = \sum_{j=1}^n |y_j|^2 = 1$, the conclusion follows.

Conversely, if $\sum_{j=1}^n \alpha_j = 1$ and $\alpha_j \geq 0$ for each j, and if $\sum_{j=1}^n \alpha_j \mu_j(\lambda_0) = 0$, $\lambda_0 \in \mathbb{R}$, define $y = [\sqrt{\alpha}_1, \dots, \sqrt{\alpha}_n]^T$ and $x = U(\lambda_0)y \neq 0$. Then

$$x^* L(\lambda) x = y^* D(\lambda_0) y = \sum_{j=1}^{n} \alpha_j \mu_j(\lambda_0) = 0.$$

□

Notice that, when the coefficients of $L(\lambda)$ can be simultaneously diagonalized by a unitary similarity: i.e. there is a unitary V such that VA_jV^* is diagonal, $j = 0, 1, \ldots, \ell$. Then we may take $U(\lambda) \equiv V$ in (2.1) and the eigenfunctions are real polynomials with degree not exceeding ℓ.

For small values of n some feeling for the number of connected components in $w(L(\lambda)) \cap \mathbb{R}$ for a QHP can be obtained from Proposition 4.1 and simple counting arguments:

Example 4.4. Let $L(\lambda)$ be a QHP with $\det A_\ell \neq 0$, distinct eigenvalues, and $n = 2$. Then the number of connected components in $w(L(\lambda)) \cap \mathbb{R}$ is ℓ or $\ell + 1$ when A_ℓ is definite or indefinite, respectively.

Example 4.5. Let $L(\lambda)$ be a QHP with $A_\ell > 0$, distinct eigenvalues, and $n = 3$. When $\ell = 2$ or 3 a QHP $L(\lambda)$ can be found for which $w(L(\lambda)) \cap \mathbb{R}$ has just one component (in contrast with the preceding example). This can be done by choosing eigenfunctions $\mu_1(\lambda)$, $\mu_2(\lambda)$, $\mu_3(\lambda)$ as polynomials of degree ℓ with ℓ distinct real zeros which interlace suitably.

Proposition 4.6. *Let $L(\lambda)$ be a regular selfadjoint matrix polynomial. If λ_0 is a singleton of $w(L(\lambda)) \cap \mathbb{R}$ then either*

(a) $L(\lambda) > 0$ *(or $L(\lambda) < 0$) in some real deleted neighbourhood of λ_0 and there exist one or more eigenfunctions with even multiplicity at λ_0, or*

(b) $L(\lambda)$ *is positive and negative definite on opposite sides of λ_0 and all eigenfunctions have odd multiplicities at λ_0. In this case*

$$L(\lambda) = (\lambda - \lambda_0)L_1(\lambda) \tag{4.2}$$

for some selfadjoint matrix polynomial $L_1(\lambda)$.

If, in addition, $L(\lambda)$ is a QHP, *then Case* (a) *cannot arise, Case* (b) *does apply, all eigenfunctions have a simple zero at λ_0, and $L_1(\lambda)$ in* (4.2) *is a* QHP.

Proof. Statements (a) and (b) follow immediately from Proposition 4.1. (Simple illustrations are given by

$$L(\lambda) = \begin{bmatrix} \lambda^2 & 0 \\ 0 & (\lambda-1)^2 \end{bmatrix}, \qquad L(\lambda) = \begin{bmatrix} \lambda^3 & 0 \\ 0 & \lambda(\lambda^2+1) \end{bmatrix},$$

respectively, with $\lambda_0 = 0$.)

When $L(\lambda)$ is a QHP λ_0 must have definite type and this means that $\mu_j'(\lambda_0) > 0$ (or < 0) for all j. Thus, Case (a) cannot arise and, in Case (b) all eigenfunctions have *simple* zeros at λ_0. To show that $L_1(\lambda)$ is a QHP, let $\hat{\lambda}$ be any real eigenvalue of $L_1(\lambda)$. Then $\hat{\lambda} \neq \lambda_0$ and it is to be shown that $\hat{\lambda}$ has definite type with respect to $L_1(\lambda)$. Clearly, $\hat{\lambda} \in \sigma(L(\lambda))$ as well. Indeed, if $L_1(\hat{\lambda})x = 0$, $x \neq 0$, then $L(\lambda)x = 0$ and, because

$$L'(\hat{\lambda}) = (\hat{\lambda} - \lambda_0)L_1'(\hat{\lambda}) + L_1(\hat{\lambda})$$

$0 \neq x^*L'(\hat{\lambda})x = (\hat{\lambda} - \lambda_0)x^*L_1'(\hat{\lambda})x$. Thus, $\hat{\lambda}$ is a definite eigenvalue of $L_1(\lambda)$ (and has the same or opposite type with respect to $L_1(\lambda)$ and $L(\lambda)$ according as $\hat{\lambda} > \lambda_0$ or $\hat{\lambda} < \lambda_0$). □

In the converse direction we have:

Proposition 4.7. *If $L(\lambda)$ is an $n \times n$ selfadjoint matrix polynomial and λ_0 is a real eigenvalue of definite type and multiplicity n, then there is a factorization* (4.1) *and λ_0 is a singleton of $w(L(\lambda)) \cap \mathbb{R}$.*

Proof. The "definite" property of λ_0 ensures that all eigenfunctions have simple zeros at λ_0 with slopes at λ_0 of the same sign. Then the conditions of part (b) of Proposition 4.6 hold. □

Theorem 4.8. *Let $L(\lambda)$ be a quadratic selfadjoint matrix polynomial with $A_2 > 0$ and at least one non-real point in $w(L(\lambda))$. Then $w(L(\lambda))$ is connected if and only if $\sigma(L(\lambda)) \cap \mathbb{R} \neq \emptyset$.*

In particular, combining the theorem with the preceding proposition we find that a QHP which is not weakly hyperbolic has a connected numerical range. Example 3.2 illustrates this statement.

Proof. It follows from Corollary 2.4 of [9] that $w(L(\lambda))$ has at most two components. So suppose that there are two such components C_1 and C_2 and that $\lambda_0 \in w(L(\lambda)) \cap \mathbb{R}$.

Let $\lambda_1 \in w(L(\lambda))$ with $\lambda_1 \neq \bar{\lambda}_1$ and, as noted above, $\bar{\lambda}_1 \in w(L(\lambda))$ as well. By Proposition 2.5 of [9], we may assume $\lambda_1 \in C_1$ and $\bar{\lambda}_1 \in C_2$ and, without loss of generality, $\lambda_0 \in C_1$. Then there is a path $s(t)$ in C_1 such that $s(0) = \lambda_0$, $s(1) = \lambda_1$. But then the path $\overline{s(t)}$ connects λ_0 with $\bar{\lambda}_1$ and a path from λ_1 to $\bar{\lambda}_1$ is obtained, which is a contradiction. Thus, it follows that $w(L(\lambda))$ has just one connected component.

Conversely, if $\sigma(L) \cap \mathbb{R}$ is empty then $L(\lambda)$ has constant signature on $\mathbb{R}$; namely, that of A_2. By Proposition 4.1 it follows that $w(L(\lambda)) \cap \mathbb{R}$ is also empty and, as $w(L(\lambda)) \neq \emptyset$ it follows from Corollary 2.4 of [9] that $w(L(\lambda))$ has exactly two connected components. □

Corollary 4.9. *Let $L(\lambda) = \lambda^2 I + \lambda B + C$ where $C > 0$ and let $B^* = B$ be indefinite (as in equation* (3.3)*). Then $w(L(\lambda))$ is connected if and only if $L(\lambda)$ has a real eigenvalue.*

Proof. Since B is indefinite there is an x such that $x^*Bx = 0$ and $\|x\| = 1$. Then

$$x^*L(\lambda)x = \lambda^2 I + x^*Cx$$

and, as $x^*Cx > 0$, $x^*L(\lambda)x = 0$ implies λ is pure imaginary and is in $w(L(\lambda))$. Now apply the theorem. □

Note, in particular, that a GS system (i.e. $L(\lambda)$ as in the Corollary together with condition (3.4)) necessarily has a connected numerical range since then $\sigma(L(\lambda)) \subset \mathbb{R}$.

Let $0 < c_1 \leq c_2 \leq \cdots \leq c_n$ be the eigenvalues of C. Then, since $c_1 \leq x^*Cx \leq c_n$ for any x with $\|x\| = 1$, it follows that, under the hypotheses of the Corollary, the non-real points of $w(L(\lambda))$ lie in the annulus

$$c_1^{1/2} \leq |\lambda| \leq c_n^{1/2}.$$

The following important special case indicates that, in general, $w(L(\lambda))$ does not include the whole of this annulus.

Example 4.10. In equation (3.3) let $B = iG$ where G is a nonsingular, real, skew-symmetric matrix (so that n is necessarily even). Also let $C \in \mathbb{R}^{n \times n}$ with $C > 0$ (see [1] for physical origins of this case). Note that Example 3.2 is of this kind. In the context of this example the two following propositions hold.

Proposition 4.11. *The line segment* $i[c_1^{1/2}, c_n^{1/2}]$ *is contained in* $w(L(\lambda))$.

Proof. Let $\mu^2 \in [c_1, c_n]$ and $x \in \mathbb{R}^n$ with $\|x\| = 1$. Then, since G is real and skew-symmetric, $x^T Gx = 0$. Thus,

$$x^*L(i\mu)x = x^T(-\mu^2 I - \mu G + C)x = -\mu^2 + x^T Cx.$$

As $\mu^2 \in [c_1, c_n]$, x can be chosen so that $x^T Cx = \mu^2$ and $x^*L(i\mu)x = 0$, i.e. $i\mu \in w(L(\lambda))$. □

Proposition 4.12. *Let* c_n *have algebraic multiplicity one as an eigenvalue of* C. *Then the circular arc* $\lambda = c_n^{1/2}e^{i\theta}$, $0 < \theta < \pi/2$ *does not belong to* $w(L(\lambda))$.

Proof. Let $Cu_n = c_n u_n$, $\|u_n\| = 1$, $u_n \in \mathbb{R}^n$. If $\lambda_0 = c_n^{1/2}e^{i\theta_0}$ where $\theta_0 \in (0, \pi/2)$ and $\lambda_0 \in w(L(\lambda))$, then there is an $x \in \mathbb{C}^n$ with $\|x\| = 1$ and

$$\lambda_0^2 + i(x^*Gx)\lambda_0 + x^*Cx = 0.$$

Since x^*Gx is pure-imaginary, λ_0 is a non-real root of a quadratic equation with real coefficients. Hence

$$|\lambda_0|^2 = c_n = x^*Cx, \tag{4.3}$$

$$-2 \text{ Re } (\lambda_0) = -2c_n^{1/2}\cos\theta_0 = i(x^*Gx). \tag{4.4}$$

Since c_n is a simple eigenvalue of C it follows from (4.3) that $x = \alpha u_n$ where $|\alpha| = 1$. Then $x^*Gx = |\alpha|^2 u_n^T G u_n = 0$ since $G^T = -G$, and (4.4) gives $\theta_0 = \pi/2$ contradicting the assumption that $\theta_0 \in (0, \pi/2)$. □

Similarly, if c_1 is a simple eigenvalue of C then the arc $\lambda = c_1^{1/2} e^{i\theta}$, $\theta \in (0, \pi/2)$ is not in $w(L(\lambda))$ (see Figure 1, for example).

Under the hypotheses of these Propositions assume, in addition, that $|G|$ is large enough that $L(\lambda)$ has at least one pair of real eigenvalues (the condition $|G| > kI + k^{-1}C$ for some $k > 0$ is certainly sufficient for this to be the case). Then Corollary 4.9 applies and the non-real points of $w(L(\lambda))$ belong to an annulus with a non-circular boundary intersecting the imaginary axis in the segment $i[c_1^{1/2}, c_n^{1/2}]$ and the positive real axis in a segment $[\xi_1, \xi_2]$ where $c_1^{1/2} < \xi_1 \le \xi_2 < c_n^{1/2}$. (The points of $w(L(\lambda)) \cap \mathbb{R}$ are, of course, described in (4.1). See also Example 3.2 and Figure 1.)

5 The numerical range of a linearization

The spectral analysis of matrix polynomials frequently uses the notion of "linearization". When $\det A_\ell \neq 0$ selfadjoint polynomials have a selfadjoint linear pencil as a linearization. This is formed as follows: Let

$$(5.1)\quad \mathcal{A} = \begin{bmatrix} A_1 & A_2 & \dots & & A_\ell \\ A_2 & & & A_\ell & 0 \\ \vdots & & \cdot^{\cdot^{\cdot}} & & \\ & A_\ell & & & \\ A_\ell & 0 & & \dots & 0 \end{bmatrix}, \quad \mathcal{B} = \begin{bmatrix} -A_0 & 0 & \dots & & 0 \\ 0 & A_2 & A_3 & \dots & A_\ell \\ 0 & A_3 & & & 0 \\ \vdots & \vdots & \cdot^{\cdot^{\cdot}} & & \\ 0 & A_\ell & 0 & \dots & 0 \end{bmatrix}.$$

Then $\sigma(\lambda\mathcal{A} - \mathcal{B}) = \sigma(L(\lambda))$ when $\det A_\ell \neq 0$.

Proposition 5.1. *Let $\mathcal{A}$, $\mathcal{B}$ be defined as above. Then for any polynomial $L(\lambda)$,*

$$w(L(\lambda)) \subseteq w(\lambda\mathcal{A} - \mathcal{B}).$$

Proof. If $\lambda_0 \in w(L(\lambda))$ then $x^* L(\lambda_0) x = 0$ for some $x \neq 0$. Construct the vector

$$x_e = \begin{bmatrix} x \\ \lambda_0 x \\ \vdots \\ \lambda_0^{\ell-1} x \end{bmatrix} \in \mathbb{C}^{\ell n}.$$

Then $x_e \neq 0$ and it can be verified that

$$x_e^* (\lambda_0 \mathcal{A} - \mathcal{B}) x_e = x^* L(\lambda_0) x = 0.$$

Hence $\lambda_0 \in w(\lambda_0 \mathcal{A} - \mathcal{B})$. □

Notice also that when $\ell \geq 2$, $w(\lambda\mathcal{A} - \mathcal{B})$ is unbounded.

The next proposition shows that, when examining the numerical range of a QHP $L(\lambda)$, the numerical range of its linearization $\lambda\mathcal{A} - \mathcal{B}$ (of (5.1)) is generally not informative.

Proposition 5.2. *Let* $L(\lambda)$ *be a* QHP *with* $\det A_\ell \neq 0$ *and variation* v*, and let* $\lambda\mathcal{A}-\mathcal{B}$ *be defined by equations* (5.1). *Then* $w(\lambda\mathcal{A} - \mathcal{B}) = \mathbb{C}$ *if* $v > 1$ *and* $w(\lambda\mathcal{A} - \mathcal{B}) \subseteq \mathbb{R}$ *if* $v = 1$.

Proof. Since $\lambda\mathcal{A}-\mathcal{B}$ is a linearization of $L(\lambda)$, $\sigma(\lambda\mathcal{A}-\mathcal{B}) = \sigma(L(\lambda))$. Also, if λ_0 is an eigenvalue of $L(\lambda)$ with eigenvector x then λ_0 is an eigenvalue of $\lambda\mathcal{A} - \mathcal{B}$ with eigenvector $x_e = \operatorname{col}[\lambda_0^j x]_{j=0}^{\ell-1}$, and it follows that

$$x_e^* \mathcal{A} x_e = x^* L'(\lambda_0) x. \tag{5.2}$$

Consequently, the eigenvalue types for $L(\lambda)$ and $\lambda\mathcal{A}-\mathcal{B}$ correspond so that $\lambda\mathcal{A}-\mathcal{B}$ also has variation v.

When $v = 1$ Theorem 1.2 of [8] implies that $\lambda\mathcal{A} - \mathcal{B}$ is a "definite" pencil (i.e. there is a pair of real numbers α, β such that $\alpha A + \beta B > 0$) and, consequently, there is no nonzero vector u such that

$$u^* \mathcal{A} u = u^* \mathcal{B} u = 0. \tag{5.3}$$

Since $\mathcal{A}$ and $\mathcal{B}$ are both hermitian and $\mathcal{A}$ is indefinite it follows (as in Theorem 4.1 of [9]) that $w(\lambda\mathcal{A} - \mathcal{B}) \subseteq \mathbb{R}$.

If $v > 1$ then, using Theorem 1.2 of [8] once more, $\lambda\mathcal{A}-\mathcal{B}$ is not a definite pencil and so there is a nonzero u such that (5.3) holds. Then it follows immediately that $w(\lambda\mathcal{A} - \mathcal{B}) = \mathbb{C}$. □

Remark. It is apparent that, when $\ell > 2$, $\mathcal{A}$ and $\mathcal{B}$ of (5.1) have common isotropic vectors $u = [0 \ 0 \ \ldots \ w]^T$ for any nonzero $w \in \mathbb{C}^n$, i.e. for which (5.3) is satisfied. Consequently, the hypothesis $v = 1$ implies that $\ell \leq 2$.

Corollary 5.3. *If* $L(\lambda)$ *is a* QHP *with* $\det A_\ell \neq 0$ *and* $\ell > 2$*, then* $w(\lambda\mathcal{A} - \mathcal{B}) = \mathbb{C}$.

Proof. It follows from Proposition 3.1 that $v \geq \ell - 1 > 1$. Now apply Proposition 5.2. □

Corollary 5.4. *If* $L(\lambda)$ *is a* QHP *with* $\det A_\ell \neq 0$ *and there is a* $z \in w(L(\lambda))$ *with* $z \notin \mathbb{R}$*, then* $w(\lambda\mathcal{A} - \mathcal{B}) = \mathbb{C}$.

Proof. By Proposition 5.1 $z \in w(\lambda\mathcal{A} - \mathcal{B})$ and the result follows from Proposition 5.2. □

Note also that, if $\lambda A - B$ is a selfajoint pencil with A nonsingular and indefinite and if $w(\lambda A - B) \neq \mathbb{C}$, then there is no $x \neq 0$ such that $x^*Ax = x^*Bx = 0$. Consequently (see [8], for example), $\lambda A - B$ is a definite pencil and, because A is indefinite it has variation $v = 1$. Applying this to a linearization (5.1) the first sentence of the following Proposition is obtained:

Proposition 5.5. *Let $\lambda\mathcal{A} - \mathcal{B}$ be the linearization of a selfadjoint polynomial $L(\lambda)$ with $\det A_\ell \neq 0$ and $\ell \geq 2$. If $w(\lambda\mathcal{A} - \mathcal{B}) \neq \mathbb{C}$ then $L(\lambda)$ is a QHP with variation $v = 1$ and, in fact, $\ell = 2$.*

Proof. The last statement follows from Proposition 3.1. □

This leads to a new characterization of quadratic hyperbolic polynomials:

Corollary 5.6. *If $\ell = 2$ and $A_2 > 0$ then $L(\lambda)$ is hyperbolic if and only if $w(\lambda\mathcal{A} - \mathcal{B}) \neq \mathbb{C}$.*

Proof. If $L(\lambda)$ is hyperbolic then $\ell = 2$ implies $v = 1$ and $\lambda\mathcal{A} - \mathcal{B}$ is a definite pencil. By Proposition 5.2 $w(\lambda\mathcal{A} - \mathcal{B}) \subseteq \mathbb{R}$.

Given that $w(\lambda\mathcal{A} - \mathcal{B}) \neq \mathbb{C}$, Proposition 5.5 says that $L(\lambda)$ is a QHP with $v = 1$. Thus, there are only two root zones and each must contain exactly n eigenvalues, i.e. $L(\lambda)$ is hyperbolic. □

It is clear (from Theorem 4.1 of [9]) that the hypothesis $w(\lambda\mathcal{A} - \mathcal{B}) \neq \mathbb{C}$ in this context implies that $w(\lambda\mathcal{A} - \mathcal{B}) \subseteq \mathbb{R}$. Notice also that "overdamped" systems of the theory of vibrations are included in Corollary 5.6. They have the additional properties that $A_1 > 0$, $A_0 > 0$ (see [1]).

6 The numerical range of real symmetric matrix polynomials

In this section we investigate the properties of matrix polynomials whose coefficients A_i are real symmetric matrices for all i. We define the *real numerical range* as

$$w_{\mathbb{R}}(L) = \{\lambda \in \mathbb{C} \mid x^*L(\lambda)x = 0 \text{ for some nonzero } x \in \mathbb{R}^n\}.$$

Our main result of this section states that for $n \geq 3$ and polynomials L as above, the real numerical range of L coincides with the numerical range of L. Moreover, in the case of $n = 2$ we show that $\partial w(L)$, the boundary of the numerical range of L, is a subset of $w_{\mathbb{R}}(L)$. The topological reason for this distinction ($n \geq 3$ and $n = 2$) lies in the fact that the real unit sphere is not simply connected only in the case when $n = 2$.

For the purposes of this section, we denote by $F(A)$ the numerical range of a $n \times n$ complex matrix A, $F(A) = \{(Ax, x) \mid x \in \mathbb{C}^n, (x, x) = 1\}$ and by $F_{\mathbb{R}}(A)$ the real numerical range of A, $F_{\mathbb{R}}(A) = \{(Ax, x) \mid x \in \mathbb{R}^n, (x, x) = 1\}$. The following is an easy but important observation involving the numerical range of L.

$$\lambda \in w(L) \text{ if and only if } 0 \in F(L(\lambda)) \textbf{ and } \lambda \in w_{\mathbb{R}}(L) \text{ if and only if } 0 \in F_{\mathbb{R}}(L(\lambda)).$$

Lemma 6.1. *(Brickman) Let $C \in \mathbb{C}^{n\times n}$ be a (complex) symmetric matrix (i.e $C^T = C$). Then*

$$\text{if } n \geq 3 \quad F_{\mathbb{R}}(C) = F(C) \textbf{ and } \text{if } n = 2 \quad \partial F(C) = F_{\mathbb{R}}(C)$$

Remark. When $n = 2$, $F_{\mathbb{R}}(C)$ is an ellipse, line segment, or singleton and $F(C)$ is the convex hull of $F_{\mathbb{R}}(C)$. The case $n = 1$ is obvious. This lemma can be found in [3].

The following lemma can be found in [11].

Lemma 6.2. *(Maroulas/Psarrakos) If $\lambda \in \partial w(L)$ then $0 \in \partial F(L(\lambda))$.*

Our main result of this section is as follows.

Theorem 6.3. *Let $L(\lambda)$ be a matrix polynomial whose coefficients are real symmetric matrices. Then*

$$\text{if } n \geq 3 \quad w_{\mathbb{R}}(L) = w(L) \textbf{ and } \text{if } n = 2 \quad \partial w(L) \subset w_{\mathbb{R}}(L).$$

Proof. Part a). Clearly $w_{\mathbb{R}}(L) \subset w(L)$. If $\lambda \in w(L)$ then $0 \in F(L(\lambda))$ and $L(\lambda)$ is complex symmetric. Hence by part a) of Lemma 6.1 $0 \in F_{\mathbb{R}}(L(\lambda))$. In other words, $\lambda \in w_{\mathbb{R}}(L)$.

Part b). If $\lambda \in \partial w(L)$ then by Lemma 6.2, $0 \in \partial F(L(\lambda))$. Part b) of Lemma 6.1 shows that $0 \in F_{\mathbb{R}}(L(\lambda))$ and, in other words, $\lambda \in w_{\mathbb{R}}(L)$. Thus $\partial w(L) \subset w_{\mathbb{R}}(L)$. □

Remark. For completeness we give an example of a 2×2 quadratic matrix polynomial for which the real numerical range is a strict subset of the numerical range.

Example 6.4. Let $L(\lambda) = \lambda^2 I + \lambda B + C$, where

$$B = \begin{bmatrix} 1 & 0 \\ 0 & -1 \end{bmatrix} \quad \text{and} \quad C = \begin{bmatrix} 2 & 1 \\ 1 & 2 \end{bmatrix}.$$

Since $\sigma(L) \cap \mathbb{R} = \emptyset$, the Theorem 4.8 implies that $w(L)$ is disconnected and hence has two connected components. Straightforward calculations show that the spectral points of L (which are the zeros of $\lambda^4 + 3\lambda^2 + 3$) do not belong to the real numerical range of L.

References

[1] Barkwell, L. and Lancaster, P. Overdamped and gyroscopic vibrating systems. *J. Appl. Mech.* **59** (1992), 176–181.

[2] Barkwell, L., Lancaster, P. and Markus, A.S. Gyroscopically stabilized systems: A class of quadratic eigenvalue problems with real spectrum. *Canad. J. Math.* **44** (1992), 42–53.

[3] Brickman, L. On The Field Of Values Of A Matrix. *Proceedings of the American Mathematical Society* **12**, 1961.

[4] Crawford, C.R. A stable generalized eigenvalue problem. *SIAM Journal on Numerical Analysis* **13** (1976), 854–860.

[5] Gohberg, I., Lancaster, P. and Rodman, L. Spectral analysis of selfadjoint matrix polynomials. *Annals of Math.* (1980), 33–71.

[6] Gohberg, I., Lancaster, P. and Rodman, L. **Matrix Polynomials**. Academic Press, New York, 1982.

[7] Lancaster, P., Markus, A. and Matsaev, V. Perturbations of G-selfadjoint operators and operator polynomials with real spectrum, *Operator Theory and its Applications*, Birkhäuser Verlag), vol. **87** (1996), 207–221.

[8] Lancaster, P. and Ye, Q. Variational properties and Rayleigh quotient algorithms for symmetric matrix pencils. In **Operator Theory: Advances and Applications**, vol. 40, pp. 247–278, Birkhäuser Verlag, 1989.

[9] Li, C.K. and Rodman, L. Numerical range of matrix polynomials. *SIAM Journal on Matrix Analysis and Applications*, **15** (1994), 1256–1265.

[10] Markus, A. **Introduction to the Spectral Theory of Polynomial Operator Pencils**, Vol. 71, Translations of Math Monographs, American Math. Soc., Providence, 1988.

[11] Maroulas, J. and Psarrakos, P. Geometrical Properties of Numerical Range of Matrix Polynomials. *Computers Math. Applic.* **31**, No. 4/5, pp. 41–47, 1996.

Department of Mathematics and Statistics
University of Calgary
Calgary, Alberta T2N 1N4
Canada

Department of Mathematics
National Technical University of Athens
Zografou
Athens, Greece

Department of Mathematics and Statistics
University of Calgary
Calgary, Alberta T2N 1N4
Canada

1991 Mathematics Subject Classification. Primary 15A60; Secondary 15A22, 47A56

Operator Theory:
Advances and Applications, Vol. 106

Spectral properties of a matrix polynomial connected with a component of its numerical range

A. Markus, J. Maroulas, and P. Psarrakos

Dedicated to Professor Heinz Langer, with admiration

We prove that any bounded component of the numerical range of a matrix polynomial $L(\lambda)$ contains at least one eigenvalue of $L(\lambda)$. Moreover, the set of corresponding Jordan chains is complete.

1. Introduction

Let

$$L(\lambda) = \sum_{k=0}^{m} \lambda^k A_k$$

be a matrix polynomial, where the coefficients A_k are complex $n \times n$ matrices. A number $\lambda_0 \in \mathbb{C}$ is called an *eigenvalue* of $L(\lambda)$ if the equation $L(\lambda_0)f_0 = 0$ has a nonzero solution $f_0 \in \mathbb{C}^n$. Such a vector f_0 is called an *eigenvector* of $L(\lambda)$ corresponding to λ_0. Vectors $f_1, f_2, \ldots, f_l$ are said to be associated with the eigenvector f_0, if

$$\sum_{j=0}^{k} \frac{1}{j!} L^{(j)}(\lambda_0) f_{k-j} = 0 \quad (k = 1, 2, \ldots, l).$$

The collection of vectors $f_0, f_0, \ldots, f_l$ is called a *Jordan chain* for $L(\lambda)$ corresponding to the eigenvalue λ_0.

The set of all eigenvalues is called the *spectrum* of $L(\lambda)$:

$$\sigma(L) = \{\lambda \in \mathbb{C} : \det L(\lambda) = 0\}.$$

Clearly, $\sigma(L)$ either consists of not more then mn points or coincides with the whole plane $\mathbb{C}$.

The set

$$W(L) = \{\lambda \in \mathbb{C} : (L(\lambda)f, f) = 0 \text{ for some } f \neq 0\}$$

is known as *numerical range* of $L(\lambda)$. Evidently, $W(L)$ is always closed and

$$\sigma(L) \subset W(L).$$

Generally speaking, the set $W(L)$ is not connected, and it is bounded if and only if the origin is not contained in the numerical range of the leading coefficient A_m [5]. In this paper we are interested in *components* of $W(L)$, i.e. maximal connected subsets of $W(L)$, and especially in bounded components. They play an important role in factorization problems (see, e.g., [5]; [6], §§ 26, 27, 31; [7]), and some properties of these components are studied in [5], [7].

We prove that any bounded component $F(\neq \emptyset)$ of $W(L)$ contains a non-empty part of the spectrum $\sigma(L)$ (Theorem 2.2). Moreover, the set of all Jordan chains corresponding to eigenvalues from F is complete, i.e., their span coincides with $\mathbb{C}^n$ (Theorem 3.2). We give also a generalization of Theorem 2.2 to the case of infinite-dimensional Hilbert space (Theorem 2.3).

We would like to note that Theorems 2.2 and 3.2 (resp. Theorem 2.3) without essential changes in the proofs can be extended to an analytic matrix (resp. operator) function $L(\lambda)$.

2. Non-emptiness of a part of the spectrum

Let F $(\neq \emptyset)$ be a bounded component of $W(L)$. Consider a bounded domain G with smooth boundary Γ such that

$$F \subset G, \quad W(L)\backslash F \subset \mathbb{C}\backslash \overline{G}.$$

If $f \in \mathbb{C}^n$, $f \neq 0$, then $(L(\lambda)f, f) \neq 0$ for $\lambda \in \Gamma$ because $\Gamma \cap W(L) = \emptyset$. Hence we can consider the index $\mathrm{ind}_\Gamma(L(\lambda)f, f)$ (for definition and main properties see, e.g., [6], p. 131). It is a simple observation that this index does not depend on f ([6], Lemma 26.8), and we will denote it by $c(F)$. Clearly, $c(F) > 0$.

The following statement is essentially known, and we prove it for the convenience of the reader. In what follows we exclude the trivial case $n = 1$.

Lemma 2.1. *If $f(\lambda)$, $\lambda \in \Gamma$, is a non-vanishing continuous vector function with values in $\mathbb{C}^n$ then*

$$\mathrm{ind}_\Gamma(L(\lambda)f(\lambda), f(\lambda)) = c(F).$$

Proof. Since the unit sphere in $\mathbb{R}^k$ is simply connected for $k > 2$ (see, e.g., [2], p. 43), every continuous mapping f from Γ to $\mathbb{C}^n\backslash\{0\}$ $(n > 1)$ is homotopic to a constant mapping. Hence $(L(\lambda)f(\lambda), f(\lambda))$ is homotopic to $(L(\lambda)g, g)$ $(g \neq 0)$. □

Theorem 2.2. *The component F contains at least one eigenvalue of $L(\lambda)$.*

Proof. Let h be a nonzero vector and $f(\lambda) = L^{-1}(\lambda)h$. Then $(L(\lambda)f(\lambda), f(\lambda)) = (h, L^{-1}(\lambda)h)$, and it follows from Lemma 2.1 that $\mathrm{ind}_\Gamma(h, L^{-1}(\lambda)h)) = c(F)$. Since $\mathrm{ind}_\Gamma \overline{u(\lambda)} = -\mathrm{ind}_\Gamma u(\lambda)$ for any continuous function $u(\lambda)(\neq 0)$, then

$$\mathrm{ind}_\Gamma(L^{-1}(\lambda)h, h) = -c(F) < 0,$$

and hence the function $(L^{-1}(\lambda)h, h)$ cannot be holomorphic everywhere inside Γ. But all singularities of this function are eigenvalues of $L(\lambda)$. □

It is easy to check that the proof of Theorem 2.2 gives also the following infinite-dimensional version of this theorem.

Theorem 2.3. *Let $L(\lambda)$ be an operator polynomial in a Hilbert space $\mathcal{H}$. If F $(\neq \emptyset)$ is a bounded component of $\overline{W(L)}$, the closure of the numerical range of $L(\lambda)$, then $\sigma(L) \cap F \neq \emptyset$.*

It should be noted that I. Krupnik and P. Lancaster obtained Theorem 2.2 independently. They use a different method of proof which, as it seems, does not work for the operator case.

3. Completeness of a part of Jordan chains

We will use the following well known result.

Lemma 3.1. *If a vector h is orthogonal to all Jordan chains of a matrix polynomial $L(\lambda)$ corresponding to eigenvalues from a domain G, then the function $(L^{-1}(\lambda)f, h)$ is analytic in G for any vector $f \in \mathbb{C}^n$.*

For proof of this statement (in a more general situation) see, e.g., [6], Lemma 18.7. A direct proof of Lemma 3.1 can be obtained from [1], Corollary 2.5.

Theorem 3.2. *If F $(\neq \emptyset)$ is a bounded component of $W(L)$ then the set of all Jordan chains of $L(\lambda)$ corresponding to eigenvalues from F is complete in $\mathbb{C}^n$.*

Proof. Let a vector h be orthogonal to all Jordan chains of $L(\lambda)$ corresponding to eigenvalues from F. Then by Lemma 3.1 the function $(L^{-1}(\lambda)h, h)$ is analytic in G, and hence $\mathrm{ind}_\Gamma(L^{-1}(\lambda)h, h) \geq 0$ (we use notations from Section 2). But we have shown in the proof of Theorem 2.2 that

$$\mathrm{ind}_\Gamma(L^{-1}(\lambda)h, h) = -c(F) < 0$$

for any nonzero vector h. Hence $h = 0$. □

Remark 3.3. In the case $c(F) = 1$, Theorem 3.2 is a consequence of a known result about existence of a linear divisor (see [6], Theorem 26.19; [7], Lemma 3.4). On the other hand, using some criteria for existence of a linear divisor ([4]; [1], p. 125) one can easily obtain above-mentioned result (for the matrix case) as a corollary of Theorem 3.2.

Remark 3.4. In the case $c(F) \geq 2$, Theorem 3.2 does not follow from known factorization theorems. We want to note also that in this case the set of Jordan chains under consideration is not always $c(F)$-fold complete (for definition of k-fold completeness, or multiple completeness see [6], p. 65). This follows from [1], Theorem 3.12, Corollary 1.14 and [3], Section 4.

Remark 3.5. In contrast to Theorem 2.2, it is impossible to generalize Theorem 3.2 to infinite-dimensional Hilbert space $\mathcal{H}$. The main reason is that the operator polynomial $L(\lambda)$ can have continuous spectrum, but of course there are also other obstacles. It is possible to obtain some results in this direction only under some strong additional assumptions (see, e.g., [6], Corollary 26.20, Theorems 30.11 and 30.12).

References

[1] I. Gohberg, P. Lancaster and L. Rodman, Matrix Polynomials, Academic Press, 1982.

[2] S.-T. Hu, Homotopy Theory, Academic Press, 1959.

[3] I. Krupnik, A. Markus and V. Matsaev, Factorization of matrix functions and characteristic properties of the circle, Int. Eq. Oper. Th. 17 (1993), 554 - 566.

[4] H. Langer, Über Lancaster's Zerlegung von Matrizen-Scharen, Arch. Rat. Mech. Anal. 29 (1968), 75 - 80.

[5] C.-K. Li and L. Rodman, Numerical range of matrix polynomials, SIAM J. Matrix Anal. Appl. 15 (1994), 1256 - 1265.

[6] A. Markus, Introduction to Spectral Theory of Polynomial Operator Pencils, Amer. Math. Soc., 1988.

[7] A. Markus and L. Rodman, Some results on numerical ranges and factorizations of matrix polynomials, Linear and Multilinear Algebra (to appear).

Department of Mathematics
and Computer Science
Ben-Gurion University of
the Negev
Beer-Sheva 84105
Israel

Department of Mathematics
National Technical University
Zografou Campus
Athens 15773
Greece

Department of Mathematics
National Technical University
Zografou Campus
Athens 15773
Greece

1991 Mathematics Subject Classification. Primary 47A56; Secondary 15A22, 15A60

Operator Theory:
Advances and Applications, Vol. 106

Lyapunov stability of a perturbed multiplication operator

S.N. NABOKO and C. TRETTER

Dedicated to Heinz Langer on the occasion of his sixtieth birthday

The perturbation of the multiplication operator in the space $L_2(0,1)$ by a Volterra operator with degenerate kernel is a particular case of the so-called "Friedrichs model". We characterize the point spectrum of such a perturbation and establish a sharp result on the Lyapunov stability in the case that the kernel vanishes on the diagonal.

1. Introduction

In this note a special case of the so-called "Friedrichs model" is considered. We study the operator of multiplication in $L_2(0,1)$ perturbed by a Volterra operator with degenerate kernel:

$$(1.1)\qquad (Ly)(x) := xy(x) + \int_0^x \varphi(x)\psi(s)y(s)\,\mathrm{d}s, \qquad x \in [0,1],$$

for $y \in L_2(0,1)$, with functions $\varphi, \psi \in L_2(0,1)$. The operator L is a model for a non-self-adjoint compact perturbation of a self-adjoint operator with nonempty essential spectrum where the perturbed spectrum remains entirely real. More complicated operators of this type arise for example in polymerisation chemistry, describing the motion of a marked monomer in a system of reacting polymers at equilibrium (see [K]).

With regard to such applications the structure of the spectrum and the Lyapunov stability of the operator L are of interest. The latter means that the group of operators $\mathrm{e}^{\mathrm{i}Lt}$, $t \in \mathbb{R}$, generated by L is bounded (see [N1], [C]). For the space $L_2(\mathbb{R})$ these problems have been studied in [N2], for the space $L_2(0,1)$ the particular case that φ and ψ are constant has been considered in [V]. The Lyapunov stability of the Friedrichs model with a perturbation of finite rank has been investigated in [FN]. In the present paper we characterize the point spectrum of the operator L in the general case (1.1) and we solve the question of the Lyapunov stability of L if the product $\varphi\psi$ vanishes.

In Section 2 an explicit formula for the resolvent of L is given. It is shown that the spectrum of L is purely essential and, thus, coincides with the interval $[0,1]$. If $\Re(\varphi\psi)$ fulfills a Lipschitz condition of order α, necessary and sufficient conditions for a point $\lambda \in [0,1]$ to be an eigenvalue of L in terms of the functions φ and ψ are given. For continuous φ these conditions amount to $\lambda \neq 1$ and $\Re(\varphi\psi)(\lambda) < -\frac{1}{2}$.

In Section 3 we first generalize a criterion for the Lyapunov stability of an operator T given in [N1] (see also [C]) which involves certain estimates for the resolvent of L near the real axis. Together with the explicit formula for the resolvent of L this allows us to prove that the operator L is Lyapunov stable if the functions φ and ψ belong to certain Lipschitz classes $\operatorname{Lip}(\omega_1)$ and $\operatorname{Lip}(\omega_2)$, respectively, and for some $\delta \in (0,1)$,

$$\int_0^\delta \frac{\omega_1(\tau)\omega_2(\tau)}{\tau} < \infty. \tag{1.2}$$

Section 4 contains an example of a pair of Lipschitz classes such that condition (1.2) is fulfilled and a counter-example which shows that the result on the Lyapunov stability of L proved in Section 3 is indeed sharp. We construct functions $\varphi \in \operatorname{Lip}(\omega_1)$, $\psi \in \operatorname{Lip}(\omega_2)$ with ω_1, ω_2 for which the condition (1.2) is violated and such that the corresponding operator L is not Lyapunov stable.

2. Resolvent and spectrum

The operator L given by (1.1) is a perturbation of the multiplication operator by an integral operator with degenerate kernel. Therefore the resolvent $(L-\lambda)^{-1}$ can be calculated explicitly.

Theorem 2.1. *For $\lambda \in \mathbb{C}\setminus[0,1]$ the inverse $(L-\lambda)^{-1}$ is given by*

$$\left((L-\lambda)^{-1}f\right)(x) = \frac{f(x)}{x-\lambda} - \frac{\varphi(x)}{x-\lambda}\,\mathrm{e}^{-G(x,\lambda)} \int_0^x \frac{\psi(t)f(t)}{t-\lambda}\,\mathrm{e}^{G(t,\lambda)}\,\mathrm{d}t \tag{2.1}$$

for $x \in [0,1]$, $f \in L_2(0,1)$, where

$$G(x,\lambda) := \int_0^x \frac{\varphi(y)\psi(y)}{y-\lambda}\,\mathrm{d}y, \qquad x \in [0,1].$$

Proof. Let $f \in L_2(0,1)$. If we denote the right hand side of (2.1) by $u(x)$, then

$$\begin{aligned}((L-\lambda)u)(x) = f(x) - \varphi(x)\,\mathrm{e}^{-G(x,\lambda)} \int_0^x \frac{\psi(t)f(t)}{t-\lambda}\,\mathrm{e}^{G(t,\lambda)}\,\mathrm{d}t + \varphi(x)\int_0^x \frac{\psi(s)f(s)}{s-\lambda}\,\mathrm{d}s \\ - \varphi(x)\int_0^x \frac{\psi(s)\varphi(s)}{s-\lambda}\,\mathrm{e}^{-G(s,\lambda)} \int_0^s \frac{\psi(t)f(t)}{t-\lambda}\,\mathrm{e}^{G(t,\lambda)}\,\mathrm{d}t\,\mathrm{d}s.\end{aligned}$$

Using $-\dfrac{\varphi(s)\psi(s)}{s-\lambda}\,\mathrm{e}^{-G(s,\lambda)} = \dfrac{\mathrm{d}}{\mathrm{d}s}\mathrm{e}^{-G(s,\lambda)}$ and integrating by parts, we find

$$\begin{aligned}((L-\lambda)u)(x) = f(x) - \varphi(x)\,\mathrm{e}^{-G(x,\lambda)} \int_0^x \frac{\psi(t)f(t)}{t-\lambda}\,\mathrm{e}^{G(t,\lambda)}\,\mathrm{d}t \\ + \varphi(x)\left[\mathrm{e}^{-G(s,\lambda)} \int_0^s \frac{\psi(t)f(t)}{t-\lambda}\,\mathrm{e}^{G(t,\lambda)}\,\mathrm{d}t\right]_0^x = f(x)\end{aligned}$$

for $x \in [0,1]$. Thus $L-\lambda$ is right invertible and its right inverse is given by the right hand side of (2.1). The proof that $L-\lambda$ is also left invertible is similar. □

Corollary 2.2. *For the spectrum of L we have $\sigma(L) = \sigma_{\text{ess}}(L) = [0,1]$.*

Proof. Theorem 2.1 implies $\sigma(L) \subset [0,1]$. By Weyl's Theorem $\sigma_{\text{ess}}(L) = \sigma_{\text{ess}}(M)$ $= [0,1]$ where M is the multiplication operator in $L_2(0,1)$, $(My)(x) = xy(x)$ for $x \in [0,1]$. This proves the assertion. □

The point spectrum of L can be characterized in terms of the real part of the function $\varphi\psi$. We start with the particular case $\varphi\psi \equiv 0$.

Proposition 2.3. *If $\varphi\psi \equiv 0$, then the operator L has no eigenvalues: $\sigma_p(L) = \emptyset$.*

Proof. Assume that $\varphi\psi \equiv 0$, and let $\lambda \in \sigma_p(L)$. Then there exists a function $y_\lambda \in L_2(0,1), y_\lambda \not\equiv 0$, such that $(L-\lambda)y_\lambda = 0$, that is,

$$(x-\lambda)y_\lambda(x) = -\int_0^x \varphi(x)\psi(s)y_\lambda(s)\,\mathrm{d}s, \qquad x \in [0,1]. \tag{2.2}$$

If we multiply this equation by $\psi(x)$ and define $v_\lambda := \psi y_\lambda$, then $v_\lambda \in L_1(0,1)$ satisfies

$$(x-\lambda)v_\lambda(x) = -\int_0^x \varphi(x)\psi(x)v_\lambda(s)\,\mathrm{d}s = 0, \qquad x \in [0,1],$$

as $\varphi\psi \equiv 0$. This implies $\psi y_\lambda = 0$ almost everywhere and hence $y_\lambda \equiv 0$ because of (2.2), a contradiction. □

Theorem 2.4. *Let $\Re(\varphi\psi) \in \operatorname{Lip}\alpha$ with $\alpha > 0$, and let $\lambda \in [0,1]$.*
i) Then $\lambda \in \sigma_p(L)$ if and only if

$$\lambda \neq 1, \qquad \Re(\varphi\psi)(\lambda) < 0, \qquad \int_\lambda^1 \frac{|\varphi(x)|^2}{(x-\lambda)^{2(1+\Re(\varphi\psi)(\lambda))}}\,\mathrm{d}x < \infty. \tag{2.3}$$

ii) Suppose in addition φ, $\psi \in C[0,1]$. Then $\lambda \in \sigma_p(L)$ if and only if

$$\lambda \neq 1, \qquad \Re(\varphi\psi)(\lambda) < -\frac{1}{2}. \tag{2.4}$$

Proof. Let $\lambda \in [0,1]$, $\lambda \in \sigma_p(L)$. Then there exists a function $y_\lambda \in L_2(0,1), y_\lambda \not\equiv 0$, such that (2.2) holds. As above, we define $v_\lambda := \psi y_\lambda \in L_1(0,1)$ and obtain

$$(x-\lambda)v_\lambda(x) = -\varphi(x)\psi(x)\int_0^x v_\lambda(s)\,\mathrm{d}s, \qquad x \in [0,1].$$

If we set $w_\lambda(x) := \int_0^x v_\lambda(s)\,\mathrm{d}s$ for $x \in [0,1]$, then $w_\lambda(0) = 0$, $w_\lambda' = v_\lambda$ which implies $w_\lambda \in W_1^1(0,1)$, and w_λ fulfills the differential equation

$$(x-\lambda)w_\lambda'(x) = -\varphi(x)\psi(x)w_\lambda(x), \qquad x \in [0,1]. \tag{2.5}$$

Hence there exists a constant C such that

$$w_\lambda(x) = \begin{cases} 0, & 0 \le x < \lambda, \\ C\mathrm{e}^{\int_x^1 \frac{\varphi(t)\psi(t)}{t-\lambda}\,\mathrm{d}t}, & \lambda < x \le 1, \end{cases}$$

as (2.5) is non-singular on $[0,\lambda)$ and $w_\lambda(0) = 0$. Since $w_\lambda \not\equiv 0$ on $[0,1]$ (otherwise $v_\lambda \equiv 0$ and thus $y_\lambda \equiv 0$ by (2.2)), it follows that $\lambda \neq 1$ and $C \neq 0$. The continuity of ω_λ implies that

$$\lim_{x \searrow \lambda} \int_x^1 \frac{\Re(\varphi\psi)(t)}{t-\lambda}\,\mathrm{d}t = -\infty.$$

A necessary (and sufficient) condition for this is that $\Re(\varphi\psi)(\lambda) < 0$ due to the smoothness of this function. Indeed, for $\Re(\varphi\psi)(\lambda) > 0$ the limit would be $+\infty$, and $\Re(\varphi\psi)(\lambda) = 0$ would imply

$$\left|\int_x^1 \frac{\Re(\varphi\psi)(t)}{t-\lambda}\,\mathrm{d}t\right| \le \int_x^1 \frac{|\Re(\varphi\psi)(t) - \Re(\varphi\psi)(\lambda)|}{|t-\lambda|}\,\mathrm{d}t \le M\int_x^1 \frac{1}{(t-\lambda)^{1-\alpha}}\,\mathrm{d}t$$

with a constant $M > 0$ as $\Re(\varphi\psi) \in \operatorname{Lip}\alpha$. Hence in this case the limit for $x \searrow \lambda$ would exist.

From the equation (2.2) for the eigenfunction y_λ and the definition of w_λ it follows that $y_\lambda(x) = -\frac{\varphi(x)}{x-\lambda}w_\lambda(x)$, and hence

$$y_\lambda(x) = \begin{cases} 0, & 0 \le x < \lambda, \\ -C\dfrac{\varphi(x)}{x-\lambda}\mathrm{e}^{\int_x^1 \frac{\varphi(t)\psi(t)}{t-\lambda}\,\mathrm{d}t}, & \lambda < x \le 1. \end{cases} \tag{2.6}$$

We are going to show that for the function y_λ given by (2.6)

$$y_\lambda \in L_2(0,1) \iff \int_\lambda^1 \frac{|\varphi(x)|^2}{(x-\lambda)^{2(1+\Re(\varphi\psi)(\lambda))}}\,\mathrm{d}x < \infty. \tag{2.7}$$

For $\lambda < x \le 1$ we have

$$\begin{aligned} \left|\mathrm{e}^{\int_x^1 \frac{\varphi(t)\psi(t)}{t-\lambda}\,\mathrm{d}t}\right| &= \left|\mathrm{e}^{\int_x^1 \frac{\varphi(t)\psi(t)-\varphi(\lambda)\psi(\lambda)}{t-\lambda}\,\mathrm{d}t}\right|\left|\mathrm{e}^{\varphi(\lambda)\psi(\lambda)\int_x^1 \frac{1}{t-\lambda}\,\mathrm{d}t}\right| \\ &= \mathrm{e}^{\int_x^1 \frac{\Re(\varphi\psi)(t)-\Re(\varphi\psi)(\lambda)}{t-\lambda}\,\mathrm{d}t}(1-\lambda)^{\Re(\varphi\psi)(\lambda)}(x-\lambda)^{-\Re(\varphi\psi)(\lambda)}. \end{aligned}$$

It follows that

$$\|y_\lambda\|^2 = C^2K_\lambda^2 \int_\lambda^1 \frac{|\varphi(x)|^2}{(x-\lambda)^{2(1+\Re(\varphi\psi)(\lambda))}}\,\mathrm{e}^{2\int_x^1 \frac{\Re(\varphi\psi)(t)-\Re(\varphi\psi)(\lambda)}{t-\lambda}\,\mathrm{d}t}\,\mathrm{d}x$$

with $K_\lambda := (1-\lambda)^{\Re(\varphi\psi)(\lambda)} > 0$. As $\Re(\varphi\psi) \in \operatorname{Lip}\alpha$ by assumption,

$$\begin{aligned}\left|\int_x^1 \frac{\Re(\varphi\psi)(t) - \Re(\varphi\psi)(\lambda)}{t-\lambda}\,\mathrm{d}t\right| &\leq M\int_x^1 \frac{1}{(t-\lambda)^{1-\alpha}}\,\mathrm{d}t \leq M\int_\lambda^1 \frac{1}{(t-\lambda)^{1-\alpha}}\,\mathrm{d}t \\ &\leq M\int_0^1 \frac{1}{\tau^{1-\alpha}}\,\mathrm{d}\tau = \frac{M}{\alpha} =: M_\alpha\end{aligned}$$

for $\lambda < x \leq 1$. Therefore,

$$(2.8)\quad \widetilde{C}\mathrm{e}^{-2M_\alpha}\int_\lambda^1 \frac{|\varphi(x)|^2}{(x-\lambda)^{2(1+\Re(\varphi\psi)(\lambda))}}\,\mathrm{d}x \leq \|y_\lambda\|^2 \leq \widetilde{C}\mathrm{e}^{2M_\alpha}\int_\lambda^1 \frac{|\varphi(x)|^2}{(x-\lambda)^{2(1+\Re(\varphi\psi)(\lambda))}}\,\mathrm{d}x$$

with $\widetilde{C} = C^2K_\lambda^2 \neq 0$ which proves (2.7). It is shown now that $\lambda \in \sigma_p(L)$ implies the conditions (2.3).

Conversely, let $\lambda \in [0,1]$ fulfill the conditions (2.3). If we define the function y_λ by (2.6) with $C = -1$, say, then $y_\lambda \in L_2(0,1)$ according to (2.7), $y_\lambda \not\equiv 0$ as $\lambda \neq 1$, $((L-\lambda)y_\lambda)(x) = 0$ for $0 \leq x < \lambda$ and

$$\begin{aligned}((L-\lambda)y_\lambda)(x) &= \varphi(x)\,\mathrm{e}^{\int_x^1 \frac{\varphi(t)\psi(t)}{t-\lambda}\,\mathrm{d}t} + \varphi(x)\int_\lambda^x \frac{\varphi(s)\psi(s)}{s-\lambda}\mathrm{e}^{\int_s^1 \frac{\varphi(t)\psi(t)}{t-\lambda}\,\mathrm{d}t}\,\mathrm{d}s \\ &= \varphi(x)\,\mathrm{e}^{\int_x^1 \frac{\varphi(t)\psi(t)}{t-\lambda}\,\mathrm{d}t} + \varphi(x)\left[-\mathrm{e}^{\int_s^1 \frac{\varphi(t)\psi(t)}{t-\lambda}\,\mathrm{d}t}\right]_\lambda^x \\ &= \varphi(x)\lim_{s\searrow\lambda}\mathrm{e}^{\int_s^1 \frac{\varphi(t)\psi(t)}{t-\lambda}\,\mathrm{d}t}\end{aligned}$$

for $\lambda < x \leq 1$. Using the Lipschitz condition for $\Re(\varphi\psi)$ again, we see

$$\left|\mathrm{e}^{\int_s^1 \frac{\varphi(t)\psi(t)}{t-\lambda}\,\mathrm{d}t}\right| \leq \mathrm{e}^{M_\alpha}K_\lambda(s-\lambda)^{-\Re(\varphi\psi)(\lambda)} \longrightarrow 0$$

for $s \searrow \lambda$ as $-\Re(\varphi\psi)(\lambda) > 0$ by assumption. Hence y_λ is an eigenfunction corresponding to λ, and the first part of the theorem is proved.

Now assume φ, $\psi \in C[0,1]$. Then $\Re(\varphi\psi)(\lambda) < 0$ implies that $\varphi(\lambda) \neq 0$, and it follows from (2.8) that

$$\Re(\varphi\psi)(\lambda) < 0, \quad \int_\lambda^1 \frac{|\varphi(x)|^2}{(x-\lambda)^{2(1+\Re(\varphi\psi)(\lambda))}}\,\mathrm{d}x < \infty \quad \Longleftrightarrow \quad \Re(\varphi\psi)(\lambda) < -\frac{1}{2}$$

for $\lambda \neq 1$. This completes the proof. □

Remark 2.5. The statements of Theorem 2.4 continue to hold if we only suppose $\Re(\varphi\psi) \in \operatorname{Lip}(\omega)$ (see Definition 3.3) with $\omega(t) = \frac{1}{\ln^{1+\varepsilon} 1/t}$, $t \in [0,\infty)$, for some $\varepsilon > 0$.

Theorem 2.6. *Let Λ be an arbitrary nonempty open subset of $[0,1)$. Then there exists a non-self-adjoint operator L in $L_2(0,1)$ of the form (1.1) with smooth φ and ψ such that*

$$\sigma_p(L) = \Lambda.$$

Proof. The operator L to be constructed will be of the form (1.1) with $\varphi \equiv 1$ and ψ constructed as follows. If $\Lambda \subset [0,1)$, $\Lambda \neq \emptyset$, is open, then Λ is the union of at most countably many open disjoint intervals,

$$\Lambda = \bigcup_{n=1}^{\infty} (\alpha_n, \beta_n),$$

where $\alpha_n, \beta_n \in [0,1]$, $\alpha_n < \beta_n$ and $(\alpha_n, \beta_n) \cap (\alpha_m, \beta_m) = \emptyset$ if $n \neq m$. We set

$$\Delta_n := \beta_n - \alpha_n, \qquad n = 1, 2, \ldots,$$

assuming without loss of generality $\Delta_1 \geq \Delta_2 \geq \cdots$. Further, we choose a function $\psi_0 \in C^\infty(\mathbb{R})$, $\operatorname{supp} \psi_0 = [0,1]$, $\psi_0(x) > 0$ for $x \in (0,1)$, and define

$$\psi(x) := -\sum_{n=1}^{\infty} \mathrm{e}^{-\frac{1}{\Delta_n}} \psi_0\left(\frac{x - \alpha_n}{\Delta_n}\right) - \frac{1}{2}, \qquad x \in [0,1]. \tag{2.9}$$

We are going to show that $\psi \in C^\infty[0,1]$. By definition of Δ_n we have $\sum_{n=1}^{\infty} \Delta_n \leq 1$ and hence $n\Delta_n \to 0$. Therefore

$$\sum_{n=1}^{\infty} \mathrm{e}^{-\frac{1}{\Delta_n}} \frac{1}{\Delta_n^s}$$

converges for any $s \in \mathbb{N}_0$. This and the continuity of $\psi_0^{(s)}$ on $[0,1]$ for any $s \in \mathbb{N}_0$ imply that the series

$$\sum_{n=1}^{N} \left(\mathrm{e}^{-\frac{1}{\Delta_n}} \psi_0\left(\frac{x-\alpha_n}{\Delta_n}\right)\right)^{(s)} = \sum_{n=1}^{N} \mathrm{e}^{-\frac{1}{\Delta_n}} \frac{1}{\Delta_n^s} \psi_0^{(s)}\left(\frac{x-\alpha_n}{\Delta_n}\right)$$

converges uniformly for $x \in [0,1]$ as $N \to \infty$. Now the statement of Theorem 2.6 immediately follows from Theorem 2.4 ii). □

3. Lyapunov stability

An operator T in a Hilbert space is said to be Lyapunov stable if the group $\mathrm{e}^{\mathrm{i}Tt}$, $t \in \mathbb{R}$, generated by T is bounded, that is,

$$\sup_{t \in \mathbb{R}} \|\mathrm{e}^{\mathrm{i}Tt}\| < \infty.$$

Equivalent is that T is similar to a self-adjoint operator ([DK], Theorem 6.3). The following result is a modification of the integral criterion for the similarity to a self-adjoint operator established in [N1] (see Theorem 2 and Remark 1 therein), and also in [C].

Theorem 3.1. *Let H be a Hilbert space and let $\mathcal{D}, \mathcal{D}'$ be dense subsets of H. A bounded operator T in H is Lyapunov stable if and only if*

i) $\sigma(T) \subset \mathbb{R}$,

ii) $\limsup_{\varepsilon \searrow 0} \varepsilon \int_{\mathbb{R}} \|(T-(k+\mathrm{i}\varepsilon))^{-1} f\|^2 \,\mathrm{d}k \le C\,\|f\|^2, \qquad f \in \mathcal{D}$,

iii) $\liminf_{\varepsilon \searrow 0} \varepsilon \int_{\mathbb{R}} \|(T^* - (k+\mathrm{i}\varepsilon))^{-1} g\|^2 \,\mathrm{d}k \le C'\|g\|^2, \qquad g \in \mathcal{D}'$.

Proof. The necessity of i), ii) and iii) for the Lyapunov stability of T is immediate from [N1], Theorem 2. Conversely, assume that i), ii) and iii) are fulfilled. Let $t \in \mathbb{R}$, $t \neq 0$, be fixed, and let $\varepsilon > 0$ be arbitrary. Since the spectrum of T is real by i), we have

$$
\begin{aligned}
\mathrm{e}^{\mathrm{i}Tt} = \lim_{N\to\infty} \Bigg(& -\frac{1}{2\pi \mathrm{i}} \int_{-N}^{N} \mathrm{e}^{\mathrm{i}(k+\mathrm{i}\varepsilon)t} (T-(k+\mathrm{i}\varepsilon))^{-1} \,\mathrm{d}k \\
& + \frac{1}{2\pi \mathrm{i}} \int_{-N}^{N} \mathrm{e}^{\mathrm{i}(k-\mathrm{i}\varepsilon)t} (T-(k-\mathrm{i}\varepsilon))^{-1} \,\mathrm{d}k \Bigg)
\end{aligned} \tag{3.1}
$$

as $\|(T-\lambda)^{-1}\| \le \frac{M}{|\lambda|}$ with some $M > 0$ for $|\lambda|$ sufficiently large. Note that the first integral tends to 0 if $t > 0$, while the second tends to 0 if $t < 0$. Now let $t > 0$ and $f \in \mathcal{D}$, $g \in \mathcal{D}'$. If we add and subtract

$$
\int_{-N}^{N} \mathrm{e}^{\mathrm{i}(k+\mathrm{i}\varepsilon)t} (T-(k-\mathrm{i}\varepsilon))^{-1} \,\mathrm{d}k
$$

on the right hand side of (3.1) and use the first resolvent equation, we obtain

$$
\begin{aligned}
2\pi \mathrm{i}\left(\mathrm{e}^{\mathrm{i}Tt} f, g\right) = & -2\mathrm{i}\varepsilon \liminf_{N\to\infty} \int_{-N}^{N} \mathrm{e}^{\mathrm{i}(k+\mathrm{i}\varepsilon)t} \left((T-(k+\mathrm{i}\varepsilon))^{-1} f, (T^*-(k+\mathrm{i}\varepsilon))^{-1} g\right) \mathrm{d}k \\
& + \liminf_{N\to\infty} \int_{-N}^{N} \mathrm{e}^{\mathrm{i}kt} (\mathrm{e}^{\varepsilon t} - \mathrm{e}^{-\varepsilon t}) \left((T-(k-\mathrm{i}\varepsilon))^{-1} f, g\right) \mathrm{d}k.
\end{aligned}
$$

Since $\varepsilon > 0$ was arbitrary, equality still holds if we take the limit $\varepsilon \searrow 0$ on the right hand side. The Cauchy–Schwarz inequality, the Hölder inequality and the assumptions ii) and iii) yield

$$\begin{aligned}
&\liminf_{\varepsilon\searrow 0}\left| 2\mathrm{i}\varepsilon \lim_{N\to\infty}\int_{-N}^{N} \mathrm{e}^{\mathrm{i}(k+\mathrm{i}\varepsilon)t}\left((T-(k+\mathrm{i}\varepsilon))^{-1}f,(T^*-(k+\mathrm{i}\varepsilon))^{-1}g\right)\mathrm{d}k\right| \\
&\le \liminf_{\varepsilon\searrow 0}\left(2\mathrm{e}^{-\varepsilon t}\left(\varepsilon\int_{\mathbb{R}}\|(T-(k+\mathrm{i}\varepsilon))^{-1}f\|^2\mathrm{d}k\right)^{\frac{1}{2}}\left(\varepsilon\int_{\mathbb{R}}\|(T^*-(k+\mathrm{i}\varepsilon))^{-1}g\|^2\mathrm{d}k\right)^{\frac{1}{2}}\right) \\
&\le 2\sqrt{CC'}\,\|f\|\,\|g\|.
\end{aligned}$$

Now let $N_0 > \|T\|$ be fixed. Then, again using the Cauchy–Schwarz inequality, the Hölder inequality and the assumption ii), we get

$$\begin{aligned}
&\liminf_{\varepsilon\searrow 0}\left|\int_{-N_0}^{N_0}\mathrm{e}^{\mathrm{i}kt}(\mathrm{e}^{\varepsilon t}-\mathrm{e}^{-\varepsilon t})\left((T-(k-\mathrm{i}\varepsilon))^{-1}f,g\right)\mathrm{d}k\right| \\
&\le \liminf_{\varepsilon\searrow 0}\left(\sqrt{\varepsilon}\left|\frac{\mathrm{e}^{\varepsilon t}-\mathrm{e}^{-\varepsilon t}}{\varepsilon}\right|\left(\varepsilon\int_{-N_0}^{N_0}\|(T-(k-\mathrm{i}\varepsilon))^{-1}f\|^2\,\mathrm{d}k\right)^{\frac{1}{2}}\sqrt{2N_0}\,\|g\|\right) \\
&\le \lim_{\varepsilon\searrow 0}\left(\sqrt{\varepsilon}\left|\frac{\mathrm{e}^{\varepsilon t}-\mathrm{e}^{-\varepsilon t}}{\varepsilon}\right|\right)\sqrt{C}\,\sqrt{2N_0}\,\|f\|\,\|g\| \\
&= 0.
\end{aligned}$$

By means of the second resolvent equation, for $N \ge N_0$,

$$\begin{aligned}
&\int_{N_0}^{N}\mathrm{e}^{\mathrm{i}kt}(\mathrm{e}^{\varepsilon t}-\mathrm{e}^{-\varepsilon t})\left((T-(k-\mathrm{i}\varepsilon))^{-1}f,g\right)\mathrm{d}k \\
&= \int_{N_0}^{N}\mathrm{e}^{\mathrm{i}kt}(\mathrm{e}^{\varepsilon t}-\mathrm{e}^{-\varepsilon t})\frac{1}{k-\mathrm{i}\varepsilon}\left(T(T-(k-\mathrm{i}\varepsilon))^{-1}f,g\right)\mathrm{d}k \\
&\quad - \int_{N_0}^{N}\mathrm{e}^{\mathrm{i}kt}(\mathrm{e}^{\varepsilon t}-\mathrm{e}^{-\varepsilon t})\frac{1}{k-\mathrm{i}\varepsilon}(f,g)\,\mathrm{d}k.
\end{aligned}$$

Now we observe

$$|\mathrm{e}^{\varepsilon t}-\mathrm{e}^{-\varepsilon t}| = \left|\int_{-\varepsilon}^{\varepsilon} t\mathrm{e}^{st}\,\mathrm{d}s\right| \le t\,2\varepsilon\,\mathrm{e}^{|\varepsilon t|}$$

and further

$$\int_{N_0}^{N}\frac{1}{k-\mathrm{i}\varepsilon}\|T(T-(k-\mathrm{i}\varepsilon))^{-1}f\|\,\mathrm{d}k \le M\,\|T\|\,\|f\|\int_{N_0}^{\infty}\frac{1}{k^2}\,\mathrm{d}k \le M'\,\|T\|\,\|f\|$$

for the first term and

$$\lim_{N\to\infty}\left|\int_{N_0}^{N}\frac{\mathrm{e}^{\mathrm{i}kt}}{k-\mathrm{i}\varepsilon}\,\mathrm{d}k\right| = \lim_{N\to\infty}\left|\left[\frac{1}{\mathrm{i}t}\frac{\mathrm{e}^{\mathrm{i}kt}}{k-\mathrm{i}\varepsilon}\right]_{N_0}^{N} + \int_{N_0}^{N}\frac{1}{\mathrm{i}t}\frac{\mathrm{e}^{\mathrm{i}kt}}{(k-\mathrm{i}\varepsilon)^2}\,\mathrm{d}k\right| \le \frac{1}{t}M''\frac{1}{N_0}$$

for the second term. Then

$$\begin{aligned}\liminf_{\varepsilon\searrow 0}&\left|\lim_{N\to\infty}\int_{N_0}^{N} \mathrm{e}^{\mathrm{i}kt}(\mathrm{e}^{\varepsilon t}-\mathrm{e}^{-\varepsilon t})\left((T-(k-\mathrm{i}\varepsilon))^{-1}f,g\right)\,\mathrm{d}k\right|\\ &\le\left(\lim_{\varepsilon\searrow 0}\left(\varepsilon\mathrm{e}^{\varepsilon t}\right)2t\,M'\,\|T\|+\lim_{\varepsilon\searrow 0}\left(\varepsilon\mathrm{e}^{\varepsilon t}\right)2M''\frac{1}{N_0}\right)\|f\|\,\|g\|\\ &=0.\end{aligned}$$

In an analogous way it can be shown that the limit for the integral from $-N$ to $-N_0$ vanishes. Altogether we have proved that

$$2\pi\left|\left(\mathrm{e}^{\mathrm{i}Tt}f,g\right)\right|\le 2\sqrt{CC'}\,\|f\|\,\|g\|,\qquad f\in\mathcal{D},\ g\in\mathcal{D}',\tag{3.2}$$

for $t>0$; the proof of (3.2) for $t<0$ is similar. As $\mathcal{D}$ and $\mathcal{D}'$ are dense in H, the above estimate also holds for $f,g\in H$, and hence $\|\mathrm{e}^{\mathrm{i}Tt}\|\le\sqrt{CC'}/\pi$ for $t\in\mathbb{R}$, $t\ne 0$. Taking the supremum over all $t\in\mathbb{R}$, we see that $\mathrm{e}^{\mathrm{i}Tt}$, $t\in\mathbb{R}$, is a bounded group and hence T is Lyapunov stable. □

In general, it is difficult to estimate the resolvent of the operator L given by (1.1). Therefore we assume from now on that $\varphi\psi\equiv 0$. In this case the foregoing theorem yields the following necessary and sufficient criterion for the Lyapunov stability of the operator L.

Proposition 3.2. *Let $\varphi\psi\equiv 0$ and let $\mathcal{D},\mathcal{D}'\subset L_2(0,1)$ be dense. Then L is Lyapunov stable if and only if*

$$\limsup_{\varepsilon\searrow 0}\ \varepsilon\int_{\mathbb{R}}\left\|\frac{\varphi(x)}{x-(k+\mathrm{i}\varepsilon)}\int_0^x\frac{\psi(t)f(t)}{t-(k+\mathrm{i}\varepsilon)}\,\mathrm{d}t\right\|^2\mathrm{d}k\le C\,\|f\|^2,\qquad f\in\mathcal{D},\tag{3.3}$$

$$\liminf_{\varepsilon\searrow 0}\ \varepsilon\int_{\mathbb{R}}\left\|\frac{\overline{\psi(x)}}{x-(k+\mathrm{i}\varepsilon)}\int_x^1\frac{\overline{\varphi(t)}g(t)}{t-(k+\mathrm{i}\varepsilon)}\,\mathrm{d}t\right\|^2\mathrm{d}k\le C'\|g\|^2,\qquad g\in\mathcal{D}'.\tag{3.4}$$

Proof. By Corollary 2.2 we have $\sigma(L)\subset\mathbb{R}$. By Theorem 2.1

$$(L-\lambda)^{-1}=(M-\lambda)^{-1}-R_\lambda,\qquad \lambda\in\mathbb{C}\setminus[0,1],$$

where M is the operator of multiplication by the independent variable and, in the case $\varphi\psi\equiv 0$,

$$(R_\lambda f)(x)=\frac{\varphi(x)}{x-\lambda}\int_0^x\frac{\psi(t)}{t-\lambda}f(t)\,\mathrm{d}t,\qquad x\in[0,1],\ f\in L_2(0,1).$$

Since M is self-adjoint, $(L-\lambda)^{-1}$ fulfills the conditions ii) and iii) of Theorem 3.1 if and only if this is true for the operator R_λ. This proves the proposition. □

In order to obtain a result on the Lyapunov stability of L, we have to require certain smoothness conditions for φ and ψ. The following definition of the Lipschitz spaces Lip(ω), where ω is a modulus of continuity, can be found e.g. in [S] (see Chapter VI, §2, therein).

Definition 3.3. Let ω be a modulus of continuity, that is,

i) $\omega : [0,\infty) \to \mathbb{R}$ is nondecreasing and continuous,

ii) $\lim_{\tau \searrow 0} \omega(\tau) = 0$,

iii) $\omega(\tau_1 + \tau_2) \le \omega(\tau_1) + \omega(\tau_2), \qquad \tau_1, \tau_2 \in [0,\infty)$.

We define

$$\operatorname{Lip}(\omega) := \big\{ f : [0,1] \to \mathbb{C},\ |f(x)| \le C,\ |f(x) - f(y)| \le C\omega\big(|x-y|\big),\ x,y \in [0,1] \big\}.$$

Theorem 3.4. *Let* $\varphi\psi \equiv 0$ *and assume* $\varphi \in \operatorname{Lip}(\omega_1)$, $\psi \in \operatorname{Lip}(\omega_2)$ *on* $[0,1]$. *If for some* δ, $0 < \delta < 1$,

$$\int_0^\delta \frac{\omega_1(\tau)\omega_2(\tau)}{\tau} < \infty, \tag{3.5}$$

then L *is Lyapunov stable.*

Proof. In order to prove the Lyapunov stability of L, we use Proposition 3.2. We are going to show that the estimate (3.3) is valid on some dense subset $\mathcal{D}$ of $L_2(0,1)$. The proof of (3.4) with lim sup instead of lim inf is analogous because the assumptions on φ and ψ are symmetric, and so (3.4) follows. We define

$$J := \{ x \in (0,1) : \psi(x) \ne 0 \}.$$

As ψ is continuous, the set J can be written as a countable union $J = \bigcup_{n=1}^\infty J_n$ of disjoint open intervals J_n. The set

$$\mathcal{D} := \Big\{ f \in C_0^\infty(0,1) : \operatorname{supp} f \cap J \text{ compact},\ \operatorname{supp} f \cap J \subset \bigcup_{n=1}^N J_{m_n},\ N \in \mathbb{N},\ m_1,\dots,m_N \in \mathbb{N} \Big\} \tag{3.6}$$

is a dense subset of $L_2(0,1)$. Now let $f \in \mathcal{D}$. Then $\operatorname{supp} f \cap J \subset \bigcup_{n=1}^N J_{m_n}$ for some $N \in \mathbb{N}$. We define $J_{m_n} =: (a_n, b_n)$ for $n = 1,2,\dots,N$. Without loss of generality, we may assume that $0 \le a_1 < b_1 \le a_2 < b_2 \le \dots \le a_N < b_N \le 1$.
First we prove that the limit for $\varepsilon \searrow 0$ on the left hand side of (3.3) exists. As $\varphi\psi \equiv 0$, we have $\varphi(x) = 0$ for $x \in (a_n, b_n)$, $n = 1,2,\dots,N$. Since $f \in \mathcal{D}$, we have $\psi(t)f(t) = 0$ for $t \in [b_n, a_{n+1}]$, $n = 1,2,\dots,N$, where $a_{N+1} := 1$. Hence

$$\begin{aligned} I_\varepsilon(f) &:= \varepsilon \int_{\mathbb{R}} \left\| \frac{\varphi(x)}{x - (k + \mathrm{i}\varepsilon)} \int_0^x \frac{\psi(t)f(t)}{t - (k+\mathrm{i}\varepsilon)}\, \mathrm{d}t \right\|^2 \mathrm{d}k \\ &= \varepsilon \int_{\mathbb{R}} \int_0^1 \frac{|\varphi(x)|^2}{(x-k)^2 + \varepsilon^2} \left| \int_0^x \frac{\psi(t)f(t)}{t - (k+\mathrm{i}\varepsilon)}\, \mathrm{d}t \right|^2 \mathrm{d}x\, \mathrm{d}k \\ &= \varepsilon \int_{\mathbb{R}} \sum_{\mu=1}^N \int_{b_\mu}^{a_{\mu+1}} \frac{|\varphi(x)|^2}{(x-k)^2 + \varepsilon^2} \left| \sum_{n=1}^\mu \int_{a_n}^{b_n} \frac{\psi(t)f(t)}{t-(k+\mathrm{i}\varepsilon)}\, \mathrm{d}t \right|^2 \mathrm{d}x\, \mathrm{d}k. \end{aligned}$$

We choose $\gamma > 0$ such that $\operatorname{supp} f \cap J_{m_n} \subset [a_n + \gamma, b_n - \gamma]$ for all $n = 1, 2, \ldots, N$. Further we define $c_n := a_n + \frac{\gamma}{2}$, $d_n := b_n - \frac{\gamma}{2}$ for $n = 1, 2, \ldots, N$, and finally $d_0 := 0$, $c_{N+1} := 1$. Then

$$\begin{aligned} I_\varepsilon(f) = {} & \varepsilon \int_{-\infty}^{0} \sum_{\mu=1}^{N} \int_{b_\mu}^{a_{\mu+1}} \frac{|\varphi(x)|^2}{(x-k)^2+\varepsilon^2}\, dx \left| \sum_{n=1}^{\mu} \int_{a_n}^{b_n} \frac{\psi(t) f(t)}{t-(k+i\varepsilon)}\, dt \right|^2 dk \\ & + \varepsilon \sum_{r=1}^{N+1} \int_{d_{r-1}}^{c_r} \sum_{\mu=1}^{N} \int_{b_\mu}^{a_{\mu+1}} \frac{|\varphi(x)|^2}{(x-k)^2+\varepsilon^2}\, dx \left| \sum_{n=1}^{\mu} \int_{a_n}^{b_n} \frac{\psi(t) f(t)}{t-(k+i\varepsilon)}\, dt \right|^2 dk \\ & + \varepsilon \sum_{r=1}^{N} \int_{c_r}^{d_r} \sum_{\mu=1}^{N} \int_{b_\mu}^{a_{\mu+1}} \frac{|\varphi(x)|^2}{(x-k)^2+\varepsilon^2}\, dx \left| \sum_{n=1}^{\mu} \int_{a_n}^{b_n} \frac{\psi(t) f(t)}{t-(k+i\varepsilon)}\, dt \right|^2 dk \\ & + \varepsilon \int_{1}^{\infty} \sum_{\mu=1}^{N} \int_{b_\mu}^{a_{\mu+1}} \frac{|\varphi(x)|^2}{(x-k)^2+\varepsilon^2}\, dx \left| \sum_{n=1}^{\mu} \int_{a_n}^{b_n} \frac{\psi(t) f(t)}{t-(k+i\varepsilon)}\, dt \right|^2 dk. \end{aligned}$$

We apply the estimate

$$\int_{b_\mu}^{a_{\mu+1}} \frac{|\varphi(x)|^2}{(x-k)^2+\varepsilon^2}\, dx \le \widetilde{C} \int_{b_\mu}^{a_{\mu+1}} |\varphi(x)|^2\, dx \le \widetilde{C} \|\varphi\|^2 \tag{3.7}$$

whenever $|x-k| > \tilde{\gamma}$ for $x \in [b_\mu, a_{\mu+1}]$ with some $\tilde{\gamma} > 0$ not depending on x, k and ε. Further we use that, by the Hoelder inequality,

$$\begin{aligned} \left| \int_{a_n}^{b_n} \frac{\psi(t)}{t-(k+i\varepsilon)} f(t)\, dt \right|^2 & \le \left(\int_{a_n}^{b_n} \frac{1}{|t-(k+i\varepsilon)|} |\psi(t) f(t)|\, dt \right)^2 \\ & \le \widehat{C}_f \|\psi\|^2 \|f\|^2 \end{aligned} \tag{3.8}$$

if $|t-(k+i\varepsilon)| > \widehat{\gamma}_f$ for $t \in [a_n, b_n] \cap \operatorname{supp} f$ with some $\widehat{\gamma}_f > 0$ independent of t, k and ε. Combining these two estimates, we find

$$\begin{aligned} I_\varepsilon(f) = {} & O(\varepsilon) + \varepsilon \sum_{r=2}^{N+1} \int_{d_{r-1}}^{c_r} \int_{b_{r-1}}^{a_r} \frac{|\varphi(x)|^2}{(x-k)^2+\varepsilon^2}\, dx \left| \sum_{n=1}^{r-1} \int_{a_n}^{b_n} \frac{\psi(t) f(t)}{t-(k+i\varepsilon)}\, dt \right|^2 dk \\ & + \varepsilon \sum_{r=1}^{N} \int_{c_r}^{d_r} \sum_{\mu=r}^{N} \int_{b_\mu}^{a_{\mu+1}} \frac{|\varphi(x)|^2}{(x-k)^2+\varepsilon^2}\, dx \left| \int_{a_r}^{b_r} \frac{\psi(t) f(t)}{t-(k+i\varepsilon)}\, dt \right|^2 dk \\ & + \varepsilon \int_{1}^{\infty} \int_{b_N}^{1} \frac{|\varphi(x)|^2}{(x-k)^2+\varepsilon^2}\, dx \left| \sum_{n=1}^{N} \int_{a_n}^{b_n} \frac{\psi(t) f(t)}{t-(k+i\varepsilon)}\, dt \right|^2 dk \\ =: {} & O(\varepsilon) + I_\varepsilon^1(f) + I_\varepsilon^2(f) + I_\varepsilon^3(f). \end{aligned}$$

The second term, $I_\varepsilon^2(f)$, is of order $O(\varepsilon)$ as

$$\int_{a_r}^{b_r} \frac{\psi(t) f(t)}{t-(k+i\varepsilon)}\, dt \to -\text{p.v.} \int_{-\infty}^{\infty} \frac{\chi_{[a_r,b_r]}(t) \psi(t) f(t)}{t-k}\, dt + i\pi\, \chi_{[a_r,b_r]}(k)\, \psi(k) f(k)$$

for $\varepsilon \searrow 0$ in $L_2(\mathbb{R})$ as a Hilbert transform (see [A], Chapter 12, and [S]) and because of (3.7). The third term, $I_\varepsilon^3(f)$, tends to 0 for $\varepsilon \searrow 0$ because of (3.8) and since

$$\varepsilon \int_{b_N}^{1} \frac{|\varphi(x)|^2}{(x-k)^2+\varepsilon^2}\,\mathrm{d}x \longrightarrow \pi\,\chi_{[b_N,1]}(k)\,|\widetilde{\varphi}(k)|^2 = 0, \qquad 1<k<\infty,$$

in $L_1(\mathbb{R})$ for $\varepsilon \searrow 0$ as a Poisson integral (see [S], Chapter III, §2, for example), where $\widetilde{\varphi}$ denotes the extension (by zero) of φ to the whole axis. Finally, for the first term we apply (3.8) and

$$\varepsilon \int_{b_{r-1}}^{a_r} \frac{|\varphi(x)|^2}{(x-k)^2+\varepsilon^2}\,\mathrm{d}x \longrightarrow \pi\,\chi_{[b_{r-1},a_r]}(k)\,|\widetilde{\varphi}(k)|^2$$

in $L_1(\mathbb{R})$ for $\varepsilon \searrow 0$, and we obtain

$$\begin{aligned} \tilde{I}_\varepsilon^1(f) &\le \widehat{C}_f \|\psi\|^2\|f\|^2\,\varepsilon \sum_{r=2}^{N+1} \int_{d_{r-1}}^{c_r} \int_{b_{r-1}}^{a_r} \frac{|\varphi(x)|^2}{(x-k)^2+\varepsilon^2}\,\mathrm{d}x\,\mathrm{d}k \\ &\to \widehat{C}_f \|\psi\|^2\|f\|^2\,\pi \sum_{r=2}^{N+1} \int_{b_{r-1}}^{a_r} |\varphi(k)|^2\,\mathrm{d}k \le \widehat{C}_f\|\varphi\|^2\|\psi\|^2\|f\|^2 \end{aligned}$$

for $\varepsilon \searrow 0$. This proves the existence of the limit of $I_\varepsilon(f)$, that is, of the left hand side of (3.3) for $\varepsilon \searrow 0$.

It remains to be proved that there exists a constant $C>0$ not depending on f such that

$$\lim_{\varepsilon \searrow 0} I_\varepsilon(f) \le C\|f\|^2. \tag{3.9}$$

Since the limit on the left hand side exists, it coincides with the formal limit for $f \in \mathcal{D}$:

$$\begin{aligned} \lim_{\varepsilon\to 0} I_\varepsilon(f) &= \pi \int_{\mathbb{R}} |\widetilde{\varphi}(k)|^2 \left| \int_0^k \frac{\psi(t)f(t)}{t-k}\,\mathrm{d}t \right|^2 \mathrm{d}k \\ &= \pi \int_0^1 |\varphi(k)|^2 \left| \int_0^k \frac{\psi(t)-\psi(k)}{t-k} f(t)\,\mathrm{d}t \right|^2 \mathrm{d}k, \end{aligned}$$

where the last equality holds because of the assumption $\varphi\psi \equiv 0$. Hence the estimate (3.9) is equivalent to the fact that the integral operator $\mathcal{K}$ given by

$$\mathcal{K}(f)(k) := \int_0^1 K(k,t)f(t)\,\mathrm{d}t, \qquad 0\le k\le 1,\ f\in L_2(0,1),$$

with kernel

$$K(k,t) := \sqrt{\pi}\,\varphi(k)\,\frac{\psi(t)-\psi(k)}{t-k}\,\chi_{[0,\infty)}(k-t), \qquad 0\le k,t\le 1,$$

is bounded in $L_2(0,1)$. It is well-known that $\mathcal{K}$ is bounded if

$$\text{ess sup}_{k\in[0,1]} \int_0^1 |K(k,t)|\,\mathrm{d}t < \infty, \tag{3.10}$$

$$\text{ess sup}_{t\in[0,1]} \int_0^1 |K(k,t)|\,\mathrm{d}k < \infty, \tag{3.11}$$

(see, e.g., [W], Satz 6.24 and Folgerung 4). As $\varphi\psi \equiv 0$, the estimate

$$\begin{aligned}
|K(k,t)| &\le \sqrt{\pi}\left|\varphi(k)\frac{\psi(t)-\psi(k)}{t-k}\right| \\
&= \sqrt{\pi}\,\frac{|\varphi(k)||\psi(t)|}{|t-k|} \\
&\le \sqrt{\pi}\,\frac{|\varphi(k)||\psi(t)| + |\varphi(t)||\psi(k)|}{|t-k|} \\
&= \sqrt{\pi}\,\frac{(|\varphi(k)| - |\varphi(t)|)(|\psi(t)| - |\psi(k)|)}{|t-k|} =: \widetilde{K}(k,t)
\end{aligned}$$

holds for $k,t \in [0,1]$. Since, in addition, $\widetilde{K}$ is symmetric, it is sufficient to prove (3.10) for $\widetilde{K}$ instead of K.

The assumptions $\varphi \in \text{Lip}\,(\omega_1), \psi \in \text{Lip}\,(\omega_2)$ imply $|\varphi| \in \text{Lip}\,(\omega_1), |\psi| \in \text{Lip}\,(\omega_2)$, that is,

$$\begin{aligned}
\big|\,|\varphi(k)| - |\varphi(t)|\,\big| &\le C_\varphi\,\omega_1(|t-k|), \\
\big|\,|\psi(t)| - |\psi(k)|\,\big| &\le C_\psi\,\omega_2(|t-k|)
\end{aligned}$$

for $k,t \in [0,1]$ with some constants $C_\varphi, C_\psi > 0$. Now let $k \in [0,1]$, and let $\delta \in (0,1)$, $\delta < \min\{k, 1-k\}$, be such that ω_1 and ω_2 fulfill the assumption (3.5). As φ and ψ are continuous on $[0,1]$, we then have

$$\begin{aligned}
&\int_0^1 |\widetilde{K}(k,t)|\,\mathrm{d}t \\
&= \int_0^{k-\delta} |\widetilde{K}(k,t)|\,\mathrm{d}t + \int_{k+\delta}^1 |\widetilde{K}(k,t)|\,\mathrm{d}t + \int_{k-\delta}^{k+\delta} |\widetilde{K}(k,t)|\,\mathrm{d}t \\
&\le \frac{4\sqrt{\pi}}{\delta}\max_{x\in[0,1]}|\varphi(x)|\max_{x\in[0,1]}|\psi(x)| + \sqrt{\pi}\,C_\varphi C_\psi \int_{k-\delta}^{k+\delta} \frac{\omega_1(|t-k|)\omega_2(|t-k|)}{|t-k|}\,\mathrm{d}t \\
&= \frac{4\sqrt{\pi}}{\delta}\max_{x\in[0,1]}|\varphi(x)|\max_{x\in[0,1]}|\psi(x)| + 2\sqrt{\pi}\,C_\varphi C_\psi \int_0^{\delta} \frac{\omega_1(\tau)\omega_2(\tau)}{\tau}\,\mathrm{d}\tau.
\end{aligned}$$

By assumption (3.5) the right hand side is finite and its value does not depend on k, which implies (3.10). This completes the proof of the theorem. □

To study the Lyapunov stability of the operator L given by (1.1) in the case $\varphi\psi \not\equiv 0$ is much more involved. A detailed discussion is left for a further occassion. At present the following statement can be shown.

Proposition 3.5. *Let $\Re(\varphi\psi) \in \operatorname{Lip}\alpha$ with $\alpha > 0$, and suppose φ, $\psi \in C[0,1])$. If the set $\{\lambda \in [0,1] : \Re(\varphi\psi)(\lambda) < -\frac{1}{2}\}$ is uncountable, then L is not Lyapunov stable.*

Proof. The assertion follows from Theorem 2.4 and from the fact that a self-adjoint operator cannot have an uncountable point spectrum. □

4. A counter-example

The following theorem provides an example of an operator L of the form (1.1) which is Lyapunov stable according to Theorem 3.4 and an example of an operator L fulfilling all assumptions of Theorem 3.4 but (3.5) which is not Lyapunov stable.

Theorem 4.1. *Let δ_1, $\delta_2 > 0$, and let ω_1, ω_2 be moduli of continuity such that for $0 < \tau < \frac{1}{2}$,*

$$\omega_i(\tau) = \frac{1}{\ln^{\delta_i} \frac{1}{\tau}}, \qquad i = 1, 2.$$

If $\delta_1 + \delta_2 > 1$, then for all functions $\varphi \in \operatorname{Lip}(\omega_1)$, $\psi \in \operatorname{Lip}(\omega_2)$ such that $\varphi\psi \equiv 0$ the corresponding operator L given by (1.1) is Lyapunov stable.
If $\delta_1 + \delta_2 < 1$, then condition (3.5) is not fulfilled and there exist functions $\varphi \in \operatorname{Lip}(\omega_1)$, $\psi \in \operatorname{Lip}(\omega_2)$ with $\varphi\psi \equiv 0$ such that the corresponding operator L is not Lyapunov stable.

Proof. First we show that for ω_1, ω_2 as in the theorem condition (3.5) is fulfilled if and only if $\delta_1 + \delta_2 > 1$. Due to Theorem 3.4 this implies the first part of the theorem. Indeed, for any $\delta \in (0,1)$,

$$\int_0^\delta \frac{\omega_1(\tau)\omega_2(\tau)}{\tau}\, d\tau = \int_{1/\delta}^\infty \frac{1}{t \ln^{\delta_1+\delta_2} t}\, dt$$

which is finite if and only if $\delta_1 + \delta_2 > 1$.
It remains to construct functions $\varphi \in \operatorname{Lip}(\omega_1)$, $\psi \in \operatorname{Lip}(\omega_2)$ in the case $\delta_1 + \delta_2 < 1$ with $\varphi\psi \equiv 0$ such that the corresponding operator L is not Lyapunov stable. To this end we define a sequence of positive numbers

$$\widetilde{\Delta}_k := \frac{1}{k^2 \ln^{2-(\delta_1+\delta_2)} k}, \qquad k = 2, 3, \ldots.$$

Then the series $\sum_{k=2}^\infty 4k\widetilde{\Delta}_k$ is convergent as $\delta_1 + \delta_2 < 1$. If we let

$$\Delta := \sum_{k=2}^\infty 4k\widetilde{\Delta}_k, \qquad \Delta_k := \frac{\widetilde{\Delta}_k}{\Delta + 1}, \qquad k = 2, 3, \ldots,$$

and

$$\gamma_n := \sum_{k=2}^{n} 4k\Delta_k, \qquad n = 2, 3, \dots,$$

then

$$\gamma := \lim_{n\to\infty} \gamma_n = \frac{\Delta}{\Delta + 1} < 1.$$

For $n = 2, 3, \dots$, we define intervals $\Gamma_n \subset [0,1]$ of length $4n\Delta_n$ by

$$\Gamma_n := [\gamma - \gamma_n, \gamma - \gamma_{n-1}], \qquad \gamma_1 := 0,$$

and divide the intervals Γ_n into $4n$ subintervals $\gamma_{n,1}, \dots, \gamma_{n,4n}$ of length Δ_n given by

$$\gamma_{n,j} := [\gamma_{n,j}^1, \gamma_{n,j}^2] := [\gamma - \gamma_n + (j-1)\Delta_n, \gamma - \gamma_n + j\Delta_n], \qquad j = 1, 2, \dots, 4n,$$

Now we choose piecewise linear functions $\varphi, \psi \in C[0,1]$ such that

$$\begin{aligned} \varphi &\equiv 0 \quad \text{on} \quad \gamma_{n,2j}, \\ \psi &\equiv 0 \quad \text{on} \quad \gamma_{n,2j-1}, \end{aligned}$$

and

$$\varphi(t) = \begin{cases} 0, & t \in [\gamma_{n,2j-1}^1, \gamma_{n,2j-1}^1 + \frac{\Delta_n}{8}], \\ \varphi_n, & t \in [\gamma_{n,2j-1}^1 + \frac{\Delta_n}{4}, \gamma_{n,2j-1}^2 - \frac{\Delta_n}{4}] =: \widehat{\gamma}_{n,2j-1}, \\ 0, & t \in [\gamma_{n,2j-1}^2 - \frac{\Delta_n}{8}, \gamma_{n,2j-1}^2], \end{cases} \qquad \text{on} \quad \gamma_{n,2j-1},$$

$$\psi(t) = \begin{cases} 0, & t \in [\gamma_{n,2j}^1, \gamma_{n,2j}^1 + \frac{\Delta_n}{8}], \\ \psi_n, & t \in [\gamma_{n,2j}^1 + \frac{\Delta_n}{4}, \gamma_{n,2j}^2 - \frac{\Delta_n}{4}] =: \widehat{\gamma}_{n,2j}, \\ 0, & t \in [\gamma_{n,2j}^2 - \frac{\Delta_n}{8}, \gamma_{n,2j}^2], \end{cases} \qquad \text{on} \quad \gamma_{n,2j},$$

for all $j = 1, 2, \dots, 2n,\ n = 2, 3, \dots$, where

$$\varphi_n := \frac{1}{\ln^{\delta_1} n}, \qquad \psi_n := \frac{1}{\ln^{\delta_2} n}. \tag{4.1}$$

Obviously, $\varphi\psi \equiv 0$. We are going to prove now that the functions φ, ψ belong to $\mathrm{Lip}\,(\omega_1)$ and $\mathrm{Lip}\,(\omega_2)$, respectively. We consider the function φ; the proof for ψ is analogous. As $\varphi_n \to 0$ for $n \to \infty$, φ is bounded on $[0,1]$ and $\varphi(0) = 0$. It is sufficient to show that

$$|\varphi(x) - \varphi(y)| \le C\omega_1(|x-y|) = C\frac{1}{\ln^{\delta_1} \frac{1}{|x-y|}}$$

for $|x - y| \le h_0$ for some h_0, $0 < h_0 \le 1$. First we consider $x = 0$. Let $\varepsilon := 1 - (\delta_1 + \delta_2)$, choose $n_0 \in \mathbb{N}$ such that

$$\frac{\ln^{\varepsilon}(n+1)}{n} \le \frac{4}{(\Delta+1)\varepsilon}, \qquad n \ge n_0,$$

and let $h_0 := \sum_{k=n_0}^{\infty} 4k\Delta_k$. Then for all $h > 0$, $h \le h_0$, $h \in \Gamma_n$ with $n \ge n_0$,

$$\begin{aligned} h &\ge \sum_{k=n+1}^{\infty} 4k\Delta_k = \frac{4}{\Delta+1} \sum_{k=n+1}^{\infty} \frac{1}{k \ln^{1+\varepsilon} k} \\ &\ge \frac{4}{\Delta+1} \int_{n+1}^{\infty} \frac{1}{t \ln^{1+\varepsilon} t} \, dt = \frac{4}{\Delta+1} \frac{1}{\varepsilon \ln^{\varepsilon}(n+1)} \quad \ge \frac{1}{n}, \end{aligned}$$

and thus, as ω_1 is nondecreasing,

$$|\varphi(h)| \le \varphi_n = \frac{1}{\ln^{\delta_1} n} = \omega_1\left(\frac{1}{n}\right) \le \omega_1(h).$$

Now let $x \ne 0$, $x \in \Gamma_n$. Consider first the case that h is so small that also $x+h \in \Gamma_n$. Then

$$\frac{|\varphi(x) - \varphi(x+h)|}{h} \le \frac{\varphi_n}{\frac{\Delta_n}{8}} = \frac{(\Delta+1)8n^2 \ln^{1+\varepsilon} n}{\ln^{\delta_1} n},$$

where again $\varepsilon := 1 - (\delta_1 + \delta_2)$. Consequently,

$$|\varphi(x+h) - \varphi(x)| \le C \frac{1}{\ln^{\delta_1} \frac{1}{h}}$$

with a constant $C > 0$ for some $h_0 > 0$ if

$$h \ln^{\delta_1} \frac{1}{h} \le C \frac{\ln^{\delta_1} n}{(\Delta+1)8n^2 \ln^{1+\varepsilon} n}, \qquad h \le h_0. \tag{4.2}$$

For the function $g(t) = t \ln^{\delta_1} \frac{1}{t}$, $t \in (0,1)$, we have $\lim_{t \searrow 0} g(t) = 0$ and

$$g'(t) = \ln^{\delta_1} \frac{1}{t} \left(1 - \delta_1 \frac{1}{\ln \frac{1}{t}}\right) > 0, \qquad t < \frac{1}{\mathrm{e}}.$$

Let $n_0 \in \mathbb{N}$, $n_0 \ge \Delta + 1$, be such that

$$\ln^{1+\varepsilon} n \le n, \qquad n \ge n_0.$$

For $h \le 2\Delta_n = \frac{2}{(\Delta+1)n^2 \ln^{1+\varepsilon} n}$, we then have

$$h \ln^{\delta_1} \frac{1}{h} \le \frac{2 \ln^{\delta_1} \left(\frac{1}{2}(\Delta+1)n^2 \ln^{1+\varepsilon} n\right)}{(\Delta+1)n^2 \ln^{1+\varepsilon} n} \le \frac{8 \ln^{\delta_1} n}{(\Delta+1)n^2 \ln^{1+\varepsilon} n}, \qquad n \ge n_0,$$

if $h < \frac{1}{\mathrm{e}}$ which proves (4.2) for $n \ge n_0$. If $n \le n_0$, then there exists an h_0, $0 < h_0 < \frac{1}{\mathrm{e}}$, such that

$$h \ln^{\delta_1} \frac{1}{h} \le \frac{1}{8n_0^2 \ln^{1+\varepsilon-\delta_1} n_0} \le \frac{1}{8n^2 \ln^{1+\varepsilon-\delta_1} n}, \qquad h \le h_0,$$

as $1 + \varepsilon - \delta_1 = 2 - 2(\delta_1 + \delta_2) + \delta_2 > 0$. Thus also in this case (4.2) holds. Now let $h \ge 2\Delta_n$, $x + h \in \Gamma_k$ with $k \le n$. As $\varphi_n \le \varphi_k$, there exist $\widetilde{x} \in \Gamma_k$ and $\widetilde{h} \le h$, $\widetilde{h} \le 2\Delta_k$, such that also $\widetilde{x} + \widetilde{h} \in \Gamma_k$ and $|\varphi(x+h) - \varphi(x)| = |\varphi(\widetilde{x}+\widetilde{h}) - \varphi(\widetilde{x})|$.

Then

$$|\varphi(x+h)-\varphi(x)| = |\varphi(\widetilde{x}+\widetilde{h})-\varphi(\widetilde{x})| \le C\frac{1}{\ln^{\delta_1}\frac{1}{\widetilde{h}}} \le C\frac{1}{\ln^{\delta_1}\frac{1}{h}}$$

for $h \le h_0$ by what we have proved before. This proves $\varphi \in \mathrm{Lip}\,(\omega_1)$.
It remains to be proved that the operator L corresponding to the functions φ, ψ defined above is not Lyapunov stable. For this purpose we choose a sequence $(f_m)_2^\infty \subset C_0^\infty(0,1)$ such that $|f_m(t)| \le 1$, $t \in [0,1]$, and

$$f_m(t) = \begin{cases} 0, & t \in [\gamma^1_{n,2j}, \gamma^1_{n,2j} + \frac{3\Delta_n}{16}], \\ 1, & t \in \widehat{\gamma}_{n,2j}, \\ 0, & t \in [\gamma^2_{n,2j} - \frac{3\Delta_n}{16}, \gamma^2_{n,2j}], \end{cases} \qquad j = 1,2,\ldots,2n,\ n = 2,3,\ldots,m.$$

Then $(f_m)_2^\infty \subset \mathcal{D}$ where $\mathcal{D}$ is given by (3.6), and $\|f_m\| \le \gamma < 1$.
Suppose now that L were Lyapunov–stable. Then, by Proposition 3.2, there exists a uniform constant $C > 0$ such that

$$\limsup_{\varepsilon \searrow 0} \varepsilon \int_{\mathbb{R}} \left\| \frac{\varphi(x)}{x-(k+i\varepsilon)} \int_0^x \frac{\psi(t) f_m(t)}{t-(k+i\varepsilon)}\, \mathrm{d}t \right\|^2 \mathrm{d}k \le C, \qquad m = 2,3,\ldots.$$

According to the proof of Theorem 3.4, the limit on the left hand side exists and is equal to the formal limit. Therefore, as $\operatorname{supp} f_m \subset \bigcup_{n=2}^m \Gamma_n$,

$$\begin{aligned} C &\ge \pi \int_0^1 |\varphi(k)|^2 \left| \int_0^k \frac{\psi(t) f_m(t)}{t-k}\, \mathrm{d}t \right|^2 \mathrm{d}k \\ &= \pi \sum_{n=2}^m \int_{\Gamma_n} |\varphi(k)|^2 \left| \int_0^k \frac{\psi(t) f_m(t)}{t-k}\, \mathrm{d}t \right|^2 \mathrm{d}k \\ &\ge \pi \sum_{n=2}^m \sum_{\substack{j=2n+1 \\ j \text{ odd}}}^{4n} \int_{\widehat{\gamma}_{n,j}} \varphi_n^2 \left(\sum_{\substack{i=1 \\ i \text{ even}}}^{j-1} \int_{\widehat{\gamma}_{n,i}} \frac{\psi_n}{k-t}\, \mathrm{d}t \right)^2 \mathrm{d}k \\ &\ge C_1 \sum_{n=2}^m \sum_{\substack{j=2n+1 \\ j \text{ odd}}}^{4n} \int_{\widehat{\gamma}_{n,j}} \varphi_n^2 \psi_n^2 \left(\sum_{\substack{i=1 \\ i \text{ even}}}^{j-1} \frac{1}{j+1-i} \right)^2 \mathrm{d}k \\ &\ge C_2 \sum_{n=2}^m n\,\Delta_n\,\varphi_n^2\,\psi_n^2 \left(\sum_{\mu=1}^{2n-1} \frac{1}{\mu} - 1 \right)^2 \\ &\ge C_3 \sum_{n=2}^m n\,\Delta_n\,\varphi_n^2\,\psi_n^2 \ln^2 n \\ &\ge C_3 \sum_{n=2}^m \frac{1}{n \ln^{\delta_1+\delta_2} n} \end{aligned}$$

with some positive constants C_1, C_2, C_3 by definition of Δ_n, φ_n and ψ_n. However, obviously, the last series is divergent as $\delta_1 + \delta_2 < 1$, a contradiction. Thus L is not Lyapunov stable, and the theorem is proved. □

Acknowledgements

The first author gratefully acknowledges the support of the Deutsche Forschungsgemeinschaft, DFG. The authors thank Prof. Dr. R. Mennicken for initiating their cooperation and Prof. Dr. J. Zemanek for useful discussions in the beginning of this work.

References

[A] Akhiezer, N.I.: Lectures on Integral Transforms; AMS Transl. Math. Monographs 70 (1988).

[C] Casteren, J.A. van: Operators similar to unitary or self-adjoint ones; Pac. J. Math. 104:1 (1983), 241–255.

[DK] Daleckii, Ju.L., Krein, M.G.: Stability of Solutions of Differential Equations in Banach Space; AMS Transl. Math. Monographs 43 (1974).

[FN] Faddeev, M.M., Naboko, S.N.: Friedrichs model operators similar to self-adjoint ones; Vestnik Leningrad Univ. Phys. 26:4 (1990), 78–92.

[K] Kokholm, N.J.: Spectral analysis of perturbed multiplication operators occurring in polymerisation chemistry; Proc. Roy. Soc. Edinburgh Sect. A 113 (1989), 119–148.

[N1] Naboko, S.N.: Conditions for similarity to unitary and self-adjoint operators; Functional Anal. Appl. 18:1 (1984), 13–22.

[N2] Naboko, S.N.: Uniqueness theorems for operator-valued functions with positive imaginary part, and the singular spectrum in the self-adjoint Friedrichs model; Ark. Mat. 25:1 (1987), 115–140.

[S] Stein, E.M.: Singular Integrals and Differentialbility Properties of Functions; Princeton University Press, Princeton 1970.

[SzN] Sz.-Nagy, B.: On uniformly bounded linear transformations in Hilbert space; Acta. Sci. Math. (Szeged) 11:3 (1947), 152–157.

[V] Veselov, V.F.: On some model for the operator similarity problem; Vestnik Leningrad. Univ. Math. 18:4 (1985), 62–66.

[W] Weidmann, J.: Lineare Operatoren in Hilbertraeumen; Teubner Verlag, Stuttgart 1976.

Department of Mathematical Physics
St.-Petersburg University
198904 St. Petersburg
Russia

Department of Mathematics
University of Regensburg
93040 Regensburg
Germany

1991 Mathematics Subject Classification. Primary 34D20, 47D03; Secondary 47G10

Operator Theory:
Advances and Applications, Vol. 106

Multiplicative perturbations of positive operators in Krein spaces

B. NAJMAN[1] AND K. VESELIĆ

Dedicated to Heinz Langer on the occasion of his 60th birthday

1. Introduction

Let A be a positive operator with a nonempty resolvent set in a Krein space $\mathcal{K}$. Then A has a spectral function with 0 and ∞ being the only possible critical points, see [9]; if neither of these points is a singular critical point then A is similar to a Hilbert space selfadjoint operator, that is, it is a scalar operator with real spectrum (see [9] for the definition and properties of the Krein space operators).

The problem of the persistence of nonsingularity of critical points has been started by one of the authors of the present note ([13, 14]) and later continued by a number of authors ([6, 2, 3, 10] etc.). All these references deal with additive perturbations $A+V$ of A; such results are insufficient in some cases, e.g. in the case of elliptic operators with indefinite weights, where

$$A = \frac{1}{\rho} L,$$

L is an elliptic operator and ρ is a real valued function which is not of constant sign. If the weight ρ is perturbed into $\tilde{\rho}$ with

$$|\tilde{\rho} - \rho| \leq \varepsilon |\rho|, \; \varepsilon < 1$$

we have to investigate multiplicative perturbation $(I+V)A$ of A. We prove a result which ensures the persistence of nonsingularity of critical points if the perturbation is sufficiently small as well as the analyticity of certain operators associated with the analytic family $(I+\varepsilon X)A$. The proof goes via the construction of the signum operator, similar to that in [13, 14], our situation is much more singular so that the above estimate for the perturbed weight has to be completed by an additional one for the derivative.

We mention the recent note [5] where the persistence of nonsingularity of the critical point ∞ under multiplicative perturbations is considered. This is in contrast with the present work where we prove the regularity of both critical points 0 and ∞.

[1]Branko Najman passed away before this manuscript was completed.

2. The results

Let $(\mathcal{K}, [\cdot|\cdot])$ be a Krein space, J a fundamental symmetry in $\mathcal{K}$. Then $\mathcal{H} = (\mathcal{K}, [J\cdot|\cdot])$ is a Hilbert space; let $(u|v) = [Ju|v]$ be the corresponding scalar product and $\|u\|$ the corresponding norm. Let A be a selfadjoint operator in $\mathcal{K}$ such that the following assumption is satisfied:

(A1) A is a strictly positive operator in $\mathcal{K}$ with a nonempty resolvent set.

It follows from (A1) that the form $a = [A\cdot|\cdot]$ defined on $\mathcal{D}(A) \times \mathcal{D}(A)$ has a closure a in $\mathcal{K}$; its domain $\mathcal{D}(a)$ coincides with the domain $\mathcal{D}((JA)^{1/2})$ of

(2.1) $$M := (JA)^{1/2}$$

in $\mathcal{H}$. As noted in the Introduction, from (A1) it follows that A has a spectral function E in the sense of [9] with the only possible critical points being 0 and ∞. The next assumption excludes the possibility that either of these points is a singular critical point:

(A2) 0 and ∞ are not singular critical points of A.

It follows from (A2) that the projections $E[0,\infty)$ and $E(-\infty,0)$ are well defined, and hence also

$$P = sgnA = E[0,\infty) - E(-\infty,0).$$

Then (see [13] or [6])

(2.2) $$P = -\frac{1}{i\pi}\, w - \lim_{\substack{r\to 0\\ R\to\infty}} \left(\int_r^R dt + \int_{-R}^{-r} dt\right)(it - A)^{-1}.$$

Moreover, P commutes with A and it is a uniformly positive bounded operator, i.e. the space $(\mathcal{K}, <\cdot|\cdot>)$ with

$$<u|v> = [Pu|v]$$

is a Hilbert space with the norm $|||\cdot|||$, equivalent to $\|\cdot\|$. Then A is selfadjoint in $(\mathcal{K}, <\cdot|\cdot>)$; the operator

$$\hat{A} = (JP)^{1/2} A (JP)^{-1/2}$$

is selfadjoint in $\mathcal{H}$. Note also that by $P^2 = 1$ we have $\|P\| \geq 1$.

Lemma 2.1. *For any bounded measurable function f we have*

$$\|f(A)\| \leq \|f\|_\infty \|P\|.$$

Proof. Since $\tilde{A}$ is selfadjoint in $\mathcal{H}$ we have

$$\|f(A)\| = \|(JP)^{-1/2}f(\tilde{A})(JP)^{1/2}\| \le \|(JP)^{-1/2}\|\|f(\tilde{A})\|\|(JP)^{1/2}\|$$

$$\le \|f\|_\infty \|JP\|^{1/2}\|P^{-1}J\|^{1/2} \le \|f\|_\infty \|P\|^{1/2}\|P^{-1}\|^{1/2} = \|f\|_\infty\|P\|$$

where we have used the unitarity of J and the fact that $P^{-1} = P$. □

Let X be an operator in $\mathcal{K}$ with the properties

(X1) X is selfadjoint and bounded in $\mathcal{K}$,
(X2) $JX = XJ$.

From (X1) and (X2) it follows that X is selfadjoint also in $\mathcal{H}$. The next assumptions connect A, M (from (2.1)) and X :

(AX1) $X\mathcal{D}(M) \subset \mathcal{D}(M)$.
(AX2) There exists a $C > 0$ such that

$$\|MXu - XMu\| \le C\|Mu\| \qquad (u \in \mathcal{D}(M)).$$

Since A is positive, M is injective and (AX2) is implied by (X1) and

Proposition 2.2. *Let*

$$A_1 = (1 + X)A$$

where A, X are as above and

$$\|X\| < 1/\|P\|.$$

Then A_1 has a non-void resolvent set and is a positive selfadjoint operator in the Krein space $(\mathcal{K}, [\cdot|\cdot]_1)$ where

$$[\cdot|\cdot]_1 = [(1+X)^{-1}\cdot|\cdot] = (J(1+X)^{-1}\cdot|\cdot).$$

Proof. $1 + X$ is bicontinuous in $\mathcal{H}$ and thus A_1 is selfadjoint and positive in $(\mathcal{K}, [\cdot|\cdot]_1)$. Also, by Lemma 2.1, for any real $\eta \neq 0$ we have

$$\|X(A(i\eta - A)^{-1}\| \le \|X\| \sup_{t\in\mathbf{R}} |\frac{t}{i\eta - t}| = \|X\|\|P\| < 1$$

and $i\eta$ belongs to the resolvent set of A for all $\eta \in \mathcal{R}\backslash\{0\}$. □

Our results are summarized in the following theorem.

Theorem 2.3. *Assume (A1), (A2),(X1),(X2),(AX1) and (AX2). Let also $\|X\| + C < 1$. Then for $|\varepsilon| \le \|X\| + C$ the operator $1 + \varepsilon X$ has a bounded inverse, the operator A_1 is a positive selfadjoint operator in $\mathcal{K}_1$ with a nonempty resolvent set.*

Neither 0 nor ∞ is a singular critical point of A_1, the operator A_1 is similar to a selfadjoint operator in $\mathcal{H}$. The operator

$$(2.3)\qquad P_1 = -\frac{1}{i\pi}\, w - \lim_{\substack{r\to 0\\ R\to\infty}} \left(\int_r^R dt + \int_{-R}^{-r} dt\right)(it - A_1)^{-1}$$

is a positive definite operator in $\mathcal{K}_1$ and $P_1^2 = I$. The space $(\mathcal{K}, < \cdot|\cdot >_1)$, where

$$< u|v >_1 = [P_1 u|v]_1,$$

is a Hilbert space. The operator A_1 is selfadjoint in $(\mathcal{K}, < \cdot|\cdot >_1)$. The operator $J(I+X)^{-1}P_1$ is boundedly invertible and positive definite in $\mathcal{H}$. The operator $(J(I+X)^{-1}P_1)^{1/2}A_1(J(I+X)^{-1}P_1)^{-1/2}$ is selfadjoint in $\mathcal{H}$.

Proof. We embed the operator A_1 in the family

$$A(\varepsilon) = (I + \varepsilon X)A$$

and accordingly define $\mathcal{K}_\varepsilon$, $[\cdot|\cdot]_\varepsilon$, $< \cdot|\cdot >_\varepsilon$ and $P(\varepsilon)$. For $\eta \in \mathbf{R}\backslash\{0\}$ set

$$(2.4)\qquad F(\eta,\varepsilon) := (i\eta - \varepsilon A)^{-1} - (i\eta A)^{-1} = \sum_{k=0}^{\infty} \varepsilon^k F_k(\eta,\varepsilon),$$

$$(2.5)\qquad F_k(\eta) := (i\eta - A)^{-1}[XA(i\eta - A)^{-1}]^k \quad (\eta \in \mathbf{C}\backslash\mathbf{R}).$$

This series converges whenever

$$(2.6)\qquad |\varepsilon|\|XA(i\eta - A)^{-1}\| < 1,\ \eta \in \mathbf{R}\backslash\{0\}.$$

Moreover, it follows from (2.6) and (2.4) that F is analytic in η and ε for $\eta \in \mathbf{C}\backslash\mathbf{R}$ and ε such that (2.4) holds. From the selfadjointness of A and (X2) it follows

$$\begin{aligned}(F_{k+1}(\eta)x|y) &= ((i\eta - A)^{-1}X[A(i\eta - A)^{-1}X]^k A(i\eta - A)^{-1}x|y)\\ &= (X[A(i\eta - A)^{-1}X]^k A(i\eta - A)^{-1}x|J(-i\eta - A)^{-1}Jy).\end{aligned}$$

From $A = JM^2$ we find $[A(i\eta - A)^{-1}X]^k = J[M^2(i\eta - JM^2)^{-1}XJ]^k J$ and therefore by (X2)

$$\begin{aligned}&(F_{k+1}(\eta)x|y) =\\ &(X[M^2(i\eta - JM^2)^{-1}XJ]^k JA(i\eta - A)^{-1}x|(-i\eta - A^{-1})Jy)\\ &= ([M^2(i\eta - JM^2)^{-1}XJ]^k M^2(i\eta - A)^{-1}x|X(-i\eta - A)^{-1}Jy).\end{aligned}$$

From $M\mathcal{D}(A) \subset \mathcal{D}(M)$ it follows that $\mathcal{R}(M(i\eta - A)^{-1}) \subset \mathcal{D}(M)$ and consequently

$$\begin{aligned}&[M^2(i\eta - JM^2)^{-1}XJ]^k M^2(i\eta - A)^{-1} =\\ &M[M(i\eta - JM^2)^{-1}XJM]^k M(i\eta - A)^{-1}.\end{aligned}$$

This implies

$$(F_{k+1}(\eta)x|y) = \tag{2.7} (G(\eta)^k M(i\eta - A)^{-1}x|(MX(-i\eta - A)^{-1}Jy)$$

with

$$G(\eta) = M(i\eta - JM^2)^{-1}JXM.$$

Note that $G(\eta)$ is defined on $\mathcal{D}(M)$ and leaves that space invariant. Our goal is to estimate the norm of $G(\eta)$.

Lemma 2.4. *For every $x \in \mathcal{D}(M)$ and every $\eta \in \mathbf{R}\backslash\{0\}$ we have*

$$\|MJ(i\eta - M^2J)^{-1}Mx\| \le \|x\|. \tag{2.8}$$

Proof. Set $M_\lambda = MJ(i\eta - M^2J)^{-1}M$. If M is bounded, then $M_\lambda = MJM(i\eta - MJM)^{-1}$ and the assertion follows immediately from the spectral calculus of the selfadjoint operator MJM. Our proof will follow the same pattern and will, in fact, construct a "selfadjoint realization" of the formal expression MJM. We thus consider the operator

$$R_{i\eta} = M(i\eta - JM^2)^{-1}M^{-1} \,:\, \mathcal{D}(M^{-1}) \to \mathcal{D}(M^{-1}),\ \eta \in \mathbf{R}\backslash\{0\}.$$

It is immediately verified that $R_{i\eta}$ satisfies the resolvent equation. For $x \in \mathcal{D}(M^{-1})$:

$$(R_{i\eta} - R_{i\eta'})x = i(\eta - \eta')R_{i\eta}R_{i\eta'}x. \tag{2.9}$$

Furthermore, for $x \in \mathcal{D}(M^{-1})$, $y \in \mathcal{D} = \mathcal{D}(M) \cap \mathcal{D}(M^{-1})$ and $y' = M^{-1}y$ we have

$$\begin{aligned}(R_{i\eta}x|y) &= (M^{-1}x|(-i\eta - M^2J)^{-1}My) = \\ &(M^{-1}x|(-i\eta - M^2J)^{-1}M^2y') = \\ &(M^{-1}x|J(-i\eta - JM^2)^{-1}JM^2y') = \\ &(M^{-1}x|M^2(-i\eta - JM^2)^{-1}y') = (x|R_{-i\eta}y).\end{aligned} \tag{2.10}$$

In particular, all $R_{i\eta}$ leave $\mathcal{D}$ invariant and commute there. We can set

$$A_\eta = \frac{R_{i\eta} + R_{-i\eta}}{2},\ B_\eta = \frac{R_{i\eta} - R_{-i\eta}}{2i}.$$

Obviously these are commuting symmetric operators in $\mathcal{H}$, defined on the dense subspace $\mathcal{D}$ and leaving this subspace invariant. For $x \in \mathcal{D}$ the resolvent equation (2.9) gives

$$B_\eta x = \eta(A_\eta - iB_\eta)(A_\eta + iB_\eta)x = \eta(A_\eta^2 + B_\eta^2)x. \tag{2.11}$$

Taking e.g. $\eta > 0$ and applying the Schwarz inequality, this gives

$$\eta(B_\eta x|B_\eta x) \le (B_\eta x|x) \le \|B_\eta x\|,$$

we see that A_η is bounded, and similarly for $\eta < 0$. By (2.11) the same follows for B_η and then also for $R_\eta|_\mathcal{D}$ whose closure $\tilde{R}_{i\eta}$ is a pseudoresolvent and its null space $\mathcal{N}$ is known to be independent of η. By (2.10) we have $\tilde{R}^*_{i\eta} = \tilde{R}_{-i\eta}$ and thus $\tilde{R}_{i\eta}$ leaves invariant both $\mathcal{N}$ and $\mathcal{N}^\perp$. Thus, there is a unique selfadjoint operator H_0 in the Hilbert space $\mathcal{N}^\perp \subset \mathcal{H}$ such that

$$R_{i\eta} = (i\eta - H_0)^{-1}P, \tag{2.12}$$

where P is the orthogonal projection onto $\mathcal{N}^\perp$.

We now connect $R_{i\eta}$ with M_η. For $x, y \in \mathcal{D}(M)$ we have

$$(M_\eta x|y) = (x|M_{-\eta}y)$$

thus, M_η is closable on $\mathcal{D}(M)$. For $x \in \mathcal{D}$ we have

$$\begin{aligned} M_\eta x &= MJ(i\eta - M^2J)^{-1}x - i\eta M(i\eta - JM^2)^{-1}M^{-1}x + i\eta R_{i\eta}x \\ &= M(i\eta - jM^2)^{-1}(JMx - i\eta M^{-1}x) + i\eta R_{-i\eta}x. \end{aligned}$$

By setting $x = Mx'$ we have $x' \in \mathcal{D}(M^2)$ and

$$M_\eta = M(i\eta - JM^2)^{-1}(JM^2 - i\eta)x' i\eta R_{-i\eta}x =$$

$$(-1 + i\eta\tilde{R}_{i\eta})x = (-1 + i\eta(i\eta - H_0^{-1})P)x$$

Thus,

$$\begin{aligned} \|M_\eta x\|^2 &= \|(1-P)x\|^2 + \|[1 - i\eta(i\eta - H_0)^{-1}]\|^2 \\ &= \|(1-P)x\|^2 + \|H_0(i\eta - H_0)^{-1}Px\|^2 \\ &\le \|(1-P)x\|^2 + \|Px\|^2 = \|x\|^2. \end{aligned}$$

Since M_η is closable on $\mathcal{D}(M)$ and $\mathcal{D} \subset \mathcal{D}(M)$ is dense in $\mathcal{H}$, the inequality above extends to all $x \in \mathcal{D}(M)$. □

Lemma 2.5. *For all $\eta \in \mathbf{R}\backslash\{0\}$ and all $x \in \mathcal{D}(M)$ we have*

$$\|G(\eta)x\| \le (\|X\| + C)\|x\|.$$

Proof. It is obviously enough to prove the same identity for the formal adjoint

$$\overline{G}(\eta) = MXJ(-i\eta - M^2J)^{-1}M \, : \, \mathcal{D}(M) \to \mathcal{D}(M)$$

which has the property

$$(G(\eta)x|y) = (x|\overline{G}(\eta)y), \; x, y \in \mathcal{D}(M).$$

We have by Lemma 2.4 and (AX2)

$$\begin{aligned} \|\overline{G}(\eta)\| &\le \|(MX - XM)J(-i\eta - M^2J)^{-1}My\| + \\ &\|XMJ(-i\eta - M^2J)^{-1}My\| \le (C + \|X\|)\|y\|. \end{aligned}$$

□

We now continue with the proof of our theorem. From (2.5) and Lemma 2.5 it follows

$$\begin{aligned}|(F_{k+1}(\eta)x|y)| &\leq \|G(\eta)^k\| \, \|M(i\eta - A)^{-1}x\| \, \|MX(-i\eta - A)^{-1}Jy\| \\ &\leq \gamma^k \|M(i\eta - A)^{-1}x\| \, \|MX(-i\eta - A)^{-1}Jy\|\end{aligned}$$

with $\gamma = \|X\| + C$. Therefore

$$|\int (F_{k+1}(\eta)x|y)d\eta|^2 \leq$$

$$\gamma^{2k} \int \|M(i\eta - A)^{-1}x\|^2 d\eta \int \|MX(-i\eta - A)^{-1}Jy\|^2 d\eta.$$

To evaluate the first integral on the RHS, we note the identity

$$\int \|M(i\eta - A)^{-1}x\|^2 d\eta = \int ((-i\eta - A)^{-1}M^2(i\eta - A)^{-1}x|x)d\eta$$

$$= \int [A(i\eta - A)^{-1}x|(i\eta - A)^{-1}x]d\eta,$$

i.e.

$$\int \|M(i\eta - A)^{-1}x\|^2 d\eta =$$

$$\int \|M(-i\eta - A)^{-1}x\|^2 d\eta = \int [A(\eta^2 + A^2)^{-1}x|x]d\eta.$$

The assumption (AX2) yields

$$\|MX(-i\eta - A)^{-1}Jy\| \leq (\|X\| + C)\|M(-i\eta - A)^{-1}Jy\|,$$

hence

$$\int \|MX(-i\eta - A)^{-1}Jy\|^2 d\eta$$

$$\leq (\|X\| + C)^2 \int \|M(-i\eta - A)^{-1}Jy\|^2 d\eta$$

$$= (\|X\| + C)^2 \int [A(\eta^2 + A^2)^{-1}Jy|Jy]d\eta.$$

For a measurable set $S \subset \mathbf{R}$, $z \in \mathcal{H}$, we set $I_s(z) = \int_S [A(y^2 + A^2)^{-1}z|z]d\eta$. We have proved

$$|\int_S (F_{k+1}(\eta)x|y)d\eta|^2 \leq \gamma^{2k}(\|X\| + C)^2 I_S(X) I_S(Jy),$$

hence

(2.13) $$|\int_S (F_{k+1}(y)x|y)d\eta| \leq \gamma^k(\|X\| + C)(I_S(x) + I_S(Jy)).$$

Inserting this into (7), we obtain

$$|\int_S (F(\eta,\varepsilon)x|y)d\eta| \leq \sum_{k=1}^{\infty} \varepsilon^k \gamma^{k-1}(\|X\|+C)(I_S(x)+I_S(Jy))$$
$$= \frac{\varepsilon}{1-\varepsilon\gamma}(\|X\|+C)(I_S(x)+I_s(Jy))$$

as soon as $|\varepsilon| < \dfrac{1}{K(\|X\|+C)}$ and in particular for $|\varepsilon| \leq 1$.

We use this estimate for $S = [-R,-r] \cup [r,R]$. Spectral calculations then yield

$$I_S(z) = [f(r,R;A)z|z]$$

where

$$f(r,R;t) = 2(arctg\,\frac{R}{|t|} - arctg\,\frac{r}{|t|})sgn\,t$$

It follows

$$|\lim_{\substack{r\to 0\\ R\to\infty}} (\int_r^R d\eta + \int_{-R}^{-r} d\eta)([(i\eta - A(\varepsilon))^{-1} - (i\eta - A)^{-1}]x|y)|$$
$$\leq C\varepsilon(\|x\|^2 + \|y\|^2).$$

This proves:

1. The existence of the limit (2.3) and its analyticity (as a bounded operator) in ε for $|\varepsilon| \leq \gamma$.
2. $P(\varepsilon)$ is selfadjoint in $\mathcal{K}_\varepsilon$, $[\cdot|\cdot]_\varepsilon$ for ε real and $|\varepsilon| \leq \gamma$.
3. By continuity $(J(1+\varepsilon X)P(\varepsilon)$ is positive definite in $\mathcal{H}$ for ε real and $|\varepsilon|$ small enough. For such ε the operator

   $$(J(I+X)^{-1}P_1)^{1/2}A_1(J(I+X)^{-1}P_1)^{-1/2}$$

   is selfadjoint in $\mathcal{H}$ and thus $P(\varepsilon) = 1$.
4. By the analyticity $P(\varepsilon) = 1$ extends to all ε for $|\varepsilon| \leq \gamma$. The positive definiteness of $(J(1+\varepsilon X)P(\varepsilon)$ thus extends to all such real ε. Those include $\varepsilon = 1$.

□

3. Application

Let $\mathcal{H} = L^2(\mathbf{R}^n), \mathcal{D}(A) = H^2(\mathbf{R}^n)$, and $Au = -sgn\, x_n\, \Delta u,\ u \in \mathcal{D}(A)$. Set $Ju = (sgn\, x_n)u,\ [u|v] = \int_{\mathbf{R}^n} u\bar{v} sgn\, x_n dx$. Then A is similar to a selfadjoint operator and (A1), (A2) are satisfied, see [4]. Let X be the operator of multiplication by a measurable real valued function ρ on $\mathbf{R}^n$ such that

$$\rho \in L^\infty(\mathbf{R}^n). \tag{3.1}$$

Then (X1) and (X2) are satisfied. Note that $\mathcal{D}(M) = \{u \in L^2 : \text{grad } u \in L^2\} = H^1(\mathbf{R}^n)$. If

$$\partial_j \rho \in L^n(\mathbf{R}^n), \quad j = 1, 2, 3, \tag{3.2}$$

then

$$\|\partial_j(\rho u)\|_2 \le \|\rho\|_\infty \|\partial_j u\|_2 + \|u(\partial_j \rho)\|_2 \le \|\rho\|_\infty \|\partial_j u\|_2 + \|\partial_j \rho\|_n \|u\|_{2n/(n-2)}.$$

Since $\|u\|_{2n/(n-2)} \le K \|\text{grad } u\|_{L^2}$ (see [1]), we obtain

$$\|\partial_j(\rho u)\|_2 \le (\|\rho\|_\infty + \|\partial_j \rho\|_n) \|\text{grad } u\|_2.$$

From $\frac{1}{K}\|Mu\| \le \|\text{grad } u\| \le K\|Mu\|$ and by (3.1) and (3.2), it follows that $XH^1(\mathbf{R}^n) \subset H^1(\mathbf{R}^n)$ and $\|MXu\| \le K(\|\rho\|_{L^\infty} + \|\text{grad } \rho\|_n)\|Mu\|$, hence (AX1) and (AX2) are also satisfied.

Hence for ε real, $|\varepsilon|$ sufficiently small, the operator $A(\varepsilon) = -(1+\varepsilon\rho)\,\text{sgn}\, x_n\, \Delta$ is similar to a selfadjoint operator in $L^2(\mathbf{R}^n)$; moreover, the similarity operator can be chosen to be analytic in ε.

If $n = 3$, instead of (3.2) one can assume

$$|\partial_j \rho(x)| \le \frac{K}{|x|}, \quad j = 1, 2, 3, \quad x \in \mathbf{R}^n \backslash \{0\}, \tag{3.3}$$

since the estimate (4.6) in [7, ∮ VI. 4] or ([11], the lemma after T. X.18) imply $\|u(\partial_j \rho)\|_2 \le K\|u\|_2$.

Further sufficient conditions for (AX1) and (AX2) can be deduced from [8, pp. 275-277]. In fact, if (16) holds then (AX1) and (AX2) are satisfied if for all $u \in H^1(\mathbf{R}^n)$, we have $\|(\partial_j \rho) M^{-1} u\| \le K\|u\|$ where $M = (-\Delta)^{1/2}$. It is sufficient that for all $j, k \in \{1, \dots, n\}$ we have $\|(\partial_j \rho) M^{-2} (\partial_k \rho)\| \le K$. In [8] sufficient conditions on the multiplication operators V and W are given in order that $V\Delta^{-1}W$ be a bounded operator in $L^2(\mathbf{R}^n)$.

Our result above is certainly not optimal. For instance, let ρ allow the application of Theorem 2.3 and set $\tilde{\rho} = \alpha + \rho$, where α is a large positive constant. Then our estimates above will not apply directly to $\tilde{A} = (1 + \alpha + \rho) sgn\, x_n\, L$ but they do apply to $(1 + \frac{\rho}{1+\alpha}) sgn\, x_n\, L$, which is proportional to $\tilde{A}$. We see that it is *the variation* of the function ρ rather than the magnitude itself that decides. We will omit here further details leaving them for forthcoming research.

References

[1] Bakry, D., Coulhon, T., Ledoux, M., Salof-Coste, L.: Sobolev inequalities in disguise, Indiana Univ. Math. J., 44(1995), 1033-1074.

[2] Ćurgus, B.: On the regularity of the critical point infinity of definitizable operators. Integral Equations Operator Theory 8 (1985), 462-488.

[3] Ćurgus,B., Najman, B.: A Krein space approach to elliptic eigenvalue problems with indefinite weights. Differential Integral Equations, 7(1994), 1241-1252.

[4] Ćurgus,B., Najman, B.: Positive differential operators in the Krein space $L^2(\mathbf{R}^n)$, this volume.

[5] Fleige, A., Najman, B.: Perturbations of Krein spaces preserving the nonsingularity of critical point ∞. Preprint.

[6] Jonas, P.: Compact perturbations of definitizable operators. II. J. Operator Theory 8 (1982), 3-18.

[7] Kato, T.: Perturbation Theory for Linear Operators. 2nd ed. Springer-Verlag, Berlin, 1976.

[8] Kato, T.: Wave operators and similarity for some non-selfadjoint operators. Math. Ann. 162 (1966), 258-279.

[9] Langer, H.: Spectral function of definitizable operators in Krein spaces. Functional Analysis, Proceedings, Dubrovnik 1981. Lecture Notes in Mathematics 948, Springer-Verlag, Berlin, 1982, 1-46.

[10] Pyatkov, S. G.: Some properties of eigenfunctions of linear pencils. (Russian) Sibirsk. Math. Zh. 30 (1989), 111-124, translation in Siberian Math. J. 30 (1989), 587-597.

[11] Reed M., Simon, B.: Methods of Modern Mathematical Physics II: Fourier Analysis, Self-adjointness. Academic Press, New York, 1975.

[12] Triebel, H.: Interpolation Theory, Function Spaces, Differential Operators. VEB Deutscher Verlag der Wissenschaften, Berlin, 1978.

[13] Veselić, K.: On spectral properties of a class of J-selfadjoint operators. I. Glasnik Mat. Ser. III 7(27) (1972), 229-248.

[14] Veselić, K.: On spectral properties of a class of J-selfadjoint operators. II. Glasnik Mat. Ser. III 7(27) (1972), 249-254.

Lehrgebiet Mathematische Physik
Fernuniversitaet Hagen
Postfach 940
D-58084 Hagen, Germany

1991 Mathematics Subject Classification. Primary 58F19; Secondary 47B50

Operator Theory:
Advances and Applications, Vol. 106

On the number of negative squares of certain functions

ZOLTÁN SASVÁRI

Dedicated to Heinz Langer on the occasion of his 60th birthday

The aim of the present paper is to compute the number of negative squares of certain functions, especially of some polynomials of several variables.

1. Introduction

We start with a brief survey of the theory of functions with a finite number of negative squares. These functions are the invention of M. G. Krein. Let G be an arbitrary group and k be a nonnegative integer. A complex-valued hermitian function[1] f on G is said to have k *negative squares* if the hermitian matrix

$$A = \left(f(x_i^{-1}x_j)\right)_{i,j=1}^n$$

has at most k negative eigenvalues (counted with their multiplicities) for any choice of n and $x_1, \ldots, x_n \in G$, and for some choice of n and $x_1, \ldots, x_n$ the matrix A has exactly k negative eigenvalues.

We denote by $P_k(G)$ the set of all functions on G with k negative squares. Thus, $P_0(G)$ is the set of positive definite functions on G. If G is a topological group then the symbol $P_k^c(G)$ denotes the set of continuous functions $f \in P_k(G)$. The above definition makes sense also for semigroups with an involution $*$ if we replace x_i^{-1} by x_i^*. Functions with k negative squares defined only on a subset $V \subset G$ can be introduced in the same way by restricting the definition to $x_1, \ldots, x_n \in V$ with $x_i^{-1}x_j \in V$ $(i, j = 1, \ldots, n)$.

Functions with k negative squares on a commutative group are *definitizable* in the sense that certain linear combinations of their translations are positive definite. The proof of this fact uses the theory of unitary operators in Pontryagin spaces. In [12] Krein proved the definitizability of real-valued functions in $P_1(\mathbb{Z})$ ($\mathbb{Z}$ denotes the additive group of integers) and $P_1^c(\mathbb{R})$ and gave integral representations for these functions. The definitizablitity of functions in $P_k(\mathbb{Z})$ has been proved by Iohvidov; integral representation appears in Iohvidov [9] (see also Iohvidov and Krein [10]). In [13] Krein proved that every function $f \in P_k^c(\mathbb{R})$ is definitizable in the following sense: there exists a polynomial Q of degree k such that the inequality

$$\int_{-\infty}^{\infty}\int_{-\infty}^{\infty} f(x-y)Q\left(-\mathrm{i}\frac{\mathrm{d}}{\mathrm{d}y}\right)h(y)\,\overline{Q\left(-\mathrm{i}\frac{\mathrm{d}}{\mathrm{d}x}\right)h(x)}\,\mathrm{d}y\,\mathrm{d}x \geq 0$$

[1] That is, $f(x^{-1}) = \overline{f(x)}$ for all $x \in G$.

holds for every infinitely differentiable function h with compact support. He obtained the integral representation

$$f(x) = p(x) + \int_{-\infty}^{\infty} \frac{\mathrm{e}^{\mathrm{i}tx} - S(x,t)}{|Q_0(t)|^2}\, \mathrm{d}\mu(t) \tag{1.1}$$

where p is a hermitian solution of the differential equation

$$\overline{Q}\left(-\mathrm{i}\frac{\mathrm{d}}{\mathrm{d}x}\right) Q\left(-\mathrm{i}\frac{\mathrm{d}}{\mathrm{d}x}\right) p(x) = 0 \qquad \left(\overline{Q}(t) = \overline{Q(\overline{t})}\right)$$

Q_0 is a polynomial that obtains by deleting the non-real zeros of Q, S is a regularizing correction compensating for the real zeros of Q, and μ is a nonnegative measure satisfying

$$\int_{-\infty}^{\infty} \frac{1}{(1+t^2)^m}\, \mathrm{d}\mu(t) < \infty$$

where m denotes the degree of Q_0. Gorbachuk [8] generalized (1.1) to functions of several variables. The definitizability of a function $f \in P_k(G)$ where G is an arbitrary commutative group, as well as the analogue of (1.1) have been proved in Sasvári [19] (see also Sasvári [23], Theorems 5.5.2 and 6.4.7).

Gorbachuk [7] has shown that every continuous function with k negative squares on $(-a, a)$ can be extended to a function in $P_k^c(\mathbb{R})$. The case $k = 1$ has been investigated earlier by Krein[12]. For more information on the extension problem for continuous functions on $(-a, a)$ we refer to Krein and Langer [16] and [15]. Langer [17] has extended Gorbachuk's result to measurable functions which are locally bounded on $(-a, a)$, while Sasvári[20] has shown that measurability implies local boundedness.[2]

Functions with k negative squares on a semigroup S have been studied in Berg and Sasvári [3] and in Thill [24]. Much more is known in the special case where S is the additive semigroup of nonnegative integers. Applying their earlier work connected with extensions of operators in spaces with an indefinite metric Krein and Langer [14] developed a theory for sequences with finitely many negative squares. These sequences are indefinite analogues of Hamburger moment sequences, and they are related to the theory of operators in Pontryagin spaces in the same way as Hamburger moment sequences are related to operators in Hilbert spaces. See also: Berg, Christensen and Maserick [2], Berg [1] and Langer and Sasvári [18].

The aim of the present paper is to compute the number of negative squares of certain functions, especially of some polynomials of several variables. The motivation is twofold. Although the analogue of the integral representation (1) holds for commutative groups ([23], Theorem 6.4.7), at present there is no method for determining the number of negative squares from the integral representation. This problem is open even for the groups $G = \mathbb{R}^n$ $(n \geq 2)$. Iohvidov and Krein [10]

[2]It is an open question whether or not an arbitrary function with k negative squares on $(-a, a)$ can be extended to a function in $P_k(\mathbb{R})$.

contains a detailed discussion of the case $G = \mathbb{Z}$ while (6.4.11) in [23] treats the case $G = \mathbb{R}$.

On the other hand, by recent results of T. M. Bisgaard and of the author, functions with a finite number of negative squares and definitizable functions are closely related to some embedding problems. We mention here the following result (see Bisgaard [4] for more details): If $p > 0$ is not an even integer then a normed real vector space $(E, \|\cdot\|)$ can be embedded in an L^p-space if and only if $(-1)^k\|\cdot\|^p \in P(\mathbf{1}, k)$ (see definitions below) where $k = \lceil p/2 \rceil$. Thus, it is of interest to develop methods to decide whether or not the function $(-1)^k\|\cdot\|^p$ is definitizable or has a finite number of negative squares, where $\|\cdot\|$ is a given norm (c.f. Corollary 3.21).

2. Notation and preliminaries

We will assume familiarity with basic information about π_k-spaces as found in [5], [11] or [23]. Throughout the rest of the paper the symbols G, G_1 and G_2 denote commutative groups. For the readers convenience we now list some definitions, notations and results from [23].

The *translation operator* E_x is defined by

$$E_x h(y) := h(y-x), \quad x, y \in G$$

where h is an arbitrary complex-valued function on G. Since $E_{x+y} = E_x E_y$ the complex linear span $\mathcal{A}$ of these operators is an algebra. The operators E_x $(x \in G)$ are linearly independent and therefore each $A \in \mathcal{A}$ has a unique representation

$$A = \sum_{i=1}^{n} c_i E_{x_i}$$

with some $c_i \in \mathbb{C} \setminus \{0\}$ and mutually different $x_i \in G$. We write

$$A^* := \sum_{i=1}^{n} \bar{c}_i E_{-x_i}.$$

It is easy to see that $(AB)^* = A^*B^*$, $A^{**} = A$ and $E_x^* = E_{-x}$ hold for all $A, B \in \mathcal{A}$ and $x \in G$.

We denote by $T(f)$ the complex linear space generated by translates of h. Obviously, $T(h) = \{Ah : A \in \mathcal{A}\}$, and $T(h)$ is invariant under each operator in $\mathcal{A}$. We write $\mathcal{A}_0 := \{A \in \mathcal{A} : \ A\mathbf{1} = 0\}$, where the function $\mathbf{1}$ is defined by $\mathbf{1}(x) := 1$ $(x \in G)$. Note that if $A = \sum c_j E_{x_j} \in \mathcal{A}$ then $A \in \mathcal{A}_0$ if and only if $\sum c_j = 0$. For a nonnegative integer l denote by $P(\mathbf{1}, l)$ the set of hermitian functions f on G such that

$$A_1 A_1^* \cdots A_l A_l^* f$$

is positive definite whenever $A_i \in \mathcal{A}_0$ $(i = 1, \ldots, l)$. Functions that are in $P(\mathbf{1}, l)$ for some l are special cases of *definitizable functions* (see [23], Chapter 6). If $f \in P(\mathbf{1}, 1)$ then f has at most one negative square.

Now let $f \in P_k(G)$. We construct an inner product $(g, h)_f = (g, h)$ on $T(f)$ by the formula

$$(g, h) := \sum_{i=1}^{n} \sum_{j=1}^{m} f(y_j - x_i) a_i \bar{b}_j$$

where

$$g = \sum_{i=1}^{n} a_i E_{x_i} f \quad \text{and} \quad h = \sum_{j=1}^{m} b_j E_{y_j} f.$$

$T(f)$, endowed with this inner product, is a pre-π_k-space that can be completed to a π_k-space $\Pi_k(f)$ such that $\Pi_k(f)$ consists of functions on G. Moreover, $\Pi_k(f)$ is translation invariant and

$$g(x) = (g, E_x f), \quad g \in \Pi_k(f),\ x \in G$$

(see [23], Section 5.1 for more details). Denoting by U_x the restriction of E_x to $\Pi_k(f)$, $(U_x) := \{U_x,\ x \in G\}$ is a commutative group of unitary operators in $\Pi_k(f)$. Hence there exists a k-dimensional, (U_x)-invariant, nonpositive subspace ([23], (B.7)). Note that any common eigenvector γ of the operators U_x is a multiplicative function, i.e., $\gamma(x + y) = \gamma(x)\gamma(y)$ holds for all $x, y \in G$.

The next 3 theorems are special cases of (6.6.2), (6.5.2) and (5.2.2), respectively, in [23].

Theorem 2.1. *If $f \in P_k(G) \cap P(\mathbf{1}, l)$ then $\mathbf{1} \in \Pi_k(f)$ and $(\mathbf{1}, \mathbf{1}) \leq 0$. Moreover, $\mathbf{1}$ is the only multiplicative function in $\Pi_k(f)$ which is a nonpositive vector.*

Theorem 2.2. *If $f \in P(\mathbf{1}, l)$ and f is bounded then $f = r + f_0$ where $r \in \mathbb{R}$ and f_0 is a positive definite function.*

Theorem 2.3. *If f is a bounded hermitian function on G then $f \in P_k(G)$ if and only if there exist measures μ_+, $\mu_- \in M^+(\Gamma)$ and mutually distinct characters $\gamma_1, \ldots, \gamma_k \in \Gamma$ satisfying $\mu_+(\{\gamma_i\}) = 0$ $(i = 1, \ldots, k)$, $\operatorname{supp}(\mu_-) = \{\gamma_1, \ldots, \gamma_k\}$ and such that*

$$f = (\mu_+ - \mu_-)\check{}.$$

Here Γ denotes the character group of G (considered with the discrete topology).

If f is a hermitian polynomial on $\mathbb{R}$ then $T(f)$ is finite dimensional and hence $\Pi_k(f) = T(f)$. Moreover, we have

Lemma 2.4. *([23], (5.4.13) and (5.5.3)) Let k be a nonnegative integer and $a_0, a_1, \ldots, a_{2k-1} \in \mathbb{R}$. Then the hermitian polynomial*

$$p(x) = (-1)^k x^{2k} + \mathrm{i} a_{2k-1} x^{2k-1} + a_{2k-2} x^{2k-2} + \cdots + \mathrm{i} a_1 x + a_0$$

has k negative squares, while the number of negative squares of the polynomial

$$q(x) = (-1)^{k+1}x^{2k} + \mathrm{i}a_{2k-1}x^{2k-1} + a_{2k-2}x^{2k-2} + \cdots + \mathrm{i}a_1 x + a_0$$

is equal to $k+1$. Besides, $\dim T(p) = \dim T(q) = 2k+1$ *and $p, q \in P(\mathbf{1}, k)$.*

Another example of functions in $P_k(\mathbb{R}) \cap P(\mathbf{1}, l)$ is given by the following result:

Lemma 2.5. *([23], (6.4.10)) Let k be a positive integer, $a \in (2k-2, 2k]$ and set $f_a(x) := (-1)^k |x|^a$ ($x \in \mathbb{R}$). Then $f_a \in P_k(\mathbb{R}) \cap P(\mathbf{1}, k)$.*

Next we prove three lemmas, not contained in [23], that will be needed later on. The first one follows immediately from the definitions of $P_k(G)$ and $P(\mathbf{1}, l)$, we omit the proof.

Lemma 2.6. *Let f be a complex-valued function on G_2 and let h be a group homomorphism from G_1 into G_2.*

1. *If $f \in P_k(G_2)$ then the function $f \circ h$ has at most k negative squares. If h maps G_1 onto G_2 then $f \circ h$ has exactly k negative squares.*
2. *If $f \in P(\mathbf{1}, l)$ then $f \circ h \in P(\mathbf{1}, l)$.*

Lemma 2.7. *If $f \in P_k(G) \cap P(\mathbf{1}, l)$ where $k \geq 1$ then the following conditions are equivalent:*

1. *$(\mathbf{1}, \mathbf{1}) < 0$;*
2. *f is bounded;*
3. *$k = 1$ and $f = r + f_0$ where $f_0 \in P_0(G)$ and r is a negative real number.*

Proof. That $\mathbf{1} \in \Pi_k(f)$ and $(\mathbf{1}, \mathbf{1}) \leq 0$ follows from Theorem 2.1.

1. $\Rightarrow$ 3.: If $(\mathbf{1}, \mathbf{1}) < 0$ then

$$\Pi_k(f) = \mathbb{C}\mathbf{1} \oplus (\mathbb{C}\mathbf{1})^{\perp} \tag{2.1}$$

where $(\mathbb{C}\mathbf{1})^{\perp}$ is a (U_x)-invariant π_{k-1}-space. In the case $k-1 > 0$ this subspace would contain a common nonpositive eigenvector of the operators U_x, $x \in G$. By Theorem 2.1, this eigenvector would be a constant multiple of $\mathbf{1}$; a contradiction. Thus, $k = 1$ and $(\mathbb{C}\mathbf{1})^{\perp}$ is a Hilbert space. The decomposition $f = r + f_0$ follows now from (2.1) by standard arguments (see e.g. (5.1.9) in [23]).

3. $\Rightarrow$ 2.: Follows from the fact that positive definite functions are bounded.

2. $\Rightarrow$ 1.: Since $f \in P(\mathbf{1}, l)$, Theorem 2.2 shows that $f = r + f_0$ where $r \in \mathbb{R}$ and $f_0 \in P_0(G)$. Consequently, $k = 1$. That $(\mathbf{1}, \mathbf{1}) < 0$ follows now from [22], Theorem 3. □

For $f \in P_k(G)$ let $L_0(f) := (\mathcal{A}_0 f)^{-} \subset \Pi_k(f)$. If $g = Af$ $(A \in \mathcal{A})$ then $g \in \mathcal{A}_0 f$ if and only if $(g, \mathbf{1}) = 0$. Moreover, $L_0(f)$ is a closed linear subspace of $\Pi_k(f)$.

Lemma 2.8. *Let f be as in Lemma 2.7 so that $\mathbf{1} \in \Pi_k(f)$. Then $(\mathbb{C}\mathbf{1})^{\perp} = L_0(f)$. If $(\mathbf{1},\mathbf{1}) < 0$ then*

$$\Pi_k(f) = \mathbb{C}\mathbf{1} \oplus L_0(f)$$

and $L_0(f)$ is a π_{k-1}-space. If $\mathbf{1}$ is a neutral vector then $\mathbb{C}\mathbf{1}$ is the isotropic subspace of $L_0(f)$ and the factor space $L_0(f)/\mathbb{C}\mathbf{1}$ is a π_{k-1}-space.

Proof. First we show that $(\mathbb{C}\mathbf{1})^{\perp} = L_0(f)$. It follows from the definition of $L_0(f)$ that $L_0(f) \subset (\mathbb{C}\mathbf{1})^{\perp}$. To prove the reversed inclusion let $g \in (\mathbb{C}\mathbf{1})^{\perp}$ be arbitrary and let $\{g_n\}_1^{\infty} \subset \mathcal{A}f$ be a sequence tending to g. Setting $g_n' := g_n - (g_n, \mathbf{1})f \in \mathcal{A}f$ we have $(g_n', \mathbf{1}) = (g_n, \mathbf{1}) - (g_n, \mathbf{1})(f, \mathbf{1}) = 0$ and hence $g_n' \in \mathcal{A}_0 f$. Since $\lim_n (g_n, \mathbf{1}) = (g, \mathbf{1}) = 0$ we have $\lim_n g_n' = g$, i.e., $g \in L_0(f)$.
If $(\mathbf{1},\mathbf{1}) < 0$ then

$$\Pi(f) = \mathbb{C}\mathbf{1} \oplus (\mathbb{C}\mathbf{1})^{\perp} = \mathbb{C}\mathbf{1} \oplus L_0(f)$$

and therefore $L_0(f)$ is a π_{k-1}-space.
If $(\mathbf{1},\mathbf{1}) = 0$ then $L_0(f) \cap L_0(f)^{\perp} = L_0(f) \cap \mathbb{C}\mathbf{1} = \mathbb{C}\mathbf{1}$ is the isotropic subspace of $L_0(f)$. By (A.19) in [23], the factor space $L_0(f)/\mathbb{C}\mathbf{1}$ is a π_{k-1}-space. □

3. The main results

Theorem 3.1. *Let $f_i \in P_{k_i}(G_i) \cap P(\mathbf{1}, l_i)$ where $k_i \geq 1$ $(i = 1, 2)$. Define the function f on the product group $G = G_1 \times G_2$ by*

$$f(x, y) := f_1(x) + f_2(y), \quad (x, y) \in G.$$

Then $f \in P_{k_1+k_2-1}(G)$.

Proof. Assume first that f is bounded. Then f_1 and f_2 are bounded, as well. By Theorems 2.3 and 2.2, $f_i = \check{\sigma}_i - p_i$ where $\sigma_i \in M^{+}(\hat{G}_i)$, $\sigma_i(\{\mathbf{1}\}) = 0$ and $p_i > 0$ (here $\hat{G}_i$ denotes the character group of the (discrete) group G_i). Consequently, $k_1 = k_2 = 1$ in view of Theorem 2.3. On the other hand, as $f(x, y) = \check{\sigma}_1(x) + \check{\sigma}_2(y) - (p_1 + p_2)$, the function f can be represented in the form $f = \check{\sigma} - (p_1 + p_2)$ where σ is a nonnegative measure on the character group of G satisfying $\sigma(\{\mathbf{1}\}) = 0$ (note that $\sigma = \sigma_1 \times \delta_0 + \delta_0 \times \sigma_2$ where δ_0 is the one-point measure at $\mathbf{1}$). Applying again Theorem 2.3, we see that $k = 1$ and therefore $k = k_1 + k_2 - 1$.

Assume now that f is unbounded. In the rest of the proof we will consider f_1 and f_2 as functions on G by setting $f_1(x, y) := f_1(x)$ and $f_2(x, y) := f_2(y)$, $(x, y) \in G$. Then, by Lemma 2.6, $f_i \in P_{k_i}(G) \cap P(\mathbf{1}, l_i)$, $i = 1, 2$. Consequently, $f \in P_k(G) \cap P(\mathbf{1}, l)$ where $0 \leq k \leq k_1 + k_2$ and $l = \max(l_1, l_2)$. Since f is unbounded we must

have $k \geq 1$. Lemma 2.7 then shows that $\mathbf{1} \in \Pi_k(f)$ and $(\mathbf{1},\mathbf{1})_f = 0$. Moreover, $\mathbb{C}\mathbf{1}$ is the isotropic subspace of $L_0(f)$ and $L_0(f)/\mathbb{C}\mathbf{1}$ is a π_{k-1}-space.
We write

$$\mathcal{A}_0^1 := \{A \in \mathcal{A} : \ A = \sum c_j E_{(x_j,0)}, \ A\mathbf{1} = 0\}$$

$$\mathcal{A}_0^2 := \{A \in \mathcal{A} : \ A = \sum c_j E_{(0,y_j)}, \ A\mathbf{1} = 0\}$$

and $L_0^i := (\mathcal{A}_0^i f)^{-}$, $i = 1, 2$. Note that L_0^i is a closed linear subspace. If $A_i \in \mathcal{A}_0^i$ then

$$A_i^* \in \mathcal{A}_0^i, \ \ A_1 f_2 = 0, \ A_2 f_1 = 0, \ A_i f = A_i f_i \ \text{ and } \ A_1 A_2^* f = 0.$$

Since $(A_1 f, A_2 f)_f = A_1 A_2^* f(0) = 0$ the subspaces L_0^1 and L_0^2 are orthogonal. If $A = \sum c_j E_{(x_j,y_j)} \in \mathcal{A}_0$ then $A_1 := \sum c_j E_{(x_j,0)} \in \mathcal{A}_0^1$, $A_2 := \sum c_j E_{(0,y_j)} \in \mathcal{A}_0^2$ and $Af = A_1 f + A_2 f = A_1 f_1 + A_2 f_2$. Thus, $\mathcal{A}_0 f = \mathcal{A}_0^1 f + \mathcal{A}_0^2 f$ which implies that

$$L_0(f) = L_0^1 + L_0^2. \tag{3.1}$$

If $A_i, B_i \in \mathcal{A}_0^i$ then

$$(A_i f, B_i f)_f = A_i B_i^* f(0) = A_i B_i^* f_i(0) = (A_i f_i, B_i f_i)_{f_i}.$$

This, together with $\mathcal{A}_0^i f = \mathcal{A}_0 f_i$ shows that L_0^i and $L_0(f_i)$ are identical (as inner product spaces).

If f_1 and f_2 are both unbounded then, in view of Lemma 2.7 and Lemma 2.8, $\mathbb{C}\mathbf{1}$ is the isotropic subspace of $L_0^i = L_0(f_i)$ and $L_0^i/\mathbb{C}\mathbf{1}$ is a π_{k_i-1}-space $(i = 1, 2)$. Since

$$L_0^1 \cap L_0^2 = \mathbb{C}\mathbf{1} \quad \text{and} \quad L_0^1 \perp L_0^2$$

from (3.1) we obtain that the Pontryagin spaces $L_0(f)/\mathbb{C}\mathbf{1}$ and $L_0^1/\mathbb{C}\mathbf{1} \oplus L_0^2/\mathbb{C}\mathbf{1}$ are isomorphic. This shows that $k - 1 = k_1 - 1 + k_2 - 1$.

Assume finally that f_1 is bounded and f_2 is unbounded (the case where f_2 is bounded can be treated in the same way). Applying again Lemma 2.7 and Lemma 2.8 we see that L_0^1 is a π_{k_1-1}-space and $L_0^2/\mathbb{C}\mathbf{1}$ is a π_{k_2-1}-space. Observing that $L_0^1 \cap L_0^2 = \{0\}$, equation (3.1) shows that $L_0(f)/\mathbb{C}\mathbf{1}$ and $L_0^1 \oplus L_0^2/\mathbb{C}\mathbf{1}$ are isomorhic. Consequently, $k - 1 = k_1 - 1 + k_2 - 1$ and the theorem is proved. □

The next corollary follows immediately from Theorem 3.1 and Lemma 2.5 by induction on n.

Corollary 3.2. *Let $p > 0$ and $k = \lceil p/2 \rceil$. The function*

$$f(x_1, \ldots, x_n) := (-1)^k (|x_1|^p + \cdots + |x_n|^p), \quad (x_1, \ldots, x_n) \in \mathbb{R}^n$$

has $(k-1)n + 1$ negative squares.

From Corollary 3.2 and Lemma 2.6 we obtain:

Corollary 3.3. *Let E be a real linear space, p and k be as in Corollary 1 and let $l_1, \dots, l_n \in E^*$ be linearly independent. Then the function*

$$f(x) := (-1)^k(|l_1(x)|^p + \cdots + |l_n(x)|^p) \quad (x \in E)$$

has $(k-1)n + 1$ negative squares.

Theorem 3.4. *Let $f_j \in P_{k_j}(G_j)$ be such that $d_j := \dim T(f_j) < \infty$ $(j = 1, 2)$. Then the function*

$$f(x, y) := f_1(x) f_2(y), \quad (x, y) \in G_1 \times G_2$$

has $k = k_1(d_2 - k_2) + k_2(d_1 - k_1)$ negative squares and $\dim T(f) = d_1 d_2$.

Proof. The statement of the theorem trivially holds if $d_1 = 0$ or $d_2 = 0$, so we may assume that $d_j > 0$ $(j = 1, 2)$. It is not hard to check that $T(f)$ is the linear span of functions of the form $g(x, y) = g_1(x) g_2(y)$ with $g_j \in T(f_j)$. Moreover,

$$(g, g)_f = (g_1, g_1)_{f_1} (g_2, g_2)_{f_2}.$$

Consequently, if $\{e_i^j\}_1^{d_j}$ is an orthonormal basis of $T(f_j)$ then the functions

$$g_{i,m}(x, y) := e_i^1(x) e_m^2(y) \ (i = 1, \dots, d_1; \ m = 1, \dots, d_2)$$

form an orthonormal basis of $T(f)$ containing exactly $k = k_1(d_2 - k_2) + k_2(d_1 - k_1)$ negative vectors. Thus, $T(f)$ is a π_k-space and therefore f has k negative squares (c.f. [23], Theorem (5.1.2)). Moreover, $\dim T(f) = d_1 d_2$.[3] □

The function $g(x) := \pm ix$ $(x \in \mathbb{R})$ has one negative square and $\dim T(g) = 2$. Applying this and Theorem 3.4 we obtain by induction:

Corollary 3.5. *The function*

$$f(x) := \pm i^n x_1 \cdot x_2 \cdots x_n, \quad x = (x_1, \dots, x_n) \in \mathbb{R}^n$$

has 2^{n-1} negative squares and $\dim T(f) = 2^n$.

The next Corollary obtains from Theorem 3.4 by induction, using Lemma 2.4.

Corollary 3.6. *For each $j = 1, \dots, n$ let*

$$p_j(y) = (-1)^m y^{2m} + ia_{2m-1,j} y^{2m-1} \cdots + ia_{1,j} y + a_{0,j}$$

be a hermitian polynomial where m is a nonnegative integer and $a_{0,j}, \dots, a_{2m-1,j} \in \mathbb{R}$. The function

$$f(x) := p_1(x_1) \cdot p_2(x_2) \cdots p_n(x_n), \quad x = (x_1, \dots, x_n) \in \mathbb{R}^n$$

[3] We note that Theorem 3.4 can also be proved by using tensor product of Pontryagin spaces.

has $\frac{1}{2}((2m+1)^n - 1)$ *negative squares, while the number of negative squares of* $-f$ *is equal to* $\frac{1}{2}((2m+1)^n + 1)$.

Remark 3.7. If $f : \mathbb{R}^n \longrightarrow \mathbb{C}$ is a hermitian polynomial then $T(f)$ is finite dimensional and hence f has a finite number of negative squares. Since $T(f)$ consists of polynomials, $\mathbf{1}$ is the only multiplicative function which is contained in $T(f)$. This shows that $f \in P(\mathbf{1}, l)$ with some l (c.f. Theorem 2.1).

For a quadratic matrix A we will denote by $\kappa^+(A)$ ($\kappa^-(A)$) the number a positive (negative, respectively) eigenvalues of A counted with their multiplicities.

Theorem 3.8. *Let* $A = (a_{ij})_{i,j=1}^n \neq 0$ *be a symmetric real matrix,* $b_0, b_1, \ldots, b_n \in \mathbb{R}$ *and*

$$f(x) = \sum_{i,j=1}^n a_{ij}x_i x_j + \mathrm{i}\sum_{j=1}^n b_j x_j + b_0, \quad x = (x_1, \ldots, x_n) \in \mathbb{R}^n.$$

Then f *has* $k = \kappa^+(A) + 1$ *negative squares and* $\dim T(f) = \operatorname{rank}(A) + 2$.

Proof. As is well known from linear algebra, f can be written in the form

$$f(x) = \sum_{j=1}^n (r_j l_j(x)^2 + \mathrm{i}q_j l_j(x) + q_j')$$

where $l_1, \ldots, l_n$ are linearly independent linear functionals on $\mathbb{R}^n$, $q_j, q_j' \in \mathbb{R}$ and $r_1, \ldots, r_n$ are the eigenvalues of A. By Lemma 2.6 and Lemma 2.4, the function $r_j l_j(x)^2 + \mathrm{i}q_j l_j(x) + q_j'$ has 2 negative squares if $r_j > 0$, it has 1 negative square if $r_j < 0$ and it has at most one negative square if $r_j = 0$. Applying now Theorem 3.1 and Lemma 2.6 we easily see that f has $k = \kappa^+(A) + 1$ negative squares.

Denoting by k^- the number of negative squares of $-f$ we have $\dim T(f) = k + k^-$ (see (5.4.12) in [23]). Hence we obtain $\dim T(f) = \kappa^+(A) + 1 + \kappa^+(-A) + 1 = \kappa^+(A) + \kappa^-(A) + 2 = \operatorname{rank}(A) + 2$. □

Lemma 3.9. *Let* $f : \mathbb{R}^n \longrightarrow \mathbb{C}$ *be a hermitian polynomial such that* $\bar{f} = -f$. *Then* $f \in P_k(\mathbb{R}^n)$ *where* $k = \frac{1}{2}\dim T(f)$.

Proof. Denote by k and k^- the numbers of negative squares of f and $-f$, respectively. Then $k + k^- = \dim T(f)$ in view of (5.4.12) in [23]. The functions f and $\bar{f}$ have obviously the same number of negative squares. Since $\bar{f} = -f$ we must have $k = k^-$ and the Lemma follows. □

Definition 3.10. Let $x = (x_1, \ldots, x_n) \in \mathbb{R}^n$ and let $\alpha = (\alpha_1, \ldots, \alpha_n)$ be an n-tuple of nonnegative integers. Then we set

$$x^\alpha := x_1^{\alpha_1} \cdot x_2^{\alpha_2} \cdots x_n^{\alpha_n}.$$

When $\alpha = (0, \ldots, 0)$, note that $x^\alpha = 1$. We write $|\alpha| = \alpha_1 + \cdots + \alpha_n$. An arbitrary complex-valued polynomial P on $\mathbb{R}^n$ can be written in the form

$$P(x) = \sum_\alpha c_\alpha x^\alpha, \quad c_\alpha \in \mathbb{C}$$

where the sum is over a finite number of n-tupels $\alpha = (\alpha_1, \ldots, \alpha_n)$. The *degree* of P, denoted by $\deg(P)$, is the maximum $|\alpha|$ such that the coefficient c_α is nonzero. We write

$$D_\alpha := \frac{\partial^{|\alpha|}}{\partial x_1^{\alpha_1} \cdots \partial x_n^{\alpha_n}}.$$

Note that $D_\alpha D_\beta P = D_{\alpha+\beta} P$ and, by definition, $D_\alpha P = P$ when $\alpha = (0, \ldots, 0)$.

Lemma 3.11. *If f is a hermitian polynomial on $\mathbb{R}^n$ then D_α maps $T(f)$ into $T(f)$ and $D_\alpha^* = (-1)^{|\alpha|} D_\alpha$.*[4] *Moreover, $T(f)$ is the linear span of the set*

$$\{D_\alpha f : 0 \leq |\alpha| \leq \deg(f)\}.$$

Proof. If $|\alpha| = 0$ then the first statement is trivial. Assume next that $\alpha = (1, 0, \ldots, 0)$ and let $A_\epsilon := \frac{1}{\epsilon}(E_{(-\epsilon,0,\ldots,0)} - I) \in \mathcal{A}$ $(\epsilon \in \mathbb{R} \setminus \{0\})$. If $g \in T(f)$ then $A_\epsilon g \in T(f)$ and

$$A_\epsilon g(x) = \frac{1}{\epsilon}(g(x_1 + \epsilon, x_2, \ldots, x_n) - g(x_1, x_2, \ldots, x_n)).$$

Moreover, $\lim_{\epsilon \to 0} A_\epsilon g(x) = D_\alpha g(x)$, $x \in \mathbb{R}^n$. Since $T(f)$ is finite dimensional it is closed under pointwise convergence. Thus, $D_\alpha T(f) \subset T(f)$.

We have $A_\epsilon^* = A_{-\epsilon}$ and $\lim_{\epsilon \to 0} A^* g(x) = -D_\alpha g(x)$ $(g \in T(f),\ x \in \mathbb{R}^n)$. From this we conclude that $D_\alpha^* = -D_\alpha$.

In the same way we see that $D_\alpha T(f) \subset T(f)$ and $D_\alpha^* = -D_\alpha$ when $|\alpha| = 1$. The statement about D_α, where α is arbitrary, follows from the fact that this operator is a finite product of operators D_{α_j} with $|\alpha_j| = 1$.

To prove the second statement let $g \in T(f)$ be orthogonal to all of the functions $D_\alpha f$ $(0 \leq |\alpha| \leq \deg(f))$. Then

$$0 = (g, D_\alpha f) = (-1)^{|\alpha|}(D_\alpha g, f) = (-1)^{|\alpha|} D_\alpha g(0), \quad 0 \leq |\alpha| \leq \deg(f).$$

Since g is a polynomial with $\deg(g) \leq \deg(f)$ we must have $g = 0$. □

[4] D_α^* denotes the adjoint of D_α regarded as an operator on the Pontryagin space $T(f)$.

Example 3.12. Let

$$f(x_1, x_2) = \mathrm{i}(x_1^3 + x_1^2 x_2 + x_1 x_2^2 + x_2^3).$$

Using the second statement of Lemma 3.11 it is not hard to check that $\dim T(f) = 6$. Lemma 3.9 now shows that f has 3 negative squares.

Definition 3.13. We denote by s_r^n $(1 \le r \le n)$ the elementary symmetric polynomial of degree r in n real variables:

$$\begin{aligned} s_1^n(x) &= x_1 + x_2 + \cdots + x_n \\ s_2^n(x) &= x_1x_2 + x_1x_3 + \cdots + x_{n-1}x_n \\ &\vdots \\ s_r^n(x) &= x_1 \cdots x_r + \cdots + x_{n-r+1} \cdots x_n \\ &\vdots \\ s_n^n(x) &= x_1x_2 \cdots x_n, \quad x = (x_1, \ldots, x_n). \end{aligned}$$

We write $d_r^n := \dim T(s_r^n)$.

Theorem 3.14. *We have*

1. $d_1^n = 2, \ \pm \mathrm{i} s_1^n \in P_1(\mathbb{R}^n)$;
2. $d_2^n = n + 2, \ s_2^n \in P_2(\mathbb{R}^n), \ -s_2^n \in P_n(\mathbb{R}^n)$;
3. $d_3^n = 2n + 2, \ \pm \mathrm{i} s_3^n \in P_{n+1}(\mathbb{R}^n)$;
4. $d_n^n = 2^n, \ \pm \mathrm{i}^n s_n^n \in P_{2^{n-1}}(\mathbb{R}^n)$.

Proof. 1. Trivial.

2. The eigenvalues of the $n \times n$ matrix

$$A = \begin{bmatrix} 0 & 1 & 1 & \cdots & 1 \\ 1 & 0 & 1 & \cdots & 1 \\ 1 & 1 & 0 & \cdots & 1 \\ \vdots & \vdots & \vdots & & \vdots \\ 1 & 1 & 1 & \cdots & 0 \end{bmatrix}$$

are -1 and $n-1$ with multiplicity $n-1$ and 1, respectively. Applying now Theorem 3.8 we obtain 2.

3. In view of Lemma 3.9 it suffices to show that $d_3^n = 2n + 2$. Write $f = \mathrm{i} s_3^n$ and let $a_1, \ldots, a_n \in \mathbb{C}$. Then

$$f(x_1 - a_1, \ldots, x_n - a_n) = f(x) + b_0 \mathbf{1} - \mathrm{i} \sum_{j=1}^{n} a_j f_{x_j}(x) + \sum_{j=1}^{n} b_j x_j$$

with some constants $b_0, \ldots, b_n \in \mathbb{C}$, showing that $d_3^n \le 2n+2$. Using Lemma 3.11 we see that the functions $f + i\mathbf{1}, f_{x_1}, \ldots, f_{x_n}$ span a nullspace $L \subset T(f)$. Next we show that these functions are linearly independent. Assume that

$$c_0(f + i\mathbf{1}) + c_1 f_{x_1} + \cdots + c_n f_{x_n} = 0$$

holds with some $c_0, \ldots, c_n \in \mathbb{C}$. This equation obviously implies that $c_0 = 0$. Differentiating with respect to x_1 we obtain

$$c_2 f_{x_1 x_2} + c_3 f_{x_1 x_3} + \cdots + c_n f_{x_1 x_n} =$$

$$c_2(x_3 + \cdots + x_n) + c_3(x_2 + x_4 + \cdots + x_n) + \cdots + c_n(x_2 + \cdots + x_{n-1}) = 0.$$

It is not hard to check that this implies $c_2 = \cdots = c_n = 0$ and hence also $c_1 = 0$.[5] Thus, dim $L = n+1$ and so f has at least $n+1$ negative squares. By Lemma 3.9, we must have $d_3^n \ge 2n+2$ and therefore $d_3^n = 2n+2$.

4. See Theorem 3.4. □

Remark 3.15. We were not able to compute d_r^n for arbitrary n and r, but we conjecture[6] that

$$d_r^n = \begin{cases} 2\sum_{j=0}^{\lceil r/2 \rceil} \binom{n}{j} & \text{if } r \text{ is odd} \\ 2\sum_{j=0}^{r/2} \binom{n}{j} - \binom{n}{r/2} & \text{if } r \text{ is even.} \end{cases}$$

In the next 2 Lemmas $\{l_1, \ldots, l_m\}$ will be a set of real linear functionals on $\mathbb{R}^n$.

Lemma 3.16. *Let $r \ge 2$ be an integer. If the functions $l_1^r, \ldots, l_m^r$ are linearly dependent then so are the functions $l_1^q, \ldots, l_m^q$ for any $q = 1, \ldots, r-1$.*

Proof. We may assume that $q = r-1$ and $l_j \ne 0$ $(j = 1, \ldots, m)$. Let

$$l_j(x_1, \ldots, x_n) = \sum_{i=1}^{n} a_{ji} x_i$$

and suppose that a nontrivial linear combination $l = \sum_1^m c_j l_j^r$ is equal to zero. Choosing j_0 and i_0 such that $c_{j_0} \ne 0$, $a_{j_0 i_0} \ne 0$ and diffrentiating l with respect to x_{i_0} we obtain

$$\sum_{j=1}^{m} c_j a_{j i_0} m l_j^{m-1} = 0.$$

Since $a_{j_0 i_0} \ne 0$, the functions $l_1^{r-1}, \ldots, l_m^{r-1}$ are dependent. □

[5] Actually, up to the order, we have to do with the same matrix as in the proof of 2.
[6] Independently, T. M. Bisgaard came to the same conjecture.

Lemma 3.17. *Let $h_1, \dots, h_m$ be complex-valued polynomials on $\mathbb{R}^n$ satisfying the equation*

$$\sum_{j=1}^{m} h_j(l_j(x)) = 0$$

for all $x \in \mathbb{R}^n$. If $l_1^2, \dots, l_m^2$ are linearly independent, then each polynomial h_j is linear.

Proof. Let $r = \max\{\deg h_j : j = 1, \dots, m\}$,

$$h_j(t) = \sum_{i=0}^{r} a_{ji} t^i \quad \text{and} \quad P_i = \sum_{j=1}^{m} a_{ji} l_j^i$$

$(a_{ij} \in \mathbb{C},\ t \in \mathbb{R})$. Then

$$0 = \sum_{j=1}^{m} h_j(l_j) = \sum_{i=0}^{r} P_i. \tag{3.2}$$

Since P_i is either identically zero or a homogeneous polynomial of degree i, equation (3.2) implies that $P_1 = \cdots = P_r = 0$. If $i \geq 2$ then the functions $l_1^i, \dots, l_m^i$ are independent in view of Lemma 3.16. Thus, $a_{ji} = 0$ $(i \geq 2,\ j = 1, \dots, m)$ and therefore $r < 2$. □

Example 3.18. It is not hard to check that

$$x_1, x_2, x_1 + x_2 \quad (n = 2, m = 3)$$

or

$$x_1, x_2, x_3, x_1 + x_2, x_1 + x_3, x_2 + x_3 \quad (n = 3, m = 6)$$

are examples of dependent linear functionals with independent squares.

Theorem 3.19. *Let f_j be a hermitian polynomial on $\mathbb{R}$ of degree $n_j \geq 3$ and with k_j negative squares and let $l_1, \dots, l_m$ be real linear functionals on $\mathbb{R}^n$ with* $\operatorname{rank}\{l_1, \dots, l_m\} = d$. *If $l_1^2, \dots, l_m^2$ are linearly independent then the function*

$$f(x) := \sum_{j=1}^{m} f_j(l_j(x)), \quad x \in \mathbb{R}^n$$

has $k = 1 + d - 2m + \sum_{j=1}^{m} k_j$ negative squares and $\dim T(f) = 2 + 2d - 3m + \sum_{j=1}^{m} n_j$.

Proof. By Theorem 3.1, the function

$$g(x_1, \dots, x_m) := \sum_{j=1}^{m} f_j(x_j)$$

has $l = 1 - m + \sum_1^m k_j$ negative squares. In view of Lemma 2.6, the restriction of g to the linear subspace $L := \{(l_1(x), \dots, l_m(x)) : x \in \mathbb{R}^n\} \subset \mathbb{R}^m$ has the same number k of negative squares as f. Let $g_0(y) := g(y_L)$ $(y \in \mathbb{R}^m)$ where y_L denotes the orthogonal projection of y onto L. Applying again Lemma 2.6 we see that $g_0 \in P_k(\mathbb{R}^m)$. Besides, $g = g_0$ on L. First we show some facts concerning $T(g)$.

The function $q_j(x_1, \dots, x_m) := x_j$ is in $T(g)$ and $(q_i, q_j)_g = 0$ $(i, j = 1, \dots m)$. Indeed, since f_j is a polynomial of degree $n_j \geq 3$ we have $D_{\alpha_j} g = c_j q_j + d_j \mathbf{1}$ with some $c_j, d_j \in \mathbb{C}$, $c_j \neq 0$, where the jth coordinate of α_j is equal to $n_j - 1$ and the other coordinates are equal to zero. Since $\mathbf{1} \in T(g)$, Lemma 3.11 shows that $q_j \in T(g)$. If $i \neq j$ then $D_{\alpha_i} D_{\alpha_j} g = 0$ by the definition of g. Since $n_j \geq 3$ we have $2(n_j - 1) \geq n_j + 1$ and therefore $D_{2\alpha_j} g = 0$. Applying now Lemma 3.11 we obtain

$$(c_i q_i, c_j q_j)_g = (D_{\alpha_i} g - d_i \mathbf{1}, D_{\alpha_j} g - d_j \mathbf{1})_g = 0.$$

We introduce the linear subspace

$$T_0 := \left\{ Ag : A = \sum c_j E_{y_j}, \; c_j \in \mathbb{C}, \; y_j \in L \right\} \subset T(g).$$

Then $h \in T_{00} := T_0^\perp$ if and only if $h(y) = (h, E_y g) = 0$ for all $y \in L$. Since h can be written in the form $h(x_1, \dots, x_m) = \sum_1^m h_j(x_j)$ where h_j is a polynomial with $h_j(0) = 0$, we have

$$h_1(l_1(x)) + \cdots + h_m(l_m(x)) = 0, \quad x \in \mathbb{R}^n.$$

By Lemma 3.17, each function h_j must be linear. Therefore, using the fact that $q_j \in T(g)$, we obtain

$$T_{00} = \left\{ \sum_{j=1}^m a_j q_j : \; a_j \in \mathbb{C}, \; \sum_{j=1}^m a_j l_j = 0 \right\}.$$

Since the functions $q_1, \dots, q_m$ are linearly independent we conclude that $\dim T_{00} = m - d$. Besides, T_{00} is a neutral space and hence $T_{00} \subset T_{00}^\perp = T_0$. Applying (A.19) in [23] we see that $T_0/T_{00} = T_{00}^\perp/T_{00}$ is a π_{l+d-m}-space.

Since $g = g_0$ on L, the mapping $Ag \longrightarrow Ag_0$ from T_0 onto $T(g_0)$ preserves the inner product and $Ag_0 = 0$ if and only if $Ag \in T_{00}$. This shows that T_0/T_{00} is a π_k-space. Consequently, $k = l + d - m = 1 + d - 2m + \sum_1^m k_j$.

To prove the last statement denote by k^- and k_j^- the number of negative squares of the functions $-f$ and $-f_j$, respectively. Since $\dim T(f_j) = n_j + 1 = k_j + k_j^-$, we obtain

$$\dim T(f) = k + k^- = 2 + 2d - 4m + \sum_{j=1}^m (k_j + k_j^-) = 2 + 2d - 3m + \sum_{j=1}^m n_j.$$

□

Example 3.20. Applying the theorem above we obtain that the function

$$f(x_1,x_2) := x_1^4 + x_2^4 + (x_1+x_2)^4, \quad (x_1,x_2) \in \mathbb{R}^2$$

has 3 negative squares.

Corollary 3.21. *Let $p = 2k \geq 4$ be an even integer and let L be an n-dimensional subspace of $\mathbf{l}^p_{n+1}$ given in the form*

$$L = \{(l_1(x), \ldots, l_{n+1}(x)) : x \in \mathbb{R}^n\}$$

where l_j is a linear functional on $\mathbb{R}^n$ and $\operatorname{rank}\{l_1, \ldots, l_{n+1}\} = n$. *Then the following conditions are equivalent:*

1. *L can be embedded into $\mathbf{l}^p_n$;*
2. *the function $(-1)^{k+1}\|\cdot\|_p^p$ has $nk+1$ negative squares on L;*
3. *$l_1^2, \ldots, l_{n+1}^2$ are linearly dependent;*
4. *$l_i = al_j$ for some $i \neq j$ and $a \in \mathbb{R}$.*

Proof. 1. $\Longrightarrow$ 2.: Follows from the fact that, in view of Theorem 3.1, the function $(-1)^{k+1}\|\cdot\|_p^p$ has $nk+1$ negative squares on $\mathbf{l}^p_n$.

2. $\Longrightarrow$ 3.: If $l_1^2, \ldots, l_{n+1}^2$ were independent then, by Theorem 3.19, the function $(-1)^{k+1}\|\cdot\|$ would have $(n+1)k > nk+1$ negative squares on L.

3. $\Longrightarrow$ 4.: Assume that no l_i is a scalar multiple of some l_j $(j \neq i)$. We show that $l_1^2, \ldots, l_{n+1}^2$ are linearly independent.

Without loss of generality we may assume that $l_1, \ldots, l_n$ are independent and $l_{n+1} = a_1 l_1 + \cdots + a_n l_n$ $(a_j \in \mathbb{R})$ where at least two of the a_j's are different from zero. By independce, there exists a linear isomorpism $A : \mathbb{R}^n \longrightarrow \mathbb{R}^n$ such that $l_j(Ax) = x_j$ $(x = (x_1, \ldots, x_n)$, $j = 1, \ldots, n)$. It is clear that $l_1^2, \ldots, l_{n+1}^2$ and $(l_1 \circ A)^2, \ldots, (l_{n+1} \circ A)^2$ are at the same time dependent or independent. To show the independence of the last system assume that

$$c_1 x_1^2 + \cdots + c_n x_n^2 + c_{n+1}(a_1 x_1 + \cdots + a_n x_n)^2 = 0$$

with some contants c_j. Since at least two of the a_j's are different from zero, the last term (and only the last) contains mixed products of the x_j's. Therefore we must have $c_{n+1} = 0$ and hence also $c_1 = \ldots = c_n = 0$.

4. $\Longrightarrow$ 1.: Assume for example that $l_{n+1} = al_1$. Then the mapping

$$(l_1(x), \ldots, l_n(x), al_1(x)) \longrightarrow ((1+a^p)^{1/p} l_1(x), \ldots, l_n(x))$$

is an embedding of L into $\mathbf{l}^p_n$. □

References

[1] BERG, C., On the uniqueness of minimal definitizing polynomials for a sequence with finitely many negative squares, in: Harmonic Analysis. Proceedings, Luxembourg 1987. Eds. P. Eymard and J.-P. Pier, Lecture Notes in Math. 1359, 93–99, Berlin-Heidelberg-New York, Springer-Verlag, 1988.

[2] BERG, C., CHRISTENSEN, J. P. R., MASERICK, P. H., Sequences with finitely many negative squares, J. Funct. Anal. 73, 260–287(1988).

[3] BERG, C., SASVÁRI, Z., Functions with a finite number of negative squares on semigroups, Monatshefte für Math. 107, 9–34(1989).

[4] BISGAARD, T. M., Embeddability in L^p-spaces, Acta Sci. Math. (to appear).

[5] BOGNÁR, J., Indefinite Inner Product Spaces, Berlin-New York, Springer-Verlag, 1974.

[6] GELFAND, I. M., VILENKIN, N. J., Verallgemeinerte Funktionen (Distributionen) IV, Berlin, Deutscher Verlag der Wissenschaften, 1964.

[7] GORBACHUK, V. I. (PLYUSHČEVA, V. I.), Integral representation of continuous Hermitian-indefinite kernels (Russian), Dokl. Akad. Nauk. SSSR, 145, 534–537(1962).

[8] ———: On integral representations of Hermitian-indefinite forms (the case of several variables) (Russian), Ukrain. Mat. Ž. 16, 232–236(1964).

[9] IOHVIDOV, I. S., On the theory of indefinite Toeplitz forms, (Russian) Dokl. Akad. Nauk SSSR, 101(2), 213–216(1955).

[10] IOHVIDOV, I. S., KREIN, M. G., Spectral theory of operators in spaces with an indefinite metric II (Russian), Trudy Moskov. Mat. Obšč. 8, 413–496(1959). English translation: Amer. Math. Soc. Translations 2(3), 283–373(1963).

[11] IOHVIDOV, I. S., KREIN, M. G., LANGER, H., Introduction to the Spectral Theory of Operators in Spaces with an Indefinite Metric, Berlin, Akademie-Verlag, 1982.

[12] KREIN, M. G., Screw lines in infinite-dimensional Lobachevski space and the Lorentz transformation (Russian), Usp. Mat. Nauk 3(3), 158–160(1948).

[13] ———, On the integral representation of a continuous Hermitian-indefinite function with a finite number of negative squares (Russian), Dokl. Akad. Nauk SSSR, 125(1), 31–34(1959).

[14] KREIN, M. G., LANGER, H., On some extension problems which are closely related with the theory of hermitian operators in a space Π_k. III. Indefinite analogues of the Hamburger and Stieltjes problems. Part (I), Beiträge zur Analysis 14, 25–40(1979).

[15] ———, On some continuation problems which are closely related to the theory of operators in spaces Π_k. IV: Continuous analogues of orthogonal polynomials on the unit circle with respect to an indefinite weight and related continuation problems for some classes of functions, J. Operator Theory, 13, 299–417(1985).

[16] ———, Continuation of Hermitian positive definite functions and related questions, (in preparation).

[17] LANGER, H., On measurable Hermitian indefinite functions with a finite number of negative squares, Acta Sci. Math. 45, 281–292(1983).

[18] LANGER, H., SASVÁRI, Z., Definitizing polynomials of unitary and Hermitian operators in Pontrjagin spaces, Math. Ann. 288, 231–243(1990).

[19] SASVÁRI, Z., Indefinite functions on commutative groups, Monatshefte für Math. 100, 223–238(1985).

[20] ———, Measurable functions with a finite number of negative squares, Acta Sci. Math. 50(3–4), 359–363(1986).

[21] ———, Definisierbare Funktionen auf Gruppen, Dissertationes Math. CCLXXXI, 1–83(1989).

[22] ———, Conditionally positive definite functions and unitary group representations in π_1-spaces, Math. Nachr. 146, 69–75(1990).

[23] ———: Positive Definite and Definitizable Functions, Berlin, Akademie Verlag, 1994.

[24] THILL, M., Exponentially bounded indefinite functions, Math. Ann. 285, 297–307(1989).

Department of Mathematics
Technical University of Dresden
Mommsenstrasse 13
01062 Dresden
Germany

1991 Mathematics Subject Classification. Primary 46C20; Secondary 43A35

Operator Theory:
Advances and Applications, Vol. 106

Factorization of elliptic pencils and the Mandelstam hypothesis

A.A. SHKALIKOV

Dedicated to Heinz Langer on the occasion of his 60th birthday

Equations of the form

$$\mathcal{A}(u) = -F\frac{d^2u}{dy^2} + iG\frac{du}{dy} + (H - \omega^2 R)u = 0$$

with symmetric operator coefficients on a Hilbert space (ω is the frequency parameter) arise often in problems of mathematical physics. We study strongly elliptic equations whose symbols

$$T_\omega(\lambda) = \lambda^2 F + \lambda G + H - \omega^2 R$$

satisfy the condition $T_\omega(\lambda) \geq \varepsilon(\lambda^2 + H - V)$ for all $\lambda \in \mathbb{R}$ provided ε is sufficiently small and V is a nonnegative H-compact positive operator. We prove that strongly elliptic equations may have only a finite number of outgoing waves. This enables us to pose the half range Cauchy problem with the radiation conditions at infinity and to prove its unique solvability. This is related to the Mandelstam hypothesis. The key tools (and the main results of the paper) are the factorization theorems which claim a representation of the symbol $T_\omega(\lambda)$ as a product of two linear divisors with special properties of involved operators.

Key words: Operator differential equations, factorization, elliptic pencils.

0. Introduction

Some problems of mathematical physics (one of them will be discussed below) can be written abstractly in the form

$$\mathcal{A}(u) = -F\frac{d^2u}{dy} + iG\frac{du}{dy} + (H - \omega^2 R)u = 0. \tag{0.1}$$

Here F, G, H, and R are symmetric operators on a suitable Hilbert space $\mathcal{H}$ satisfying certain additional conditions which ensure the elliptic nature of this equation, and ω is a physical parameter (frequency) which appears after the separation of the time variable.

Physical meaning have solutions of equation (0.1) which are bounded as $y \to \infty$ and satisfy the so-called radiation principle. Different approaches to formulate the radiation principle have been widely discussed in physical and mathematical literature (see, for example, SVESHNIKOV [Sv], the books of ZILBERGLEIT and KOPILEVICH [ZK], VOROVICH and BABESHKO [VB]). The formulation of the radiation principles is based on the preliminary spectral analysis of the pencil

$$T_\omega(\lambda) = \lambda^2 F + \lambda G + H - \omega^2 R. \tag{0.2}$$

We say $\{\lambda_k, v_k\}$ with $v_k \neq 0$ is an eigenpair of the pencil $T_\omega(\lambda)$ if $T_\omega(\lambda_k)v_k = 0$. Any eigenpair $\{\lambda_k, v_k\}$ generates the solution

$$u_k = e^{-i\lambda_k y} v_k \tag{0.3}$$

of equation (0.1). Those solutions which correspond to the real eigenvalues λ_k are of particular interest, they are called propagating waves. Among propagating waves there are the outgoing and incoming ones. It was understood after the author's discussions with physicists, that the Mandelstam hypothesis can be formulated as follows (see [BS], [ZK], although the problem is not clearly formulated there): *given an element $x \in \mathcal{H}$ there is a unique solution $u(y)$ of equation (0.1) such that $u(0) = x$, and as $y \to \infty$ the solution $u(y)$ asymptotically coincides with a linear combination of outgoing waves.*

This problem is also related to those settled by REYLEIGH on the wave diffraction on a periodic surface. Some of them are treated in the book of WILCOX [W]. This connection, however, is not easily seen, and its demonstration is left for a future occasion.

Our first aim is to define an abstract model of strongly elliptic equations in waveguide domains whose symbols are quadratic selfajoint pencils. The main goal is to prove factorization theorems for these pencils and investigate the properties of a right divisor. The results obtained enable us, in particular, to approve the Mandelstam hypothesis.

Our starting point was a celebrated paper of KREIN and LANGER [KL] which deals with pencils of the form

$$L(\lambda) = I + \lambda B + \lambda^2 C.$$

Here I is the identity operator, B is bounded and selfadjoint, while C is positive and compact. The fundamental theorem of [KL] yields the factorization

$$L(\lambda) = (I - \lambda Z_1)(I - \lambda Z).$$

Among possible divisors there is an operator Z whose spectrum $\sigma(Z)$ lies in the closed upper (or lower) half plane and coincides with the spectrum of $L(\lambda)$ in the open half plane. A further analysis of an operator Z occuring in this factorization was given in the papers of KOSTYUCHENKO and ORAZOV [KO1] and

KOSTYUCHENKO and SHKALIKOV [KS]. However, while attempting to apply the method of KREIN and LANGER to attack the factorization problem for quadratic pencils with *unbounded coefficients*, one faces new serious obstacles. Moreover, *a further analysis of divisors* has to be carried out after the factorization is already proved. In particular, to prove the Mandelstam hypothesis we have to show that *among the possible factorizations*

$$T_\omega(\lambda) = (\lambda - Z_1)F(\lambda - Z),$$

there is only one operator Z *which generates a* C_0 *(or holomorphic) semigroup in an appropriate Hilbert space.*

The plan of this paper is the following. In Section 1 we define strongly elliptic pencils as relatively compact pertubations of uniformly positive ones. For pencils with discrete spectrum our definition is equivalent to the asymptotic inequality

$$T_\omega(\lambda) \geq \varepsilon(\lambda^2 + H), \qquad \text{for } \lambda \in \mathbb{R},\ |\lambda| > r_0,$$

provided r_0 is sufficiently large. This assumption can be easily checked for concrete elliptic systems, since it is equivalent to the Gårding inequality (this is shown in Section 3). Following the paper [S1] we define the "classical" and the "generalized" spectra of $T_\omega(\lambda)$. We show that the classical and generalized spectra of a strongly elliptic pencil coincide in the union of a ball centered in the origin and a sufficiently small double sector containing the real axis. Moreover, in this domain the spectrum consists of finitely many normal eigenvalues. For large values of $|\lambda|$ inside a double sector we prove the resolvent estimates which play an important role in the sequel. They look similar to the classical a priori estimates for regular elliptic boundary value problems obtained by AGMON, DOUGLAS and NIRENBERG [ADN], [AN] and AGRANOVICH and VISHIK [AV]. Nevertheless, the estimates obtained in Section 1 are of a different nature, in particular, they can be used for elliptic systems on non-smooth domains. One can feel the difference while considering the example in Section 6.

In Section 2 we give more details about the real spectrum of $T_\omega(\lambda)$. In particular, we show that the outgoing waves correspond to those eigenpairs which have the positive sign characteristics

$$\varepsilon_k = (T'_\omega(\lambda_k)v_k, v_k).$$

In Section 3 we prove a factorization theorem for positive strongly elliptic pencils. We could obtain this theorem (although this is not easy) using classical results on the factorization of non-negative operator functions on the real line (see the exposition of this theory in the books of FOIAS and NAGY [FN] and of ROSENBLUM and ROVNJAK [RR]). However, we preferred to give a new approach based on semigroup theory, as it seems more natural for the problem in question. Moreover, we believe that this method can be modified to fit arbitrary strongly elliptic pencils not positive ones only.

In Section 4 we prove a factorization theorem for strongly elliptic pencils (not necessarily positive) under an additional assumption (the so-called Keldysh-Agmon condition). The proof is based on the preliminary analysis of the half-range completeness and minimality problem for the pencil $T_\omega(\lambda)$. To solve these problems we borrow the ideas from the papers [KS] and [SS]. In this exposition, however, we get rid of some superfluous assumptions and present the material in a different and shorter way. In particular, in contrast to the cited papers, we now can apply our results in the case where the operator H is generated by an elliptic operator (or system) on a nonsmooth domain.

The results of Section 3 and 4 are used in Section 5 to approve the Mandelstam hypothesis. Finally, in Section 6, we demonstrate how the obtained results can be applied to the elliptic system of differential equations of elasticity theory.

The second part of the paper (Sections 4–6) is a revised version of results on elliptic pencils presented by the author in the unpublished manuscripts [S2], [S3]. Using the opportunity the author thanks Professors V. A. Kondratiev, Yu. I. Kopilevich, A. G. Kostyuchenko, P. Lancaster and A. S. Zilbergleit for valuable discussions. I am also indebted to dr. R. O. Hryniv who took the job of looking through the manuscript and correcting mistakes.

1. Elliptic pencils and their spectrum

Definition of regular elliptic and strongly elliptic pencils. In what follows we always assume that the coefficients of equation (0.1) or a quadratic pencil $T_\omega(\lambda)$ of the form (0.2) are operators on a separable Hilbert space $\mathcal{H}$ having the following properties (we borrow the terminology from the book of Kato [Ka]):

F is a bounded and uniformly positive operator ($0 \ll F \ll \infty$)*;*

H is a selfadjoint uniformly positive operator with domain $\mathcal{D}(H) \subset \mathcal{H}$ ($H = H^* \gg 0$);

G is a symmetric operator ($G \subset G^*$) with domain $\mathcal{D}(G) \supset \mathcal{D}(H^{1/2})$;

R is an H-compact positive operator (i.e. $R > 0$, $\mathcal{D}(R) \supset \mathcal{D}(H)$ and RH^{-1} is compact on $\mathcal{H}$), and the closure of the operator $H^{-1/2}RH^{-1/2}$ has trivial kernel.

It is worth noting that for any symmetric H-bounded operator R the closure of $H^{-1/2}RH^{-1/2}$ exists and is a bounded operator on $\mathcal{H}$ (see the remark explaining the boundedness of the operator C defined in (1.7)).

The parameter ω plays a role in the sequel only in cases when we appeal to physical considerations. For fixed ω it will be convenient to denote $S = H - \omega^2 R$ and consider the pencil

$$T(\lambda) = \lambda^2 F + \lambda G + S \tag{1.1}$$

implying that

$$\left\{ \begin{array}{l} 0 \ll F \ll \infty, \quad G \subset G^*, \quad \mathcal{D}(G) \supset \mathcal{D}(H^{1/2}), \\ S = S^* \text{ is a relatively compact perturbation of } \quad H = H^* \gg 0. \end{array} \right. \tag{1.2}$$

We use the scale of Hilbert spaces $\mathcal{H}_\theta$ generated by the "main" operator H. Namely, for $\theta \geq 0$ the space $\mathcal{H}_\theta$ coincides with $\mathcal{D}(H^{\theta/2})$ endowed with the norm $\|x\|_\theta = \|H^{\theta/2}x\|$, while $\mathcal{H}_{-\theta}$ is the dual space to $\mathcal{H}_\theta$ with respect to $\mathcal{H}$. The following fact will be used in the sequel: If $S \gg 0$ then the scale of Hilbert spaces generated by S coincides with $\mathcal{H}_\theta$ for $0 \leq \theta \leq 2$. This fact follows from the assumption $\mathcal{D}(S) = \mathcal{D}(H)$ and the interpolation theorem (see, e.g., [LM], Ch 1).

Further, by writing $T(\lambda)$ instead of $T_\omega(\lambda)$ we always assume that $T(\lambda)$ is of the form (1.1) with coefficients satisfying conditions (1.2).

The definition of a regular elliptic boundary value problem (see [AN], [AV], [LM]) is expressed algebraically in terms of principle symbols of a differential equation and boundary operators (the so-called ellipticity condition for the equation and the complementing Lopatinskii condition for boundary operators). Suppose that we consider a regular elliptic problem in a wave-guide domain $\Omega \times \mathbb{R}$ (Ω is a smooth bounded domain in $\mathbb{R}^n$) and write it in abstract form (0.1) (homogeneous boundary conditions are included in the domain of the main operator H). It follows from the results of [AN] and [AV]: a problem is regular elliptic if and only if $T(\lambda)$ is invertible for $\lambda \in \mathbb{R}$ and $|\lambda| > r_0$, with r_0 large enough and for these values of λ

$$\|HT^{-1}(\lambda)\| + |\lambda|\ \|H^{1/2}T^{-1}(\lambda)\| + |\lambda|^2\|T^{-1}(\lambda)\| \leq const. \tag{1.3}$$

These arguments lead to the folowing definition (as we agreed the parameter ω is omitted).

Definition 1.1. *A pencil $T(\lambda)$ or equation $T\left(i\frac{d}{dy}\right)u(y) = 0$ is said to be regular elliptic if estimate (1.3) holds for $\lambda \in \mathbb{R}$, $|\lambda| > r_0$.*

In this paper, however, we deal mostly with equations which are abstract generalizations of strongly elliptic equations (see, e.g., the book of Fichera [Fi]).

Definition 1.2. *A pencil $T(\lambda)$ is said to be uniformly positive if there exists a number $\varepsilon > 0$ such that*

$$T(\lambda) \geq \varepsilon(\lambda^2 + H) \qquad \text{for all } \lambda \in \mathbb{R}. \tag{1.4}$$

It follows from the definition that $S \gg 0$ if $T(\lambda)$ is uniformly positive. As $\mathcal{D}(S) = \mathcal{D}(H)$, both the operators SH^{-1} and HS^{-1} are defined on the whole $\mathcal{H}$, and it follows from the definition that they are closed. Hence, by the closed graph theorem these operators are bounded and then there exist positive constants c_0, c_1 such that

$$c_0\|Hx\| \leq \|Sx\| \leq c_1\|Hx\|, \qquad x \in \mathcal{D}(H).$$

By virtue of the Heinz inequality (see [Ka], Ch.5.4) we have $c_0H \leq S \leq c_1H$. Therefore, (1.4) implies also

$$T(\lambda) \geq \varepsilon_1(\lambda^2 + S), \qquad \lambda \in \mathbb{R},$$

with $\varepsilon_1 = \varepsilon/c_1$. Actually, we have just showed that the operator H in Definition 1.2 can be replaced by any operator $S = S^* \gg 0$ such that $\mathcal{D}(S) = \mathcal{D}(H)$.

Definition 1.3. *A pencil $T(\lambda)$ of the form (1.1) is said to be strongly elliptic if there exists an H-compact positive operator V such that $T(\lambda) + V$ is uniformly positive.*

Proposition 1.4. *Let $T(\lambda)$ be a strongly elliptic pencil. Then there exist numbers $\varepsilon > 0$ and $r_0 > 0$ such that*

$$T(\lambda) \geq \varepsilon(\lambda^2 + H) \qquad \text{for all } \lambda \in \mathbb{R} \text{ and } |\lambda| > r_0. \tag{1.5}$$

Proof. By the definition we have

$$T(\lambda) \geq \varepsilon(\lambda^2 + H) - V, \tag{1.6}$$

where V is an H-compact positive operator. Obviously, if VH^{-1} is compact in $\mathcal{H}$ then V is H-bounded with zero H-bound, i.e. for any $\varepsilon > 0$ there exist $c = c(\varepsilon)$ such that

$$||Vx|| \leq \varepsilon||Hx|| + c||x||, \quad c = c(\varepsilon), \quad x \in \mathcal{D}(H).$$

By virtue of the Heinz inequality we have

$$V \leq \varepsilon H + cI,$$

where I is the identity operator. Taking in the last inequality $\varepsilon/2$ instead of ε we obtain (1.5) from (1.6). □

The inverse assertion of Proposition 1.4, generally, is not true. Examples can be easily given by considering bounded operators G and S on $\mathcal{H}$. However, we can invert the statement of Proposition 1.4 assuming that H has discrete spectrum or, equivalently, the identity operator is H-compact.

Proposition 1.5. *Let H^{-1} be compact in $\mathcal{H}$. Then condition* 1.5 *implies that $T(\lambda)$ is strongly elliptic.*

Proof. According to (1.2) G is $H^{1/2}$-bounded operator and $S - H$ is H-compact. Hence, G is H-compact and for any $\lambda \in \mathbb{R}$, we have

$$T(\lambda) = \lambda^2 F + H + K(\lambda),$$

where $K(\lambda)H^{-1}$ is compact. Given $\varepsilon > 0$ there exists $c = c(\lambda) > 0$ such that

$$|(K(\lambda)x, x)| \leq \varepsilon(Hx, x) + c(x, x).$$

If $\varepsilon = 1/2$ and c_0 is the maximum of $c(\lambda)$ on the interval $(-r_0, r_0)$ then

$$T(\lambda) + c_0 I \geq \frac{1}{2}(\lambda^2 F + H), \qquad \text{for } \lambda \in (-r_0, r_0),$$

and together with (1.5) this implies that $T(\lambda) + c_0 I$ is uniformly positive. □

Location of the spectrum and the resolvent estimates. In [S1] three different approaches are proposed to define the spectrum of a pencil with unbounded coefficients. In particular, the "classical" and the "generalized" spectrum of $T(\lambda)$ are defined as follows. We say μ *belongs to the classical spectrum of the pencil* $T(\lambda)$ *if* $T(\mu)$ *is not boundedly invertible in* $\mathcal{H}$. This concept is natural but not always convenient (see [S1]). To define the generalized spectrum, let us consider F, G and S as the operators acting on the space $\mathcal{H}_{-1}$ with domain $\mathcal{D} = \mathcal{H}_1$ (recall that $\mathcal{H}_\theta = \mathcal{D}(H^{\theta/2})$ is the scale of Hilbert spaces generated by the operator H). Since all these operators are H-bounded and symmetric, they are well defined in $\mathcal{H}_{-1}$ with domain $\mathcal{H}_1$ (see details in [S1]). Now we can consider $T(\lambda)$ as an operator function in the space $\mathcal{H}_{-1}$ defined on the domain $\mathcal{D}(T) = \mathcal{H}_1$. We say that μ *belongs to the generalized spectrum of the pencil* $T(\lambda)$ *if* $T(\mu)$ *is not boundedly invertible in* $\mathcal{H}_{-1}$. The complement of the generalized spectrum is said to be the generalized resolvent set of $T(\lambda)$. It can be easily checked (see [S1]) that μ belongs to the generalized spectrum of $T(\lambda)$ if and only if μ belongs to the spectrum of the pencil

$$L(\lambda) = \lambda^2 A + \lambda B + C,$$

with bounded in $\mathcal{H}$ coefficients

$$(1.7) \qquad A = H^{-1/2} F H^{-1/2}, \quad B = H^{-1/2} G H^{-1/2}, \quad C = H^{-1/2} S H^{-1/2}.$$

We have to explain why C is bounded. The operator SH^{-1} is defined on the whole $\mathcal{H}$ and is closed. Then SH^{-1} and its adjoint $H^{-1}S$ are bounded, and according to the interpolation theorem the operator C is bounded, too.

Generally, we can not claim that the generalized and the classical spectra of $T(\lambda)$ coincide. In the subsequent theorems we clarify the relationship between these concepts.

Theorem 1.6. *Let* $\rho_{cl}(T)$ *and* $\rho_{gen}(T)$ *be the classical and the generalized resolvent sets of a strongly elliptic pencil* $T(\lambda)$. *Then*

$$\rho_{cl}(T) \subset \rho_{gen}(T).$$

The real line belongs to $\rho_{cl}(T) \cap \rho_{gen}(T)$ *with the possible exception of finitely many normal eigenvalues whose algebraic multiplicity coincide in both sences. If* H^{-1} *is compact then the classical and the generalized spectra coincide in the whole* $\mathbb{C}$ *and consist of normal eigenvalues.*

Proof. The last assertion of the theorem and the coincidence of the algebraic multiplicities of the normal eigenvalues in both sences are proved in [S1], §3.

Let $\lambda \in \rho_{cl}(T)$. Then

$$T(\lambda) : \mathcal{H}_2 \to \mathcal{H} \quad \text{and} \quad T(\overline{\lambda}) : \mathcal{H}_2 \to \mathcal{H}$$

are isomorphisms, hence, so are the operators

$$T^*(\lambda) = T(\overline{\lambda}) : \mathcal{H} \to \mathcal{H}_{-2} \quad \text{and} \quad T^*(\overline{\lambda}) = T(\lambda) : \mathcal{H} \to \mathcal{H}_{-2}.$$

From the interpolation theorem (see [LM], Ch1) we obtain that $T(\lambda) : \mathcal{H}_1 \to \mathcal{H}_{-1}$ is an isomorphism, i.e. $\lambda \in \rho_{gen}(T)$.

Let us prove the second statement. The assumption $\mathcal{D}(G) \supset \mathcal{D}(H^{1/2})$ implies that $GH^{-1/2}$ is defined on the whole $\mathcal{H}$ and it follows from the definition that it is closed. Hence, $GH^{-1/2}$ is bounded and its norm $\leq c$. Then for any $\varepsilon > 0$ we have

$$\|Gx\|^2 \leq c^2 \|H^{1/2}x\| \leq c^2(Hx, x) \leq 2c^2(\varepsilon\|Hx\|^2 + \varepsilon^{-1}\|x\|^2), \quad x \in \mathcal{D}(\mathcal{H}).$$

This means that G is H-bounded with zero H-bound, and so is the operator

$$K(\lambda) = \lambda^2 F + \lambda G + S - H + V$$

for any H-compact operator V and $\lambda \in \mathbb{C}$. We can choose a positive operator V such that

$$T(\lambda) + V \geq \varepsilon^2(\lambda^2 + H), \quad \lambda \in \mathbb{R}.$$

It follows from the stability Theorem V.4.11 of [Ka] that $T(\lambda) + V = H + K(\lambda) \gg 0$ is selfadjoint for any fixed $\lambda \in \mathbb{R}$. Therefore, $T(\lambda) + V$ is boundedly invertible in $\mathcal{H}$ for all $\lambda \in \mathbb{R}$ (and, hence, in a neighbourhood of any point $\lambda \in \mathbb{R}$). We have the representation

$$T(\lambda) = [I - V(T(\lambda) + V)^{-1}](T(\lambda) + V),$$

where $V(T(\lambda) + V)^{-1}$ is a holomorphic operator function in a neighbourhood of $\mathbb{R}$ whose values are compact operators. It follows from the theorem on holomorphic operator function (see [GGK], Ch. XI) that the spectrum of $T(\lambda)$ in a neighbourhood of $\mathbb{R}$ consists of finitely many isolated eigenvalues of finite algebraic multiplicity. According to Proposition 1.4 all the real eigenvalues are located in a finite interval $[-r_0, r_0]$. This ends the proof. □

For $\delta > 0$ and $0 < \varphi \leq \pi/2$ we denote

$$B_\delta = \{\lambda : |\lambda| \leq \delta\}, \quad \Lambda_\varphi^- = \{\lambda : |arg\lambda| < \varphi\}, \quad \Lambda_\varphi^- = \{\lambda : |\pi - arg\lambda| < \varphi\}$$

and $\Lambda_\varphi = \Lambda_\varphi^+ \cup \Lambda_\varphi^-$.

Theorem 1.7. *Let $T(\lambda)$ be strongly elliptic. Then there exist positive numbers φ and δ such that the union $\Lambda_\varphi \cup B_\delta$ with the possible exeption of finitely many normal eigenvalues belongs to the classical resolvent set of $T(\lambda)$ (and hence to $\rho_{gen}(T)$). Moreover, the estimate*

$$|\lambda|^2 \, \|T^{-1}(\lambda)\| + |\lambda| \, \|H^{1/2}T^{-1}(\lambda)\| + \|H^{1/2}T^{-1}(\lambda)H^{1/2}\| \leq const \tag{1.8}$$

holds for all $\lambda \in \Lambda_\varphi, |\lambda| > r_0$ if r_0 is large enough.

Proof. Let us prove (1.8) for $\lambda \in \Lambda_\varphi^+$, the same arguments can be applied for $\lambda \in \Lambda_\varphi^-$. If $\lambda = re^{i\theta}$, then

$$T(\lambda) = T(r) + r^2(e^{2i\theta} - 1)F + r(e^{i\theta} - 1)G. \tag{1.9}$$

This equality and Proposition 1.4 yield the estimate

$$Re\,(T(\lambda)x, x) \geq \varepsilon(r^2(x,x) + (Hx,x)), \quad x \in \mathcal{H}_2,\ \lambda \in \Lambda_\varphi,\ |\lambda| > r_0, \tag{1.10}$$

for sufficiently small φ and large r_0. We noticed already that the coefficients of the pencil $L(\lambda) = H^{-1/2}T(\lambda)H^{-1/2}$ are bounded operators. From (1.9) we have

$$\|L(\lambda)\|\,\|y\| \geq Re(L(\lambda)y, y) \geq \varepsilon r^2(y,y), \quad y \in \mathcal{H}_1,\ \lambda \in \Lambda_\varphi,\ |\lambda| > r_0.$$

By continuity this inequality holds for all $y \in \mathcal{H}$ and implies that zero does not belong to the numerical range of $L(\lambda)$. Then $L(\lambda)$ is invertible, and

$$||L^{-1}(\lambda)|| \leq \varepsilon^{-1}. \tag{1.11}$$

From this we have that

$$T(\lambda) = H^{1/2}L^{-1}(\lambda)H^{1/2} : \mathcal{H}_1 \to \mathcal{H}_{-1}$$

is an isomorphism, and $T^{-1}(\lambda)$ exists in $\mathcal{H}_{-1}$. Now, $L^{-1}(\lambda) = H^{1/2}T^{-1}(\lambda)H^{1/2}$ and from (1.11) we obtain the estimate of the third term in (1.7).

It follows from Proposition 1.5 that

$$\|T^{1/2}(r)x\| \geq \varepsilon\|(r^2 + H)^{1/2}x\|, \quad r > r_0, \quad x \in \mathcal{D}(H).$$

By virtue of Theorem 1.6 $T(r)$ is invertible for $r > r_0$, hence,

$$\begin{aligned} &\|T^{-1/2}(r)\| \leq \varepsilon^{-1}\|(r^2 + H)^{-1/2}\| \leq \varepsilon^{-1}r, \\ &\|GT^{-1/2}(r)\| \leq c\|H^{1/2}T^{-1/2}(r)\| \leq c\varepsilon^{-1}. \end{aligned} \tag{1.12}$$

We have

$$T(\lambda) = T^{1/2}(r)(I + G(\lambda))T^{1/2}(r), \quad G(\lambda) = T^{-1/2}(r)(T(\lambda) - T(r))T^{-1/2}(r).$$

It follows from representation (1.9) and estimates (1.12) that $\|G(\lambda)\| \leq 1/2$ if $\lambda \in \Lambda_\varphi^+$ and φ is sufficently small. Hence, $T(\lambda)$ is invertible in $\mathcal{H}$ for $\lambda \in \Lambda_\varphi^+$, and

$$\|T^{-1}(\lambda)\| \leq 2\|T^{-1/2}(r)\|^2 \leq 2\varepsilon^{-2}r^2,$$

$$\|H^{1/2}T^{-1}(\lambda)\| \leq 2\|H^{1/2}T^{-1/2}(r)\|\,\|T^{-1/2}(r)\| \leq 2\varepsilon^{-2}r^{-1}.$$

This completes the proof. □

Remark 1.8. We say $T(\lambda)$ is *positive* if $T(\lambda) > 0$ for all $\lambda \in \mathbb{R}$. We claim: *A positive strongly elliptic pencil is uniformly positive.* Indeed, if $T(\lambda) > 0$ for all $\lambda \in \mathbb{R}$ then λ is not an eigenvalue of $T(\lambda)$, and according to Theorem 1.6 $T(\lambda)$ is boundedly invertible in $\mathcal{H}$ as well as in $\mathcal{H}_{-1}$. Therefore $T(\lambda) : \mathcal{H}_1 \to \mathcal{H}_{-1}$ is a continuous bijection for $\lambda \in [-r_0, r_0]$, hence, so is $T^{-1}(\lambda) : \mathcal{H}_{-1} \to \mathcal{H}_1$. This yields the estimate $\|H^{1/2}T^{-1}(\lambda)H^{1/2}\| \leq const$, which implies $T(\lambda) \geq \varepsilon H$. Bearing in mind Proposition 1.4, we find that $T(\lambda)$ is uniformly positive.

2. The real spectrum of a strongly elliptic pencil

We noticed in the Introduction, that the real eigenvalues of a pencil $T(\lambda)$ play a significant role in physical considerations, as they correspond to waves propagating the energy at the infinity (or from the infinity). We already proved that strongly elliptic pencils may have only finitely many real eigenvalues. In this section we obtain additional valuable information.

First, recall that a point $\mu \in \mathbb{C}$ is said to be a normal eigenvalue of $T(\lambda)$ if it is an isolated point of the spectrum of $T(\lambda)$ and the principal part of the Laurent expansion of the resolvent $T^{-1}(\lambda)$ in a neighbourhood of μ admits a representation of the form

$$\sum_{k=1}^{N}\sum_{s=0}^{p_k} \frac{(\cdot, z_k^{p_k-s})\, x_k^s}{(\lambda-\mu)^{p_k+1-s}}. \tag{2.1}$$

Here

$$x_k^0, \dots, x_k^{p_k}, \qquad k = 1, \dots, N, \tag{2.2}$$

is a canonical system of eigen and associated elements of $T(\lambda)$ and

$$z_k^0, \dots, z_k^{p_k}, \qquad k = 1, \dots, N, \tag{2.3}$$

is the adjoint canonical system which is uniquely defined by the choice of system (2.2).

Let $\mu \in \mathbb{R}$. Since the classical and the generalized spectra of $T(\lambda)$ coincide in a neighbourhood of $\mathbb{R}$, the elements of systems (2.2) and (2.3) belong to $\mathcal{H}_2$. It follows from [KS], Lemma 2.1 that there exists a canonical system (2.2) such that

$$x_k^s = \varepsilon_k z_k^s, \quad k = 1, \dots, N, \quad s = 0, \dots, p_k,$$

where $\varepsilon_k = \pm 1$. Such a canonical system is called *normal* and the numbers ε_k are called *the sign characteristics* of the corresponding Jordan chains.

A real eigenvalue μ is said to be of *positive (negative) type* if

$$(T'(\mu)y, y) > 0 \ (< 0) \quad \text{for all } y \in Ker\, T(\mu).$$

Proposition 2.1. *If μ is a semisimple real eigenvalue of $T(\lambda)$ and $V = \sum \varepsilon(\cdot, y_k^0)\, y_k^0$ is the residue operator of $T^{-1}(\lambda)$ at the pole μ then*

$$(L'(\mu)Vy, Vy) = (y, Vy) \qquad \text{for all } y \in Ker\, T(\mu). \tag{2.4}$$

In particular, μ is of positive (negative) type if and only if all the sign characteristics are positive (negative).

Proof. For $y \in Ker\, T(\mu)$ we have $Vy \in KerT(\mu)$ and

$$y = T(\lambda)T^{-1}(\lambda)y = (T(\mu) + (\lambda - \mu)T'(\mu) + \ldots)(V(\lambda - \mu)^{-1} + R(\mu) + \ldots)y =$$
$$= T'(\mu)Vy + T(\mu)R(\mu)y + o(1),$$

where $o(1) \to 0$ as $\lambda \to \mu$ and $R(\mu)$ is a bounded operator on $\mathcal{H}$. Taking the scalar product with Vy and letting $\lambda \to \mu$, we obtain (2.4). □

Theorem 2.2. *Let a pencil $T_\omega(\lambda)$ of the form (0.2) be strongly elliptic. Then for all $\omega > 0$ with possible exception of some values $\omega_k \to \infty$ (the so-called resonant frequences) there is an even number, say 2κ, of real eigenvalues of $T_\omega(\lambda)$ counting geometric multiplicities. They all are of definite type and exactly κ of them are of positive (negative) type.*

Proof. Consider the pencil

$$L_\omega(\lambda) = H^{-1/2}T_\omega(\lambda)H^{-1/2} = L(\lambda) - \omega^2 R_0, \quad R_0 = H^{-1/2}RH^{-1/2}.$$

The assumptions on the operators (see Section 1) ensure us that the coefficients of $L_\omega(\lambda)$ are bounded operators on $\mathcal{H}$, moreover, $R_0 > 0$. By virtue of Theorem 1.6 there are finitely many normal eigenvalues of $L_\omega(\lambda)$ on the real axis. To prove that they are of definite type with possible exception of isolated values $\omega_k \to \infty$ we apply the known results of pertubation operator theory which are based on theorems due to RELLICH and NAGY (see Ch 9 of [RN]), KREIN and LYUBARSKII [KL], KOSTYUCHENKO and ORAZOV [KO2]. A concentrated exposition of this material can be found in the paper of SHKALIKOV and HRINIV [SH], Propositions 1.6-1.9. The only reservation: the condition $R_0 \gg 0$ assumed in [SH] can be replaced by $R_0 > 0$ provided the coeficients of the pencil $L_\omega(\lambda)$ are bounded. The main idea of proving this result is the following. Let μ be a real eigenvalue of the pencil $L_\omega(\lambda)$ with fixed $\omega = \omega_0$, and let $\theta_0 = \omega_0^2$. We notice that $L_\omega(\lambda)$ is a linear selfadjoint pencil with respect to the parameter $\theta = \omega^2$ and its eigenvalues $\theta_j(\lambda)$ according to the Rellich-Nagy theorem depend analytically on λ in a neighbourhood of an eigenvalue $\lambda = \mu$, namely,

$$\theta_j(\lambda) = \theta_0 + a_j(\lambda - \mu)^{p_j} + \ldots$$

with some $0 \neq a_j \in \mathbb{R}$ and integer $p_j > 0$. Then $\lambda_j(\theta)$ represent the branches of the inverse algebraic functions

$$\lambda_{j,k}(\theta) = \mu + (a_j^{-1}(\theta - \theta_0))^{1/p_j} + \ldots, \quad k = 0, \ldots, p_j - 1,$$

and p_j coincide with the lengths of the corresponding Jordan chains. Hence, $\lambda_j(\theta)$ move locally either in the complex plane or leave on the real axis depending monotonically on θ, moreover, the condition $R_0 > 0$ implies that the real branches $\lambda_j(\theta)$ are strictly monotone functions. Thus, all the real eigenvalues in a small punctured neighbourhood of μ are semisimple. Further, it turns out (see Proposition 2.3 below) that the sign characteristics of the real eigenvalues $\lambda_j(\theta)$ coincide with $sign\, \lambda_j'(\theta)$. Taking into account Proposition 2.1, we obtain that all the real eigenvalues of $T_\omega(\lambda)$ in a small right (left) neighbourhood of μ are of positive (negative) type. Hence, the resonant frequences are isolated points.

Let us prove the other statements. Fix a non-resonant frequency ω, and fix a positive H-compact operator V such that $T_\omega(\lambda)+V$ is uniformly positive. Consider the pencil

$$T_\omega(\lambda) + \rho(V + I), \qquad 0 \le \rho \le 1. \tag{2.5}$$

Obviously, the closure of $H^{-1/2}(V + I)H^{-1/2}$ is positive in $\mathcal{H}$. Now apply Proposition 1.9 from [SH] which says:

$$E^+(\rho) + E^-(\rho) = const,$$

where $E^+(\rho)$ and $E^-(\rho)$ are the number of real eigenvalues of positive and negative type, respectively. This equality holds also for the resonant values of θ if the numbers $E^\pm(\rho)$ are defined as in [SH]. Since for $\rho = 1$ the pencil (2.5) is uniformly positive, we have

$$E^+(1) - E^-(1) = 0, \quad \text{hence} \quad E^+(0) = E^-(0).$$

This ends the proof. □

Let $\{\mu, f_0\}$ be a normal eigenpair corresponding to a simple or semisimple eigenvalue of $T_\omega(\lambda)$ with a fixed $\omega = \omega_0 > 0$. As we mentioned above the eigenvalue $\theta_0 = \omega_0^2$ admits an analytic continuation $\theta_j(\lambda) = \omega^2$ when λ runs in a neighbourhood of μ. The value $\theta'(\lambda)\big|_{\lambda=\mu}$ is called the group velocity (see, for example, [ZK] or [VB]) of the wave solution

$$u(y) = e^{-i\mu y} f_0.$$

Proposition 2.3. *If μ is a definite type eigenvalue of a pencil $T_\omega(\lambda)$ and $\{\mu, f_0\}$ is a corresponding normal eigenpair then*

$$(T_\lambda'(\mu)f_0, f_0) = \theta_\lambda'(\mu)(Rf_0, f_0), \quad \theta = \omega^2, \tag{2.6}$$

i.e. the sign characteristic of an eigenpair coincides with the sign of its group velocity.

Proof. Let $f(\lambda) = f_0 + (\lambda - \mu)f_1 + \dots$ be the eigenelement of $T_\omega(\lambda)$ corresponding to $\lambda = \lambda(\theta)$. Denoting $\theta_0 = \omega_0^2$ we obtain

$$[T(\mu) - \theta_0 + T'(\mu)(\lambda - \mu) - (\theta - \theta_0)R + \dots][f_0 + (\lambda - \mu)f_1 + \dots] = 0,$$

therefore

$$(T'(\mu)f_0, f_0) + o(1) = \frac{\theta - \theta_0}{\lambda - \mu}(Rf_0, f_0).$$

Letting $\lambda \to \mu$, we get (2.6). □

Corollary 2.4. *For any non-resonant frequency ω equation (0.1) with strongly elliptic symbol (0.2) possesses finitely many, say $2\kappa \geq 0$, propagating waves and exactly κ of them are outgoing (incoming), i.e. have positive (negative) group velocity or the sign characteristics.*

3. Factorization of positive strongly elliptic pencils

In this section we use abstract Sobolev spaces. Namely, by $W_m(\mathbb{R}^+, \mathcal{H})$ we denote the space consisting of $\mathcal{H}_m$-valued functions $u(y)$ defined on $\mathbb{R}^+$, such that $u^{(j)}(y)$ exist in the generalized sense for $j \leq m$ as $\mathcal{H}$-valued functions and the integral

$$\int_0^\infty \left(\|u^{(m)}(y)\|_0^2 + \|u(y)\|_m^2 \right) dy =: \|u\|_{W_m}^2$$

converges. The detailed information on abstract Sobolev spaces can be found in the book of Lions and Magenes [LM]. We recall here some facts which we need below. According to *the theorem on intermediate* derivatives we have

$$u^{(j)}(y) \in L_2(\mathbb{R}^+, \mathcal{H}_{m-j}), \quad j = 1, \ldots, m, \quad \text{if } u(y) \in W_m(\mathbb{R}^+, \mathcal{H}).$$

An important role in the sequel plays the trace theorem which we formulate (as it needed) in the case $m = 1$.

Trace Theorem. *A function $u(y) \in W_1(\mathbb{R}^+, \mathcal{H})$ is continuous and uniformly bounded on $\mathbb{R}^+$ an $\mathcal{H}_{1/2}$-valued function and the trace operator*

$$\mathcal{T}_r : W_1(\mathbb{R}^+, \mathcal{H}) \to \mathcal{H}_{1/2}, \qquad \mathcal{T}_r u = u(r), \tag{3.1}$$

is bounded for any fixed $r \in \mathbb{R}^+$, moreover, $\|\mathcal{T}_r\| \leq c$ with a constant c not depending on $r \in \mathbb{R}^+$.

If $u(y) \in W_2(\mathbb{R}^+, \mathcal{H})$ then $u'(y) \in W_2(\mathbb{R}^+, \mathcal{H}_1)$. As $\mathcal{D}(G) \supset \mathcal{H}_1$, we have $Gu'(y) \in L_2(\mathbb{R}^+, \mathcal{H})$. Therefore

$$\mathcal{A}u = T\left(i\frac{d}{dy}\right)u = -Fu''(y) + iGu'(y) + Su(y)$$

is well defined in the space $L_2(\mathbb{R}^+, \mathcal{H})$ with domain $\mathcal{D}(\mathcal{A}) = W_2(\mathbb{R}^+, \mathcal{H})$. Let $\mathcal{A}_0$ be the restriction of $\mathcal{A}$ on the domain

$$\mathcal{D}(\mathcal{A}_0) = \overset{\circ}{\mathrm{W}}_1(\mathbb{R}^+, \mathcal{H}) := \{y \mid y \in W_1(\mathbb{R}^+, \mathcal{H}),\ y(0) = 0\}.$$

Lemma 3.1. *Let $T(\lambda)$ be strongly elliptic. Then there exist a number $\varepsilon > 0$ and an H-compact selfadjoint operator $V \geq 0$, such that*

$$\varepsilon \|u\|_{W_1} - \|V^{1/2}u\|_{L_2} \leq (\mathcal{A}_0 u,\, u) \leq \varepsilon^{-1}\|u\|_{W_1}, \quad u \in \mathcal{D}(\mathcal{A}_0). \tag{3.2}$$

If in addition $T(\lambda)$ is positive then the left hand side estimate holds with $V = 0$.

Proof. Denote

$$\hat{u}(\lambda) = \int_0^\infty u(y)e^{i\lambda y}\, dy, \qquad \lambda \in \mathbb{R}.$$

It follows from the Plancherel theorem that

$$\Big(\hat{f}(\lambda), \hat{g}(\lambda)\Big) = \Big(f(y), g(y)\Big), \qquad \text{for } f, g \in L_2(\mathbb{R}^+, \mathcal{H}),$$

where the scalar product $\Big(,\Big)$ is taken in $L_2(\mathbb{R}^+, \mathcal{H})$. As $T(\lambda)$ is strongly elliptic, there is an H-compact operator V such that

$$T(\lambda) + V \geq \varepsilon(\lambda^2 + H), \quad \lambda \in \mathbb{R}.$$

We can suppose that $V = V^*$, otherwise the Fridrichs extension of V should be considered. Bearing in mind that for all functions $u(y) \in \overset{\circ}{\mathrm{W}}_1(\mathbb{R}^+, \mathcal{H}) = \mathcal{D}(\mathcal{A}_0)$

$$-i\lambda\hat{u}(\lambda) = \int_0^\infty u'(y)\, e^{i\lambda y}\, dy,$$

we find that

$$\begin{aligned}
\Big((\mathcal{A}_0 + V)u,\, u\Big) &= \Big(Fu',\, u'\Big) - i\Big(Gu,\, u'\Big) + \Big((S+V)u,\, u\Big) \\
&= \Big((T(\lambda)+V)\hat{u}(\lambda),\, \hat{u}(\lambda)\Big) \geq \varepsilon\Big((\lambda^2 + H)\hat{u}(\lambda),\, \hat{u}(\lambda)\Big) \\
&= \varepsilon\Big[\Big(u',\, u'\Big) + \Big(H^{1/2}u,\, H^{1/2}u\Big)\Big] = \varepsilon\|u\|^2_{W_1}, \quad u \in \mathcal{D}(\mathcal{A}_0).
\end{aligned}$$

This implies the left hand side estimate of (3.2). The right one is trivial and follows from the inequality

$$|\Big(Gu,\, u'\Big)| \leq c\Big(H^{1/2}u,\, H^{1/2}u\Big)\Big(u',\, u'\Big) \leq c\|u\|^2_{W_1}.$$

To get the last statement of Lemma, recall Remark 1.8. □

Let $\overset{\circ}{\mathrm{W}}_{-1}(\mathbb{R}^+, \mathcal{H})$ be the dual space to $\overset{\circ}{\mathrm{W}}_1(\mathbb{R}^+, \mathcal{H})$ with respect to $L_2(\mathbb{R}^+, \mathcal{H})$. For any $v \in \overset{\circ}{\mathrm{W}}_1(\mathbb{R}^+, \mathcal{H})$ and $u \in W_2(\mathbb{R}^+, \mathcal{H})$

$$(\mathcal{A}u,\, v) = (Fu',\, v') - i(Gu, v') + (KH^{1/2}u,\, H^{1/2}v), \tag{3.3}$$

where $K = H^{-1/2}SH^{-1/2} = K^*$ is a bounded operator on $\mathcal{H}$. For any fixed $u \in W_1(\mathbb{R}^+, \mathcal{H})$ the right hand side of (3.3) represents a continuous linear functional on $\overset{\circ}{W}_1(\mathbb{R}^+, \mathcal{H})$. According to the definition of a dual space, any such a functional admits a representation (f, v), with $f \in \overset{\circ}{W}_{-1}(\mathbb{R}^+, \mathcal{H})$. Hence $\mathcal{A}$ admits the extension

$$\mathcal{A} : W_1(\mathbb{R}^+, \mathcal{H}) \to \overset{\circ}{W}_{-1}(\mathbb{R}^+, \mathcal{H}). \tag{3.4}$$

A function $u(y) \in W_1(\mathbb{R}^+, \mathcal{H})$ is called a generalized solution of the equation

$$T\left(i\frac{d}{dy}\right) u(y) = 0 \tag{3.5}$$

if $u(y)$ belongs to the kernel of operator (3.4).

Lemma 3.2. *Let $T(\lambda)$ be a positive strongly elliptic pencil. Then for any $x \in \mathcal{H}_{1/2}$ there is a unique generalized solution $u(y)$ of equation (3.5) such that $u(0) = x$.*

Proof. This statement is familiar from PDO theory; its abstract version is proved in the same way, one should use only the Friedrichs theorem instead of the Lax-Milgram lemma. Namely, taking into account Lemma 3.1 and the Friedrichs theorem (see [RN], Ch8), we obtain that $\mathcal{A}_0$ admits the only selfadjoint extention $\mathcal{A}_F \gg 0$ such that $\mathcal{D}(\mathcal{A}_F^{1/2}) = \overset{\circ}{W}_1(\mathbb{R}^+, \mathcal{H})$. Hence,

$$\mathcal{A}_F : \overset{\circ}{W}_1(\mathbb{R}^+, \mathcal{H}) \to \overset{\circ}{W}_{-1}(\mathbb{R}^+, \mathcal{H}) \tag{3.6}$$

is an isomorphism. Since the trace operator $\mathcal{T}_0$ defined in (3.1) is surjective, for any $x \in \mathcal{H}_{1/2}$ there is a function $v_1(y) \in W_1(\mathbb{R}^+, \mathcal{H})$ such that $v_1(0) = x$. Then $\mathcal{A}v_1(y) \in \overset{\circ}{W}_{-1}(\mathbb{R}^+, \mathcal{H})$ and taking into account that mapping (3.6) is an isomorphism, we find a function $v_2(y) \in \overset{\circ}{W}_1(\mathbb{R}^+, \mathcal{H})$ such that $\mathcal{A}_F v_2(y) = \mathcal{A}v_1(y)$. Hence, the function $u(y) = v_1(y) - v_2(y)$ is a generalized solution of the equation $\mathcal{A}u(y) = 0$ and $u(0) = v_1(0) = x$. The uniqueness follows from the condition $Ker\, \mathcal{A}_F = 0$. □

Lemma 3.3. *Let $u(y)$ be a solution of equation (3.5) on the semiaxis $\mathbb{R}^+$ in the following sense:*

$$u(y), u'(y) \in C(\mathbb{R}^+, \mathcal{H}_1), \quad u''(y) \in C(\mathbb{R}^+, \mathcal{H}),$$

and equation (3.5) holds as an equality in $\mathcal{H}$. If $u(y) \in W_1(\mathbb{R}^+, \mathcal{H})$, then

$$\|S^{1/2}u(y)\| = \|F^{1/2}u'(y)\|, \quad y \geq 0. \tag{3.7}$$

Proof. Consider in $\mathcal{H}^2 = \mathcal{H} \times \mathcal{H}$ the operator

$$\mathbf{T} = \begin{pmatrix} -F^{-1/2}GF^{-1/2} & -F^{-1/2}S^{1/2} \\ S^{1/2}F^{-1/2} & 0 \end{pmatrix}, \tag{3.8}$$

acting in $\mathcal{H}^2 = \mathcal{H} \times \mathcal{H}$ (the linearization of $T(\lambda)$). Obviously, $\mathbf{T}$ is symmetric (and even selfadjoint) in the Krein space $\mathcal{K} = \{\mathcal{H}^2, \mathbf{J}\}$ with the fundamental symmetry $\mathbf{J} = \begin{pmatrix} I & 0 \\ 0 & -I \end{pmatrix}$. It is easy to see that equation (3.5) is equivalent to the following one

$$\mathbf{T}\,\mathbf{u}(y) = i\mathbf{u}'(y), \qquad \mathbf{u}(y) = \begin{pmatrix} F^{1/2}u'(y) \\ -iS^{1/2}u(y) \end{pmatrix}.$$

Using this equation we find (differentiation is allowed by our assumptions)

$$(\mathbf{Ju}(y),\, \mathbf{u}(y))' = (\mathbf{Ju}'(y),\, \mathbf{u}(y)) + (\mathbf{Ju}(y),\, \mathbf{u}'(y)) =$$

$$= -i(\mathbf{JTu}(y),\, \mathbf{u}(y)) + i(\mathbf{u}(y),\, \mathbf{JTu}(y)) = 0.$$

Therefore, $(\mathbf{Ju}(y),\, \mathbf{u}(y)) = const.$ The condition $u(y) \in W_1(\mathbb{R}^+, \mathcal{H})$, obviously, implies $(\mathbf{Ju}(y),\, \mathbf{u}(y)) = 0$ and (3.7) follows. □

Theorem 3.4. *Let $T(\lambda)$ be a strongly elliptic positive pencil. Then there exists a closed operator Z in the space $\mathcal{H}$ with domain $\mathcal{D}(Z) \subset \mathcal{H}_1$, such that*

$$T(\lambda)x = (F\lambda - Z_1)(\lambda - Z)x \qquad \textit{for all } x \in \mathcal{D}(Z), \tag{3.9}$$

where $Z_1 = -(G + FZ)$ and the equality is understood in $\mathcal{H}_{-1}$. Moreover,

(a) Z has a representation $Z = KS^{1/2}$ where K is a partial isometry in $\mathcal{H}$ whose image $\Re(K) = \mathcal{H}$;

(b) -iZ generates a holomorphic semigroup in the spaces $\mathcal{H}_\theta, 0 \le \theta \le 1/2$;

(c) the generalized solutions of equation (3.5) satisfy the equation

$$u'(y) = -iZu(y).$$

Factorization (3.9) with these properties is unique.

Proof. Let $x \in \mathcal{H}_{1/2}$. By virtue of Lemma 3.2 there is a generalized solution of equation (3.5) such that $u_x(0) = x$. Define the operator function $U(t)$ on $\mathbb{R}$ as follows

$$U(t)x = u_x(t), \qquad t \ge 0.$$

Note that according to Lemma 3.2 the restriction of the trace operator $\mathcal{T}_0$ to $Ker\mathcal{A} \in W_1(\mathbb{R}^+, \mathcal{H})$ is a bounded isomorphism onto $\mathcal{H}_{1/2}$. Hence the inverse operator

$$\mathcal{T}_0^{-1} : \mathcal{H}_{1/2} \to Ker\mathcal{A}, \qquad \mathcal{T}_0^{-1}x = u_x(t)$$

is bounded, as well as the operator $U(t) = \mathcal{T}_t \mathcal{T}_0^{-1}$ acting in $\mathcal{H}_{1/2}$ (for any $t \geq 0$). It follows from the definition of the operator $U(t)$ and from the trace theorem (see the formulation at the beginning of this section) that

$$U(t+s) = U(t)U(s), \qquad U(0) = I, \qquad \|U(t)\| \leq const,$$

$$\text{s-}\lim_{t\to s} U(t) = U(s), \quad 0 \leq s \leq t,$$

where the strong limit is understood in $\mathcal{H}_{1/2}$. This means that $U(t)$ is a uniformly bounded C_0-semigroup in the space $\mathcal{H}_{1/2}$ (see, e.g., [Yo]). If $U(t) = e^{-iZt}$ where $-iZ$ is the generator of $U(t)$, then property (c) of Theorem 3.4 is satisfied, and by Lemma 3.2 it defines Z uniquely.

It is known from semigroup theory that Z and Z^2 (as well as the other powers) are closed operators in $\mathcal{H}_{1/2}$ whose domains $\mathcal{D}(Z)$ and $\mathcal{D}(Z^2)$ are densely defined in $\mathcal{H}_{1/2}$.

Let $x \in \mathcal{D}(Z^2) \subset \mathcal{H}_{1/2}$ and $u_x(t)$ is the corresponding generalized solution of (3.5). In view of the semigroup properties the functions $u_x'(t), u_x''(t)$ are continuous in $\mathcal{H}_{1/2}$ on $\mathbb{R}^+$ and

$$u_x'(t) = iZu_x(t) \qquad u_x''(t) = -Z^2 u_x(t).$$

The operator $G : \mathcal{H} \to \mathcal{H}_{-1}$ is bounded, therefore, $Gu_x'(t)$ is continuous in $\mathcal{H}_{-1}$. Since $u_x(t)$ is a generalized solution, we have the equality

$$-Fu_x''(t) + iGu_x'(t) = -Su_x(t) \tag{3.10}$$

which is understood as an equality in $\overset{\circ}{\mathrm{W}}_{-1}(\mathbb{R}^+, \mathcal{H})$. The left hand side is a continuous function in $\mathcal{H}_{-1}$, hence, so is the function $Su_x(t)$. Equivalently, $u_x(t)$ is continuous in $\mathcal{H}_1$. In particular, $x = u_x(0) \in \mathcal{H}_1$ and (3.10) gives

$$(FZ^2 + GZ + S)x = 0, \qquad \text{for } x \in \mathcal{D}(Z^2), \tag{3.11}$$

where the equality is understood in $\mathcal{H}_{-1}$.

Our further aim is to extend (3.11) to a larger domain. Notice, if $x \in \mathcal{D}(Z^2)$ then the conditions of Lemma 3.3 are fulfilled and we have

$$\|F^{1/2}u_x'(t)\| = \|S^{1/2}u_x(t)\|, \qquad t \geq 0.$$

In particular, we have the equality $\|F^{1/2}Zx\| = \|S^{1/2}x\|$ which gives (for $x \in \mathcal{D}(Z^2)$) the representation $Z = KS^{1/2}$, where K is a partial isometry in $\mathcal{H}$. Since $\mathcal{D}(Z^2)$ is dense in $\mathcal{H}_{1/2}$ and $\mathcal{H}_{1/2}$ is dense in $\mathcal{H}$, we have $\Re(K) = \mathcal{H}$. Hence, Z is boundedly invertible in $\mathcal{H}$ and $Z^{-1} = S^{-1/2}K^*$. This enables us to extend Z from $\mathcal{H}_{1/2}$ onto $\mathcal{H}$ with domain $D_{\mathcal{H}}(Z) = \Re(S^{-1/2}K^*)$. Further (and in (3.9)) we omit the index $\mathcal{H}$ and imply that Z acts in $\mathcal{H}$ and its domain $\mathcal{D}(Z)$ is understood as described. Certainly, $\mathcal{D}(Z) \subset \mathcal{H}_1$ and it coincides with $\mathcal{H}_1$ if and only if K is a unitary operator. Now, both terms Gx and Sx are in $\mathcal{H}_{-1}$ for $x \in \mathcal{D}(Z)$, so equality (3.11) can be extended to all $x \in \mathcal{D}(Z)$. This is equivalent to the factorization (3.9), moreover, for Z_1 we have the representation $Z_1 = S^{1/2}K^*$ as well as $Z_1 = -(G + FZ)$. Then we obtain

$$(\lambda - Z)^{-1} = T^{-1}(\lambda)(F\lambda - S^{1/2}K^*)$$

where the both sides are understood as operators in $\mathcal{H}$. Applying Theorem 1.7 to the right hand side of the last identity we obtain the right hand side of the last identity we obtain

$$\|(\lambda - Z)^{-1}x\| \leq \frac{c\|x\|}{1 + |\lambda|} \tag{3.12}$$

in a double sector Λ_φ containing the real axis. Let us prove that (3.12) holds also for all λ from the upper half plane $\mathbb{C}^+$. Since $-iZ$ is a generator of a C_0-semigroup in $\mathcal{H}_{1/2}$ we have (see [Yo, Ch. 9])

$$\|(\lambda - Z)^{-1}x\| \leq \|(\lambda - Z)^{-1}x\|_{1/2} \leq \frac{c_x}{1 + |\lambda|} \quad \text{for all } x \in \mathcal{H}_{1/2} \text{ and } \lambda \in \mathbb{C}^- \setminus \Lambda_\varphi,$$

with a constant c_x depending on x, and the estimate holds in the whole upper half plane $\mathbb{C}^+$ outside an arbitrary small double sector Λ_φ containing the real axis. Applying the Phragmen-Lindelöf theorem (see [Bo], for example) we obtain estimate (3.12) for all $\lambda \in \mathbb{C}^-$ and $x \in \mathcal{H}_{1/2}$ with the same constant c as it was in (3.12). By continuity (3.12) can be extended for all $x \in \mathbb{R}$. This implies that $-iZ$ generates a holomorphic semigroup in $\mathcal{H}$.

Actually, $-iZ$ generates a holomorphic semigroup in the space $\mathcal{H}_{1/2}$, too. To prove this, we consider the pencil

$$T_\varphi(\lambda) = T(e^{i\varphi}\lambda) \quad \text{and} \quad \mathcal{A}^\varphi u = T_\varphi\left(i\frac{d}{dy}\right).$$

For sufficiently small $|\varphi|$ we can reprove Lemma 3.1 changing $(\mathcal{A}_0 u, u)$ in (3.2) by $Re(\mathcal{A}_0^\varphi u, u)$. This is possible, since the Friedrichs extension exists for the sectorial operators (see [Ka], Ch. 6). Repeating the arguments we find that there exists an operator $-iZ_\varphi$ which generates a C_0-semigroup in $\mathcal{H}_{1/2}$, Z_φ possesses property (c) and realizes a factorization of the form (3.9) for the pencil $T_\varphi(\lambda)$. From this we obtain $Z_\varphi = e^{i\varphi}Z$. Then the minimal resolvent growth estimate of the form (3.12) holds for Z in a small double sector containing the real axis. Hence, the C_0-semigroup generated by $-iZ$ is, actually, a holomorphic semigroup. Now, applying the interpolation theorem we get assertion (b). This ends the proof. □

4. Elliptic pencils satisfying the Keldysh-Agmon condition

4.1. The resolvent growth condition

In this section we will use the condition which in general form can be formulated as follows.

The resolvent growth condition. *Assuming that* $T^{-1}(\lambda)H^{1/2}x(\lambda)$ *is holomorphic in the upper (lower) half plane* $\mathbb{C}^+$ ($\mathbb{C}^-$) *where*

$$x(\lambda) = x_0 + \lambda x_1 + \ldots + \lambda^n x_n$$

is an $\mathcal{H}$*-valued polynomial, we have*

$$\|H^{1/2}T^{-1}(\lambda)H^{1/2}x(\lambda)\| \le C|\lambda|^m \quad \textit{for all } \lambda \in \mathbb{C}^+(\mathbb{C}^-), \quad |\lambda| > r_0, \tag{4.1}$$

with some constants c and m.

This condition is by no means obvious to verify and we formulate the other one which can be checked out more easily.

Keldysh-Agmon condition. $T(\lambda)$ *is of the form (1.1) and*

(a) the operator H has discrete spectrum (i.e. H^{-1} *is compact and its eigenvalues are subject to the estimates*

$$\lambda_j(H) \ge cj^p, \quad j = 1, 2, \ldots, \tag{4.2}$$

with some constants c and p;

(b) either $p \ge 2$ *or* $p < 2$ *but there are rays* $\gamma_j = \{\lambda|\ \arg\lambda = \theta_j\}, j = 1, \ldots, N$, *in the upper (lower) half plane* $\mathbb{C}^+(\mathbb{C}^-)$ *such that*

$$0 < \theta_j < \theta_{j+1} < 2\pi/p,\ j = 1, \ldots, N-1; \quad \max(\theta_1, \theta_{j+1} - \theta_j, \pi - \theta_N) < 2\pi/p,$$

and

$$\|H^{1/2}T^{-1}(\lambda)H^{1/2}\| \le c(1 + |\lambda|^m), \quad \textit{for } \lambda \in \gamma_j,$$

with some constants c and m.

Proposition 4.1. *If* $T(\lambda)$ *is a strongly elliptic pencil then the Keldysh-Agmon condition implies the resolvent growth condition, moreover, one can take in (4.1)* $m = n$.

Proof. First, notice that (4.2) implies that the generalized and the classical spectra of $T(\lambda)$ coincide (Theorem 1.6). The essense of the matter is that condition (a) together with $\mathcal{D}(G) \supset \mathcal{D}(H^{1/2})$ imply that $H^{1/2}T^{-1}(\lambda)H^{1/2}$ is an $\mathcal{H}$-valued meromorphic operator function of order $2/p$. The proof is based on the results of Keldysh [Ke], Agmon [Ag], Matsaev [Mat] et. al. (see historical remarks and details in [S3], §2). Now, if

$$F(\lambda) = H^{1/2}T^{-1}(\lambda)H^{1/2}x(\lambda)$$

is holomorphic in $\mathbb{C}^+$ and $x(\lambda)$ is a polynomial then condition (b), on account of the Phragmen-Lindelöf theorem, implies that $F(\lambda)$ has a polynonial growth in $\mathbb{C}^+$. According to Theorem 1.7

$$|F(\lambda)| < c(1+|\lambda|^n), \qquad \lambda \in \mathbb{R},\ |\lambda| > r_0,\ n = \deg x(\lambda). \tag{4.3}$$

Since $F(\lambda)$ is of order zero in $\mathbb{C}^+$, by virtue of the Phragmen-Lindelöf theorem the estimate (4.3) holds asymptotically for all $\lambda \in \mathbb{C}^+$. □

4.2. Half-range completeness and minimality

In what follows we consider for simplicity a generic situation when $T(\lambda)$ has only semisimple real eigenvalues of definite type. For a pencil of the form (0.2) this is true according to Theorem 2.2 for all values of ω with the possible exception of isolated resonant frequences $\omega_k \to \infty$.

Let $T(\lambda)$ have discrete spectrum and let the eigenvalues of $T(\lambda)$ be numerated according to their geometric multiplicity (i.e. every eigenvalue λ_k is repeated $n = \operatorname{nul} \mathrm{T}(\lambda_k)$ times). In this case we have a one-to-one correspondence between the eigenvalues λ_k and canonical Jordan chains of the form (2.2). As we agreed, all the real eigenvalues are supposed to be semisimple. The eigenelements corresponding to every real eigenvalue are assumed to form a normal canonical system (see Section 2). Take all the chosen Jordan chains of $T(\lambda)$ corresponding to the eigenvalues from the open upper (lower) half-plane and all the eigenelements corresponding to the real eigenvalues of positive (negative) type. Denote the system consisting of all these elements by $E^+(E^-)$ and call it the first (second) half of the root elements of $T(\lambda)$.

Let us recall the well-known definitions. A system $\{e_k\}_1^\infty$ is said to be *minimal* in Hilbert space $\mathcal{H}$ if there exists an adjoint system $\{e_k^*\}_1^\infty$ such that $(e_k, e_j^*) = \delta_{kj}$, where δ_{jk} is the Kronecker symbol. Equivalently, $\{e_k\}_1^\infty$ is minimal if any element e_k is not contained in the closed linear span of the other ones. A system $\{e_k\}_1^\infty$ is said to be *complete* in $\mathcal{H}$ if there is no non-zero element in $\mathcal{H}$ which is orthogonal to all the elements of the system.

Theorem 4.2. *The first and the second half of the root elements of a pencil $T(\lambda)$ form minimal systems in $\mathcal{H}$ provided $T(\lambda)$ is strongly elliptic and has discrete spectrum.*

Proof. Let us work with the system E^+, for example. By virtue of Propostition 1.4 there is a number $r_0 > 0$ such that the pencil $T_1(\lambda) = T(\lambda - r_0)$ has only positive eigenvalues on the real axis (to prove the minimality of E^- one should consider the pensil $T_1(\lambda) = T(\lambda + r_0)$). The Jordan chains $x_k^0, \dots, x_k^p$ of the pencil $T(\lambda)$ are changed after this transformation in the following way

$$\xi_k^0 = x_k^0\ ,\ \xi_k^1 = x_k^1 - r_0^{-1}x_k^0,\ \dots\ , \xi_k^p = x_k^p - r_0^{-1}x_k^{p-1} - \dots - r_0^{-p}x_k^0,$$

while the sign characteristics of the pairs $\{\lambda_k, y_k^0\}$ and $\{\lambda_k + r_0, y_k^0\}$ are the same. Hence, it suffices to prove the minimality for the case when $T(\lambda)$ has only positive real eigenvalues and $S > 0$. Let us consider the system

$$\mathbf{x}_k^s = \begin{pmatrix} F^{1/2}(\lambda_k x_k^s + x_k^{s-1}) \\ S^{1/2} x_k^s \end{pmatrix}, \qquad x_k^s \in E^+. \tag{4.4}$$

It is an easy exercise to show that the $\mathbf{x}_k^s$ are the root elements of the operator $\mathbf{T}$ defined by (3.8) ($\mathbf{T}$ is the linearization of $T(\lambda)$). As we mentioned $\mathbf{T}$ is a symmetric operator in the Krein space $\mathcal{K} = \{\mathcal{H}^2, \mathbf{J}\}$ with the fundamental symmetry $\mathbf{J} = \begin{pmatrix} I & 0 \\ 0 & -I \end{pmatrix}$. From this we have the biorthogonality relationships (see, e.g., [AI], Ch.1)

$$(\mathbf{J}\mathbf{x}_k^s, \mathbf{x}_j^h) = 0 \qquad \text{for all } \; j, k, s, h$$

except for the case $k = j$ and $\lambda_k \in \mathbb{R}$. For $\lambda_k \in \mathbb{R}$ we have

$$(\mathbf{J}\mathbf{x}_k, \mathbf{x}_k) = \lambda_k^2 (F x_k, x_k) - (S x_k, x_k)$$

$$= 2\lambda_k^2 (F x_k, x_k) + \lambda_k (G x_k, x_k) = \lambda_k (T'(\lambda_k) x_k, x_k) = \lambda_k \varepsilon_k,$$

where ε_k is the sign characteristic of the pair $\{\lambda_k, y_k\}$. Hence,

$$(\mathbf{J}\mathbf{x}_k^s, \mathbf{x}_j^h) = \delta_{kj} \lambda_k \varepsilon_k, \qquad x_k^s, x_k^h \in E^+, \tag{4.5}$$

where $\varepsilon_k = 0$ for the nonreal λ_k and $\lambda_k \varepsilon_k > 0$ for $\lambda_k \in \mathbb{R}$ and $x_k \in E^+$. Let x be a finite linear combination of elements (4.4)

$$\mathbf{v} := \begin{pmatrix} v_1 \\ v_2 \end{pmatrix} = \sum c_k^s \begin{pmatrix} F^{1/2}(\lambda_k x_k^s + x_k^{s-1}) \\ S^{1/2} x_k^s \end{pmatrix} =: \sum c_k^s \mathbf{x}_k^s. \tag{4.6}$$

From (4.5) we obtain $\|v_1\| \geq \|v_2\|$, therefore $\|\mathbf{v}\| \leq 2\|v_1\|$. Recall that the system $\mathbf{x}_k^s$ is minimal as the system of the root elements of the operator $\mathbf{T}$ with discrete spectrum. Then the inequality $\|v_1\| \leq \|\mathbf{v}\| \leq 2\|v_1\|$ implies (if we use the second definition of minimality) that

$$\{F^{1/2}(\lambda_k x_k^s - x_k^{s-1})\}, \qquad x_k^s \in E^+,$$

is a minimal system in $\mathcal{H}$. Hence E^+ is minimal in $\mathcal{H}$, too. □

Theorem 4.3. *The first and the second half of the root elements of a pencil $T(\lambda)$ form complete systems in $\mathcal{H}_1$ provided $T(\lambda)$ is strongly elliptic and the Keldysh-Agmon condition holds.*

Proof. As before, we deal with the system E^+. Suppose that there is an element $f \in \mathcal{H}_1$ such that

$$(f, x_k^s)_1 = (H^{1/2} f, H^{1/2} x_k^s) = 0 \qquad \text{for all } x_k^s \in E^+. \tag{4.7}$$

Choose a number r_0 such that $T(\lambda) > 0$ for $\lambda > r_0$ and consider the function

$$(4.8) \qquad F(\lambda) = \frac{1}{\lambda - r_0}(H^{1/2}T^{-1}(\lambda)H^{1/2}g, g), \quad g = H^{1/2}f \in \mathcal{H}.$$

The principal part of $F(\lambda)$ in a neighbourhood of a real pole λ_k has the representation

$$\sum \frac{\varepsilon_k(g, H^{1/2}x_k)(H^{1/2}x_k, g)}{(\lambda - r_0)(\lambda - \lambda_k)}$$

where ε_k are the sign characteristics corresponding to the eigenpair $\{\lambda_k, x_k\}$. Due to (4.7) all the terms with $\varepsilon_k > 0$ in the last expression are equal to zero. Since $\lambda_k - r_0 < 0$, all the residues of $F(\lambda)$ at the poles $\lambda_k \in \mathbb{R}$ are non-negative. The residue at the additional pole $\lambda = r_0$ is non-negative, too. Taking into account the representation (2.1) of $T^{-1}(\lambda)$ in a neighbourhood of a non-real pole $\lambda_k \in \mathbb{C}^+$ and assumption (4.7), we find that $F(\lambda)$ is holomorphic in $\mathbb{C}^+$. By the Schwarz symmetry principle it is holomorphic in $\mathbb{C}^-$. Proposition 4.1 gives us $F(\lambda) = O(\lambda^{-1})$ when $\lambda \to \infty$ uniformly in $\mathbb{C}$.

Let us show that the residue of $F(\lambda)$ at ∞ equals zero. Given $\varepsilon > 0$ we can find $g_0 \in \mathcal{H}_1$ such that $\|g - g_0\|_1 < \varepsilon$. If we put g_0 in (4.8) instead of g then by virtue of Theorem 1.7 the corresponding function vanishes at ∞ as $O(\lambda^{-3})$ when $\lambda \to \pm\infty$ uniformly in $\mathbb{C}$. Therefore, $F(\lambda) = o(\lambda^{-1})$ as $\lambda \to \infty$ uniformly in $\mathbb{C}$, i.e. the residue at ∞ is equal to zero. Now, recall that all the residues of $F(\lambda)$ at the finite poles are non-negative. This is possible only if all they are equal to zero, in particular,

$$(H^{1/2}T(r_0)H^{1/2}g,\, g) = 0.$$

This implies $g = 0$. □

Corollary 4.4. *The first and the second half of the root elements of a strongly elliptic pencil $T(\lambda)$ satisfying the Keldysh-Agmon condition form complete and minimal systems in spaces $\mathcal{H}_\theta$ for all $0 \le \theta \le 1$.*

Proof. It follows from the definitions: if a system is minimal (complete) in $\mathcal{H}(\mathcal{H}_1)$ then it has the same property in $\mathcal{H}_\theta$ for $\theta > 0$ ($\theta < 1$). Now apply Theorems 4.2 and 4.3. □

4.3. Factorization

The obtained results enable us to construct a divisor of an elliptic pencil.

Theorem 4.5. *Let $T(\lambda)$ be a strongly elliptic pencil satisfying the Keldysh-Agmon condition. Then*

$$(4.9) \qquad T(\lambda)x = (\lambda - Z_1)F(\lambda - Z)x$$

where (a) Z *and* Z_1 *admit a representation*

$$r - Z = K_0 H^{1/2}, \qquad r - Z_1 = H^{1/2} K_1$$

with bounded and boundedly invertible in $\mathcal{H}$ *operators* K_0 *and* K_1*, provided* $r \in \mathbb{R}$ *is not an eigenvalue of* $T(\lambda)$*. In particular,* Z *is a closed operator on* $\mathcal{H}$ *with domain* $\mathcal{D}(Z) = \mathcal{H}_1$ *;*
(b) *the spectra of* $T(\lambda)$ *and* $\lambda - Z$ *coincide in the upper half-plane, while on the real axis* $\lambda - Z$ *inherits only the positive type eigenpairs of* $T(\lambda)$*, i.e. the system of the root functions of* Z *coincides with the first half of the root functions of* $T(\lambda)$*;*
(c) iZ *generates a holomorphic semigroup in all spaces* $\mathcal{H}_\theta, 0 \leq \theta \leq 1$*.*
Equality (4.9) holds for all $x \in \mathcal{H}_1$ *and is understood in sense of operators acting from* $\mathcal{H}_1$ *to* $\mathcal{H}_{-1}$*. Factorization (4.9) with property (b) is unique.*

Proof. As in Theorem 4.2 we may assume that $T(\lambda)$ has only positive eigenvalues, otherwise we have to work with $T(\lambda - r_0)$, $r_0 \gg 1$.

Let us consider the set of all finite linear combinations of elements (4.4). The elements of this set have representation (4.6). If the system E^+ is complete in $\mathcal{H}_1$ then the system $H^{1/2}(E^+)$ is complete in $\mathcal{H}$. Therefore, Theorem 4.3 implies that the linear span of the elements $\{v_2\}$ in (4.6) form a dense subset in $\mathcal{H}$ as well as the elements $\{v_1\}$. Define the operator K by

$$Kv_1 = v_2. \tag{4.10}$$

It was shown in Theorem 4.2 that $\|v_2\| \leq \|v_1\|$. Hence, K is densely defined on $\mathcal{H}$ and can be extended as a contraction on the whole $\mathcal{H}$. The image of K is dense in $\mathcal{H}$.

Denote by E^0 the subsystem of E^+ consisting of all elements $x_k^s \in E^+$ corresponding to the nonreal eigenvalues. Let $\mathcal{H}_0$ be the closure in $\mathcal{H}$ of the linear span generated by E^0. Denote $\kappa = \operatorname{codim} \mathcal{H}_0$ (κ coincides with the number of positive type eigenvalues counting with geometric multiplicity). It is clear from (4.5) that $\|Kv_1\| = \|v_1\|$ for $v_1 \in \mathcal{H}_0$, hence, $K(\mathcal{H}_0)$ is a closed subspace in $\mathcal{H}$. By virtue of Corollary 4.4 the system $H^{1/2}(E^+)$ is minimal and complete in $\mathcal{H}$. This implies that $\operatorname{codim} \mathrm{K}(\mathcal{H}_0) = \kappa$. Hence, there is a unitary operator U in $\mathcal{H}$ such that the restriction of U onto $\mathcal{H}_0$ coincides with K, i.e. $U - K$ is of finite rank. We noticed already that the image of K is dense in $\mathcal{H}$. Now, it follows from the Fredholm theorem that K is boundedly invertible on $\mathcal{H}$.

Denote $Z = F^{-1/2} K^{-1} S^{1/2}$, where $S = T(0) > 0$. From (4.6) and (4.10) we have

$$Zx_k^s = \lambda_k x_k^s + x_k^{s-1}, \qquad x_k^s \in E^+. \tag{4.11}$$

Since x_k^s are the root elements of $T(\lambda)$, we have

$$(FZ^2 + GZ + S)x_k^s = 0 \qquad \text{for all } x_k^s \in E^+.$$

The linear span of E^+ is dense in $\mathcal{H}_1$, hence,

$$-(FZ+G)Z = S, \tag{4.12}$$

where the equality is understood in the sense of operators acting from $\mathcal{H}_1$ to $\mathcal{H}_{-1}$. Denoting $Z_1 = -(FZ+G)F^{-1}$ we obtain from (4.12) the factorization

$$T(\lambda) = (\lambda - Z_1)F(\lambda - Z).$$

As $\mathcal{D}(S) = \mathcal{D}(H)$ we have $S^{1/2} = K_2 H^{1/2}$ with a bounded and boundedly invertible operator K_2. Hence, $Z = K_0 H^{1/2}$ with $K_0 = F^{-1/2}K^{-1}K_2$. We have also

$$Z_1 = SH^{-1/2}K_0^{-1}F^{-1} = H^{1/2}K_2^* K_2 K_0^{-1}F^{-1} =: H^{1/2}K_1.$$

Thus (a) is proved. The assertion (b) follows from (4.11). The uniqueness of a factorization with property (b) follows from the completeness of the system E^+. It remains to prove (c). To this end we obtain from (4.9)

$$(\lambda - Z)^{-1} = T^{-1}(\lambda)(\lambda - H^{1/2}K_1)F. \tag{4.13}$$

Applying Theorem 1.7 we obtain

$$\|(\lambda - Z)^{-1}\| \le C|\lambda|^{-1}, \qquad \lambda \in \Lambda_\varphi, \quad |\lambda| > r_0. \tag{4.14}$$

Moreover, $(\lambda - Z)^{-1}$ is holomorphic in $\mathbb{C}^-$. By virtue of (4.13) and Proposition 4.1 $(\lambda - Z)^{-1}$ has a polynomial growth in C^-. Consequently, (4.14) holds for all $\lambda \in \Lambda_\varphi \cup \mathbb{C}^-, |\lambda| > r_0$. Thus, iZ generates a holomorphic semigroup in $\mathcal{H}$. Since $Z : \mathcal{H}_1 \to \mathcal{H}$ is an isomorphism, iZ possesses the same property in $\mathcal{H}_1$. Applying the interpolation theorem we obtain assertion (c). This ends the proof. □

5. The Mandelstam hypothethis

In this section we solve the problem

$$T\left(i\frac{d}{dy}\right) u(y) = 0 \tag{5.1}$$

$$u(0) = f \tag{5.2}$$

$$u(y) = u_+(y) + u_0(y), \quad u_0(y) \to 0 \text{ as } y \to \infty, \tag{5.3}$$

where $u_+(y)$ is a linear combination of outgoing waves (0.3).

Below we clarify the understanding of this problem and prove the solvability in the classical sense and the uniqueness in the generalized sense. We may say that (5.1)–(5.3) is the half-range Cauchy problem because instead of two initial conditions at $y = 0$ we set only one, but force a solution to behave at ∞ in a special way.

Further we denote by $C_2(a,b;\mathcal{H})$ the space of continuous on (a,b) $\mathcal{H}_2$-valued functions whose derivatives $v'(y)$ and $v''(y)$ exist in $\mathcal{H}_1-$ and $\mathcal{H}$-norm and belong to $C(a,b;\mathcal{H}_1)$ and $C(a,b;\mathcal{H})$, respectively (the continuity at the ends of (a,b) is not assumed!)

Theorem 5.1. *Let $T(\lambda)$ be strongly elliptic and assume that the Keldysh-Agmon condition holds. Then for any $\theta \in [0,1]$ and any $f \in \mathcal{H}_\theta$ there exists a function $u(y) \in C_2(0,\infty;\mathcal{H})$ satisfying equation (5.1), having representation (5.3) with exponentially decaying $\|u_0(y)\|_2$ and satisfying initial condition (5.2) in the following sense*

$$\lim_{y\to+0} \|u(y) - f\|_\theta = 0. \tag{5.4}$$

Proof. We find a solution of the problem in question by means of the operator Z which was constructed in Theorem 4.5. Namely, denote

$$u(y) = \frac{1}{2\pi i}\left(\int_\gamma + \int_\Gamma\right) e^{i\lambda y}(\lambda - Z)^{-1} f\, d\lambda, \tag{5.5}$$

where γ surrounds only real eigenvalues of Z, while Γ lies in the upper half-plane and is asymptotically directed along the rays $arg\,\lambda = \delta$ and $arg\,\lambda = \pi - \delta$ with sufficiently small $\delta > 0$. By virtue of Theorem 4.5 iZ generates a holomorphic semigroup in $\mathcal{H}_\theta$, hence integral (5.5) is well defined and (5.4) holds (see [Yo, Ch. 9]). Moreover, the functions $Z^k u^{(j)}(y)$ are well defined for $y > 0, k, j \geq 0$ and are continuous in $\mathcal{H}_\theta \subset \mathcal{H}$. Since $Z : \mathcal{H}_1 \to \mathcal{H}$ is an isomorphism, we obtain that $u^{(j)}(y)$ are continuous for $y > 0$ in $\mathcal{H}_1$. The equality

$$-(G + FZ)Zx = Sx$$

holds for all $x \in \mathcal{H}_1$, in particular, for $x \in \mathcal{H}_2$. As $u'(y) = iZu(y)$ we obtain that $iGu'(y) - Fu''(y) \in C(0,\infty;\mathcal{H})$, equation (5.1) is satisfied in $\mathcal{H}$ and $u(y) \in C_2(0,\infty;\mathcal{H})$. Representation (5.3) with an exponentially decaying function $u_0(y)$ follows from (5.5). □

Theorem 5.2. *A generalized solution $u(y)$ of problem (5.1)–(5.3), such that $u(y) \in L_1(0,\varepsilon;\mathcal{H}_1)$ with some $\varepsilon > 0$, is unique.*

Proof. In Section 3 we assumed that generalized solutions $u(y)$ belong to the space $W_2^1(0,\infty;\mathcal{H})$. Here our assumptions are weaker: we only assume that $u(y) \in W_2^1(\varepsilon,\infty;\mathcal{H})$ and $u(y) \in L_1(0,\varepsilon;\mathcal{H}_1)$ for any $\varepsilon > 0$. Certainly, if $u(y)$ belongs to $W_2^1(0,\infty;\mathcal{H})$ then $u(y) \in L_2(0,\varepsilon;\mathcal{H}_1)$ and $u(y) \in L_1(0,\varepsilon;\mathcal{H}_1)$. By the definition of a generalized solution, the equation

$$-Fu''(y) = -iFu'(y) + Su(y)$$

is satisfied in the sense of $\overset{\circ}{\mathrm{W}}_{-1}(\varepsilon,\infty;\mathcal{H})$. The right hand side belongs to the space $L_2(\varepsilon,\infty;\mathcal{H}_{-1})$, hence, so does the left hand side.

Suppose that $\|u(y)\| \to 0$ as $y \to 0$. For $\lambda \in \mathbb{C}^+$ we have

$$
\begin{aligned}
0 &= \int_\varepsilon^\infty T_\omega\left(i\tfrac{d}{dy}\right) u(y)e^{i\lambda y}dy = \\
&= e^{i\lambda\varepsilon}\left(Fu'(\varepsilon) - i(\lambda F - G)u(\varepsilon)\right) + T_\omega(\lambda)\hat{u}_\varepsilon(\lambda) = 0,
\end{aligned}
\tag{5.6}
$$

where

$$\hat{u}_\varepsilon(\lambda) = \int_\varepsilon^\infty u(y)e^{i\lambda y}dy.$$

We consider (5.6) as an equality in $\mathcal{H}_{-1}$. Since $u(y)$ is locally integrable at zero as a function with values in $\mathcal{H}_1$ we can take the limit as $\varepsilon \to 0$ and obtain

$$T_\omega(\lambda)\hat{u}_0(\lambda) = -Fu'(0) = g \in \mathcal{H}_{-1}.$$

Therefore, $\hat{u}_0(\lambda) = T_\omega^{-1}(\lambda)g$. Let us consider the function

$$F(\lambda) = \frac{1}{\lambda + r_0}\left(T_\omega^{-1}(\lambda)g, g\right).$$

It follows from (4.10) that $F(\lambda)$ is bounded in $\mathbb{C}^+$ and has finitely many poles on $\mathbb{R}$ with positive residues provided r_0 is sufficiently large. Repeating the arguments of Theorem 4.9 we obtain $F(\lambda) \equiv 0$. Hence, $\hat{u}(\lambda) \equiv 0$ and $u(y) \equiv 0$. □

6. Application to the Lame system of the elasticity theory

Small oscillations of an elastic medium are described by the system of equations (see the books of LANDAU and LIFSHITZ [LL] or KUPRADZE et.al. [Ku])

$$\rho\frac{\partial^2 w}{\partial t^2} + Lw = 0,$$

where $w = w(t,x) = (w_1, w_2, w_3)$ is the displacement vector, $\rho = \rho(x)$ is the density of the medium, L is the operator matrix with the entries

$$L_{kj}(D) = (\hat{\lambda} + \hat{\mu})D_kD_j + \delta_{kj}\hat{\mu}(D_1^2 + D_2^2 + D_3^2), \quad D_k = i\frac{\partial}{\partial x_k},$$

and $\hat{\lambda}, \hat{\mu}$ are the Lame constants. We suppose that the space variable $x = (x_1, x_2, x_3)$ belongs to the wave-guide domain $Q = [0,\infty) \times \Omega$ where Ω is a bounded domain in the plane (x_2, x_3). Separating the time variable $w = ue^{i\omega t}$ we obtain the stationary equation with given frequency ω

(6.1) $$(L - \omega^2 \rho)u = 0.$$

We have to impose with this equation boundary and initial conditions. We pose on the lateral surface of the half-cylinder Q homogeneous conditions, since Q is a wave-guide domain. For simplicity let us consider the Dirichlet boundary conditions

(6.2) $$u(x_1, x_2, x_3)|_{(x_2,x_3)\in\partial\Omega} = 0 \quad \forall x_1 \geq 0.$$

At the base of Q we assume that

(6.3) $$u(0, x_2, x_3) = \varphi(x),$$

where $\varphi(x)$ is a given function. We rewrite equation (6.1) in the form

(6.4) $$T_\omega\left(i\frac{d}{dy}\right)u = -F\frac{d^2u}{dy^2} + iG\frac{du}{dy} + (H - \omega^2 R)u = 0,$$

where $y = x_1$,

$$F = \begin{pmatrix} \hat\lambda + 2\hat\mu & 0 & 0 \\ 0 & \hat\mu & 0 \\ 0 & 0 & \hat\mu \end{pmatrix}, \quad G = i(\hat\lambda + \hat\mu)\begin{pmatrix} 0 & D_2 & D_3 \\ D_2 & 0 & 0 \\ D_3 & 0 & 0 \end{pmatrix},$$

$$H = \begin{pmatrix} \mu\Delta & 0 & 0 \\ 0 & \hat\mu\Delta + (\hat\lambda + \hat\mu)D_2^2 & (\hat\lambda + \hat\mu)D_2D_3 \\ 0 & (\hat\lambda + \hat\mu)D_2D_3 & \hat\mu\Delta + (\hat\lambda + \hat\mu)D_3^2 \end{pmatrix},$$

$$R = \rho(x)I, \quad \Delta = -(D_2^2 + D_3^2),$$

and I is the identity matrix. We suppose that the operators F,G,H act in the Hilbert space $\mathcal{H} = [L_2(\Omega)]^3$.

We have to specify a domain of the main operator H. Taking into account boundary conditions (6.2) we define

$$\mathcal{D}(H) = \left\{v|\ v \in \left[W_2^2(\Omega)\right]^3, v|_{\partial\Omega} = 0\right\},$$

$$\mathcal{D}(G) = \left\{v|\ v \in \left[W_2^1(\Omega)\right]^3, v|_{\partial\Omega} = 0\right\} =: \left[\overset{\circ}{\mathrm{W}}{}_2^1(\Omega)\right]^3,$$

where $\left[W_2^k(\Omega)\right]^3$ are the Sobolev spaces of vector functions on $\Omega \subset \mathbb{R}^2$.

We notice that the operator H is positive, since

$$(Hv, v) = \int_\Omega Hv(x)\overline{v(x)}dx = \hat\mu\left(\sum_{j=1}^3 \|D_2v_j\|^2 + \|D_3v_j\|^2\right) +$$

$$+(\hat\lambda + \hat\mu)\|D_2v_2 + D_3v_3\|^2,$$

where $\|f\|^2 = \int_\Omega |f|^2 dx$. Taking into account boundary condition (6.2) and the Friedrichs inequality we obtain $(Hv, v) \geq \varepsilon\|v\|^2$ with some $\varepsilon > 0$. The operator G is symmetric, as

$$(Gv, v) = 2(\hat{\lambda} + \hat{\mu})Re(iD_2v_2 + iD_3v_3, v_1).$$

The operator F, obviously, is uniformly positive and bounded provided $\rho(x) \geq \varepsilon > 0$ is a measurable bounded function on Ω.

It is well-known (see [Ag] or [Tr], Ch. 5) that $H + cI$ is invertible in $[L_2(\Omega)]^3$ provided $c \geq 0$ and Ω is a smooth domain. Therefore, H is a selfadjoint operator if Ω is smooth. This is not always true, if Ω, for instance, has corner points (see examples in the paper of KONDRATIEV and SHKALIKOV [KoS]). In this case let us consider the Friedrichs extention H_F of the operator H. It is known (see [RN], Ch. 8) that it is the only extension which possesses the property $\mathcal{D}(H_F^{1/2}) = \mathcal{D}(G)$.

Denoting $H_F = H$ we remark that all the assumptions on the operator coefficients claimed at the beginning of Section 1 are fulfilled.

We note that in the case of a non-smooth domain Ω there is no precise information on $\mathcal{D}(H_F)$, hovewer, we do know that

$$\mathcal{D}(H_F^{1/2}) = \mathcal{D}(G) = \left[\overset{\circ}{\mathrm{W}}{}_3^1(\Omega)\right]^3.$$

Actually, the domain of $H_F(= H)$ is not involved in our considerations, the knowledge of $\mathcal{D}(H^{1/2}) = \mathcal{H}_1$ is the only important information which we need.

Now let us prove that the pencil corresponding to equation (6.4) is regular elliptic in the case of a smooth domain and strongly elliptic otherwise.

Proposition 6.1. *The pencil $T_\omega(\lambda)$ generated by the Lame system and the Dirichlet boundary conditions is strongly elliptic.*

Proof. We have

$$(T_\omega(\lambda)v, v) = \lambda^2\left[(\hat{\lambda} + 2\hat{\mu})\|v_1\|^2 + \hat{\mu}(\|v_2\|^2 + \|v_3\|^2)\right] +$$

$$+2\lambda(\hat{\lambda} + \hat{\mu})Re\left[(iD_2v_2 + iD_3v_3, v_1)\right] + (\hat{\lambda} + \hat{\mu})\|D_2v_2 + D_3v_3\|^2 +$$

$$+\hat{\mu}\sum_{j=1}^{3}(\|D_2v_j\|^2 + \|D_3v_j\|^2) - \omega^2\|\rho^{1/2}v\|^2 \geq \tag{6.5}$$

$$\geq \hat{\mu}\lambda^2\sum_{j=1}^{3}(\|v_j\|^2 + \|D_2v_j\|^2 + \|D_3v_j\|^2) - \omega^2\|\rho^{1/2}v\|^2 \geq$$

$$\geq \varepsilon[\lambda^2(v, v) + (Hv, v)] - \omega^2(\rho v, v), \quad \lambda \in \mathbb{R}, \quad \|v\|^2 = \|v_1\|^2 + \|v_2\|^2.$$

It is known (see [Tr], Ch 4.10) that the embedding $I : [\overset{\circ}{W}{}_2^1(\Omega)]^3 \to [L_2(\Omega)]^3$ is compact for any bounded domain Ω (we pay attention that if we consider, say, Neuman boundary conditions, then we have to assume in addition that Ω is a Lipshitzian domain). By virtue of Proposition 1.4 we obtain that $T_\omega(\lambda)$ is strongly elliptic. □

We remark that Proposition 6.1 can be also proved in the case of an unbounded domain Ω if we assume $\rho(x) \to 0$ as $|x| \to \infty$.

Proposition 6.2. *The pencil $T_\omega(\lambda)$ is regular elliptic if a domain Ω is smooth. Moreover, estimate (1.3) holds asymptotically outside any double sector containing the imaginary axis and the Keldysh-Agmon condition holds.*

Proof. (Cf.[KO2]). Denoting $-iD_k = \xi_k$, let us calculate the principal characteristic symbol of the Lame system (the principal symbol does not depend on ω and we can assume $\omega = 0$). We have

$$\det T_0(\lambda) = det \left[\lambda^2 \begin{pmatrix} \hat\lambda + 2\hat\mu & 0 & 0 \\ 0 & \hat\mu & 0 \\ 0 & 0 & \hat\mu \end{pmatrix} + \lambda(\hat\lambda + \hat\mu) \begin{pmatrix} 0 & \xi_2 & \xi_3 \\ \xi_2 & 0 & 0 \\ \xi_3 & 0 & 0 \end{pmatrix} + \right.$$

$$\left. + \begin{pmatrix} \hat\mu|\xi|^2 & 0 & 0 \\ 0 & \hat\mu|\xi|^2 + (\hat\lambda + \hat\mu)\xi_2^2 & (\hat\lambda + \hat\mu)\xi_2\xi_3 \\ 0 & (\hat\lambda + \hat\mu)\xi_2\xi_3 & \hat\mu|\xi|^2 + (\hat\lambda + \hat\mu)\xi_3^2 \end{pmatrix} \right] = \hat\mu^2(\hat\lambda + 2\hat\mu)(\lambda^2 + |\xi|^2)^3,$$

where $|\xi|^2 = \xi_2^2 + \xi_3^2$.

Hence, the ellipticity condition in the sense of [AN] and [AV] holds for all λ not belonging to the imaginary axis. It is well known (see, e.g., [LM]) that the Dirichlet boundary condition satisfies the Lopatinskii condition for all elliptic systems. Hence, the problem (6.1), (6.2) is regular elliptic and according to the results of [AN] and [AV] estimate (1.3) holds outside arbitrary small sector containing the imaginary axis. Since $T_\omega(\lambda)$ is a seladjoint pencil, estimate (1.3) implies

$$||HT_\omega^{-1}(\lambda)|| + ||T_\omega^{-1}(\lambda)H|| \leq const,$$

and, by virtue of the interpolation theorem, we have

$$(6.6) \qquad ||H^{1/2}T_\omega^{-1}(\lambda)H^{1/2}|| \leq const, \quad |\lambda| > r_0,$$

at any ray in $\mathbb{C}$ with exception of the imaginary exis. According to the Weyl asymptotic formula for eigenvalues of the elliptic operators, we have the estimate (4.2) with $p = 1$. Hence, if Ω is a smooth domain then the Keldysh-Agmon condition for the Lame system is valid. □

In the case of a non-smooth domain we are able to prove the validity of the Keldysh-Agmon condition only under additional constraints on the Lame constants.

Proposition 6.3. *Let Ω be a bounded domain in $\mathbb{R}^2$. If $\hat{\mu} > \sqrt{2}\hat{\lambda}$ then estimate(1.9) is saisfied in a double sector Λ_φ with some $\varphi > \pi/4$ and the Keldysh-Agmon condition holds.*

Proof. Let us estimate the quadratic form $(T_\omega(\lambda)v, v)$ at the ray $\lambda = e^{i\pi/4}\zeta, \quad \zeta > 0$. Suppose $\omega = 0$. Bearing in mind (6.5) we obtain

$$Re\ e^{-i\pi/4}\left(T_0(e^{i\pi/4}\zeta)v, v\right) \geq$$

$$\frac{\sqrt{2}}{2}\hat{\mu}\left(\zeta^2\sum_{j=1}^{3}\|v_j\|^2 + \|D_2v_j\|^2 + \|D_3v_j\|^2\right) +$$

$$+\frac{\sqrt{2}}{2}(\hat{\lambda}+\hat{\mu})\left(\zeta^2\|v_1\|^2 - 2\sqrt{2}|\zeta|\ \|D_2v_2 + D_3v_3\|\ \|v_1\| + \|D_2v_2 + D_3v_3\|^2\right).$$

Taking into account the inequality

$$2(ac+bc) \leq \sqrt{2}(a^2+b^2+c^2), \quad a,b,c > 0$$

we can estimate the second summand as follows

$$\geq -(\hat{\lambda}+\hat{\mu})(2-\sqrt{2})\frac{\sqrt{2}}{2}\left(\zeta^2\|v_1\|^2 + \|D_2v_2\|^2 + \|D_3v_3\|^2\right).$$

Therefore,

$$Re\ e^{i\pi/4}\left(T_\omega(e^{i\pi/4}\zeta)v, v\right) \geq \varepsilon\left(\|v\|_1 + \zeta^2\|v\|\right) - \omega^2(\rho v, v). \tag{6.7}$$

with some $\varepsilon > 0$ provided $\hat{\mu} > \sqrt{2}\hat{\lambda}$. Obviously, a similar estimate (if $\pi/4$ is replaced by θ) holds at any ray $\lambda = re^{i\theta}$ in a double sector Λ_φ provided $0 < \varphi - \pi/4$ is small enough.

According to Theorem 1.7 estimate (6.7) gives the estimate of the resolvent (6.6). Since (4.2) holds in our case with $p = 1$, we see that the Keldysh -Agmon condition is satisfied. □

For simplicity we formulate the main result of this section not in the whole generality.

Theorem 6.4. *Let Ω be a bounded domain in $\mathbb{R}^2$ and assume that either Ω is smooth or Lame constants satisfy the condition $\hat{\mu} > \sqrt{2}\hat{\lambda}$. Then for any function*

$\varphi(x) \in \left[\overset{\circ}{\mathrm{W}}{}_2^1(\Omega)\right]^3$ *there is a unique classical solution* $u(y)$ *in the half-cylinder* $Q = \mathbb{R}^+ \times \Omega$ *of the stationary Lame system (6.4), (6.2) with given nonresonant frequency* ω*, such that this solution satisfies the Mandelstam radiation principle as* $x_1 = y \to \infty$ *and the initial condition is understood in the following sense*

$$\lim_{y\to 0} \|u(y, x_2, x_3) - \varphi(x_2, x_3)\|_1 = 0.$$

Proof. It follows from results of Section 5. □

Our conjecture (which we can not prove at the moment) is that the condition $\hat{\mu} > 2\hat{\lambda}$ in Theorem 6.4 is superfluous. This condition is used only in the proof of the existence. Apparently it is essential for the validity of the Keldysh–Agmon condition and, hence, for the half range completeness. However, it has not to be essential for the existence of a solution. The reason is that for sufficiently small frequencies ω the pencil $T_\omega(\lambda)$ is positive and Theorem 3.4 can be applied to prove the existence.

References

[ADN] S. Agmon, A. Douglas and L. Nirenberg, Estimates near the boundary for solutions of elliptic partial differential equations satisfying general boundary conditions. Comm. Pure Appl. Math. vol. **12** (1959) 623–727.

[Ag] S. Agmon, Lectures on elliptic boundary value problems, New York, 1965.

[AI] T. Ja. Azizov and I. S. Iohvidov, Linear operators in spaces with indefinite metric, John Wiley, Chichester, 1989.

[AN] S. Agmon and L. Nirenberg, Properties of solutions of ordinary differential equations in Banach space. Comm. Pure Appl. Math. Vol **16** (1963), 121–239.

[AV] M. S. Arganovich and M. I. Vishik, Elliptic problems with a parameter and parabolic problems of general type, Uspekhi Mat. Nauk **19** (1964), no. 3, (117), 53–161; English transl. in Russian Math. Surveys **19** (1964).

[Bo] R. Ph. Boas, Entire functions. New York, 1954.

[BS] B. M. Bolotovskii and S. N. Stoljarov, Modern state of the electrodynamics of moving media. In book: Einstein collection, 1974, "Nauka", Moscow, 1976.

[Fi] G. Fichera. Existance theorems in elasticity and Boundary value problems of elasticity with unilateral constraints, Handbuch der Physik, Band VIa/2, Springer–Verlag, Berlin, 1972, 347–389, 391–424.

[FN] C. Foias and B. Sz-Nagy, Analyse harmonique des operateurs de L'espace de Hilbert, Academiai Kiado, 1967.

[GGK] I. Gohberg, S. Goldberg and M. A. Kaashoek, Classes of linear operators, vol. 1. Operator theory: Adv. and Appl. vol. **49**, 1990.

[GK] I. GOHBERG AND M. G. KREIN, Introduction to the theory of linear nonselfajoint operators in Hilbert space. Moscow, 1965, "Nauka"; English transl. Amer. Math. Soc., Providence, RI, 1969.

[Ka] T. KATO, Perturbation theory for linear operators (2-nd edition), Springer-Verlag, New York, 1976.

[Ke] M. V. KELDYSH, On the completeness of eigenfunction of certain classes of non-selfajoint linear operators, Russian Math. Surveys **26** no. 4 (1971), 295–305.

[KL] M. G. KREIN AND G. K. LANGER [H. LANGER], On some mathematical principles in the linear theory of damped oscillations of continua, Appl. Theory of Functions in Continuum Mech. (Proc. Internat. Sympos. Tbilisi , 1963) Vol. II: Fluid and Gas Mech., Math. Methods, "Nauka", Moscow, 1965, 283–322; English transl., Parts I,II Integral Equations and Operator Theory **1** (1978),364–399, 539–566.

[KLu] M. G. KREIN AND G. JA. LUBARSKII, On analytical properties of the multiplicators of positive type periodic canonical differential systems. Izvestija Acad. Nauk USSR. Ser. Mathem. vol **26** (1992) no. 4, 549–572.

[KO1] A. G. KOSTYUCHENKO AND M. B. ORASOV, On certain properties of the roots of the selfajoint quadric pencil, J. Funct. Anal. and Appl. **9** (1975), 28–40.

[KO2] A. G. KOSTYUCHENKO AND M. B. ORAZOV, Vibrations of an elastic semicylinder and associated selfadjoint quadric pencils, Trudy Seminara im I. G. Petrovskogo, vol **6** (1981), 97–146. English transl. in J. Soviet Math.

[KoS] V. A. KONDRATIEV AND A. A. SHKALIKOV, Completeness of the eigenfunctions of elliptic operators on non-smooth domains. Preprint of the Potsdam University, 1998.

[KS] A. G. KOSTYUCHENKO AND A. A. SHKALIKOV, Selfajoint quadric operator pencils and elliptic problems, J. Funct. Anal. and Appl. **17** (1983), 109–128.

[Ku] V. D. KUPRADZE ET AL., Three-dimensional problems of the mathematical theory of elasticity and thermoelasticity, Izdat. Tbilis. Univ., Tbilisi, 1968; English transl. of 2nd rev. aug. ed., North-Holland, Amsterdam, 1979.

[LL] L. D. LANDAU AND E. M. LIFSHITS, Course of theoretycal physics, Vol. **7**: Theory of elasticity, 4th ed., "Nauka", Moscow, 1987; English transl., Pergamon Press, Oxford, 1986.

[LM] J. L. LIONS AND E. MAGENES, Problems aux Limites Nonhomogenes et Applications. Vol.**1**, Dunod Paris, 1968; English transl. in Springer-Verlag.

[Ma] A. S. MARKUS, Introduction to the Spectral Theory of Polynomial Operator Pencils, Amer. Math. Soc., Providence, 1988.

[Mat] V. I. MATSAEV, A method of estimating of the resolvent of non-selfadjoint operators. Dokl. Acad. Nauk SSSR, vol. **154** no. 5 (1964), 1034–1037.

[RN] F. RISS AND B. SZ-NAGGY, Lecons d'analyse fonctionnelle , Acad. Kiado, Budapest, 1972.

[RR] M. ROSENBLUM AND J. ROVNJAK, Hardy classes and operator theory, Oxford Univ. Press, New York – Oxford 1985.

[S1] A. A. SHKALIKOV, On spectral theory of operator pencils and solvability of operator-differential equations. Doctoral dissertation, Moscow State Univ., Moscow 1987 (Russian)

[S2] A. A. SHKALIKOV, Operator pencils and operator equations in Hilbert space, (Unpublished manuscript, University of Calgary), 1992.

[S3] A. A. SHKALIKOV, Elliptic equations in Hilbert space and associated apectral problems, J. Soviet Math. **51** no. 4 (1990), 2399–2467.

[S4] A. A. SHKALIKOV, Operator pencils arising in elasticity and hydrodynamics: the instability index formula. Operator Theory: Adv. and Appl. vol. bf 87, Birkhäser Verlag (1996), 358–385.

[SH] A. A. SHKALIKOV AND R. O. HRYNIV, On operator pencils arising in the problem of beam oscillation with internal damping, Matem. Zametki **56** no. 2 (1994), 114–131 (Russian); English transl. in Math. Notes **56** (1994).

[SS] A. A. SHKALIKOV AND A. V. SHKRED, The problem of steady-state oscilations of a transversally isotropic half-cylinder, Math. USSR Sbornik, vol. **73** (1992) no. 2, 579–602.

[Sv] A. G. SVESHNIKOV, On the radiation principle, Dokl. Akad. Nauk SSSR, vol. **73** (1950), no. 5, 917–920.

[Tr] H. TRIEBEL, Interpolation theory, function spaces, differential operators, VEB Deutcher Verlag Wiss., Berlin 1977; North-Holland, Amsterdam, 1978.

[VB] , I. I. VOROVICH AND V. A. BABESHKO, Mixed dynamic problems of the elasticity theory in non-classical domains. "Nauka", Moscow, 1979.

[W] C. H. WILCOX, Scattering theory for diffraction gratings. Springer-Verlag, 1984, New York Inc.

[Yo] K. YOSIDA, Functional analysis, Springer-Verlag, Berlin-Heidelberg-New York, 1978.

[ZK] A. S. ZILBERGLEIT AND YU. I. KOPILEVICH, Spectral theory of regular waveguides. Ioffe Inst. of Physics, Leningrad, 1983.

Moscow Lomonosov State University
Department of Mechanics and Mathematics
Moscow, 119899
Russia

1991 Mathematics Subject Classification. 47A70, 70J15.

Operator Theory:
Advances and Applications, Vol. 106

An inductive limit procedure within the quantum harmonic oscillator

FRANCISZEK HUGON SZAFRANIEC

To Heinz

We show how the very classical quantum oscillator acting in $\mathcal{L}^2(\mathbb{R})$ can be approximated in a sense by a finite difference one. This continues our earlier investigations ([6], [7] and [9]) of this classical object of Mathematical Physics as acting in the Hilbert space environment.

0. A few words of introduction

The quantum harmonic oscillator is a so well known object of Mathematical Physics that there is no need to give any reference (however, for those who urgently need it we suggest [4] as a quick guide). For our purpose, as for it to act in a Hilbert space, we can say that it can be represented by a single operator S[1], called the *creation* one, satisfying[2]

$$S^\times S - SS^\times = I.$$

The other members of the family: the *annihilation* operator $S^\times$, the *number* one $N = SS^\times$ as well as the *hamiltonian* (which is the begining of the whole story in Physics) can be represented by S; this is why we concentrate on the latter. The very classical model for the quantum harmonic oscillator is that which acts in $\mathcal{L}^2(\mathbb{R})$. This, when embedded in ℓ^2 in a natural way, is a limit case of the finite difference oscillator in there (as developed in [6]); the fact we consider in more detail a bit later, in Section 3. The purpose of this paper is to establish a kind of dual relation, now performing in $\mathcal{L}^2(\mathbb{R})$.

1. The quantum harmonic oscillator in an abstract set-up

Let $\mathcal{H}$ be a separable (complex) Hilbert space and let $e = \{e_n\}_{n=0}^{\infty}$ be an orthogonal basis in it. Set $\mathcal{D} = \text{lin}\{e_n\}_{n=0}^{\infty}$. Then S is called a *creation operator* with respect to the basis e if

$$\mathcal{D}(S) = \mathcal{D}, \quad Se_n = \sqrt{n+1}\, e_{n+1}, \quad n = 0, 1, \dots$$

[1] S is to remind us that it is a subnormal operator, cf. [8]
[2] more about this can be found in [7]

The *annihilation operator* $S^\times$ with respect to e is defined by

$$\mathcal{D}(S^\times) = \mathcal{D}, \quad S^\times e_n = \sqrt{n}\, e_{n-1}, \quad S^\times e_0 = 0.$$

The operators S and $S^\times$ are *formally adjoint* each to the other, that is

$$< Sf, g > = < f, S^\times g >, \quad f, g \in \mathcal{D}.$$

Set

$$\mathcal{D}_{\max} = \{f \in \mathcal{H}; \sum_{n=0}^{\infty} n |< f, e_n >|^2 < +\infty\}$$

and define

$$\mathcal{D}(S_{\max}) = \mathcal{D}_{\max}, \quad S_{\max} f = \sum_{n=0}^{\infty} \sqrt{n+1} < f, e_n > e_{n+1}, \quad f \in \mathcal{D}_{\max},$$

$$\mathcal{D}(S^\times_{\max}) = \mathcal{D}_{\max}, \quad S^\times_{\max} f = \sum_{n=1}^{\infty} \sqrt{n} < f, e_n > e_{n-1}, \quad f \in \mathcal{D}_{\max}.$$

Then it can be get by standard argumentation that

$$S^- = S_{\max}, \quad (S^\times)^- = S^\times_{\max}, \quad S^* = S^\times_{\max} \quad \text{and} \quad (S^\times)^* = S_{\max}.$$

Call $S_{\max}$ the *maximal creation operator* and $S^\times_{\max}$ the *maximal annihilation operator*, both with respect to e.

2. The quantum harmonic oscillator in $\mathcal{L}^2(\mathbb{R})$

The *very classical* model of the quantum harmonic oscillator couple, the creation and the annihilation operator, is

$$S = \frac{1}{\sqrt{2}}\left(x - \frac{\mathrm{d}}{\mathrm{d}x}\right), \quad S^\times = \frac{1}{\sqrt{2}}\left(x + \frac{\mathrm{d}}{\mathrm{d}x}\right)$$

considered in $\mathcal{L}^2(\mathbb{R})$ with $\mathcal{D}(S) = \mathcal{D}(S^\times) = \operatorname{lin}\{h_n\}_{n=0}^{\infty}$ where h_n is the n-th Hermite function

$$h_n = 2^{-n/2}(n!)^{-1/2}\pi^{-1/4}\mathrm{e}^{-x^2/2}H_n$$

with H_n, the n-th Hermite polynomial, defined as

$$H_n(x) = (-1)^n \mathrm{e}^{x^2}\frac{\mathrm{d}^n}{\mathrm{d}x^n}\mathrm{e}^{-x^2}.$$

Because of the orthogonality relation

$$\int_{-\infty}^{\infty} H_m(x)H_n(x)\mathrm{e}^{-x^2}\,\mathrm{d}x = 2^n(n!)\sqrt{\pi}\delta_{mn},$$

$\{h_n\}_{n=0}^{\infty}$ is an orthonormal sequence and in fact, what is well known, an orthonormal basis in $\mathcal{L}^2(\mathbb{R})$.

We also have $S_{\max}$ and $S_{\max}^{\times}$, both with respect to $\{h\}_{n=0}^{\infty}$, according to the previous section.

3. The quantum harmonic oscillator in ℓ^2

Charlier polynomials $C_n^{(a)}$, $n = 0, 1, \ldots$, are determined by

$$\mathrm{e}^{-az}(1+z)^x = \sum_{n=0}^{\infty} C_n^{(a)}(x)\frac{z^n}{n!},$$

with orthogonality

$$\sum_{x=0}^{\infty} C_m^{(a)}(x)C_n^{(a)}(x)\frac{\mathrm{e}^{-a}a^x}{x!} = \delta_{mn}a^n n!,\ m,n = 0,1,\ldots.$$

Define the Charlier functions $\tilde{c}_n^{(a)}$, $n = 0, 1, \ldots$, $a > 0$, as

$$\tilde{c}_n^{(a)}(x) = a^{-\frac{n}{2}}(n!)^{-\frac{1}{2}}C_n^{(a)}(x)\mathrm{e}^{-\frac{a}{2}}a^{\frac{x}{2}}\begin{cases} \Gamma(x+1)^{-\frac{1}{2}} & \text{for } x \geq 0 \\ 0 & \text{for } x < 0 \end{cases}$$

and finally Charlier sequences $c_n^{(a)}$ by

$$c_n^{(a)} \stackrel{\mathrm{df}}{=} \tilde{c}_n^{(a)}|_{\mathbb{N}}, \quad n = 0, 1, \ldots$$

So $c_n^{(a)}$'s are apparently in ℓ^2 and they form an orthonormal basis therein. On the other hand, $\tilde{c}_n^{(a)} \in \mathcal{L}^2(\mathbb{R})$, the fact which can be get this or another way (for instance with some help of the Stirling formula).

Define the operators $S_{(a)}$ and $S_{(a)}^{\times}$, by

$$\mathcal{D}(S_{(a)}) = \mathcal{D}(S_{(a)}^{\times}) = \operatorname{lin}\{c_n^{(a)}\}_{n=0}^{\infty},\ S_{(a)}f = g,\ S_{(a)}^{\times}f = h$$

where g and h are given by

$$g(x) = \sqrt{x}f(x-1) - \sqrt{a}f(x),\ x = 1, 2, \ldots \quad g(0) = -\sqrt{a}f(0),$$

$$h(x) = \sqrt{x+1}f(x+1) - \sqrt{a}f(x),\ x = 0, 1, \ldots$$

$S_{(a)}$ and $S_{(a)}^{\times}$ are the creation and the annihilation operators in ℓ^2 with respect to the orthonormal basis $\{c_n^{(a)}\}_{n=0}^{\infty}$.

So defined finite difference operator S appeared in [5] [3] and was presented in detail in [6].

[3]another (and rather formal) version of it can be found in [1] where a possibility of a related limit procedure was mentioned.

Since the canonical zero-one basis $\{e_n\}_{n=0}^{\infty}$ defined as

$$e_n = \{\delta_{n,x}\}_{x=0}^{\infty}, \quad n = 0, 1, \ldots$$

is apparently in $\mathcal{D}(S_{(a),\max})$. Then after introducing the unitary operator $V_{(a)} : \ell^2 \mapsto \ell^2$ defined as

$$V_{(a)} c_n^{(a)} = e_n, \quad n = 0, 1, \ldots,$$

this means precisely that

$$V_{(a)}^{-1} \overline{S^{(a)}} V_{(a)} c_n^{(a)} = \sqrt{n+1} c_{n+1}^{(a)} - \sqrt{a} c_n^{(a)}.$$

Passing with a to 0 we get $S^{(a)} V_{(a)} c_n^{(a)} \to S e_n$ where S is defined as in Section 1 (and which may be identified with that from Section 2). The aim of this paper is to examine what happens if $a \to +\infty$.

4. "Charlier" tend to "Hermite"

In this section we work out in detail a somehow general idea that Hermite polynomials can be obtained as a limit case of Charlier ones, or, to be a little bit more precise, that [4]

$$\lim_{a \to +\infty} (2/a)^{n/2} C_n^{(a)}(\sqrt{2a}x + a) = H_n(x)$$

where H_n is the n-th Hermite polynomial. Whatever the meaning of this convergence is we want to have it in $\mathcal{L}^2(\mathbb{R})$, so let us do this. Reminding the three term recurrence relation for Charlier polynomials

$$C_{n+1}^{(a)}(x) = (x - n - a) C_n^{(a)}(x) - an C_{n-1}^{(a)}, \; n = 0, 1, \ldots, \quad C_{-1}^{(a)}(x) = 0, \; C_0^{(a)} = 1$$

and writing it for $g_n^{(a)}(x) \stackrel{\text{df}}{=} (2/a)^{n/2} C_n^{(a)}(\sqrt{2a}x + a)$ we get

$$g_{n+1}^{(a)}(x) = (2x - n\sqrt{\frac{2}{a}}) g_n^{(a)}(x) - 2n g_{n-1}^{(a)}(x), \; n = 0, 1, \ldots, \quad g_{-1}^{(a)} = 0, \; g_0^{(a)} = 1.$$

Because the three term recurrence relation for Hermite polynomials is

$$H_{n+1}(x) = 2x H_n(x) - 2n H_{n-1}(x), \; n = 0, 1, \ldots, \; H_{-1} = 0, \; H_0 = 1,$$

an induction argument and the Lebesgue dominated convergence theorem imply that

$$g_n^{(a)} \to H_n \text{ in } \mathcal{L}^2(\mathbb{R}, \mathrm{e}^{-x^2} \mathrm{d}x). \tag{1}$$

Set

$$\widetilde{m}_a(x) \stackrel{\text{df}}{=} 2^{n/2} (n!)^{1/2} \mathrm{e}^{a/2} a^{(\sqrt{2a}x)/2} \Gamma(\sqrt{2a}x + a + 1)^{1/2}$$

[4] cf. [3], however notice that our definition of Charlier polynomials is as in [2]

and

$$m_a(x) \stackrel{\text{df}}{=} \begin{cases} \pi^{1/4}\mathrm{e}^{-a/2}a^{-(\sqrt{2a}x+a)/2}\Gamma(\sqrt{2a}x+a+1)^{-1/2}\mathrm{e}^{x^2/2} & \text{if } x > -a/\sqrt{2a} \\ 0 & \text{otherwise} \end{cases}.$$

Because

$$g_n^{(a)}(x) - H_n(x) = \widetilde{m}_a(x)(\tilde{c}_n^{(a)}(x) - m_a(x)h_n(x))$$

for $x > -a/\sqrt{2a}$, in order to get from (1) that

$$\tilde{c}_n^{(a)} - m_a h_n \to 0 \text{ in} \mathcal{L}^2(\mathbb{R})$$

it suffices to show that

$$|1/\widetilde{m}_a(x)| \leq C, \text{ with } A_n \text{ independent of } a$$

To prove the latter use the Stirling formula [5] so as to get

$$|1/\widetilde{m}_a(x)|^2 \leq B_n \exp\{(1-2a_0)(1+\log a_0) - 1\}$$

where D_n does not depend on a and a_0 is the (only) positive solution of

$$4 + 2\log a - \frac{1}{a} = 0.$$

This establishes

(2) $$\|\tilde{c}_n^{(a)} - m_a h_n\|_{\mathcal{L}^2(\mathbb{R})} \to 0 \text{ as } a \to +\infty.$$

5. Embedding ℓ^2 into $\mathcal{L}^2(\mathbb{R})$

Set

$$X_a \stackrel{\text{df}}{=} \{x;\ \sqrt{2a}x + a \in \mathbb{N}\} = \{x_k^{(a)} = \frac{k-a}{\sqrt{2a}};\ k = 0, 1, \ldots\}$$

$$(J_a f)(x) = \begin{cases} \sqrt{2a} f(x_k^{(a)}) & \text{if } x_k^{(a)} \leq x < x_{k+1}^{(a)} \\ 0 & \text{if } \frac{-a}{\sqrt{2a}} < x \end{cases}, \quad x \in \mathbb{R},\ f \in \ell^2(X_A),$$

Apparently J_a is an isometry from $\ell^2(X_a)$ onto its range $\mathcal{R}(J_a)$. For

(3) $$a = a_N = 2^{2N+1}$$

set $X_N = X_{a_N}$ and $J_N = J_{a_N}$; thus

$$X_N \subset X_M \text{ if } N < M.$$

[5] $\Gamma(x) = \sqrt{\frac{2\pi}{x}}(\frac{x}{\mathrm{e}})^x(1 + \frac{1}{12x} + \ldots)$

Moreover we complete the inductive limit construction by defining in the natural way the embeddings

$$J_{N,M} : \mathcal{R}(J_N) \mapsto \mathcal{R}(J_M),\ N < M.$$

Now after setting $c_n^{(N)} \stackrel{\text{df}}{=} c_n^{(a_N)}$ we get

$$\|J_N c_n^{(N)} - \tilde{c}_n^{(N)}\|_{\mathcal{L}^2(\mathbb{R})} \to 0 \text{ as } N \to +\infty. \tag{4}$$

Also under new notation $m_N \stackrel{\text{df}}{=} m_{a_N}$ (2) looks like

$$\|c_n^{(N)} - m_N h_n\|_{\mathcal{L}^2(\mathbb{R})} \to 0 \text{ as } N \to +\infty. \tag{5}$$

6. The limit: "finite difference" to the "very classical"

Set

$$\mathcal{D}_N \stackrel{\text{df}}{=} \operatorname{lin}\{J_N c_n^{(N)}\}_{n=0}^{\infty} \oplus (\operatorname{lin}\{J_N c_n^{(N)}\}_{n=0}^{\infty})^{\perp}$$

where $c_n^{(N)}$'s are defined as in the Section 4. Define the operator S_N by

$$\mathcal{D}(S_N) \stackrel{\text{df}}{=} \mathcal{D}_N \text{ and } S_N c_n^{(N)} \stackrel{\text{df}}{=} S_{(a_N)} J_N c_n^{(N)},\ n = 0, 1, \ldots$$

$$S_N f = 0, \quad f \in (\operatorname{lin}\{c_n^{(N)}\}_{n=0}^{\infty})^{\perp}.$$

First of all notice that, because J_N is an isometry onto $\operatorname{clolin}\{c_n^{(N)}\}_{n=0}^{\infty}$ and $S_{(a_N)}$ is injective, the operator S_N is well defined. Then we get immediatelly

$$S_N c_n^{(N)} = \sqrt{n+1}\, c_{n+1}^{(N)} \quad n = 0, 1, \ldots$$

which means that S_N, when restricted to $\operatorname{clolin}\{c_n^{(N)}\}_{n=0}^{\infty}$, is a creation operator, according to our definition. Define $S_N^{\times}$, having the same domain as S_N, by

$$S_N^{\times} c_n^{(N)} \stackrel{\text{df}}{=} \sqrt{n}\, c_{n-1}^{(N)}, \quad S_N^{\times} \stackrel{\text{df}}{=} 0 \text{ if } f \in (\operatorname{lin}\{c_n^{(N)}\}_{n=0}^{\infty})^{\perp}.$$

Now we are ready to state our conclusion

Theorem 1 *Keeping all the notations introduced so far we have, for any* $n = 0, 1, \ldots,$

$$\|S_N c_n^{(N)} - M_N S h_n\| \to 0 \text{ and } \|S_N^{\times} c_n^{(N)} - M_N S^{\times} h_n\| \to 0 \text{ as } N \to +\infty,$$

where M_N *is the (bounded) operator of multiplication by* m_N *on* $\mathcal{L}^2(\mathbb{R})$.

Proof. Apply (4) and (5). □

References

[1] N. Atakishiyev, S.K. Suslov, A model of the harmonic oscillator on the lattice (in Russian), in "Contemporary Group Analysis", Proc. VI All-Union Colloquium, pp. 17–21, Izdat. Elm, Baku ,1989

[2] T.S. Chihara, An introduction to orthogonal polynomials, Gordon and Breach, New York, N.Y., 1978

[3] R. Koekoek, R.F. Swarttouw, The Askey-scheme of hypergeometric orthogonal polynomials and its q-analogue, Delft University of Technology, Report of the Faculty of Technical Mathematics and Informatics no. 94–05

[4] W. Mlak, M. Słociński, Quantum phase and circular operators, Univ. Iagell. Acta Math., 24 (1992), 133–144

[5] F.H. Szafraniec, Orthogonal polynomials and subnormality of related shift operators, in "Orthogonal polynomials and their applications", Proc., Segovia (Spain), 1986, Monogr. Acad. Cienc. Zaragoza R-C Palacios, ed., 1, pp. 153–155, 1988

[6] F.H. Szafraniec, Yet another face of the creation operator, Operator Theory Adv. Appl., 80 (1995), 266–275

[7] F.H. Szafraniec, Analytic models of the quantum harmonic oscillator, Contemp. Math., 212 (1997), 269–276

[8] F.H. Szafraniec, Unbounded subnormal operators. Why?, Univ. Iagel. Acta Math, 34 (1996), 149–152

[9] F.H. Szafraniec, The quantum harmonic oscillator in $\mathcal{L}^2(\mathbb{R})$, in "Special Functions and Differential Equations", Proc., The Institute of Mathematical Sciences, Chennai (formerly Madras), India, January 13–23, 1997 K. Srinivasa Rao, R. Jagannathan, G. Vanden Berghe, J. Van der Jeugt, eds., pp. 206–211, Allied Publishers, New Dehli, 1997

Instytut Matematyki
Uniwersytet Jagielloński
ul. Reymonta 4, PL-30059 Kraków

1991 Mathematics Subject Classification. Primary 81Q05; Secondary 81S05, 47B20, 47B38, 47B39

Operator Theory:
Advances and Applications, Vol. 106

Canonical systems with a semibounded spectrum

HENRIK WINKLER

Dedicated to Heinz Langer on the occasion of his 60th birthday

We consider a singular two-dimensional canonical system $Jy' = -zHy$ on $[0, L)$ such that at L Weyl's limit point case holds. Here H is a real and nonnegative definite matrix function, the so-called Hamiltonian. From results of L. de Branges it follows that the correspondence between canonical systems (or their Hamiltonians H) and their Titchmarsh-Weyl coefficients is a bijection between the class of trace normed Hamiltonians H and the class of Nevanlinna functions. In this note we show that the Hamiltonian H of a canonical system with a semibounded spectrum has the property $\det H = 0$ and that its components are functions of locally bounded variation. Further, a characterization of the class of Hamiltonians corresponding to canonical systems with a finite number of negative (or positive) eigenvalues is given.

1. Introduction

In this note we consider initial value problems of the form

$$J\frac{dy(x)}{dx} = -zH(x)y(x), \quad y_1(0) = 0, \tag{1.1}$$

on $[0, L)$, $0 < L \leq \infty$, where $J = \begin{pmatrix} 0 & -1 \\ 1 & 0 \end{pmatrix}$, $z \in \mathbb{C}$, and H is a real, symmetric and nonnegative definite 2×2-matrix function on $[0, L)$, the so-called *Hamiltonian*. We assume that Weyl's limit point case prevails at L for the canonical system (1.1), which is equivalent to the relation (see [dB2]) $\int_0^L \operatorname{tr} H(x)dx = +\infty$, where *tr* denotes the trace of a matrix. Let $\mathbf{N}$ be the set of all *Nevanlinna functions*, i.e. the set of all functions which are analytic on the upper half plane $\mathbb{C}^+$ and map $\mathbb{C}^+$ into $\mathbb{C}^+ \cup \mathbb{R}$. In [dB2] it is shown that to each canonical system (1.1) (or to each Hamiltonian H) there corresponds a Titchmarsh-Weyl coefficient $Q \in \tilde{\mathbf{N}} := \mathbf{N} \cup \{f \equiv \infty\}$ (see Section 2 below). If $\operatorname{tr} H(x) = 1$, $0 \leq x < L$, then $L = \infty$ and the Hamiltonian H is called *trace normed.* In the class of trace normed Hamiltonians the inverse result of [dB4] can be formulated as follows (see [W1]): *Each function $Q \in \tilde{\mathbf{N}}$ is the Titchmarsh-Weyl coefficient of an a.e. on $[0, +\infty)$ uniquely determined trace normed Hamiltonian H.* This result holds only for two-dimensional canonical systems; for a spectral theory of canonical systems of higher dimension see e.g. [DI], [S1, S2].

The proof that there is a bijective correspondence between trace normed Hamiltonians and their Titchmarsh-Weyl coefficients is not constructive. Therefore it seems to be of interest to get a more detailed characterization of Hamiltonians corresponding to certain subclasses of **N**. In [W2] some rules are given for an explicit construction of H corresponding to a special spectral measure σ of Q. In this note we derive some properties of Hamiltonians of canonical systems whose spectrum is semibounded, that is their corresponding spectral measures have a semibounded support. In particular we characterize the canonical systems with a finite number of negative (or positive) eigenvalues.

In the case that all moments of the spectral measure exist, M.G. Krein and H. Langer gave in [KL2] explicit representations of the Hamiltonian by means of corresponding orthogonal polynomials (see also [A]). Starting from this result, it will be shown in Section 3 below that each canonical system with a semibounded spectrum corresponds to a trace normed Hamiltonian H with $\det H = 0$ whose components are functions of locally bounded variation. Using a (not trace normed) representation of H of the form $H = \begin{pmatrix} v^2 & v \\ v & 1 \end{pmatrix}$, in Theorems 3.3 and 3.5 below canonical systems with a semibounded spectrum are characterized by means of the function v. The main result of Section 4 below is Theorem 4.1, stating that the number of the "critical" points of v plus the number of H-indivisible intervals of type 0 is equal to the number of the negative eigenvalues of the corresponding canonical system. Theorem 4.3 below is the analogous result for canonical systems with a finite number of positive eigenvalues.

2. Canonical systems

Let H be a real, symmetric and nonnegative definite 2×2-matrix function given on $[0, L)$, $0 < L \leq \infty$: $H(x) = H(x)^T \geq 0$, $H(x) \neq 0$,

$$H(x) = \begin{pmatrix} h_1(x) & h_3(x) \\ h_3(x) & h_2(x) \end{pmatrix}, \qquad x \in [0, L). \tag{2.1}$$

The entries of H are assumed to be locally integrable functions and such that

$$\int_0^L \operatorname{tr} H(x)dx = +\infty.$$

Two matrix functions H_1 and H_2 are considered to be equivalent if $H_1(x) = H_2(x)$ a.e. with respect to the Lebesgue measure.

With no loss of generality we can suppose that H is defined on $[0, \infty)$ and trace normed, i.e., $\operatorname{tr} H(x) = 1$, $x \in [0, \infty)$. To justify this, let H be any Hamiltonian on $[0, L)$ and let y be a solution of the corresponding problem (1.1).

By $\hat{x} := \int_0^x \operatorname{tr} H(t)dt$ and $\hat{H}(\hat{x}) := H(x)(\operatorname{tr} H(x))^{-1}$ a trace normed Hamiltonian $\hat{H}$ on $[0,\infty)$ is defined. It follows easily that with $\hat{y}(\hat{x}) := y(x)$ the equation

$$J\frac{d\hat{y}(\hat{x})}{d\hat{x}} = -z\hat{H}(\hat{x})\hat{y}(\hat{x}), \ \hat{y}_1(0) = 0$$

is satisfied. It is shown below, that the Hamiltonians H and $\hat{H}$ correspond to the same Titchmarsh-Weyl coefficient.

The following intervals play a special role in the sequel, (see [Ka], [dB1-4]. The open interval $I \subset [0,\infty)$ is called *H-indivisible* if for some positive function k on I and the constant vector $\xi_\phi := (\cos\phi, \ \sin\phi)^T$ the relation

$$H(x) = k(x)\xi_\phi\xi_\phi^T \text{ for all } x \in I$$

holds. Here ϕ is called the type of I. In particular, $\det H(x) = 0$ if $x \in I$. Note that $k(x) = 1$ if H is trace normed. An H-indivisible interval is called *maximal*, if it is not a proper subset of another H-indivisible interval.

The following Hilbert space L^2_H is associated with the Hamiltonian H: L^2_H is the set of all equivalence classes of measurable, a.e. finite vector functions $f(x) = (f_1(x) \ f_2(x))^T$ on $[0,\infty)$ with the properties:

1. $\displaystyle\int_0^\infty f(x)^*H(x)f(x)dx < +\infty.$
2. For any H-indivisible interval I_ϕ of type ϕ there exists a constant $c_{I_\phi,f} \in \mathbb{C}$ such that $\xi_\phi^T f(x) = c_{I_\phi,f}$ for all $x \in I_\phi$.

In [Ka] and [dB5] it is shown that L^2_H is a Hilbert space with respect to the inner product

$$\langle f,g\rangle := \int_0^\infty g(x)^*H(x)f(x)dx.$$

On L^2_H we define the following linear relation A:

The pair (f,g), $f,g \in L^2_H$, belongs to A if f is identically zero in a neighbourhood of $+\infty$ and the following relations hold:

$$J\frac{df(x)}{dx} = -H(x)g(x), \text{ if } x \in [0,\infty), \quad f_1(0) = 0. \tag{2.2}$$

Let $\mathcal{D}(A)$ be the domain of A.

If the Hamiltonian H satisfies the following two conditions:

a) $\int_0^\epsilon h_2(x)dx > 0$ for every $\epsilon > 0$,

b) there exists a $\beta > 0$ such that $\operatorname{rank}\left(\int_0^\beta H(x)dx\right) = 2$,

the main result of [Ka] implies that $\mathcal{D}(A)$ is dense in L^2_H and the closure of A is equal to the adjoint A^*, i.e., A is an essentially selfadjoint operator. The condition

a) excludes that $(0,b)$ is a maximal H-indivisible interval of type $\phi=0$ for some $b>0$, that is, there is no $b>0$ such that $H(x)=\begin{pmatrix}1&0\\0&0\end{pmatrix}$ if $x\in(0,b)$. In the terminology of [Ka] this means that the "first exceptional case" is excluded. The condition b) excludes that the whole interval $(0,\infty)$ is H-indivisible. In this case for $e\in\mathbb{C}^2$ the relation $\int_0^\infty (H(t)e,e)dt=0$ implies $e=0$, that is, the canonical system (1.1) is definite.

A nonnegative measure σ on $\mathbb{R}$ is called a *spectral measure* of the operator A if there exists a linear isometric mapping $\mathbf{F}$ from L^2_H into L^2_σ with the property $\mathbf{F}A\mathbf{F}^{-1}\subset M_\lambda$, where M_λ is the operator of multiplication by the independent variable in L^2_σ: $M_\lambda(f)(\lambda):=\lambda f(\lambda)$, $f\in L^2_\sigma$. In [dB2] it is shown, that there exists a unique spectral measure σ. To this end for any Hamiltonian H the initial value problem

$$\frac{dW(x,z)}{dx}J=zW(x,z)H(x),\quad W(0,z)=\begin{pmatrix}1&0\\0&1\end{pmatrix}, \tag{2.3}$$

is considered, its solution, the 2×2 matrix function W, is said to be the *fundamental matrix* function of the canonical system (1.1).

The fact that at ∞ for the canonical system (1.1) Weyl's limit point case prevails means that for an arbitrary $t\in\tilde{\mathbf{N}}$ and $z\in C^+$ the limit

$$Q_H(z):=\lim_{x\to\infty}\frac{w_{11}(x,z)t(z)+w_{12}(x,z)}{w_{21}(x,z)t(z)+w_{22}(x,z)} \tag{2.4}$$

exists, is independent of t and, as a function of z, belongs to the class $\tilde{\mathbf{N}}$. The function Q_H is called the *Titchmarsh-Weyl* coefficient of the canonical system. If $Q_H\not\equiv\infty$, the function Q_H admits the unique spectral representation (see [AG])

$$Q_H(z)=bz+a+\int_{-\infty}^{+\infty}\left(\frac{1}{\lambda-z}-\frac{\lambda}{1+\lambda^2}\right)d\sigma(\lambda) \tag{2.5}$$

with $b\geq0$, $a\in\mathbb{R}$, and a nonnegative measure σ such that

$$\int_{-\infty}^{+\infty}\frac{d\sigma(\lambda)}{1+\lambda^2}<\infty.$$

The measure σ is called the spectral measure of the canonical system (1.1) or of the Hamiltonian H. If H satisfies the conditions a) and b), then σ is the spectral measure of the corresponding operator A.

If $(0,\infty)$ is an H-indivisible interval of type 0, then $Q=\infty$. The condition a) implies $b=0$ in the representation (2.5), indeed, it holds $b>0$ in (2.5) if and only if $(0,b)$ is a maximal H-indivisible interval of type 0 (see [W1]). The condition b) implies $d\sigma \not\equiv 0$. In the sequel we often assume that H satisfies the conditions a) and b), i.e. A is a densely defined essentially selfadjoint operator in L^2_H.

With the operator A there is associated the following Fourier transformation (see [dB3], [Ka]). Denote by $L^2_{H,0}$ the subset of L^2_H of elements which vanish identically near ∞, and define

$$F(u,z) := \int_0^\infty (w_{21}(t,z)\ w_{22}(t,z))H(t)u(t)dt. \tag{2.6}$$

Then the mapping $u \mapsto F(u,\cdot)$ is an isometry from $L^2_{H,0}$ onto a dense subset of L^2_σ. Hence it can be extended by continuity to all of L^2_H. The inverse transformation, mapping L^2_σ onto L^2_H, is given by

$$u(x) = \text{l.i.m.} \int_{-N}^{+N} (w_{21}(x,\lambda)\ w_{22}(x,\lambda))^T F(u,\lambda)d\sigma(\lambda), \quad N \to +\infty, \tag{2.7}$$

where l.i.m. denotes the limit in the norm of L^2_σ.

In the sequel we will use the following approximation procedure (see [dB2]):

Lemma 2.1. (see [dB2]) *The convergence $Q_{H_n}(z) \to Q_H(z)$, $n \to \infty$, holds locally uniformly for $z \in \mathbb{C}^+$ if and only if it holds*

$$\int_0^x H_n(t)dt \to \int_0^x H(t)dt,\ n \to \infty, \quad \textit{locally uniformly for } x \in [0,\infty). \tag{2.8}$$

Proof. In the following we write $H_n \to H$ for the convergence in the sense of (2.1). Assume that Q_{H_n} tends to a function $Q \in \tilde{\mathbf{N}}$ locally uniformly in $\mathbb{C}^+$. According to [dB2] there is a subsequence $\{H_{n_k}\}$ with $H_{n_k} \to H$. Then H is the Hamiltonian corresponding to Q. If $H_n \not\to H$ then there is a subsequence $\{H_{n_{\tilde{k}}}\}$ which converges to a Hamiltonian $\tilde{H}$ different from H, then $\tilde{H}$ also corresponds to Q, a contradiction to the uniqueness. This shows that $H_n \to H$. The converse statement can be proved by similar arguments. □

Theorem 2.2. *For a canonical system with Hamiltonian H on $[0,L)$ and spectral measure σ it holds*

$$\sigma(\{0\}) = \left(\int_0^L h_2(t)dt \right)^{-1}. \tag{2.9}$$

Proof. First we prove the theorem under the assumption that H is trace normed. An H-indivisible interval $(0,b)$ of type 0 has no influence on the relation (2.9), so we assume without loss of generality that the conditions a) and b) are satisfied. If $\sigma(\{0\}) > 0$, according to (2.7) the function

$$F(\lambda) = \begin{cases} (\sigma(\{0\})^{-1} & \text{if } \lambda = 0 \\ 0 & \text{if } \lambda \neq 0 \end{cases}$$

in L^2_σ is in correpondence with the function $u(x) = (0\ 1)^T$, $x \in [0, \infty)$ in L^2_H. Then the Fourier transformation (2.6) implies $\sigma(\{0\})^{-1} = \int_0^\infty h_2(t)dt$.

Now we assume that $\int_0^\infty h_2(t)dt < \infty$. First we consider the case that the Hamiltonian H is *discrete,* that is $[0, \infty)$ is the union of a finite number of maximal H-indivisible intervals and their boundary points. In this case $\int_0^\infty h_2(t)dt < \infty$ holds if and only if there exists an $l > 0$ such that (l, ∞) is a maximal H-indivisible of type 0, that is $h_2(x) = 0$ if $x \in (l, \infty)$. As it follows from (2.3) that the functions $w_{11}(\cdot, z)$ and $w_{21}(\cdot, z)$ are constant on (l, ∞), we get

$$Q(z) = \frac{w_{11}(l, z)}{w_{21}(l, z)}. \tag{2.10}$$

The spectral representation (2.5) of Q yields

$$\sigma(\{0\}) = \lim_{y \to +0} -iyQ(iy). \tag{2.11}$$

Using the relations $w_{21}(l, z) = -z \int_0^l w_{22}(t, z)h_2(t)dt$ and $W(x, 0) = \begin{pmatrix} 1 & 0 \\ 0 & 1 \end{pmatrix}$, $x \in [0, \infty)$, we get

$$\lim_{y \to +0} -iy(w_{21}(l, iy))^{-1} = \left(\int_0^l h_2(t)dt \right)^{-1} = \left(\int_0^\infty h_2(t)dt \right)^{-1}.$$

As $w_{11}(l, 0) = 1$ from (2.11) and (2.10) the assertion of the theorem follows in the case that H is discrete.

In general, a Hamiltonian H with $\int_0^\infty h_2(t)dt = K < \infty$ can be approximated in the sense of Lemma 2.1 by a sequence of discrete Hamiltonians H_n for which $\int_0^\infty h_{2,n}(t)dt < K + 1$ for all n. If $y > 0$ we have

$$|iyQ_n(iy)| \geq |\text{Re } iyQ_n(iy)| = \int_{-\infty}^{+\infty} \frac{y^2(1 + \lambda^2)}{\lambda^2 + y^2} d\sigma_n(\lambda) \geq \sigma_n(\{0\}) \geq \frac{1}{K + 1}.$$

As $Q_n(iy) \to Q(iy)$ if $n \to \infty$ (see Lemma 2.1), it follows that $|iyQ(iy)|) \geq (K+1)^{-1}$ and hence, by (2.11), that $\sigma(\{0\}) \geq \frac{1}{K+1} > 0$. According to the first part of the proof, (2.9) follows for trace normed Hamiltonians.

If H is not trace normed, it follows for the corresponding trace normed Hamiltonian $\hat{H}$ in the transformation above that $\int\limits_0^L h_2(t)dt = \int\limits_0^\infty \hat{h}_2(t)dt$, this proves the relation (2.9) in the general case. □

3. Semibounded canonical systems

A canonical system is called *semibounded* if its spectrum, that is the support of its spectral measure σ, is semibounded.

Theorem 3.1. *If a canonical system is semibounded, its trace normed Hamiltonian H has the property* $\det H = 0$ *on* $[0, +\infty)$ *and the components of H are functions of locally bounded variation.*

Proof. In the first part of the proof we show that $\det H = 0$ a.e. on $[0, +\infty)$. First we assume that the spectral measure σ has a bounded support with $\operatorname{supp} \sigma \subset [0, \infty)$ and that the corresponding Titchmarsh-Weyl coefficient Q_H has the representation $Q_H(z) = \int\limits_0^\infty \frac{d\sigma(\lambda)}{\lambda - z}$. Then the function $\tilde{Q}_H$ defined by $\tilde{Q}_H(z) := zQ_H(z) = -\int\limits_0^\infty d\sigma + \int\limits_0^\infty \frac{\lambda d\sigma(\lambda)}{\lambda - z}$ belongs to the Nevanlinna class $\mathbf{N}$. As $\operatorname{supp} \sigma$ is bounded, for all integers $k \geq 0$ the moments

$$s_k := \int\limits_0^\infty \lambda^k d\sigma(\lambda) < +\infty, \quad k = 0,\ 1,\ 2, \ldots,$$

exist. From $Q_H \in \mathbf{N}$ it follows that the sequence of moments $\{s_0,\ s_1,\ s_2, \ldots\}$ is nonnegative definite and as $\tilde{Q}_H \in \mathbf{N}$, the sequence of the moments corresponding to its spectral measure $d\tilde{\sigma}(\lambda) = \lambda d\sigma(\lambda)$ is equal to $\{s_1,\ s_2, \ldots\}$ and nonnegative definite. For the determinants

$$D_k := \begin{vmatrix} s_0 & s_1 & \cdots & s_k \\ s_1 & s_2 & \cdots & s_{k+1} \\ \vdots & \vdots & & \vdots \\ s_k & s_{k+1} & \cdots & s_{2k} \end{vmatrix} \quad \text{and} \quad B_k := \begin{vmatrix} s_1 & s_2 & \cdots & s_{k+1} \\ s_2 & s_3 & \cdots & s_{k+2} \\ \vdots & \vdots & & \vdots \\ s_{k+1} & s_{k+2} & \cdots & s_{2k+1} \end{vmatrix}$$

we get $D_k \geq 0$ and $B_k \geq 0$ (see [A]). Further, if $D_n = 0$ ($B_n = 0$) then $D_{n+k} = 0$ ($B_{n+k} = 0$, respectively) for all integers $k \geq 1$.

On the space of all polynomials a linear functional $\mathcal{S}$ is defined as follows: If $p(\lambda) := p_0 + p_1\lambda + \ldots + p_k\lambda^k$, then $\mathcal{S}(p) := p_0 s_0 + p_1 s_1 + \ldots + p_k s_k$. A sequence $\{P_k(\lambda)\}$ of orthogonal polynomials with respect to $\mathcal{S}$ (i.e. it holds $\mathcal{S}(P_i P_k) = \delta_{ik}$) is given by

$$P_0(\lambda) = (s_0)^{-1/2},\ P_k(\lambda) = \frac{1}{\sqrt{D_{k-1}D_k}} \begin{vmatrix} s_0 & s_1 & \cdots & s_k \\ \vdots & \vdots & & \vdots \\ s_{k-1} & s_k & \cdots & s_{2k-1} \\ 1 & \lambda & \cdots & \lambda^k \end{vmatrix} \quad \text{if } k \geq 1.$$

The polynomials $P_k(\lambda)$ are called *orthogonal polynomials* of the first kind, the orthogonal polynomials of the second kind $\{Q_k(\lambda)\}$ are given by

$$Q_k(\lambda) := \mathcal{S}\left(\frac{P_k(\lambda) - P_k(\cdot)}{\lambda - \cdot}\right).$$

With these orthogonal polynomials the Hamiltonian H corresponding to the Titchmarsh-Weyl coefficient Q_H can be constructed explicitely (see [KL2]), namely it holds

$$H(x) = \frac{1}{Q_k(0)^2 + P_k(0)^2} \begin{pmatrix} Q_k(0)^2 & -P_k(0)Q_k(0) \\ -P_k(0)Q_k(0) & P_k(0)^2 \end{pmatrix}$$
$$\text{if } \sum_{i=0}^{k-1} Q_i(0)^2 + P_i(0)^2 \leq x < \sum_{i=0}^{k} Q_i(0)^2 + P_i(0)^2.$$

With the determinants

$$C_k := \begin{vmatrix} 0 & s_0 & \cdots & s_{k-1} \\ s_0 & s_1 & \cdots & s_k \\ \vdots & \vdots & & \vdots \\ s_{k-1} & s_k & \cdots & s_{2k-1} \end{vmatrix}$$

the relations (see [A], [KL2])

$$Q_k(0)^2 = \frac{C_k^2}{D_k D_{k-1}}, \quad P_k(0)^2 = \frac{B_{k-1}^2}{D_k D_{k-1}}. \tag{3.1}$$

hold and Sylvesters rule implies

$$C_{k+1}B_{k-1} = C_k B_k - D_k^2. \tag{3.2}$$

From $D_n \neq 0$ it follows that $B_{n-1} \neq 0$, as $B_{n-1} = 0$ would imply that $B_n = 0$ and $D_n = 0$ by the relation (3.2). Again by (3.2) we get $B_n^2 + C_{n+1}^2 > 0$. If $D_{n+1} = 0$, from (3.1) it follows that $P_{n+1}(0)^2 + Q_{n+1}(0)^2 = +\infty$ and $[0, +\infty)$ consists of $n+1$ maximal H-indivisible intervals.

Let

$$v_k := \frac{-Q_k(0)}{P_k(0)}, \quad k = 0,\ 1, \ldots,$$

with $v_k := \infty$ if $P_k = 0$. Note that $v_0 = 0$. With the relations (3.1) and (3.2) we get

$$v_{k+1} = v_k + \frac{D_k^2}{B_{k-1}B_k}.$$

This shows that $\{v_k\}$ is a nondecreasing sequence. Equivalently, the function

$$v(x) := \frac{h_3(x)}{h_2(x)}, \quad x \in \mathcal{D}(v) := \{x | x \geq 0,\ h_2(x) > 0\}, \tag{3.3}$$

is a nondecreasing step function. Note that $\mathcal{D}(v)$ is a connected set. If H_1 is the Hamiltonian corresponding to the Weyl coefficient $Q_1 := a + Q$, $a \in \mathbb{R}$, for the function v_1 defined by H_1 according to the relation (3.3) it follows from Lemma 3.2 of [W1] that $v_1(x) = v(x) + a$.

Now let Q_H be a Nevanlinna function with supp $\sigma \subset [0, +\infty)$. Using the approximation principle of Lemma 2.1, we choose a sequence of Nevanlinna functions Q_{H_n} with discrete spectral functions such that $Q_{H_n} \to Q_H$ locally uniformly in $\mathbb{C}^+$.

As for each H_n the function v_n is nondecreasing, the components of H_n are functions of uniformly bounded variation. By Helly's Theorem, there exists a subsequence $\{H_{n_k}\}$ such that $H_{n_k}(x)$ converges to a matrix function $\tilde{H}(x)$ at each $x < \infty$. As the components of H_{n_k} are uniformly bounded by 1, the theorem of Lebesgue implies

$$\int_0^x H_{n_k}(t)dt \to \int_0^x \tilde{H}(t)dt, \ k \to \infty,$$

for each x. On the other hand, it holds $H_{n_k} \to H$ in the sense as stated in Lemma 2.1, and by the uniqueness theorem of de Branges it follows that $\tilde{H}$ and H are equivalent, that is $\tilde{H} = H$. As $\det H_{n_k}(x) = 0$ for all $x \in [0, \infty)$ and H_{n_k} converges to H a.e. on $[0, \infty)$, it holds $\det H = 0$ a.e. on $[0, \infty)$. Note that $v_n(x) \to v(x) := \frac{h_1(x)}{h_3(x)}$, $x < \infty$, where v is a nondecreasing function.

If supp $\sigma \subset (-l, \infty)$, $l > 0$, we use a transformation formula from [W2] which represents the Hamiltonian corresponding to σ in terms of the Hamiltonian corresponding to the spectral measure $\tilde{\sigma}$ which arises by "shifting" supp σ to the right such that supp $\tilde{\sigma} \subset [0, \infty)$: If W is the fundamental matrix of the canonical system corresponding to H, by

$$\begin{aligned}
\tilde{H}(\tilde{x})d\tilde{x} &= W(x, -l)H(x)W(x, -l)^T dx, \\
\tilde{x}(x) &= \operatorname{tr}\left(\int_0^x W(t, -l)H(t)W(t, -l)^T dt\right)
\end{aligned}$$

a Hamiltonian $\tilde{H}$ is defined whose Titchmarsh-Weyl coefficient $\tilde{Q}_{\tilde{H}}$ has the property

$$\tilde{Q}_{\tilde{H}}(z) = Q_H(z-l), \ d\tilde{\sigma}(\lambda) = d\sigma(\lambda - l).$$

As $\det W(x) = 1$, it follows that $\det H(x) = 0$ if $\det \tilde{H}(x) = 0$.

If supp $\sigma \subset (-\infty, l)$, note that $\tilde{H}(x) = DH(x)D$ with $D = \begin{pmatrix} -1 & 0 \\ 0 & 1 \end{pmatrix}$ is the Hamiltonian corresponding to the spectral measure $d\tilde{\sigma}(\lambda) := d\sigma(-\lambda)$ and again it follows that $\det H = 0$ if $\det \tilde{H} = 0$. This finishes the proof of the first statement of the theorem.

Now we show that the components of H are functions of locally bounded variation. If supp $\sigma \subset [0, \infty)$, for $x \in \mathcal{D}(v)$ the Hamiltonian H is of the form

$$H(x) = \left(1 + v(x)^2\right)^{-1} \begin{pmatrix} v(x)^2 & v(x) \\ v(x) & 1 \end{pmatrix}.$$

If $x \notin \mathcal{D}(v)$ it holds $h_2(x) = h_3(x) = 0$ and $h_1(x) = 1$. As v is a nondecreasing function, the components of H are of locally bounded variation on $[0, \infty)$.

If supp $\tilde{\sigma} \subset (-l, \infty)$ with $l > 0$, then the corresponding Hamiltonian $\tilde{H}$ has a representation

$$\tilde{H}(\tilde{x}) = W(x,l)H(x)W(x,l)^T \left(\text{tr } (W(x,l)H(x)W(x,l)^T)\right)^{-1},$$

where the spectral measure σ corresponding to H satisfies supp $\sigma \subset [0, \infty)$. Let $\tilde{E} \subset [0, \infty)$ be any finite interval, and let E be the set of all $x \in \mathbb{R}$ with $\tilde{x}(x) \in \tilde{E}$. Then E is also a finite interval (see [W2]). As the entries of $W(\cdot, l)$ and their derivatives are bounded on E, it follows that the components of $W(\cdot, l)HW(\cdot, l)^T$ are of bounded variation on E (see [N]). As $\det H = 0$, H is of the form $H(x) = \xi_{\phi(x)}\xi_{\phi(x)}^T$. Then $r(x) := \text{tr } (W(x,l)H(x)W(x,l)^T)$ is equal to $\|W(x,l)\xi_{\phi(x)}\|^2$, where $\|\cdot\|$ denotes the Euclidian norm. If λ_{min} is the smallest eigenvalue of $W(x,l)^T W(x,l)$, then $r(x) \geq \lambda_{min}$ (see [G]). Let $K(x) := \text{tr } (W(x,l)^T W(x,l))$. From $\det W(x,l) = 1$ it follows by computation that $K(x) \geq 2$. If λ_1 and λ_2 are the eigenvalues of $W(x,l)^T W(x,l)$, then $\lambda_1 \lambda_2 = 1$ and $\lambda_1 + \lambda_2 = K(x)$, that is $\lambda_{min} + \lambda_{min}^{-1} = K(x)$. This yields $\lambda_{min} \geq K(x)^{-1}$. Let $c_K < \infty$ be an upper bound of $K(\cdot)$ on E. Then it holds $r(x) \geq c_K^{-1}$ if $x \in E$. As r is on E of bounded variation, this implies that $r(x)^{-1}$, $x \in E$, is of bounded variation (see [N]). It follows that the components of $\tilde{H}$ are of bounded variation on $\tilde{E}$. □

Corollary 3.2. *If the spectral measure σ of the canonical system with Hamiltonian H has the property supp $\sigma \subset [0, \infty)$, then the set $\mathcal{D}(v) := \{x | x \geq 0, \ h_2(x) > 0\}$ is connected and the function*

$$v(x) := \frac{h_3(x)}{h_2(x)} \left(= \frac{h_1(x)}{h_3(x)} \right), \quad x \in \mathcal{D}(v), \tag{3.4}$$

is nondecreasing. If supp $\sigma \subset (0, \infty)$, then $\mathcal{D}(v) = (b, \infty)$.

Proof. The first statement of the corollary follows from the proof of Theorem 3.1. Let $\sup \mathcal{D}(v) =: d_0 < \infty$. Then $h_2(x) = 0$ if $x \in (d_0, \infty)$, it follows that $\int_0^\infty h_2(t)dt < \infty$. By Theorem 2.2, $\sigma(\{0\}) = \left(\int_0^\infty h_2(t)dt\right)^{-1} > 0$. □

At the points where $h_2 > 0$ the Hamiltonian of a semibounded canonical system is characterized by the function v. It turns out that it is sometimes more convenient to consider Hamiltonians which are normalized as follows:

$$\text{(3.5)} \qquad H(x) = \begin{cases} \begin{pmatrix} v^2(x) & v(x) \\ v(x) & 1 \end{pmatrix} & \text{if } h_2(x) \neq 0, \\ \begin{pmatrix} 1 & 0 \\ 0 & 0 \end{pmatrix} & \text{if } h_2(x) = 0. \end{cases} \qquad x \in [0, L),\ L \leq +\infty,$$

Note that the function v^2 is locally integrable on $[0, L)$ as $\int_0^x \mathrm{tr}\ H(t)dt < +\infty$ if $x < L$.

Theorem 3.3. *A canonical system which is semibounded from below has a Hamiltonian of the form* (3.5) *with the following properties: Let I_k, $k = 1, 2, \ldots$ be the maximal H-indivisible intervals of type* 0, *and let $E := [0, L) \setminus \bigcup_k \overline{I_k}$. Then the intervals I_k can only accumulate at L. There is an at most countable number of critical points $x_i \in E$, $i = 1, 2, \ldots$, whose only possible accumulation point is L such that in each interval of $E \setminus \{x_1, x_2, \ldots\}$ the function v is nondecreasing and right-continuous. A critical point $x_i \in \mathcal{D}(v)$ is either a (finite) negative jump of v, that is $v(x_i+) - v(x_i-) < 0$, or it is a singularity with $v(x_i-) = +\infty$ or $v(x_i+) = -\infty$.*

Proof. Let l be a lower bound of the spectrum of A such that $l < 0$. By $d\tilde{\sigma}(\lambda) := d\sigma(\lambda + l)$ the spectral measure $\tilde{\sigma}$ of a canonical system with Hamiltonian $\tilde{H}$ defined on $[0, \tilde{L})$ and fundamental matrix $\tilde{W}$ is given such that with

$$Z_l := \{s |\ \tilde{w}_{21}(s, -l)\tilde{v}(s) + \tilde{w}_{22}(s, -l) = 0\}$$

it holds

$$\text{(3.6)} \qquad \begin{aligned} H(x)dx &= \tilde{W}(t, -l)\tilde{H}(t)\tilde{W}(t, -l)^T dt, \\ x(t) &= \int_0^t (\tilde{w}_{21}(s, -l)\tilde{v}(s) + \tilde{w}_{22}(s, -l))^2 ds \\ &\quad + \int_0^t \chi_{Z_l}(w_{11}(s, -l)\tilde{v}(s) + \tilde{w}_{12}(s, -l))^2 ds, \\ L &= \lim_{t \to \tilde{L}} x(t) \end{aligned}$$

(see [W2]). As supp $\tilde{\sigma} \subset (0,\infty)$, the Hamiltonian $\tilde{H}$ is of the form (3.5) with a nondecreasing and right-continuous function $\tilde{v}$ defined on $[0,\tilde{L})$. From the relations (3.6) it follows that $x(t)$ is a strictly increasing function of t, in particular it follows that $x(t)$ belongs to an indivisible interval of H if and only if t belongs to an indivisible interval of $\tilde{H}$. Suppose that $\tilde{w}_{21}(\cdot,-l) = 0$ vanishes at t_0. As $\det \tilde{W} \equiv 1$, it follows that t_0 is not a zero of $\tilde{w}_{22}(\cdot,-l)$ and hence $\tilde{w}_{21}'(t_0,-l) = -l\tilde{w}_{22}(t_0,-l) \neq 0$. This shows that the zeros of $\tilde{w}_{21}(\cdot,-l)$ are isolated with the only possible accumulation point $\tilde{L}$. If $\tilde{w}_{21}(t,-l) \neq 0$, a computation implies

$$\frac{d}{dt}\left(\frac{\tilde{w}_{22}(t,-l)}{\tilde{w}_{21}(t,-l)}\right) = -l\left(\frac{\tilde{w}_{22}(t,-l)}{\tilde{w}_{21}(t,-l)} + \tilde{v}(t)\right)^2,$$

that is, $\frac{\tilde{w}_{22}(\cdot,-l)}{\tilde{w}_{21}(\cdot,-l)}$ is nondecreasing. Consequently, between two consecutive zeros of $\tilde{w}_{21}(\cdot,-l)$ there is at most one interval or point where $\tilde{w}_{21}(t,-l)\tilde{v}(t)+\tilde{w}_{22}(t,-l) = 0$ is satisfied or where $v(t-) + \frac{\tilde{w}_{22}(t,-l)}{\tilde{w}_{21}(t,-l)}$ and $v(t+) + \frac{\tilde{w}_{22}(t,-l)}{\tilde{w}_{21}(t,-l)}$ have different signs. If $\tilde{w}_{21}(t,-l)\tilde{v}(t) + \tilde{w}_{22}(t,-l) \neq 0$ it holds

$$v(x(t)) = \frac{\tilde{w}_{11}(t,-l)\tilde{v}(t) + \tilde{w}_{12}(t,-l)}{\tilde{w}_{21}(t,-l)\tilde{v}(t) + \tilde{w}_{22}(t,-l)}$$

and a short computation gives

$$dv(x(t)) = \frac{d\tilde{v}(t)}{(\tilde{w}_{21}(t,-l)\tilde{v}(t-) + \tilde{w}_{22}(t,-l))(\tilde{w}_{21}(t,-l)\tilde{v}(t+) + \tilde{w}_{22}(t,-l))}.$$

This yields that v is nondecreasing between the zeros of $\tilde{w}_{21}(\cdot,-l)\tilde{v}(\cdot) + \tilde{w}_{22}(\cdot,-l)$ up to the points where $v(t-) + \frac{\tilde{w}_{22}(t,-l)}{\tilde{w}_{21}(t,-l)}$ and $v(t+) + \frac{\tilde{w}_{22}(t,-l)}{\tilde{w}_{21}(t,-l)}$ have different signs. The intervals where the function $\tilde{w}_{21}(\cdot,-l)\tilde{v}(\cdot) + \tilde{w}_{22}(\cdot,-l)$ is equal to zero correspond by (3.6) to the H-indivisible intervals of type 0 and the isolated zeros of $\tilde{w}_{21}(\cdot,-l)\tilde{v}(\cdot) + \tilde{w}_{22}(\cdot,-l)$ correspond to the singularities of v. □

Remark 3.4. Let

$$\tilde{x}(x) := \int_0^x ((v^2(t) + \chi_{\{v=0\}}(t))\chi_{\mathcal{D}(v)}(t) + \chi_{\{h_2=0\}}(t))dt, \quad \tilde{L} := \lim_{x\to L} \tilde{x}(x)$$

and for $x \in [0,L)$, $L \leq +\infty$,

$$\tilde{H}(\tilde{x}) := \begin{cases} \begin{pmatrix} 1 & -v(x)^{-1} \\ -v(x)^{-1} & v^{-2}(x) \end{pmatrix} & \text{if } v(x) \neq 0, \\ \begin{pmatrix} 0 & 0 \\ 0 & 1 \end{pmatrix} & \text{if } v(x) = 0, \\ \begin{pmatrix} 1 & 0 \\ 0 & 0 \end{pmatrix} & \text{if } h_2(x) = 0, \end{cases}$$

that is, with $\mathcal{D}(m) = \{\tilde{x} | \tilde{h}_1(\tilde{x}) > 0\}$, $m(\tilde{x}) := -v(x)^{-1}$ if $v(x) \neq 0$ and $m(\tilde{x}) = 0$ if $h_2(x) = 0$ we can write for $\tilde{x} \in [0, \tilde{L})$, $\tilde{L} \leq +\infty$,

$$\tilde{H}(\tilde{x}) = \begin{cases} \begin{pmatrix} 1 & -m(\tilde{x}) \\ -m(\tilde{x}) & m(\tilde{x})^2 \end{pmatrix} & \text{if } \tilde{h}_1(\tilde{x}) > 0, \\ \begin{pmatrix} 0 & 0 \\ 0 & 1 \end{pmatrix} & \text{if } \tilde{h}_1(\tilde{x}) = 0. \end{cases} \tag{3.7}$$

Note that $\tilde{x}(x)$ is strictly increasing with respect to x and that $\tilde{H}$ and H correspond to the same Titchmarsh-Weyl coefficient Q. By the definition of m it follows that $m(\tilde{x})$ is locally nondecreasing at the points $\tilde{x}(x)$ where $v(x)$ is locally nondecreasing and different from 0. That is, $m(\tilde{x}(x))$ has possibly negative jumps or singularities at the critical points of v, at the isolated zeros of v, or at the end points of the intervals I_k. (Note that $m(\tilde{x}(x)) = 0$ if $x \in I_k$.) It follows that m is locally nondecreasing up to an at most countable number of critical points $\tilde{x}_i$, whose only possible accumulation point is L. For more results concerning semibounded canonical systems with Hamiltonians of the form (3.7) see [LW].

In the same way as Theorem 3.3 one can prove a corresponding result about canonical systems which are semibounded from above.

Theorem 3.5. *A canonical system which is semibounded from above has a Hamiltonian of the form* (3.5) *with the following properties: Let I_k, $k = 1, 2, \dots,$ be the maximal H-indivisible intervals of type* 0, *and let $E := [0, L) \setminus \bigcup_k \overline{I_k}$. Then the intervals I_k can only accumulate at L. There is an at most countable number of critical points $x_i \in E$, $i = 1, 2, \dots,$ whose only possible accumulation point is L such that in each interval of $E \setminus \overline{\{x_1, x_2, \dots\}}$ the function v is nonincreasing and right-continuous. A critical point $x_i \in \mathcal{D}(v)$ is either a (finite) positive jump of v, that is $v(x_i+) - v(x_i-) < 0$, or it is a singularity with $v(x_i-) = -\infty$ or $v(x_i+) = +\infty$.*

4. The main result

A canonical system has κ ($< \infty$) negative eigenvalues if its spectral measure σ has the property that $\operatorname{supp} \sigma \cap (-\infty, 0)$ is a set of κ points, that is, the corresponding operator A has κ negative eigenvalues. According to Theorem 3.1., the corresponding Hamiltonian H has the property that $\det H \equiv 0$, The following theorem establishes a connection between the number of negative eigenvalues of the canonical system or, equivalently, the number of negative squares of the form $(Af, f)_{L^2_H}$, and the Hamiltonian H with the corresponding function v.

Theorem 4.1. *Suppose that the canonical system has a finite number κ of negative eigenvalues. Then the Hamiltonian H is of the form* (3.5) *with the following properties.*

1. *There is a finite number κ_1 of bounded and maximal H-indivisible intervals $I_1, \ldots, I_{\kappa_1}$ of type* 0.
2. *There is a finite number κ_2 of critical points $x_i \in E := [0, L) \setminus \bigcup_k \overline{I_k}$ such that in each interval of $E \setminus \{x_1, \ldots, x_{\kappa_2}\}$ the function v is nondecreasing and right-continuous. At a critical point the function v has either a negative jump or a singularity.*
3. *If* 0 *is the left end point of an H-indivisible interval of type* 0 *it holds $\kappa = \kappa_1 + \kappa_2 - 1$, otherwise, if* 0 *is not an end point of an H-indivisible interval of type* 0, *it holds $\kappa = \kappa_1 + \kappa_2$.*

Conversely, if the Hamiltonian H of the form (3.5) has the properties 1 and 2, then the canonical system has a finite number of negative eigenvalues.

Proof. A maximal H-indivisible interval $(0, b)$ of type 0 with $b > 0$ corresponds to the linear term bz in the spectral representation formula (2.5) of the function Q_H (see [dB2], [W1]). Such an interval has no influence on the number of negative squares, so we can restrict ourselves to the case that the Hamiltonian H satisfies the conditions a) and b) of Section 2.

Assume that a canonical system with a finite number of negative eigenvalues is given. As the operator A is semibounded from below, there is a Hamiltonian H corresponding to A of the form as stated in Theorem 3.3. We are going determine the number of negative squares of the form $(Af, f)_{L^2_H}$.

The canonical system $Jf' = -H(Af)$ given at $[0, L)$, $L \leq \infty$, reads as

$$
\begin{aligned}
f_2'(x) &= h_1(x)(Af)_1(x) - h_3(x)(Af)_2(x), \\
f_1'(x) &= -h_3(x)(Af)_1(x) - h_2(x)(Af)_2(x).
\end{aligned} \tag{4.1}
$$

With $f, g \in \mathcal{D}(A)$ it follows that the form $\mathbf{t}[f, g] := (Af, g)_{L^2_H}$ can be written as

$$
\begin{aligned}
\mathbf{t}[f, g] &= \int_0^L g^*(x) H(x)(Af)(x)dx = -\int_0^L g^*(x) J f'(x) dx \\
&= \int_0^L (f_2'(x)\overline{g_1(x)} - f_1'(x)\overline{g_2(x)})dx.
\end{aligned}
$$

Let $I_k = (a_k, b_k)$, $k = 0, 1, \ldots$, be the ordered sequence of all maximal H-indivisible intervals of type 0 and let $E := [0, L) \setminus \bigcup_k \overline{I_k}$.

In the following we restrict ourselves to canonical systems with a finite number of negative eigenvalues such that the function v in (3.5) is of locally bounded variation. Then it holds $L = \infty$ and v is finite on E.

On the set E it holds $f_2'(x) = -v(x)f_1'(x)$ and $g_2'(x) = -v(x)g_1'(x)$. On (b_{k-1}, a_k), a component of E, $(b_{-1} := 0)$ integration by parts yields

$$\int_{b_{k-1}}^{a_k} (f_2'(x)\overline{g_1(x)} - f_1'(x)\overline{g_2(x)})dx = -f_1(x)(v(x)\overline{g_1(x)} + \overline{g_2(x)})|_{b_{k-1}}^{a_k} + \int_{b_{k-1}}^{a_k} f_1(x)\overline{g_1(x)}dv(x)$$

as $d(g_1(x)v(x)+g_2(x)) = g_1(x)dv(x)$. On $I_k = (a_k, b_k)$ as $f_1' = 0$ and g_1 is constant we get:

$$\int_{a_k}^{b_k} (f_2'(x)\overline{g_1(x)} - f_1'(x)\overline{g_2(x)})dx = (f_2(b_k) - f_2(a_k))\overline{g_1(a_k)}.$$

Summing over the intervals (b_{k-1}, b_k) it follows

$$\begin{aligned}(4.2)\quad \mathbf{t}[f,g] = & \int_{[0,\infty)\setminus B} f_1(x)\overline{g_1(x)}dv(x) + f_1(0)\overline{g_1(0)}v(0) \\ & + \sum_k (f_1(b_i)\overline{g_1(b_i)}(v(b_k) - v(a_k)) \\ & + \sum_k (f_2(b_k) - f_2(a_k))\overline{g_1(a_k)} + (\overline{g_2(b_k)} - \overline{f_2(a_k)})f_1(a_k).\end{aligned}$$

On the H-indivisible intervals of type 0 the functions f_2' and g_2' are constant as $f, g \in \mathcal{D}(A)$. Note that $f_2' = g_2' = 0$ on an H-indivisible interval of type 0 of the form (L, ∞). Let $z_k := \sum_{i=0}^k (a_i - b_{i-1})$ and let c_f and g_f be functions defined on $\{z_k\}$ such that $c_f(z_k) := f_2'|_{(a_k,b_k)}$ and $g_f(z_k) := g_2'|_{(a_k,b_k)}$. If we connect the components of E to an interval of length

$$l := \int_0^L \chi_E(t)dt, \ (l \leq \infty),$$

and include the jumps $v(b_k) - v(a_k)$ of v at the endpoints of the intervals I_k in the integral, after a rescaling we can write the relation (4.2) in the form

$$(4.3)\qquad \mathbf{t}[f,g] = \int_0^l f_1\overline{g_1}dv + \int_0^l (f_2\overline{c_g} + c_f\overline{g_2})d\mu,$$

where the measure μ is defined on $\{z_k\}$ by $d\mu(z_k) := b_k - a_k$. For $f \in \mathcal{D}(A)$ it holds

$$(f,f)_{L_H^2} = \int_0^l |vf_1 + f_2|^2 dx + \int_0^l |f_2|^2 d\mu,$$

and with $v(x)f_1(x) + f_2(x) = \int_0^x f_1 dv + \int_0^x c_f d\mu + f_2(0)$ it follows that

$$(4.4) \qquad (f,f)_{L^2_H} = \int_0^l \left| \int_0^x f_1 dv + \int_0^x c_f d\mu + f_2(0) \right|^2 dx + \int_0^l |f_1|^2 d\mu.$$

As A is a semibounded essentially selfadjoint operator, from [K], Chapter 6, Theorem 1.27 it follows that the form $\mathbf{t}$ is closable. The closure of $\mathbf{t}$ in L^2_H we denote with $\bar{\mathbf{t}}$ and its domain with $\mathcal{D}(\bar{\mathbf{t}})$. Let $|v|(x)$ denote the total variation of v on $[0,x)$:

$$(4.5) \qquad |v|(x) := \sup \left\{ \sum_{i=1}^{n} |v(x_{i+1}) - v(x_i)| \, |0 \le x_1 \le \ldots \le x_{n+1} < x \right\}.$$

Lemma 4.2. *Let a canonical system of the form (3.5) be given such that the corresponding operator A is semibounded from below and that v is a function of locally bounded variation. Let $f_1 \in L^2_{|v|}$ be a function with compact support such that f_1 is linear on $[0,l) \setminus$ supp $|v|$, and $\int_0^l |f_1|^2 d\mu < \infty$. Let $c_f \in L^1_\mu \cap L^2_\mu$ and $c_0 \in \mathbb{R}$ such that $F_2(x) := \int_0^x f_1 dv + \int_0^x c_f d\mu + c_0$ has the property $F_2 \in L^2_{[0,l)}$.*

With $f_2(x) := v(x)f_1(x) + \int_0^x f_1 dv + \int_0^x c_f d\mu + c_0$ the vector function $f := (f_1, f_2)^T$ belongs to $\mathcal{D}(\bar{\mathbf{t}})$ and

$$(4.6) \qquad \bar{\mathbf{t}}[f,f] = \int_0^l |f_1|^2 dv - \int_0^L (f_1 \overline{c_f} + c_f \overline{f_1}) d\mu.$$

For functions f_1, c_f and g_1, c_g with the assumed properties it holds

$$(4.7) \qquad \bar{\mathbf{t}}[f,g] = \int_0^l f_1 \overline{g_1} dv + \int_0^l (f_1 \overline{c_g} + c_f \overline{g_1}) d\mu.$$

Proof. Using [K], Chapter 6, Theorem 1.16 we have to show that f can be approximated by a sequence $\{f_n\} \subset \mathcal{D}(A)$ such that $\|f - f_n\|_{L^2_H} \to 0$ $(n \to \infty)$ and $\sup \mathbf{t}[f_n, f_n] < \infty$.

Let C^1_c be the class of all continuously differentiable functions on $[0,l)$ which vanish outside a compact set and are linear at $[0,L) \setminus$ supp $|v|$ and let $\hat{C}^1_c$ be the class of all functions $f \in C^1_c$ with $f(l) = 0$ if $l < \infty$. As $|v|(\{l\}) = 0$ the class $\hat{C}^1_c$ is dense in $L^2_{|v|}$. If $l = \infty$ this is wellknown [C], if $l < \infty$ there is either an interval

$I = (l - \epsilon, l)$, $\epsilon > 0$, where v is constant, then $\|f\|_{L^2_{|v|}}$ is independent of the values of f at I, or there is a sequence $\{x_n\} \subset \operatorname{supp} |v|$ with $x_n \to l$ $(n \to \infty)$. Then it is easy to see that the linear functional $F(f) := f(l)$, $\mathcal{D}(F) = C^1_c$, is unbounded in $L^2_{|v|}$, which implies that the kernel of F is dense in $L^2_{|v|}$.

Let $\hat{f}_1$ belong to $\hat{C}^1_c$ and let $\hat{f}_1$, $c_{\hat{f}}$ and c_0 satisfy the assumptions of the lemma. Then it holds $\hat{f}_2' = v\hat{f}_1'$ and the function $\hat{f} = (\hat{f}_1, \hat{f}_2)^T$ belongs to $\mathcal{D}(A)$.

Let f_1 and c_f be as in the lemma. If $l = \infty$ choose $K < \infty$ such that $f_1 = 0$ on the interval (K, l). If $l < \infty$ put $l = K$. Now choose $\hat{f}_1 \in \hat{C}^1_c$ with $\hat{f}_1 = 0$ on (K, l) and $\int_0^K f_1 dv = \int_0^K \hat{f}_1 dv$ and with $f_1(z_i) = \hat{f}_1(z_i)$ at finitely many points $z_i \in \operatorname{supp} \mu$, such that $\int_0^l |f_1 - \hat{f}_1|^2 d\mu < \epsilon$, $\epsilon > 0$, and the relation $\int_0^l |f_1 - \hat{f}_1|^2 d|v| < \epsilon$. holds. Then it follows from the Cauchy-Schwartz inequality

$$\int_0^K \left| \int_0^x (f_1 - \hat{f}_1) dv \right|^2 dx \leq K|v|(K) \int_0^l |f_1 - \hat{f}_1|^2 d|v| < \epsilon K |v|(K).$$

With $c_{\hat{f}} := c_f$ and $\hat{c}_0 = c_0$ we get the relations

$$(f - \hat{f}, f - \hat{f})_{L^2_H} \leq \epsilon(1 + K|v|(K)).$$

Again by the Cauchy-Schwartz inequality we get

$$|\mathbf{t}[\hat{f}, \hat{f}]| \leq \int_0^l |\hat{f}_1|^2 d|v| + 2 \left(\int_0^l |\hat{f}_1|^2 d\mu \right)^{1/2} \left(\int_0^l |c_f|^2 d\mu \right)^{1/2},$$

this relation implies

$$|\mathbf{t}[\hat{f}, \hat{f}]| \leq \epsilon + \int_0^l |f_1|^2 d|v| + 2 \left(\epsilon + \int_0^l |f_1|^2 d\mu \right)^{1/2} \left(\int_0^l |c_f|^2 d\mu \right)^{1/2}.$$

Let $\{\epsilon_n\}$ be a sequence of positive numbers with $\epsilon_n \to 0$ if $n \to \infty$ and let $\hat{f}^n \in \mathcal{D}(A)$ satisfy the above relations with ϵ_n instead of ϵ. Then it is easy to see that the sequence $\{\hat{f}^n\}$ satisfies the relations $\|f - \hat{f}^n\|_{L^2_H} \to 0$ $(n \to \infty)$ and $\sup \mathbf{t}[\hat{f}^n, \hat{f}^n] < \infty$. This shows that the assumptions of [K], Chapter 6, Theorem 1.16 are satisfied, which implies the relations (4.6) and (4.7). □

In order to determine the number of negative squares of the form (4.6) we can restrict us to consider real sample functions, since the components of the corresponding eigenfunctions of A are real too. In the sequel we denote the function

f_2 simply by f and with $\{z_k|\ k=0,1,\ldots\} = \operatorname{supp}\mu$ let $C_k = c_f(z_k)d\mu(z_k)$. Then the form (4.6) becomes

$$\int_0^l f^2 dv + 2\sum_k f(z_k)C_k. \tag{4.8}$$

If $l=\infty$ the number c_0 must be chosen such that

$$0 = \int_0^\infty f dv + \sum_k C_k + c_0.$$

As f has a compact support the integral on the right-hand side of the relation (4.4) exists.

Choosing suitable f and C_k, it is easy to see that the form (4.8) gets a negative square at each point $z_k \in \operatorname{supp}\mu$ and also each negative jump of v which does not belong to supp μ gives one additional negative square. To show this, at a point z_k we simply choose a function f with $f(z_k)=1$ and $f(x)=0$, $x \neq z_k$ and a suitable C_k such that $v(z_k) - v(z_k-) + 2C_k < 0$. This proves the assertion of the theorem under the assumption that the function v in the representation (3.5) has no singularities.

Now we consider the general case, that the function v in the representation (3.5) can also become singular. The operator A has κ negative eigenvalues by assumption and κ' denotes the number of critical points, that is the points belonging to supp μ or where v has a negative jump or a singularity. Let x_i, $i=1,\ldots,\kappa_s$ be the $\kappa_s (\leq \infty)$ singularities of v. We consider the form

$$\mathbf{t}[f,f] = \int_0^L (f_2'(x)\overline{f_1(x)} - f_1'(x)\overline{f_2(x)})dx$$

for functions $f \in \mathcal{D}(A)$ which are constant on intervals $(x_i - \epsilon_i, x_i + \epsilon_i)$ for some $\epsilon_i > 0$. Then we get

$$\mathbf{t}[f,f] = \int_{[0,L)\setminus\cup_i(x_i-\epsilon_i,x_i+\epsilon_i)} (f_2'(x)\overline{f_1(x)} - f_1'(x)\overline{f_2(x)})dx$$

and integration by parts using $vf_1' = -f_2'$ on $[0,L)\setminus \cup_i(x_i-\epsilon_i, x_i+\epsilon_i)$ implies with $\epsilon_0 = \epsilon_{\kappa_s+1} = 0$, $x_0 = 0$, $x_{\kappa_s+1} = l$,

$$\begin{aligned} \mathbf{t}[f,f] &= \sum_{i=0}^{x_{\kappa_s}} \int_{x_i+\epsilon_i}^{x_{i+1}-\epsilon_{i+1}} |f_1|^2 dv + 2\sum_k f_1(z_k)C_k \\ &+ \sum_{i=0}^{x_{\kappa_s}} |f_2(x_i+\epsilon_i)|^2 (v(x_i+\epsilon_i) - v(x_i-\epsilon_i)), \end{aligned}$$

where after a change of scale for the function f for convenience the same notation is used. Now we choose the ϵ_i in such a way that $v(x_i+\epsilon_i)-v(x_i-\epsilon_i)<0$ and that no other critical point is contained in the interval $(x_i-\epsilon_i, x_i+\epsilon_i)$. That is, for the corresponding functions f the singularities act as negative jumps. By a choice of suitable sample functions f as above, paying attention that the constancy of f on the intervals $(x_i-\epsilon_i, x_i+\epsilon_i)$ is a restriction, we get the relation $\kappa' \leq \kappa$.

On the other hand, according to Lemma 2.1 we can approximate the corresponding Titchmarsh-Weyl coefficient Q_H by a sequence Q_{H_n} where each Hamiltonian H_n has κ' critical points without singularities. Let $\{\epsilon_i^{(n)} | i=1,\ldots,\kappa_s,\ n\in \mathbb{N}\}$ be a sequence of positive numbers with $\sum_{i=1}^{\kappa_s}\epsilon_i^{(n)} \to 0$ $(n\to\infty)$ such that $v(x_i+\epsilon_i^{(n)})-v(x_i-\epsilon_i^{(n)})<0$ for each $\epsilon_i^{(n)}$. If H is the Hamiltonian of the canonical system corresponding to A, we define a sequence of Hamiltonians $\{H_n\}$ by $H_n(x)=\begin{pmatrix}1&0\\0&0\end{pmatrix}$ on the intervals $(x_i-\epsilon_i^{(n)}, x_i+\epsilon_i^{(n)})$ and $H_n(x)=H(x)$ outside of these intervals. Then it holds $H_n \to H$ $(n\to\infty)$ in the sense of Lemma 2.1, that is $Q_{H_n}\to Q_H$ $(n\to\infty)$ uniformly on compact sets of $\mathbb{C}^+$. On $(-\infty,0)$ the spectral measure of each Q_{H_n} consists of κ' point masses. It follows that σ_H has not more than κ' point masses on $(-\infty,0)$, which implies $\kappa=\kappa'$.

In order to prove the converse direction, assume that a Hamiltonian of the form (3.5) has κ critical points. If there are no singularities among the critical points, we have already shown that the corresponding canonical system has κ negative eigenvalues. If there are singularities, by an approximation corresponding to Lemma 2.1 as above it is shown that A has a finite number $\kappa' \leq \kappa$ of negative eigenvalues, and by the first direction of the proof it follows that $\kappa=\kappa'$. □

In the same way as Theorem 4.1 one can prove the following result:

Theorem 4.3. *Suppose that the canonical system has a finite number κ of positive eigenvalues. Then the Hamiltonian H is of the form* (3.5) *with the following properties:*

1. *There is a finite number κ_1 of bounded and maximal H-indivisible intervals $I_1,\ldots,I_{\kappa_1}$ of type* 0.
2. *There is a finite number κ_2 of critical points $x_i \in E := [0,L)\setminus \bigcup_k \overline{I_k}$ such that in each interval of $E\setminus\{x_1,\ldots,x_{\kappa_2}\}$ the function v is nonincreasing and right-continuous. At a critical point the function v has either a positive jump or a singularity.*
3. *If* 0 *is the left end point of an H-indivisible interval of type* 0 *it holds $\kappa=\kappa_1+\kappa_2-1$, otherwise, if* 0 *is not an end point of an H-indivisible interval of type* 0, *it holds $\kappa=\kappa_1+\kappa_2$.*

Conversely, if the Hamiltonian H of the form (3.5) *has the properties 1 and 2, then the canonical system has a finite number of positive eigenvalues.*

Acknowledgements

I want to thank Prof. Heinz Langer for fruitful and inspiring discussions.

References

[AG] N.I. Achieser and I.M. Glasmann: *Theorie der linearen Operatoren im Hilbert Raum.* Akademie-Verlag, Berlin, 1954.

[A] N.I. Akhiezer: *The Classical Moment Problem.* Oliver & Boyd, Edinburgh, 1965.

[dB1-4] L. de Branges: Some Hilbert spaces of entire functions. Trans. Amer. Math. Soc. **96** (1960), 259–295; **99** (1961), 118–152; **100** (1960), 73–115; **105** (1962), 43–83.

[dB5] L. de Branges: *Hilbert Spaces of Entire Functions.* Prentice Hall, Englewood Cliffs, N.J., 1968.

[C] D.L. Cohn: *Measure Theory.* Birkhäuser Verlag, Boston, 1980.

[DI] H. Dym and A. Iacob: Positive definite extensions, canonical equations and inverse problems. *Operator Theory: Advances and Applications,* vol.**12** Birkhäuser Verlag, Basel, 1984, pp. 141–240.

[G] F.R. Gantmacher: *Matrizentheorie.* Deutscher Verlag der Wissenschaften, Berlin, 1986.

[K] T. Kato: *Perturbation Theory for Linear Operators.* Springer-Verlag, Berlin, Heidelberg, New York, 1980.

[Ka] I.S. Kac: Linear relations, generated by a canonical differential equation on an interval with a regular endpoint, and expansibility in eigenfunctions (Russian). Deposited paper 517.9, Odessa, 1984.

[KL1] M.G. Krein and H. Langer: Über einige Fortsetzungsprobleme, die eng mit der Theorie hermitescher Operatoren im Raume Π_κ zusammenhängen. I. Einige Funktionenklassen und ihre Darstellungen. Math. Nachr. **77** (1977), 187–236.

[KL2] M.G. Krein and H. Langer: On some extension problems which are closely connected with the theory of hermitian operators in a space Π_κ. III. Indefinite analogues of the Hamburger and Stieltjes moment problems. Part (1): Beiträge zur Analysis **14** (1979), 25–40. Part (2): Beiträge zur Analysis **15** (1981), 27–45.

[L] H. Langer: Spektralfunktionen einer Klasse von Differentialoperatoren zweiter Ordnung mit nichtlinearem Eigenwertparameter. Ann. Acad. Sci. Fenn. Ser. A I Math. **2** (1976), 269–301.

[LW] H. Langer and H. Winkler: Direct and inverse spectral problems for generalized strings. (submitted).

[N] I.P. Natanson: *Theorie der Funktionen einer reellen Veränderlichen.* Akademie-Verlag, Berlin, 1981.

[S1] A.L. Sakhnovich: Spectral functions of a canonical system of order $2n$. Math. USSR Sbornik **71** (1992), 355–369.

[S2] L.A. SAKHNOVICH: The method of operator identities and problems of analysis. Algebra and Analysis **5** (1993), 4–80.

[W1] H. WINKLER: The inverse spectral problem for canonical systems. Integral Equations Operator Theory **22** (1995), 360–374.

[W2] H. WINKLER: On transformations of canonical systems. *Operator Theory: Advances and Applications,* vol. **80**, Birkhäuser Verlag, Basel, 1995, pp. 276–288.

TU Dresden
Institut für Mathematische Stochastik
D–01062 Dresden
Germany

1991 Mathematics Subject Classification. Primary 34A55, 47E05 ; Secondary 34B20, 34L05, 47B25

OPERATOR THEORY: ADVANCES AND APPLICATIONS
BIRKHÄUSER VERLAG

Edited by
I. Gohberg,
School of Mathematical Sciences, Tel-Aviv University, Ramat Aviv, Israel

This series is devoted to the publication of current research in operator theory, with particular emphasis on applications to classical analysis and the theory of integral equations, as well as to numerical analysis, mathematical physics and mathematical methods in electrical engineering.

---

81 H. Upmeier
Toeplitz Operators and Index Theory in Several Complex Variables
1996. ISBN 3-7643-5282-5

82 T. Constantinescu
Schur Parameters, Factorization and Dilation Problems
1996. ISBN 3-7643-5285-X

83 A.B. Antonevich
Linear Functional Equations. Operator Approach.
1995. ISBN 3-7643-2931-9

84 L.A. Sakhnovich
Integral Equations with Difference Kernels on Finite Intervals
1996. ISBN 3-7643-5267-1

85 Y.M. Berezansky / G.F. Us / Z.G. Sheftel
Functional Analysis. Volume I.
1996. ISBN 3-7643-5344-9

86 Y.M. Berezansky / G.F. Us / Z.G. Sheftel
Functional Analysis. Volume II.
1996. 3-7643-5345-7

87 I. Gohberg / P. Lancaster / P.N. Shivakumar (Eds)
Recent Developments in Operator Theory and Its Applications. International Conference in Winnipeg, October 2–6, 1994.
1996. ISBN 3-7643-5414-5

88 J. van Neerven (Ed.)
The Asymptotic Behaviour of Semigroups of Linear Operators
1996. ISBN 3-7643-5455-0

89 Y. Egorov / V. Kondratiev
On Spectral Theory of Elliptic Operators
1996. ISBN 3-7643-5390-2

90 A. Böttcher / I. Gohberg (Eds)
Singular Integral Operators and Related Topics. Joint German-Israeli Workshop, Tel Aviv, March 1–10, 1995.
1996. ISBN 3-7643-5466-6

91 A.L. Skubachevskii
Elliptic Functional Differential Equations and Applications.
1997, ISBN 3-7643-5404-6

92 A.Ya. Shklyar
Complete Second Order Linear Differential Equations in Hilbert Spaces
1997. ISBN 3-7643-5377-5

93 Y. Egorov / B.-W. Schulze
Pseudo-Differential Operators, Singularities, Applications
1997. ISBN 3-7643-5484-4

94 M.I. Kadets / V.M. Kadets
Series in Banach Spaces. Conditional and Unconditional Convergence.
1997. ISBN 3-7643-5401-1

95 H. Dym / V. Katsnelson / B. Fritzsche / B. Kirstein (Eds)
Topics in Interpolation Theory
1997. ISBN 3-7643-5723-1

96 D. Alpay / A. Dijksma / J. Rovnyak / H. de Snoo
Schur Functions, Operator Colligations, and Reproducing Kernel Pontryagin Spaces
1997. ISBN 3-7643-5763-0

97 M.L. Gorbachuk / V.I. Gorbachuk
M.G. Krein's Lectures on Entire Operators
1997. ISBN 3-7643-5704-5

98 I. Gohberg / Yu. Lyubich (Eds)
New Results in Operator Theory and Its Applications
The Israel M. Glazman Memorial Volume
1997. ISBN 3-7643-5775-4

99 T. Ayerbe Toledano / T. Dominguez Benavides / G. López Acedo
Measures of Noncompactness in Metric Fixed Point Theory
1997. ISBN 3-7643-5794-0

100 C. Foias / A.E. Frazho / I. Gohberg / M.A. Kaashoek
Metric Constrained Interpolation, Commutant Lifting and System
1998. ISBN 3-7643-5889-0

101 S.D. Eidelman / N.V. Zhitarashu
Parabolic Boundary Value Problems
1998. ISBN 3-7643-2972-6

102 I. Gohberg / R. Mennicken / C.Tretter (Eds)
Differential and Integral Operators. International Workshop on Operator Theory and Applications, IWOTA 95, in Regensburg, July 31–August 4, 1995.
1998. ISBN 3-7643-5890-4

103 I. Gohberg / R. Mennicken / C. Tretter (Eds)
Recent Progress in Operator Theory. International Workshop on Operator Theory and Applications, IWOTA 95, in Regensburg, July 31–August 4, 1995.
1998. ISBN 3-7643-5891-2

104 H. Bercovici / C. Foias (Eds)
Nonselfadjoint Operator Algebras, Operator Theory, and Related Topics. The Carl M. Pearcy Anniversary Volume.
1998. ISBN 3-7643-5954-4